Cell Biology: Advanced Principles

Cell Biology: Advanced Principles

Edited by Larry Madison

SYRAWOOD
PUBLISHING HOUSE
New York

Published by Syrawood Publishing House,
750 Third Avenue, 9th Floor,
New York, NY 10017, USA
www.syrawoodpublishinghouse.com

Cell Biology: Advanced Principles
Edited by Larry Madison

International Standard Book Number: 978-1-68286-871-3 (Hardback)

Cataloging-in-Publication Data

Cell biology : advanced principles / edited by Larry Madison.
     p. cm.
Includes bibliographical references and index.
ISBN 978-1-68286-871-3
1. Cytology. 2. Cells. 3. Biology. I. Madison, Larry.
QH581.2 .C45 2020
571.6--dc23

TABLE OF CONTENTS

PREFACE

This book has been an outcome of determined endeavour from a group of educationists in the field. The primary objective was to involve a broad spectrum of professionals from diverse cultural background involved in the field for developing new researches. The book not only targets students but also scholars pursuing higher research for further enhancement of the theoretical and practical applications of the subject.

Cell is considered to be the basic unit of life. The branch of biology which studies the structure and function of the cell is known as cell biology or cytology. The basic concerns of cell biology are the physiological properties, metabolic processes, life cycle, signaling pathways, chemical composition and interactions of the cell with their environment. It encompasses both prokaryotic and eukaryotic cells. Therefore, it functions on microscopic as well as molecular level. The knowledge of how cells work and its components is fundamental to all biological sciences. It is also essential to the on-going research in bio-medical. Genetics, molecular biology, immunology and biochemistry are some fields which are closely related to the research in cell biology. The various studies that are constantly contributing towards advancing technologies and evolution of the field of cell biology are examined in detail in this book. With state-of-the-art inputs by acclaimed experts of this field, this book targets students and professionals.

It was an honour to edit such a profound book and also a challenging task to compile and examine all the relevant data for accuracy and originality. I wish to acknowledge the efforts of the contributors for submitting such brilliant and diverse chapters in the field and for endlessly working for the completion of the book. Last, but not the least; I thank my family for being a constant source of support in all my research endeavours.

Editor

Thrombin-Induced Calpain Activation Promotes Protease-Activated Receptor 1 Internalization

**Alejandro Alvarez-Arce, Irene Lee-Rivera, Edith López,
Arturo Hernández-Cruz, and Ana María López-Colomé**

Instituto de Fisiología Celular, Universidad Nacional Autónoma de México, México City, Mexico

Correspondence should be addressed to Ana María López-Colomé; acolome@ifc.unam.mx

Academic Editor: Richard Tucker

The serine protease thrombin activates Protease-Activated Receptors (PARs), a family of G-protein-coupled receptors (GPCRs) activated by the proteolytic cleavage of their extracellular N-terminal domain. Four members of this family have been identified: PAR1–4. The activation of Protease-Activated Receptor 1(PAR1), the prototype of this receptor family, leads to an increase in intracellular Ca^{+2} concentration ($[Ca^{+2}]i$) mediated by $G_{q11}\alpha$ coupling and phospholipase C (PLC) activation. We have previously shown that the stimulation of PAR1 by thrombin promotes intracellular signaling leading to RPE cell transformation, proliferation, and migration which characterize fibroproliferative eye diseases leading to blindness. Within this context, the elucidation of the mechanisms involved in PAR1 inactivation is of utmost importance. Due to the irreversible nature of PAR1 activation, its inactivation must be efficiently regulated in order to terminate signaling. Using ARPE-19 human RPE cell line, we characterized thrombin-induced $[Ca^{+2}]i$ increase and demonstrated the calcium-dependent activation of μ-calpain mediated by PAR1. Calpains are a family of calcium-activated cysteine proteases involved in multiple cellular processes including the internalization of membrane proteins through clathrin-coated vesicles. We demonstrated that PAR1-induced calpain activation results in the degradation of α-spectrin by calpain, essential for receptor endocytosis, and the consequent decrease in PAR1 membrane expression. Collectively, the present results identify a novel μ-calpain-dependent mechanism for PAR1 inactivation following exposure to thrombin.

1. Introduction

The retinal pigment epithelium (RPE) is a monolayer of differentiated, quiescent cells located between the neural retina and the choroid. The RPE is the predominant component of the outer blood-retina barrier (BRB) and plays an essential role in the maintenance of the functional and structural integrity of the neural retina required for visual function. Among its functions, the RPE is involved in the transepithelial transport of nutrients, the storage and metabolism of vitamin A derivatives, the renewal of photoreceptor outer segments, and the ionic homeostasis of the subretinal space [1].

Under pathological conditions involving the alteration of the BRB due to ocular trauma, diabetic retinopathy, retinal detachment, retinal hemorrhage, or, importantly, retinal surgery, the RPE is exposed to serum-contained growth factors and proinflammatory agents including thrombin, generated by the activation of the coagulation cascade. Thrombin has been shown to promote the release of cytokines, chemokines, and growth factors by RPE cells [2]. This results in the proliferation, dedifferentiation, and migration of RPE and glia cells to the vitreous and the subsequent assembly of contractile membranes on both retinal surfaces, thus promoting retinal detachment and, eventually, the loss of vision [3]. The identification of molecular targets for the prevention of this outcome is still lacking.

Thrombin-induced intracellular signaling is triggered by Protease-Activated Receptors (PARs), a family of G-protein-coupled receptors (GPCRs) activated by the proteolytic cleavage of the extracellular N-terminal domain, which unmasks a fresh N-terminal sequence that functions as an intramolecular ligand. Four members of this family have been identified: PAR1, PAR3, and PAR4 activated by thrombin and PAR2 mainly activated by trypsin, tryptase, and other trypsin-like proteases, in addition to high concentrations of thrombin [4]. PAR1 is the prototype of this receptor family, and its cleavage by thrombin at the Arg41-Ser42 bond exposes a new N-terminus (^{42}SFLLRN^{47}) that acts as a tethered ligand which binds intramolecularly to the second extracellular loop of the receptor and triggers signaling. Synthetic ligands corresponding to the cleaved N-terminus can displace the tethered ligand from the binding site and fully activate PAR1 in an intermolecular mode. The coupling of PARs to GPCR Gα subunits G$_{q11}\alpha$, G$_{12/13}\alpha$, and G$_i\alpha$ has been linked to a wide array of physiologic responses. Particularly, PAR1 coupling to G$_{q11}\alpha$ activates phospholipase C-β (PLC-β), which catalyzes the formation of inositol 1,4,5-trisphosphate (IP$_3$) and diacylglycerol (DAG), leading to an increase in intracellular Ca^{+2} concentration ([Ca^{+2}]i) [5].

Due to the irreversible nature of PAR1 activation by thrombin, PAR1 signaling and inactivation must be tightly regulated [6, 7]. Similar to classic GPCRs, PAR1 is rapidly desensitized by G-protein-coupled receptor kinase-mediated phosphorylation [8, 9] and β-arrestin binding, which uncouples the receptor from heterotrimeric G protein signaling [10]. Activated PAR1 is then internalized from the cell surface, sorted directly to lysosomes and degraded, which prevents continued signaling by previously activated receptors [11, 12]. These findings indicate that internalization and lysosomal sorting of PAR1 are important for regulating the magnitude and duration of G protein signaling. Studies by Paing et al. (2006) on this matter have shown that although β-arrestin binding is mainly responsible for PAR1 desensitization, it is not required for receptor internalization. Moreover, the mechanisms that control constitutive internalization of native, uncleaved receptors appear to differ from those controlling the internalization of activated receptors. Nonactivated PAR1 cycles constitutively between the plasma membrane and intracellular stores, thereby replenishing the cell surface following thrombin exposure, leading to rapid resensitization to thrombin signaling independent of de novo receptor synthesis [13].

Both the constitutive internalization and the internalization of activated PAR1 proceed through clathrin-coated vesicles [13–15]. Previous studies have shown that the clathrin adaptor protein complex 2 (AP-2) is essential for constitutive receptor internalization and cellular recovery of thrombin signaling [13]. The clathrin adaptor AP-2 is a heterotetrameric complex formed by α, β2, μ2, and σ2 adaptin subunits and has critical functions in the assembly and recruitment of cargo proteins to clathrin-coated pits. The μ2-adaptin subunit of AP-2 binds directly to tyrosine-based motifs within the cytoplasmic- (C-) tail domain of a number of GPCRs including PAR1 [16] and is required for constitutive internalization and cellular resensitization to thrombin [13]. AP-2 also regulates activated PAR1 internalization via recognition of

distal C-tail phosphorylation sites rather than the canonical tyrosine-based motif and has been shown to depend on epsin-1 interaction with ubiquitinated PAR1 [13, 17]. The mechanism by which activated PAR1 is recruited to clathrin-coated pits is presently not clear, and whether internalization of PAR2, PAR3, and PAR4 proceeds through the same pathway as PAR1 or is regulated by distinct mechanisms remains to be established.

Calpains are a family of calcium-activated cysteine proteases that catalyze the limited proteolysis of a number of cellular proteins in eukaryotes [18, 19]. To date, more than a dozen calpain isoforms and multiple splice variants have been identified [20]. The best-characterized members of this family, μ-calpain (calpain 1) and m-calpain (calpain 2), are activated, respectively, by micro- and millimolar Ca^{2+} concentrations [21]. Both these calpains include a 78–80 kDa catalytic subunit encoded by the CAPN1 or CAPN2 genes [20] and share a common 28 kDa regulatory subunit (CAPNS1; formerly CAPN4) [22]. Calcium binding to EF-hand calcium binding domains present in both calpain subunits results in conformational changes which expose the catalytic domain in the large subunit, thus activating calpain [18]. μ-Calpain is present in an inactive form in the cytoplasm of non-stimulated cells. Upon stimulation, the increase in [Ca^{+2}]i within the low micromolar range induces μ-calpain activation by autoproteolysis of the N-terminus, which exposes the catalytic site [23].

Calpains are involved in multiple cellular processes such as cell migration, actin cytoskeleton remodeling, and apoptosis [18]. Among these functions, calpains regulate the internalization of membrane proteins through clathrin-coated vesicles [24], such as epidermal growth factor receptor (EGFR) [25], transferrin receptor (RTF) [26], Cystic Fibrosis Transmembrane Conductance Regulator (CFTR) [27], and Low Density Lipoprotein (LDL) [28] among others. This process has been shown to require the proteolysis of distinct structural proteins, including the cytoskeletal protein α-spectrin by calcium-activated μ-calpain [28, 29].

Using human RPE-derived ARPE-19 cell line, we analyzed thrombin-induced [Ca^{+2}]i increase and demonstrated that the calcium-dependent activation of μ-calpain and the subsequent degradation of α-spectrin by calpain significantly decrease PAR1 expression at the cell membrane. These findings indicate that PAR1 inactivation by internalization is controlled by specific mechanisms, distinct from those promoting RPE cell transformation, and further support a role for thrombin in this process.

2. Materials and Methods

2.1. Reagents. All reagents used were cell culture grade. Thrombin, PAR1 agonist (SFLLRNPNDKYEPF), and anti-α-spectrin antibody (MAB1266) were purchased from Calbiochem/EMD Millipore (Billerica, MA, USA). PAR 3 (SFNGGP-NH$_2$) and PAR4 (GYPGKF-NH$_2$) agonist peptides were from Bachem (Torrance, CA, USA). PPACK (D-phenylalanyl-prolyl-arginyl chloromethyl ketone) was from

Enzo Life Sciences (New York, NY, USA). DMEM/F12 was from Thermo Fisher Scientific (Waltham, MA, USA). Fetal bovine serum (FBS), Fluo-4 AM, Thapsigargin, EGTA, and Lanthanum (III) Chloride were purchased from Invitrogen, Life Sciences (Carlsbad, CA, USA). Calpain-Glo Protease Assay was from Promega (Madison WI, USA), and Epoxomicin and PAR1-Alexa 488 antibody (FAB3855G) were from R&D Systems, Inc. (Minneapolis, MN USA). Anti-PAR1 antibody (SC13503) was purchased from Santa Cruz Biotechnology (CA, USA). CY3 anti-mouse secondary antibody (115-165-003) was from Jackson ImmunoResearch (PA, USA). BAPTA-AM, acetyl-calpastatin, and PD150606 were from Tocris Bioscience (Minneapolis, MN, USA). All other reagents were from Sigma Aldrich (St. Louis MO, USA).

2.2. Cell Culture. ARPE-19 cell line derived from human RPE was used throughout (ATCC® CRL-2302™). Cells were grown in DMEM/F12 (Sigma Aldrich, St. Louis, MO, USA) supplemented with 15 mM HEPES, 14.2 mM NaHCO$_3$, 0.5 mM sodium pyruvate, 0.005% penicillin, streptomycin, 0.01% neomycin, and 10% fetal bovine serum (FBS) pH 7.4. Cells were subcultured as suggested by ATCC. For all assays, cultures were serum-deprived for 24 hours prior to the experiment. Unless otherwise specified, experiments were carried in DMEM/F12 medium without FBS.

2.3. Measurement of Calpain Activity. Calpain-Glo Protease Assay (Promega, Madison, WI, USA) kit was used to determine the activity of m- and μ-calpain [30]. Cells were incubated for 30 minutes with 100 nM epoxomicin in order to inhibit proteasome activity. Cultures were then washed with phosphate buffered saline (PBS) and incubated for one hour with Suc-LLVY substrate peptide diluted with Calpain-Glo Buffer at 37°C. Cells were rinsed with PBS to remove unincorporated peptide and subsequently stimulated with 10 nM thrombin or 2.5 μM PAR1-Agonist Peptide (AP) in DMEM/F12 medium at room temperature for 1 minute. The agonists were removed with PBS and cells were permeabilized using PBS 0.9% Triton X-100. Luciferin detection reagent (Promega, Madison, WI, USA) was then added for 5 min and luminescence was measured in a plate reader (Synergy HT, BioTek Instruments Inc., Winooski, VT, USA). Measurements for each experimental condition were registered at 20 s intervals for 2 minutes. Results are expressed as percent Relative Luminescence Units (% RLU). Data are the mean ± SEM of three independent experiments. Values for nonstimulated cultures maintained in serum-free DMEM/F12 (negative control) were set as 100%.

2.4. Anti-α-Spectrin Western Blot. Degradation of α-spectrin by calpain is one of the first events required for receptor internalization. Cells were stimulated for 1 minute with thrombin or PAR1-AP in serum-free DMEM/F12 at room temperature. Cultures were then rinsed with PBS and incubated for 15 minutes at 37°C in serum-free DMEM/F12. Cells were lysed in 50 mM Tris-HCl pH = 7.4, 150 mM NaCl, 10 mM EDTA, 0.1% SDS, 1% Triton X-100, 1% CHAPS, 0.5% NP40, 0.1% BSA, 10% protease inhibitor cocktail (Sigma Aldrich P8340),

40 mM β-glycerophosphate, and 10 mM sodium pyrophosphate. Protein concentration was determined using bicinchoninic acid assay (Sigma Aldrich) and 20 μg of the total protein was used for protein immunodetection analysis. The lysates were solubilized in Laemmli buffer (0.75 mM Tris-HCl; pH 8.8, 5% SDS, 20% glycerol, 0.01% bromophenol blue, 10% β mercaptoethanol), boiled for 5 min, resolved by SDS/PAGE (6.5%), and transferred onto nitrocellulose membranes (Amersham Biosciences). After blocking for 1 h at room temperature with 7.5% nonfat milk in Tween TBS, the membranes were probed with primary antibodies against α-spectrin (1 : 4000 in blocking buffer) overnight at 4°C. Secondary HRP-conjugated antibody (1 : 5000) was incubated for 1 h and membranes were developed using the Luminata Forte Western Chemiluminescent Substrate (Millipore, Billerica, MA, USA). Kodak® film images were digitized using an Alpha Digi-Doc system (Alpha Innotech, San Leandro, CA, USA), and densitometry analysis was performed using the ImageJ Software and normalized to control values. GAPDH immunodetection was used as loading control.

2.5. Intracellular Ca^{+2} Concentration ([Ca^{2+}]i) Measurement. [Ca^{+2}]i was determined as previously described [31]. Briefly, coverslips containing ARPE-19 cells were incubated with 2 μM of the cell-permeable fluorescent Ca^{2+} indicator fluo-4 AM (Molecular Probes, Eugene, OR, USA) for 35 min at room temperature in Krebs-Ringer-Bicarbonate buffer (KRB; 118 mM NaCl; 2 mM KH$_2$PO$_4$; 4.7 mM KCl; 2.5 mM CaCl$_2$; 1.4 mM MgSO$_4$; 25 mM NaHCO$_3$; 5.6 mM Glucose; pH 7.4). The coverslips were placed in a recording chamber (Mod. RC-25; Warner Instruments, Hamden, CT, USA) attached to the stage of an upright microscope (Nikon Eclipse 80i; Nikon Corp., Tokyo, Japan) and continuously superfused (3 ml/min) with KRB applied to the recording chamber by gravity-fed superfusion. Fluo-4 was excited at 488 nm with monochromatic light from an argon laser (Laser Physics, Reliant 100 s488, West Jordan, UT), coupled to a Yokogawa spin-disk confocal scan head (CSU10B, Yokogawa Electronic Co., Tokyo, Japan and Solamere Technology Group, Salt Lake city, USA). Emission light was captured with a 510 nm filter. Fluorescence images were acquired with a water-immersion, Nikon objective (20x, 0.5 NA), and a cooled digital CCD camera (Andor Technology iXon 897, Oxford Instruments, High Wycombe, UK) controlled by the iQ software (Andor iQ version 1.10.2). Fluorescence images were acquired at 10 ms exposure and 500 ms intervals. All intracellular Ca^{2+} imaging experiments were performed at room temperature (22–24°C). Image sequences were analyzed using Image J software (National Institutes of Health). The values obtained are expressed as Arbitrary Fluorescence Units (AFU). The area under each curve (AUC) was calculated using GraphPad PRISM 6.0 software (La Jolla, CA, USA). For those experiments performed in the absence of Ca^{+2}, cells were superfused with Ca^{+2}-free KRB; 0.25 mM EGTA or 100 μM LaCl$_3$ were included in some experiments as stated in the figure legends. Cells were stimulated with thrombin or PAR1, PAR3, or PAR4 APs for 1 minute.

2.6. Epifluorescence Microscopy. ARPE-19 cells were seeded onto 22 mm plates. Cells were serum-deprived for 24 hours and subsequently incubated for 30 min in the presence of BAPTA-AM (10 μM) or the calpain inhibitors: calpastatin (1 μM) or PD1506060 (100 μM), followed by stimulation with either thrombin or PAR1-AP in DMEM/F12 medium for 5 minutes at room temperature. Subsequently, cells were rinsed with PBS and incubated for 15 min at 37°C. Blocking was performed with 5% BSA for 30 minutes and incubated for 4 hrs with Santa Cruz Anti-PAR1 antibody (1 : 500). Following primary antibody incubation, cells were fixed for 10 min with 4% paraphormaldehyde at 4°C. CY3 anti-mouse antibody was incubated for 1 hr (1 : 1000). Cells were washed 2x with PBS for 5 minutes. Nuclei were stained with Hoechst and further washed as before. Images were acquired with ACT-1 software in a Nikon microscope (Eclipse TE 2000-U) with DXM1200F camera and 40x objective (0.6 NA). Corrected Total Cell Fluorescence (CTCF) was calculated with ImageJ Software and normalized as percentage of control values.

2.7. Flow Cytometry. ARPE-19 cells were seeded onto 100 mm plates and serum-deprived for 24 hours. The cells were then detached using versene (5 mM Tris-HCl, 0.13 M NaCl, 0.5 M KCl, and 1.3 mM EDTA), and 5×10^5 cells were used for each experimental condition. Suspended cells were incubated for 30 minutes with 10 μM BAPTA-AM or calpain inhibitors (1 μM calpastatin or 100 μM PD1506060); the inhibitors were present throughout the experiment. Cells were stimulated for 5 minutes at room temperature, either with thrombin or with PAR1-AP in DMEM/F12 medium, rinsed with PBS, and blocked with 5% BSA-PBS. Cells were incubated with Anti-PAR1-Alexa 488 antibody for 3 hours at 4°C, and fluorescence was measured with Attune Acoustic Focusing flow cytometer (Thermo Fisher Scientific, USA). Data were analyzed using FlowJo, LLC 10.2 software (FlowJo, USA).

2.8. Statistical Analysis. Raw data for analyses were obtained from at least three independent experiments, as specified in the figure legends. Multiple comparison ANOVA and Tukey's post hoc test was applied to all results for statistical analysis. Prism V6.0 from GraphPad (La Jolla, CA, USA) was used.

3. Results

3.1. Thrombin Induces $[Ca^{+2}]i$ Increase. The effect of thrombin stimulation on RPE cell $[Ca^{+2}]i$ was determined using fluorescence microscopy, as described in the Methods. Although thrombin-induced increase in $[Ca^{+2}]i$ in RPE cells has been reported [32, 33], the characteristics of this response as a function of the intensity and duration of the stimulus, determinant for functional outcome, have not been analyzed.

Results in Figure 1(a) show that stimulation with thrombin induces a transient, dose-dependent $[Ca^{2+}]i$ increase in RPE cells, sustained for ~3 min. $[Ca^{2+}]i$ increase was found to be dose-dependent from 100 pM thrombin concentration with maximum stimulation attained at 10 nM thrombin treatment, equivalent to calcium elevation induced by FBS (positive control). The specificity of the effect was demonstrated

by inhibition upon the addition of the thrombin catalytic inhibitor PPACK (25 μM) (Figures 1(a) and 1(b)). Calculation of the area under the curves (AUC) in (Figure 1(a)) from the stimulation time-point to the 4th minute is plotted as a function of fluorescence intensity (Arbitrary Fluorescence Units (AFU)) over time (Figure 1(b)). The Ec_{50} for thrombin effect was calculated from the logarithmic curve in (Figure 1(b)) and found to be $Ec50 = 0.55$ nM (Figure 1(c)).

3.2. Thrombin-Induced Increase in $[Ca^{2+}]i$ Is Mediated by PAR1. In order to identify the specific receptor mediating thrombin effect, we tested the effect of PAR-APs on $[Ca^{2+}]i$ using fluorescence microscopy as described in Methods. $[Ca^{2+}]i$ was plotted as a function of fluorescence intensity (AFU) over time (minutes). Results in Figure 2(a) show that only PAR1-AP (2.5 μM) induced an increase in $[Ca^{2+}]i$ comparable to that elicited by 10 nM thrombin. Figure 2(b) shows the calculated area under the curves (AUC) shown in Figure 2(a). These results demonstrate that PAR1 is responsible for $[Ca^{+2}]i$ increase induced by thrombin.

3.3. Calcium Release from Intracellular Pools Is the Main Source of Thrombin-Induced $[Ca^{2+}]i$ Increase. In order to determine the contribution of external and internal calcium pools to thrombin- and PAR1-induced $[Ca^{+2}]i$ increase, thrombin stimulation was carried in Ca^{+2} free medium containing 0.25 mM EGTA. As shown in Figure 3, thrombin and PAR1-AP responses were decreased by ~20% in this condition, indicating a minor contribution of extracellular calcium to thrombin-induced $[Ca^{2+}]i$ rise. In contrast, thrombin- and PAR1 AP-induced $[Ca^{+2}]i$ increase was inhibited by ~80% upon the inclusion of 2 μM thapsigargin, known to deplete ER calcium stores (Figures 3(a) and 3(b)). These results were confirmed by the complete inhibition of thrombin effect by the joint inclusion of EGTA and thapsigargin. On this line, the store-operated calcium entry (SOCE) channels Orai and TRPC are activated by the depletion of intracellular Ca^{2+} stores. Since thrombin induces Gq/PLCβ signaling, which is a physiological stimulus for store depletion, we tested the effect of the membrane Ca^{2+} channel inhibitor $LaCl_3$, on PAR1-mediated $[Ca^{+2}]i$ increase. Figure 3(c) shows that $LaCl_3$ had a similar effect to that of EGTA, suggesting the possible participation of SOCE in PAR1-induced $[Ca^{+2}]i$ increase. Collectively, these data indicate that Ca^{2+} release from intracellular Ca^{2+} stores is the main source of $[Ca^{2+}]i$ increase induced by thrombin.

3.4. Thrombin Promotes Calpain Activity through PAR1 Activation. The effect of thrombin stimulation on calpain activity was assessed by the degradation of the synthetic calpain substrate peptide Suc-LLVY using Calpain-Glo Protease Assay, designed for measuring the activation of calpain isoforms μ and m [34]. Our results show that thrombin stimulation increases calpain activity by ~250% over control level. Activation appeared to be thrombin-specific, since it was prevented by the catalytic thrombin inhibitor PPACK (Figure 4(a)). In order to identify PAR1 as the mediator of

(a)

(b)

(c)

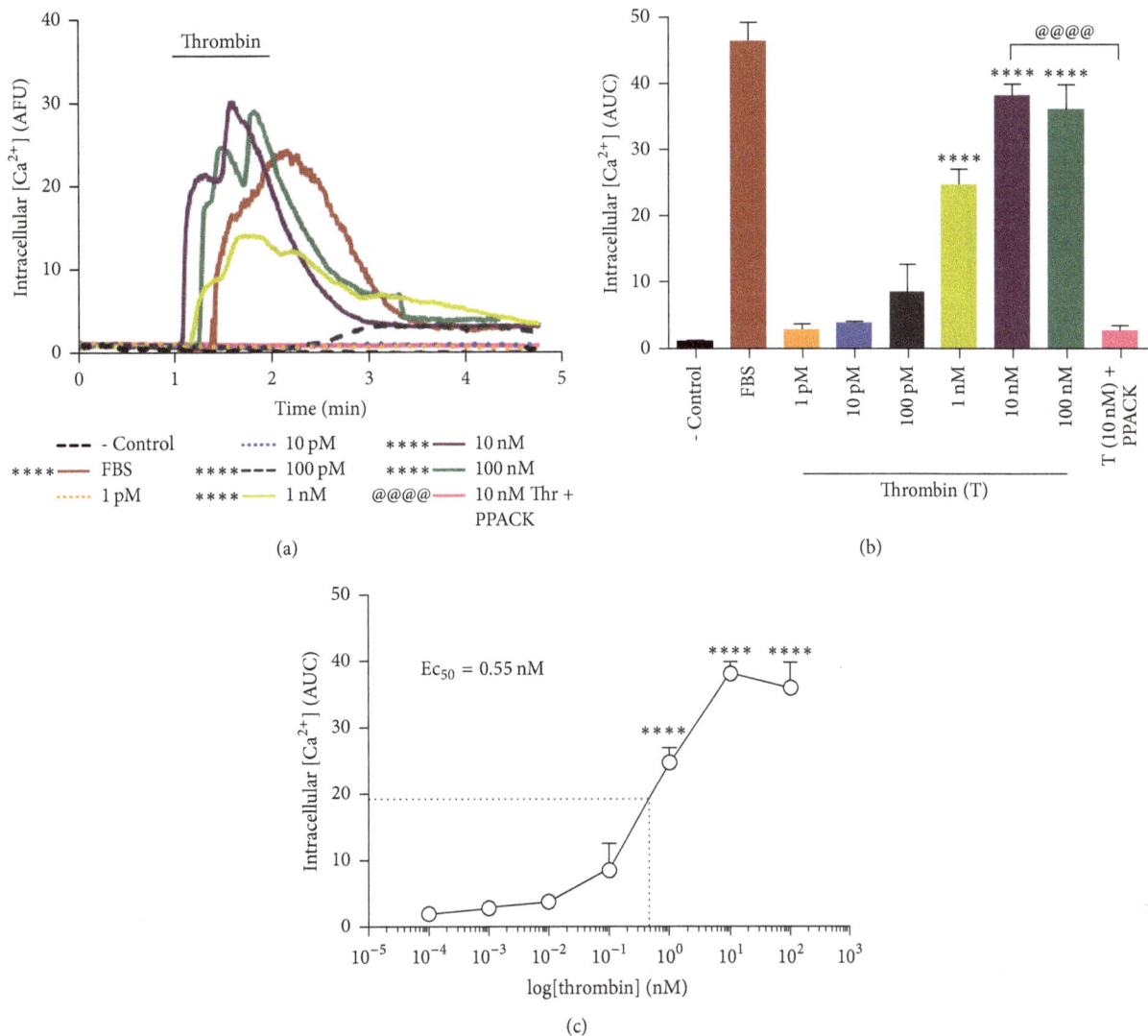

FIGURE 1: Thrombin induces a specific, dose-dependent increase in $[Ca^{2+}]i$. Cells were serum-deprived for 24 hours prior to stimulation with increasing concentrations of thrombin. $[Ca^{2+}]i$ was monitored by fluorescence microscopy as described in the Methods. (a) Thrombin induces a transient, saturable $[Ca^{2+}]i$ increase, sustained for ~3 min. $[Ca^{+2}]i$ was plotted as a function of fluorescence intensity (Arbitrary Fluorescence Units (AFU)) over time (minutes). Specificity of thrombin effect was assessed by inclusion of the thrombin inhibitor PPACK ($25\,\mu M$). 10% FBS was included as positive control. (b) The area under the curves (AUC) plotted in (a) was calculated from the stimulation point up to the 4th minute of stimulation. (c) Thrombin $Ec_{50} = 0.55\,nM$ was calculated from the logarithmic transformation of data in (b). Results are expressed as the mean ± SEM of three independent experiments compared to nonstimulated cells (- Control). Multiple comparison ANOVA and Tukey's test: $\alpha = 0.001$ (∗∗∗∗) referred to negative control or $\alpha = 0.001$ (@@@@) referred to thrombin stimulation.

thrombin effect, we tested the effect of PAR1-AP on calpain activity. Results showed that PAR1-AP stimulated calpain activity by ~170% (Figure 4(a)).

To establish if calpain activation by thrombin depends on thrombin-induced $[Ca^{+2}]i$ increase, cells were loaded with the selective cell-permeant Ca^{2+} chelator BAPTA-AM prior to thrombin stimulation. Results in Figure 4(b) show that inclusion of BAPTA-AM ($10\,\mu M$) completely prevented thrombin-induced calpain activation. Calpain activity was abolished by the calpain endogenous inhibitor calpastatin ($1\,\mu M$) and by the synthetic inhibitor PD150606 ($100\,\mu M$),

thus confirming the specificity of the effect (Figure 4(b)). These data demonstrate that thrombin-induced $[Ca^{+2}]i$ increase promotes calpain activation through the activation of PAR1.

3.5. Activation of Calpain by Thrombin Stimulation Promotes α-Spectrin Proteolysis. Spectrin, an extended rod-shaped tetramer composed of two α and two β subunits, forms a cytoskeletal meshwork with actin and other accessory proteins lining the cytoplasmic surface of most cell membranes, thus providing a mechanism for restricting the free diffusion

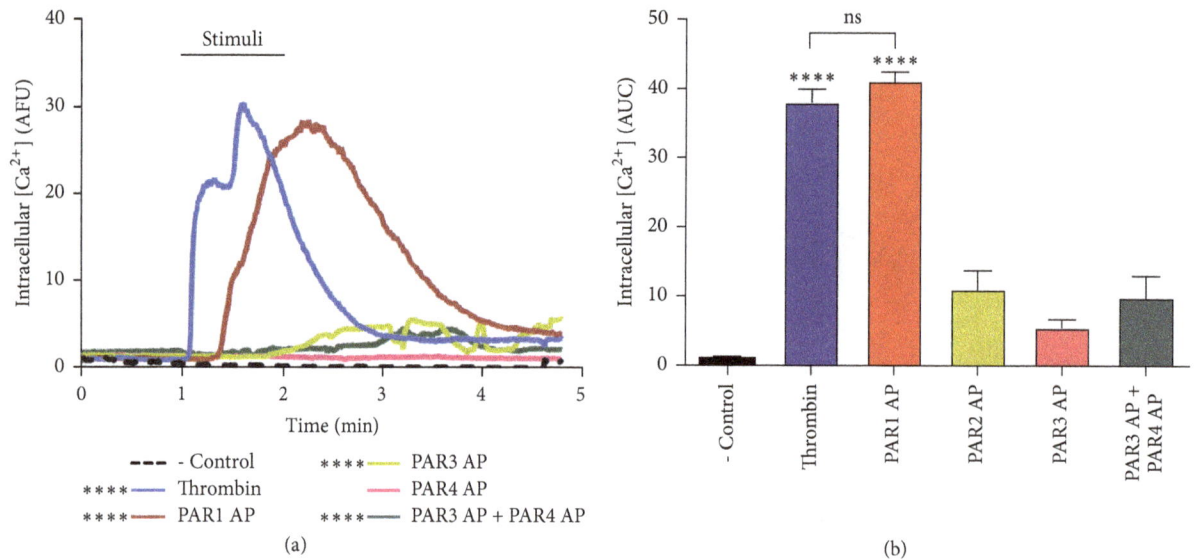

(a)

(b)

FIGURE 2: Thrombin-induced $[Ca^{+2}]i$ increase is mimicked by PAR1-AP. Cells were serum-deprived for 24 hours prior to stimulation with thrombin or PAR-APs. $[Ca^{2+}]i$ was assessed by fluorescence microscopy as described in Methods. (a) PAR1-AP (2.5 μM) induced an increase in $[Ca^{2+}]i$, comparable to thrombin. $[Ca^{2+}]i$ was plotted as a function of fluorescence intensity (Arbitrary Fluorescence Units (AFU)) over time (minutes). (b) The graph represents the area under the curves in (a). Thrombin (10 nM) stimulation was used as positive control. Results are expressed as the mean $\pm$ SEM of three independent experiments, compared to nonstimulated cells (- Control). Multiple comparison ANOVA and Tukey's test: $\alpha = 0.001$ (****) referred to negative control.

of transmembrane proteins [35]. Limited proteolysis of α-spectrin by μ-calpain [36] prevents spectrin interaction with actin [37] and also spectrin association with membrane-binding sites [38] allowing protein endocytosis [28]. Based on this evidence, we tested the effect of thrombin stimulation on α-spectrin proteolysis.

As shown in Figure 5(a), thrombin stimulation induced the proteolysis of α-spectrin 250 kDa subunit into ~150 kDa and ~120 kDa fragments. This effect appeared to be thrombin-specific, prevented by PPACK (Figure 5(a)). Thrombin-induced α-spectrin proteolysis was mimicked by PAR1-AP (Figure 5(b)). The specificity of α-spectrin degradation by calpain was confirmed by the blockage of the effect by the calpain inhibitors calpastatin and PD150606 and by the calcium chelator BAPTA-AM (Figure 5(c)). These results indicate that thrombin-induced $[Ca^{2+}]i$ increase activates calpain-mediated proteolysis of α-spectrin through the activation of PAR1.

3.6. Thrombin Decreases PAR1 Membrane Expression through PAR1-Induced Calpain Activation. The effect of thrombin-induced calpain activation on PAR1 plasma membrane expression was investigated using epifluorescence microscopy (Figure 6) and flow cytometry (Figure 7). Epifluorescence analysis in Figure 6(b) shows that stimulation with 10 nM thrombin for 5 min significantly decreased PAR1 expression at the plasma membrane; this effect was prevented by the thrombin inhibitor PPACK. To determine if thrombin

effect on PAR1 membrane expression is mediated by the activation of PAR1, we analyzed the effect of PAR1-AP on PAR1 membrane expression. Time-course analysis of PAR1-AP effect showed that although stimulation by 25 μM PAR1-AP for 5 min did not modify PAR1 membrane expression, sustained stimulation with the agonist for 60 min decreased PAR1 expression at the membrane to the same extent as 5 min application of thrombin (Figure 6(c)). Epifluorescence analysis showed that thrombin stimulation decreased PAR1 membrane expression by 89%. This effect was significantly prevented by calcium removal (BAPTA-AM plus EGTA) or by the inhibition of calpain by the inclusion of calpastatin or PD150606 (Figure 6(d)). In order to discard a nonspecific effect of fixation in epifluorescence analysis, PAR1 membrane expression was assessed in living cells using flow cytometry. Results from epifluorescence analysis were precisely confirmed by flow cytometry experiments. Figure 7 shows that stimulation by thrombin (Figure 7(b)) or by PAR1-AP (Figure 7(c)) decreases PAR1 membrane expression by 89%. This effect was partially prevented by inclusion of EGTA and BAPTA-AM or the inhibition of calpain by calpastatin or PD150606 (Figure 7(d)). The specificity of thrombin effect was confirmed by PPACK inhibition (Figure 7(b)). These results demonstrate that thrombin induces the Ca^{2+} and calpain-dependent decrease in PAR1 membrane expression and suggest the participation of distinct thrombin-activated processes in the regulation of PAR1 membrane expression.

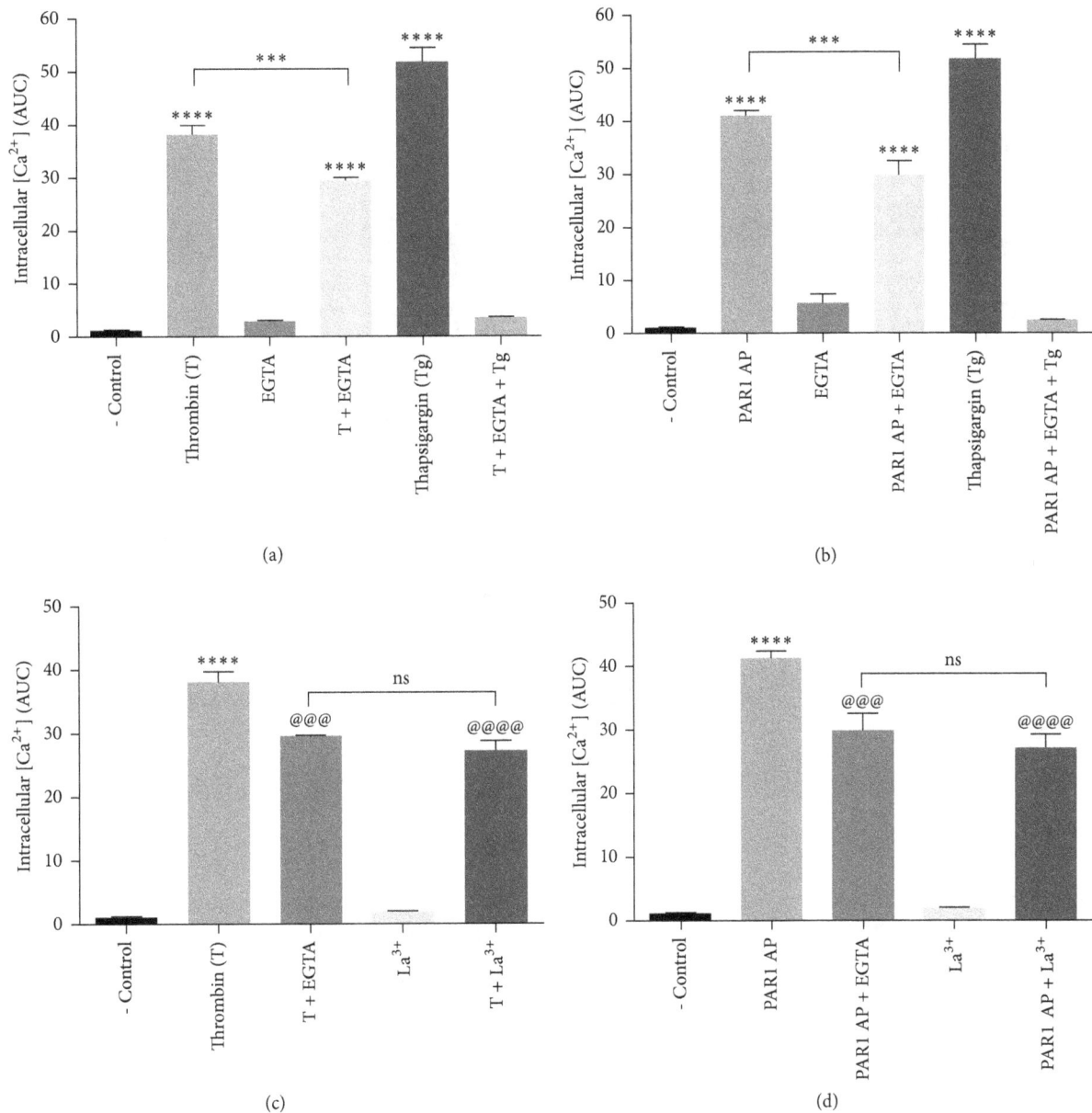

(a)

(b)

(c)

(d)

FIGURE 3: Thrombin promotes $[Ca^{2+}]i$ increase mainly by the release from intracellular pools. Serum-deprived cells were stimulated with thrombin or PAR1-AP. $[Ca^{2+}]i$ was determined by fluorescence microscopy as described in Methods. (a) Thrombin-induced $[Ca^{+2}]i$ increase is partially prevented by EGTA (0.25 mM) and abolished by thapsigargin (Tg; $2\,\mu$M) + EGTA. (b) Thrombin-induced calcium release is mimicked by PAR1-AP. (c) Blockage of plasma membrane calcium channels by Lanthanum (La^{3+}; $100\,\mu$M) decreases thrombin-induced and (d) PAR1-AP-induced calcium increase to a similar extent as EGTA. Results are expressed as the area under the curve (AUC) compared to nonstimulated (NS) cells (negative control). Data are the mean ± SEM of three independent experiments. Multiple comparison ANOVA and Tukey's test: $\alpha = 0.001$ ($****$) or $\alpha = 0.01$ ($***$) referred to negative control. And $\alpha = 0.001$ (@@@@) or $\alpha = 0.01$ (@@@) referred to thrombin stimulation (10 nM) or PAR1 peptide agonist (2.5 μM).

4. Discussion

The internalization and membrane recycling of native PARs is required for cell resensitization to thrombin, whereas the internalization and lysosomal degradation of activated PARs are important for the termination of G protein signaling [11, 12]. However, the mechanisms that control PAR1 inactivation are still not clear. Thrombin has been shown to activate PAR1 in RPE cells, which promotes the epithelial-mesenchymal transformation, proliferation, and migration of RPE cells, processes involved in the development of proliferative eye diseases leading to blindness [3]. Using human-derived ARPE-19 cells as a model for RPE, in the present study we investigated the mechanism responsible for the termination

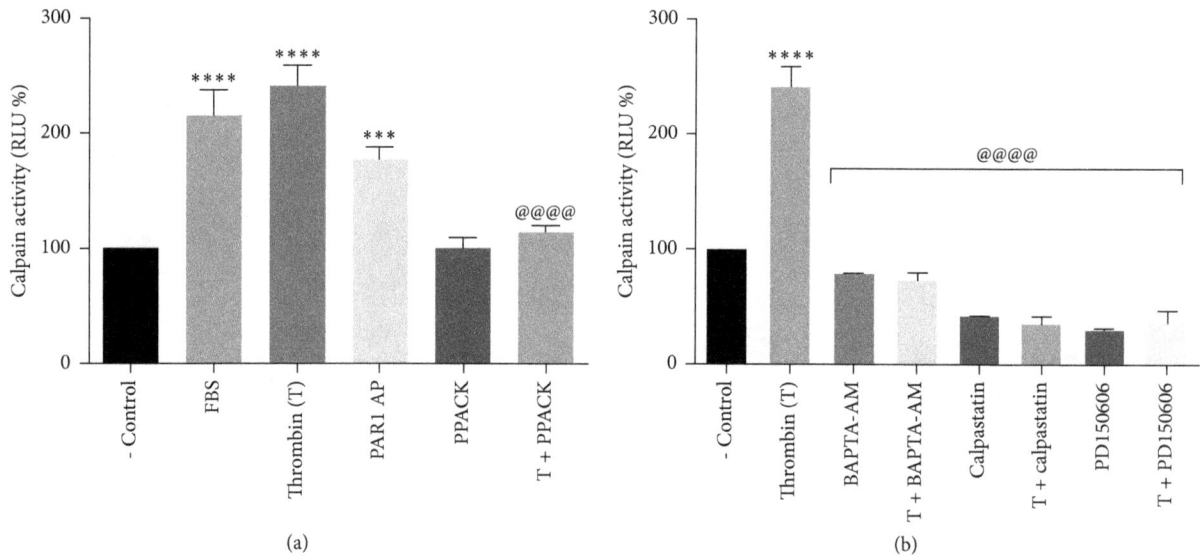

(a)

(b)

FIGURE 4: Thrombin activation of PAR1 stimulates the calcium-dependent activation of calpain. ARPE-19 cells were serum-deprived for 24 hours prior to stimulation for 1 min with 10 nM thrombin or 2.5 μM PAR1-AP. PPACK, BAPTA-AM, and calpain inhibitors were included 30 minutes prior to stimulation. Calpain activation was measured as a function of luciferase activity as described in Methods. (a) Thrombin stimulates calpain activity by ~250% in a specific manner, prevented by PPACK. 10% FBS was included as positive control. PAR1-AP stimulates calpain activity by ~170% compared to negative control (- Control). (b) Calpain activation was prevented by the inclusion of BAPTA-AM (10 μM), calpastatin (1 μM), and PD150606 (100 μM). Results in Relative Luminescence Units (RLU) are the mean $\pm$ SEM of three independent experiments. Values for nonstimulated cultures maintained in serum-free DMEM/F12 (- Control) were set as 100%. Multiple comparison ANOVA and Tukey's test: $\alpha = 0.001$ ($* * * *$) or $\alpha = 0.01$ ($* * *$) referred to negative control; $\alpha = 0.001$ (@@@@) referred to thrombin stimulation.

of thrombin-induced PAR1 signaling and demonstrated that activation of PAR1 by thrombin decreases PAR1 membrane expression through the Ca^{2+}-dependent activation of calpain. Our results suggest a novel regulatory mechanism by which activation of PAR1 controls its own availability at the plasma membrane.

We demonstrated that thrombin induces calpain activation through a specific process prevented by PPACK and identified to PAR1 as the receptor responsible for this effect (Figure 4(a)). As expected, thrombin-induced calpain activation required $[Ca^{+2}]i$ and was prevented by the specific calpain inhibitors calpastatin and PD150606 (Figure 4(b)). In order to establish a role for calpain in PAR1 membrane decrease, we examined thrombin effect on the cleavage of the endogenous μ-calpain substrate α-spectrin, required for clathrin-mediated endocytosis [39, 40]. Thrombin and PAR1-AP were shown to increase α-spectrin degradation (Figures 5(a) and 5(b)), indicating that PAR1 is responsible for thrombin-induced activation of μ-calpain and the subsequent degradation of α-spectrin (Figure 5(c)).

Thrombin-induced $[Ca^{2+}]i$ mobilization in RPE cells has been reported [32, 33]. Since calpain is a Ca^{2+} activated protease, we showed that thrombin induces the specific, dose-dependent increase in $[Ca^{2+}]i$ with an Ec_{50} of 0.55 nM (Figure 1(c)). In agreement with previous findings [8], $[Ca^{2+}]i$ elevation was mildly decreased in Ca^{2+}-free medium (20%) and significantly inhibited by thapsigargin (80%; Figure 3),

indicating that the main source of $[Ca^{2+}]i$ rise is Ca^{2+} release from internal stores, although Ca^{2+} influx is involved to a minor extent. Interestingly, La^{3+} inhibition of thrombin- and PAR1-induced $[Ca^{2+}]i$ increase to the same extent as EGTA (20%) suggests that Ca^{2+} entry through Orai-containing or voltage-gated Ca^{2+} channels could account for thrombin-induced extracellular Ca^{2+} entry (Figure 3) [41–43]. We identified PAR1 as the main receptor responsible for thrombin-induced $[Ca^{2+}]i$ increase, since PAR1-AP elicited $[Ca^{2+}]i$ increase equivalent to that induced by thrombin (Figure 2), whereas stimulation by PAR3-AP or PAR4-AP did not. These results are in agreement with findings in epithelial cells [44, 45], although in contrast with a previous report showing the increase in $[Ca^{2+}]i$ induced by PAR4 stimulation in ARPE-19 cells [33]. This discrepancy could be ascribed to the use by Narayan et al. (2010) of a synthetic PAR4-AP 10 times more potent than the endogenous agonist peptide used in the present work [46].

The main objective of this work was to investigate the effect of thrombin on PAR1 membrane expression and the role of calpain in the control of this process. Our results showed that the activation of PAR1 by thrombin induces a significant decrease in PAR1 membrane expression in RPE cells, via the Ca^{2+}-dependent activation of μ-calpain and the subsequent proteolysis of α-spectrin by this enzyme. Since calpains are involved in the formation of coated vesicles and α-spectrin degradation is essential for the endocytic

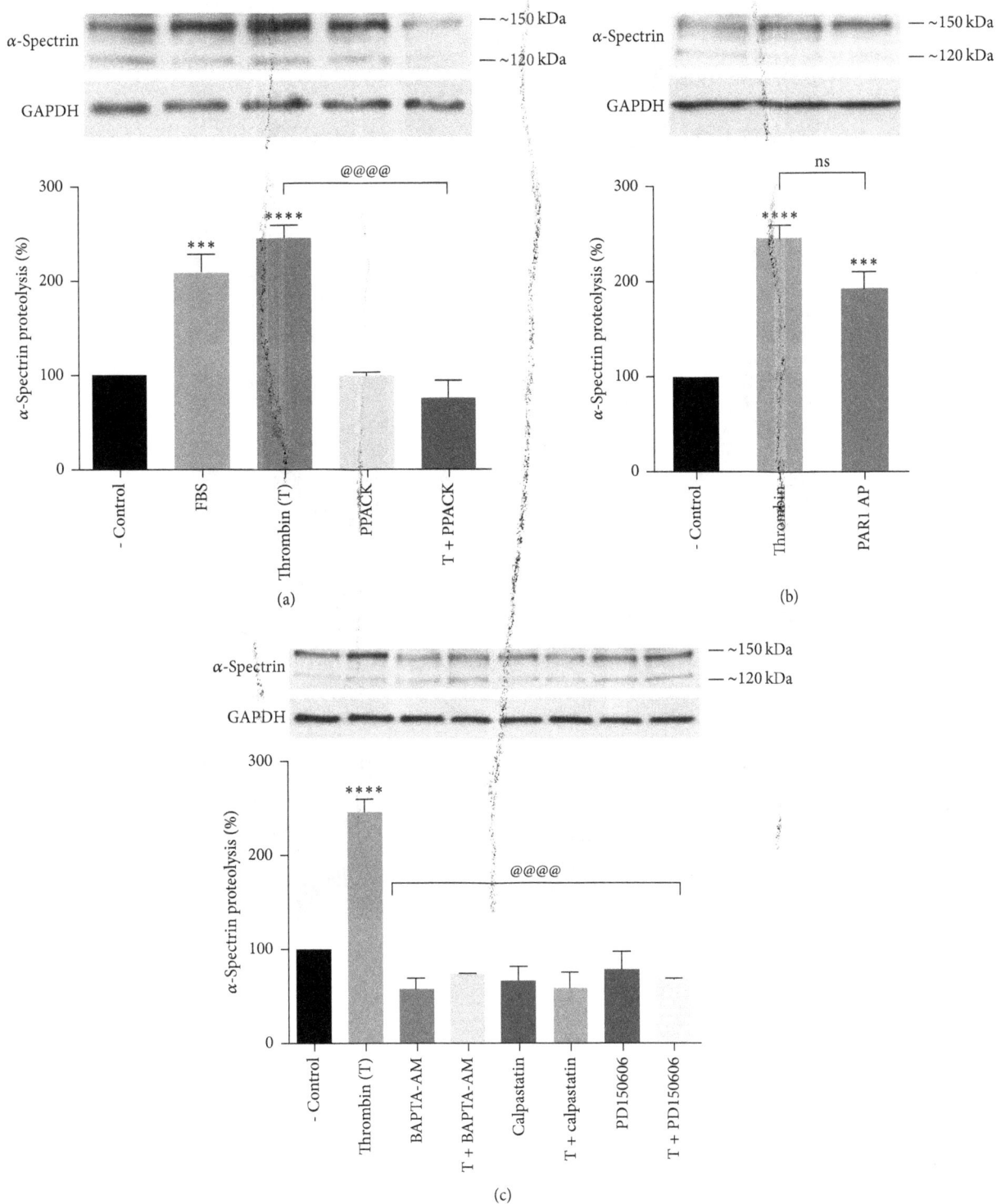

FIGURE 5: Thrombin induces α-spectrin degradation through PAR1-mediated Ca^{2+}-dependent activation of calpain. Serum-deprived ARPE-19 cells were stimulated for 1 minute with 10 nM thrombin or 2.5 μM PAR1-AP. PPACK, calcium chelators, and calpain inhibitors were included 30 minutes ahead of stimulation. α-Spectrin proteolysis into 150 kDa and 120 kDa fragments was assessed by Western blot. (a) Thrombin stimulates α-spectrin proteolysis by ~ 250% in a specific manner prevented by PPACK (25 μM). 10% FBS was included as positive control. (b) Thrombin effect was mimicked by 2.5 μM PAR1-AP. (c) Chelation of intracellular Ca^{2+} (10 μM BAPTA-AM), calpastatin (1 μM), and PD150606 (100 μM) prevented calpain activation. Results are expressed as percentage of α-spectrin proteolysis compared to nonstimulated cells (- Control). Data are the mean $\pm$ SEM of three independent experiments. Multiple comparison ANOVA and Tukey's test: $\alpha = 0.001$ ($****$) or $\alpha = 0.01$ ($***$) referred to negative control. Or $\alpha = 0.001$ (@@@@) referred to thrombin stimulation.

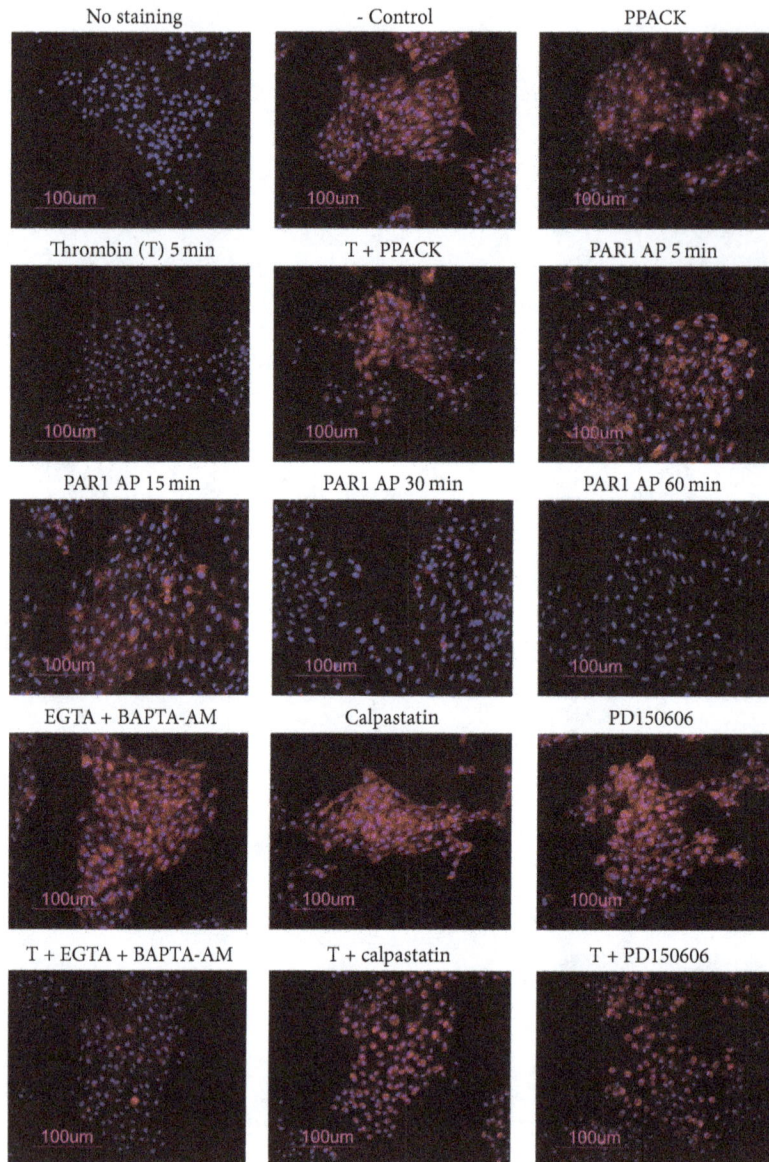

(a)

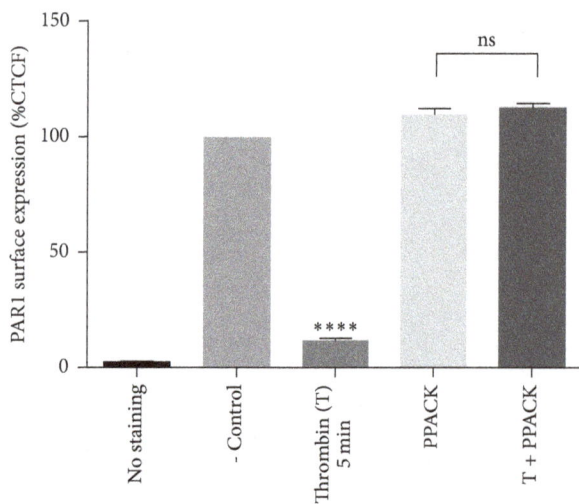

(b)

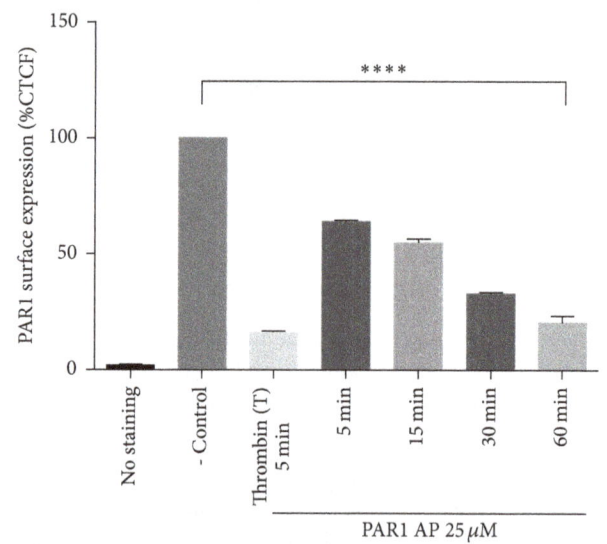

(c)

FIGURE 6: Continued.

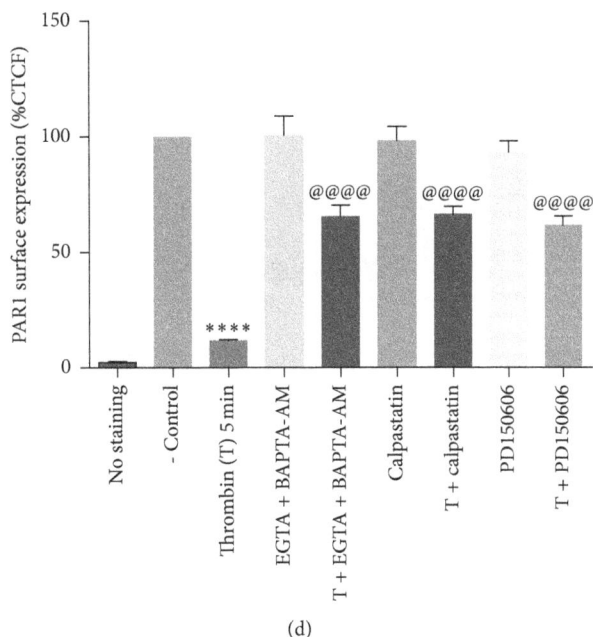

FIGURE 6: Thrombin stimulation decreases PAR1 membrane expression through calpain activation. Cells were serum-deprived for 24 hours prior to stimulation with thrombin or PAR1-AP. PAR1 membrane expression was assessed by epifluorescence microscopy. (a) PAR1 (CY3; red) and cell nuclei (Hoechst; blue) merged images. (b) Quantitative analysis of epifluorescence images shown in (a). Thrombin (10 nM) decreases PAR1 membrane expression through a specific process prevented by PPACK (25 μM). (c) Stimulation with PAR1-AP (25 μM) for 60 min decreases PAR1 membrane expression to the same extent as thrombin stimulation for 5 min. (d) Thrombin-induced decrease in PAR1 membrane expression requires calcium-dependent activation of calpain. Thrombin-induced decrease in PAR1 membrane expression is prevented by the joint addition of BAPTA-AM (10 μM) and EGTA (0.25 mM), by calpastatin (1 μM) and by PD150606 (100 μM). Results are expressed as the Corrected Total Cell Fluorescence (CTCF) compared to nonstimulated cells (- Control). Data are the mean ± SEM of three independent experiments. Multiple comparison ANOVA and Tukey's test: $\alpha = 0.001$ ($\ast\ast\ast\ast$) referred to negative control. Or $\alpha = 0.001$ ($@@@@$) referred to thrombin stimulation.

process, our data indicate that PAR1 signals to promote the internalization of PAR1 carried by clathrin-coated pits [47]. Nevertheless, epifluorescence and flow cytometry analysis in Figures 6(d) and 7(d) clearly showed that calcium withdrawal or calpain inhibition only partially prevents thrombin-induced reduction in PAR1 membrane expression, suggesting that calcium-dependent calpain activation is not the only mechanism involved in thrombin-induced decrease in PAR1 membrane expression. Among the $[Ca^{2+}]i$- and calpain-independent mechanisms shown to regulate PAR1 membrane expression, the N-linked glycosylation of PAR-1 second extracellular loop is known to regulate the internalization of this receptor [48]. Also on this line, Bicaudal D1 (BicD1) has been identified from a screen of an embryonic cDNA library as an adapter molecule directly interacting with the C-terminal cytoplasmic domain of PAR1, involved in the transport of PAR1 from the plasma membrane to endosomal vesicles. In fact, silencing of BicD1 expression impairs endocytosis of PAR1 [49]. However, the possible effect of thrombin on the activation/regulation of these processes remains to be explored.

We here identified a novel mechanism through which thrombin activation of PAR1 autoregulates its membrane expression through the increase in $[Ca^{2+}]i$, the activation of calpain by calcium, and the degradation of α-spectrin by calpain. Together with the present results, the future definition of additional thrombin-activated pathways involved in the regulation of PAR1 membrane expression may provide information aimed at preventing thrombin-mediated RPE cell transformation involved in proliferative retinopathies.

Conflicts of Interest

The authors declare that they have no conflicts of interest.

Acknowledgments

This work was partially supported by CONACyT (Grant 254333) and by PAPIIT/UNAM (Grant IN205317) to Ana María López-Colomé. Alejandro Alvarez-Arce is a doctoral student from Programa de Doctorado en Ciencias Biomédicas, Universidad Nacional Autónoma de México (UNAM), and received fellowship 573317 from CONACyT. Authors acknowledge Nicolás Jiménez-Pérez for technical assistance in Ca^{2+} imaging experiments and Dr. Laura Ongay for technical support in flow cytometry experiments.

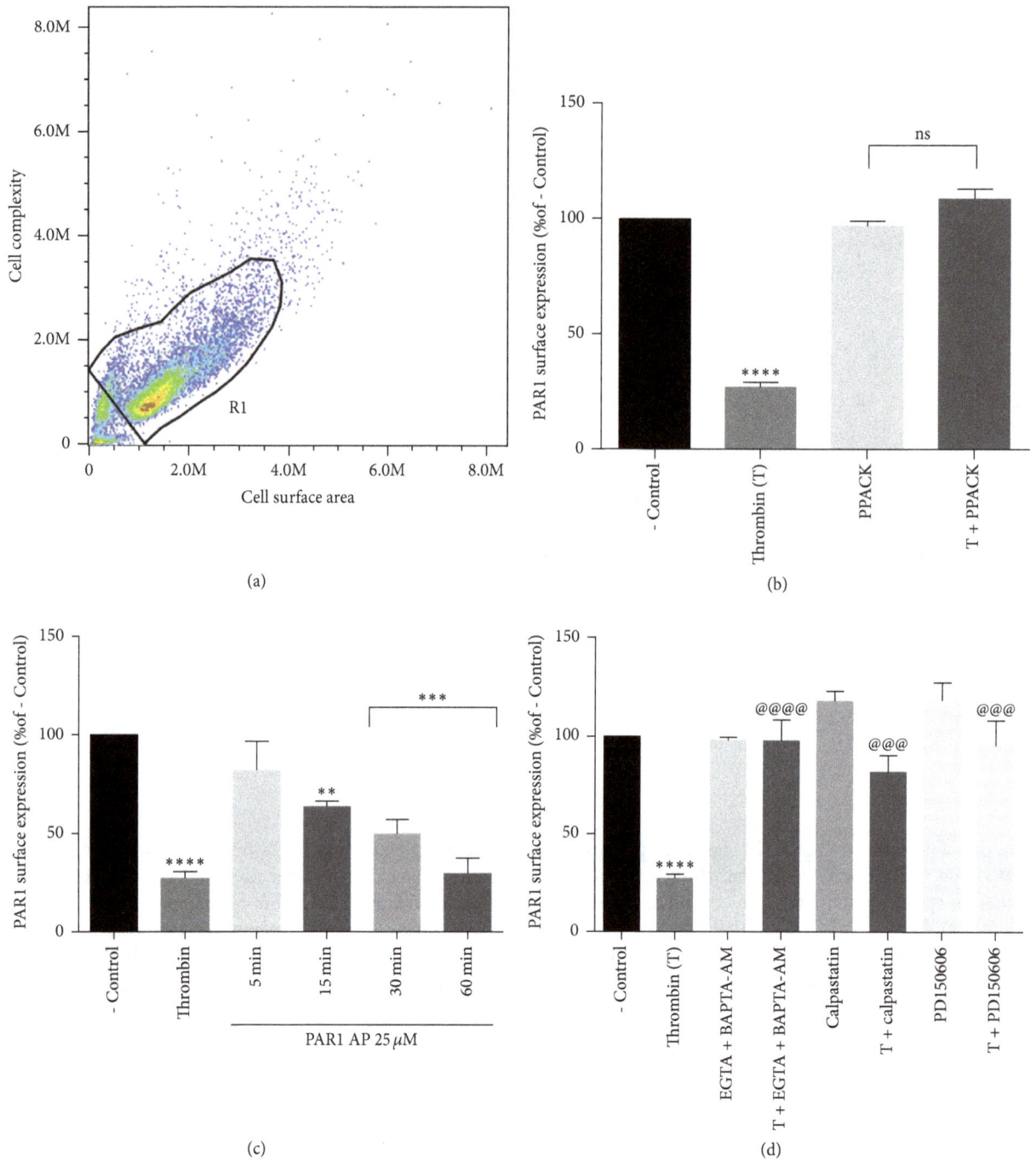

FIGURE 7: Stimulation of living cells by thrombin decreases PAR1 membrane expression through a Ca^{2+}- and calpain-dependent process. Cells were serum-deprived for 24 hours prior to stimulation with thrombin or PAR1-AP. The decrease in thrombin-induced PAR1 membrane expression measured in fixed preparations (Figure 6) was confirmed in living cells by flow cytometry. (a) Cell population selected for analysis. (b) Thrombin (10 nM) stimulation decreases PAR1 membrane expression by a specific process prevented by PPACK (25 μM). (c) PAR1-AP (25 μM) induces time-dependent decrease in PAR1 membrane expression. (d) Thrombin induces the decrease in PAR1 membrane expression through the calcium-dependent activation of calpain. This effect was prevented by the joint addition of BAPTA-AM (10 μM) and EGTA (0.25 mM), calpastatin (1 μM), or PD150606 (100 μM). Results are expressed as a percent of fluorescence in nonstimulated cells (- Control). Data are the mean ± SEM of three independent experiments. Multiple comparison ANOVA and Tukey's test: $\alpha = 0.001$ (****), $\alpha = 0.01$ (***), or $\alpha = 0.1$ (**) referred to negative control. And $\alpha = 0.001$ (@@@@) or $\alpha = 0.01$ (@@@) referred to thrombin stimulation.

References

[1] O. Strauss, "The retinal pigment epithelium in visual function," *Physiological Reviews*, vol. 85, no. 3, pp. 845–881, 2005.

[2] J. Bastiaans, J. C. van Meurs, V. C. Mulder et al., "The Role of Thrombin in Proliferative Vitreoretinopathy," *Investigative Opthalmology & Visual Science*, vol. 55, no. 7, pp. 4659–4666, 2014.

[3] H. Nagasaki, K. Shinagawa, and M. Mochizuki, "Risk factors for proliferative vitreoretinopathy," *Progress in Retinal and Eye Research*, vol. 17, no. 1, pp. 77–98, 1998.

[4] K. Mihara, R. Ramachandran, M. Saifeddine et al., "Thrombin-mediated direct activation of proteinase-activated Receptor-2: Another target for thrombin signaling," *Molecular Pharmacology*, vol. 89, no. 5, pp. 606–614, 2016.

[5] V. S. Ossovskaya and N. W. Bunnett, "Protease-activated receptors: contribution to physiology and disease," *Physiological Reviews*, vol. 84, no. 2, pp. 579–621, 2004.

[6] T.-K. H. Vu, D. T. Hung, V. I. Wheaton, and S. R. Coughlin, "Molecular cloning of a functional thrombin receptor reveals a novel proteolytic mechanism of receptor activation," *Cell*, vol. 64, no. 6, pp. 1057–1068, 1991.

[7] T.-K. H. Vu, V. I. Wheaton, D. T. Hung, I. Charo, and S. R. Coughlin, "Domains specifying thrombin-receptor interaction," *Nature*, vol. 353, no. 6345, pp. 674–677, 1991.

[8] J. Chun, T. Hla, S. Spiegel, and W. Moolenaar, "Inhibition of thrombin receptor signaling by a *G*-protein coupled receptor kinase. Functional specificity among *G*-protein coupled receptor kinases," *Journal of Biological Chemistry*, vol. 269, no. 2, pp. 1125–1130, 1994.

[9] C. Tiruppathi, W. Yan, R. Sandoval et al., "G protein-coupled receptor kinase-5 regulates thrombin-activated signaling in endothelial cells," *Proceedings of the National Acadamy of Sciences of the United States of America*, vol. 97, no. 13, pp. 7440–7445, 2000.

[10] M. J. Lohse, J. L. Benovic, J. Codina, M. G. Caron, and R. J. Lefkowitz, "β-arrestin: A protein that regulates β-adrenergic receptor function," *Science*, vol. 248, no. 4962, pp. 1547–1550, 1990.

[11] M. A. Booden, L. B. Eckert, C. J. Der, and J. Trejo, "Persistent Signaling by Dysregulated Thrombin Receptor Trafficking Promotes Breast Carcinoma Cell Invasion," *Molecular and Cellular Biology*, vol. 24, no. 5, pp. 1990–1999, 2004.

[12] J. Trejo, S. R. Hammes, and S. R. Coughlin, "Termination of signaling by protease-activated receptor-1 is linked to lysosomal sorting," *Proceedings of the National Acadamy of Sciences of the United States of America*, vol. 95, no. 23, pp. 13698–13702, 1998.

[13] M. M. Paing, C. A. Johnston, D. P. Siderovski, and J. Trejo, "Clathrin adaptor AP2 regulates thrombin receptor constitutive internalization and endothelial cell resensitization," *Molecular and Cellular Biology*, vol. 26, no. 8, pp. 3231–3242, 2006.

[14] M. M. Paing, A. B. Stutts, T. A. Kohout, R. J. Lefkowitz, and J. Trejo, "β-arrestins regulate protease-activated receptor-1 desensitization but not internalization or down-regulation," *The Journal of Biological Chemistry*, vol. 277, no. 2, pp. 1292–1300, 2002.

[15] U. J. Soh, M. R. Dores, B. Chen, and J. Trejo, "Signal transduction by protease-activated receptors," *British Journal of Pharmacology*, vol. 160, no. 2, pp. 191–203, 2010.

[16] B. Chen, D. P. Siderovski, R. R. Neubig, M. A. Lawson, and J. Trejo, "Regulation of protease-activated receptor 1 signaling by the adaptor protein complex 2 and R4 subfamily of regulator of G protein signaling proteins," *The Journal of Biological Chemistry*, vol. 289, no. 3, pp. 1580–1591, 2014.

[17] B. Chen, M. R. Dores, N. Grimsey, I. Canto, B. L. Barker, and J. Trejo, "Adaptor Protein complex-2 (AP-2) and epsin-1 mediate protease-activated receptor-1 internalization via phosphorylation- and ubiquitination-dependent sorting signals," *The Journal of Biological Chemistry*, vol. 286, no. 47, pp. 40760–40770, 2011.

[18] D. E. Goll, V. F. Thompson, H. Q. Li, W. Wei, and J. Y. Cong, "The calpain system," *Physiological Reviews*, vol. 83, no. 3, pp. 731–801, 2003.

[19] S. Ohno, Y. Emori, S. Imajoh, H. Kawasaki, M. Kisaragi, and K. Suzuki, "Evolutionary origin of a calcium-dependent protease by fusion of genes for a thiol protease and a calcium-binding protein?" *Nature*, vol. 312, no. 5994, pp. 566–570, 1984.

[20] Y. Huang and K. K. W. Wang, "The calpain family and human disease," *Trends in Molecular Medicine*, vol. 7, no. 8, pp. 355–362, 2001.

[21] J. D. Glass, D. G. Culver, A. I. Levey, and N. R. Nash, "Very early activation of m-calpain in peripheral nerve during Wallerian degeneration," *Journal of the Neurological Sciences*, vol. 196, no. 1-2, pp. 9–20, 2002.

[22] U.-J. P. Zimmerman, L. Boring, U. H. Pak, N. Mukerjee, and K. K. W. Wang, "The calpain small subunit gene is essential: Its inactivation results in embryonic lethality," *IUBMB Life*, vol. 50, no. 1, pp. 63–68, 2000.

[23] A. Khorchid and M. Ikura, "How calpain is activated by calcium," *Nature Structural & Molecular Biology*, vol. 9, no. 4, pp. 239–241, 2002.

[24] K. Sato, Y. Saito, and S. Kawashima, "Identification and Characterization of Membrane-Bound Calpains in Clathrin-Coated Vesicles from Bovine Brain," *European Journal of Biochemistry*, vol. 230, no. 1, pp. 25–31, 1995.

[25] Y. Maemoto, Y. Ono, S. Kiso et al., "Involvement of calpain-7 in epidermal growth factor receptor degradation via the endosomal sorting pathway," *FEBS Journal*, vol. 281, no. 16, pp. 3642–3655, 2014.

[26] N. Rudinskiy, Y. Grishchuk, A. Vaslin et al., "Calpain hydrolysis of α- and β2-adaptins decreases clathrin-dependent endocytosis and may promote neurodegeneration," *The Journal of Biological Chemistry*, vol. 284, no. 18, pp. 12447–12458, 2009.

[27] M. Averna, R. Stifanese, R. Grosso et al., "Role of calpain in the regulation of CFTR (cystic fibrosis transmembrane conductance regulator) turnover," *Biochemical Journal*, vol. 430, no. 2, pp. 255–263, 2010.

[28] A. Kamal, Y.-S. Ying, and R. G. W. Anderson, "Annexin VI-mediated loss of spectrin during coated pit budding is coupled to delivery of LDL to lysosomes," *The Journal of Cell Biology*, vol. 142, no. 4, pp. 937–947, 1998.

[29] A. Czogalla and A. F. Sikorski, "Spectrin and calpain: A 'target' and a 'sniper' in the pathology of neuronal cells," *Cellular and Molecular Life Sciences*, vol. 62, no. 17, pp. 1913–1924, 2005.

[30] K. Seyb, J. Ni, M. Huang, E. Schuman, M. L. Michaelis, and M. A. Glicksman, "Screen for calpain inhibitors using a cell - based, high - throughput assay calpain assay," *Cell Notes*, vol. 18, pp. 6–8, 2007.

[31] A. Hernández-Cruz, A. L. Escobar, and N. Jiménez, "Ca2+-induced Ca2+ release phenomena in mammalian sympathetic neurons are critically dependent on the rate of rise of trigger Ca2+," *The Journal of General Physiology*, vol. 109, no. 2, pp. 147–167, 1997.

[32] R.-M. Catalioto, C. Valenti, C. A. Maggi, and S. Giuliani, "Enhanced Ca^{2+} response and stimulation of prostaglandin release by the bradykinin B2 receptor in human retinal pigment epithelial cells primed with proinflammatory cytokines," *Biochemical Pharmacology*, vol. 97, no. 2, pp. 189–202, 2015.

[33] S. Narayan, G. Prasanna, K. Tchedre, R. Krishnamoorthy, and T. Yorio, "Thrombin-induced endothelin-1 synthesis and secretion in retinal pigment epithelial cells is Rho kinase dependent," *Journal of Ocular Pharmacology and Therapeutics*, vol. 26, no. 5, pp. 389–397, 2010.

[34] M. O'Brien, M. Scurria, K. Rashka, B. Daily, and T. Riss, "A bioluminescent assay for calpain activity," *Promega Notes*, vol. 91, pp. 6–9, 2005.

[35] V. Bennett, "Spectrin-based membrane skeleton: A multipotential adaptor between plasma membrane and cytoplasm," *Physiological Reviews*, vol. 70, no. 4, pp. 1029–1065, 1990.

[36] L. Löfvenberg and L. Backman, "Calpain-induced proteolysis of β-spectrins," *FEBS Letters*, vol. 443, no. 2, pp. 89–92, 1999.

[37] P. S. Becker, M. A. Schwartz, J. S. Morrow, and S. E. Lux, "Radiolabel-transfer cross-linking demonstrates that protein 4.1 binds to the N-terminal region of β spectrin and to actin in binary interactions," *European Journal of Biochemistry*, vol. 193, no. 3, pp. 827–836, 1990.

[38] R.-J. Hu and V. Bennett, "In vitro proteolysis of brain spectrin by calpain I inhibits association of spectrin with ankyrin-independent membrane binding site(s)," *The Journal of Biological Chemistry*, vol. 266, no. 27, pp. 18200–18205, 1991.

[39] M. W. Lambert, "Functional Significance of Nuclear α Spectrin," *Journal of Cellular Biochemistry*, vol. 116, no. 9, pp. 1816–1830, 2015.

[40] P. Zhang, D. Sridharan, and M. W. Lambert, "Knockdown of μ-calpain in fanconi anemia, FA-A, cells by siRNA restores αiI spectrin levels and corrects chromosomal instability and defective DNA interstrand cross-link repair," *Biochemistry*, vol. 49, no. 26, pp. 5570–5581, 2010.

[41] S. Cordeiro and O. Strauss, "Expression of Orai genes and ICRAC activation in the human retinal pigment epithelium," *Graefe's Archive for Clinical and Experimental Ophthalmology*, vol. 249, no. 1, pp. 47–54, 2011.

[42] S. Wimmers, L. Coeppicus, R. Rosenthal, and O. Strauß, "Expression profile of voltage-dependent Ca2+ channel subunits in the human retinal pigment epithelium," *Graefe's Archive for Clinical and Experimental Ophthalmology*, vol. 246, no. 5, pp. 685–692, 2008.

[43] I.-H. Yang, Y.-T. Tsai, S.-J. Chiu et al., "Involvement of STIM1 and Orai1 in EGF-mediated cell growth in retinal pigment epithelial cells," *Journal of Biomedical Science*, vol. 20, no. 1, article 41, 2013.

[44] E. Ostrowska and G. Reiser, "The protease-activated receptor-3 (PAR-3) can signal autonomously to induce interleukin-8 release," *Cellular and Molecular Life Sciences*, vol. 65, no. 6, pp. 970–981, 2008.

[45] L. Seminario-Vidal, S. Kreda, L. Jones et al., "Thrombin promotes release of ATP from lung epithelial cells through coordinated activation of Rho- and Ca^{2+}-dependent signaling pathways," *The Journal of Biological Chemistry*, vol. 284, no. 31, pp. 20638–20648, 2009.

[46] T. R. Faruqi, E. J. Weiss, M. J. Shapiro, W. Huang, and S. R. Coughlin, "Structure-function analysis of protease-activated receptor 4 thetered Ligand peptides. Determinants of specificity and utility in assays of receptor function," *The Journal of Biological Chemistry*, vol. 275, no. 26, pp. 19728–19734, 2000.

[47] K. Sato, S. Hattori, S. Irie, H. Sorimachi, M. Inomata, and S. Kawashima, "Degradation of fodrin by m-calpain in fibroblasts adhering to fibrillar collagen I gel," *The Journal of Biochemistry*, vol. 136, no. 6, pp. 777–785, 2004.

[48] A. G. Soto and J. Trejo, "N-linked glycosylation of protease-activated receptor-1 second extracellular loop: A critical determinant for ligand-induced receptor activation and internalization," *The Journal of Biological Chemistry*, vol. 285, no. 24, pp. 18781–18793, 2010.

[49] S. Swift, J. Xu, V. Trivedi et al., "A novel protease-activated receptor-1 interactor, bicaudal D1, regulates G protein signaling and internalization," *The Journal of Biological Chemistry*, vol. 285, no. 15, pp. 11402–11410, 2010.

Regeneration and Regrowth Potentials of Digit Tips in Amphibians and Mammals

Yohan Choi,[1,2] Fanwei Meng,[1,2] Charles S. Cox,[1] Kevin P. Lally,[1] Johnny Huard,[3,4,5] and Yong Li[1,2,4]

[1]*Department of Pediatric Surgery, University of Texas McGovern Medical School, Houston, TX 77030, USA*
[2]*Center for Stem Cell and Regenerative Medicine, The Brown Foundation Institute of Molecular Medicine for the Prevention of Human Diseases (IMM), The University of Texas Health Science Center at Houston (UT Health), Houston, TX 77030, USA*
[3]*Department of Orthopaedic Surgery, University of Texas McGovern Medical School, Houston, TX 77030, USA*
[4]*Center for Tissue Engineering and Aging Research, The IMM, The University of Texas Health Science Center at Houston (UT Health), Houston, TX 77030, USA*
[5]*Center for Regenerative Sports Medicine, Steadman Philippon Research Institute, Vail, CO, USA*

Correspondence should be addressed to Yong Li; yong.li.1@uth.tmc.edu

Academic Editor: Michael Peter Sarras

Tissue regeneration and repair have received much attention in the medical field over the years. The study of amphibians, such as newts and salamanders, has uncovered many of the processes that occur in these animals during full-limb/digit regeneration, a process that is highly limited in mammals. Understanding these processes in amphibians could shed light on how to develop and improve this process in mammals. Amputation injuries in mammals usually result in the formation of scar tissue with limited regrowth of the limb/digit; however, it has been observed that the very tips of digits (fingers and toes) can partially regrow in humans and mice under certain conditions. This review will summarize and compare the processes involved in salamander limb regeneration, mammalian wound healing, and digit regeneration in mice and humans.

1. Introduction

Mammalian fingertips and toes can partially regrow under certain conditions; however, regeneration is greatly limited compared to urodele amphibians such as newts and salamanders that can completely regrow an amputated limb [1–3]. The question is why there is such a difference between the regenerative potentials of mammals and amphibians. Embryonic, neonatal, and adult mice can regenerate digit tips if the amputation is midway through the third phalanx [4–6]; however, if the amputation occurs proximal to the midway point of the third phalanx in mice, regeneration of the digit tip does not typically occur [7, 8]. Similarly, young patients have also been documented to regrow the tips of amputated fingers if treated conservatively [9–11]. Although adults and even elderly individuals have potentially regenerated amputated digit tips, the regenerative process may not be as efficient as it is in younger patients and usually results in fibrous scars in adults. The regeneration process of the digit following injury may be related to the age of the host, with decreased restoration in adults compared to fetal or neonatal mammals [8, 10–12]. Injured adult mammalian tissues are usually replaced with fibrotic scar tissue, whereas scarless healing typically occurs in fetal wound healing which results in complete tissue recovery [13–15]. Stem cell activation and scarless wound healing are considered to be essential requisites for quality tissue regeneration [16–18]; however, for some regenerative processes a dedifferentiation process, but not stem cell activation, is required [19]. This review will summarize the literature in the context of amputated digit regeneration and beyond.

2. Salamander Limb Regeneration

Studies of axolotl regeneration are ongoing in order to understand the differences between regenerating and

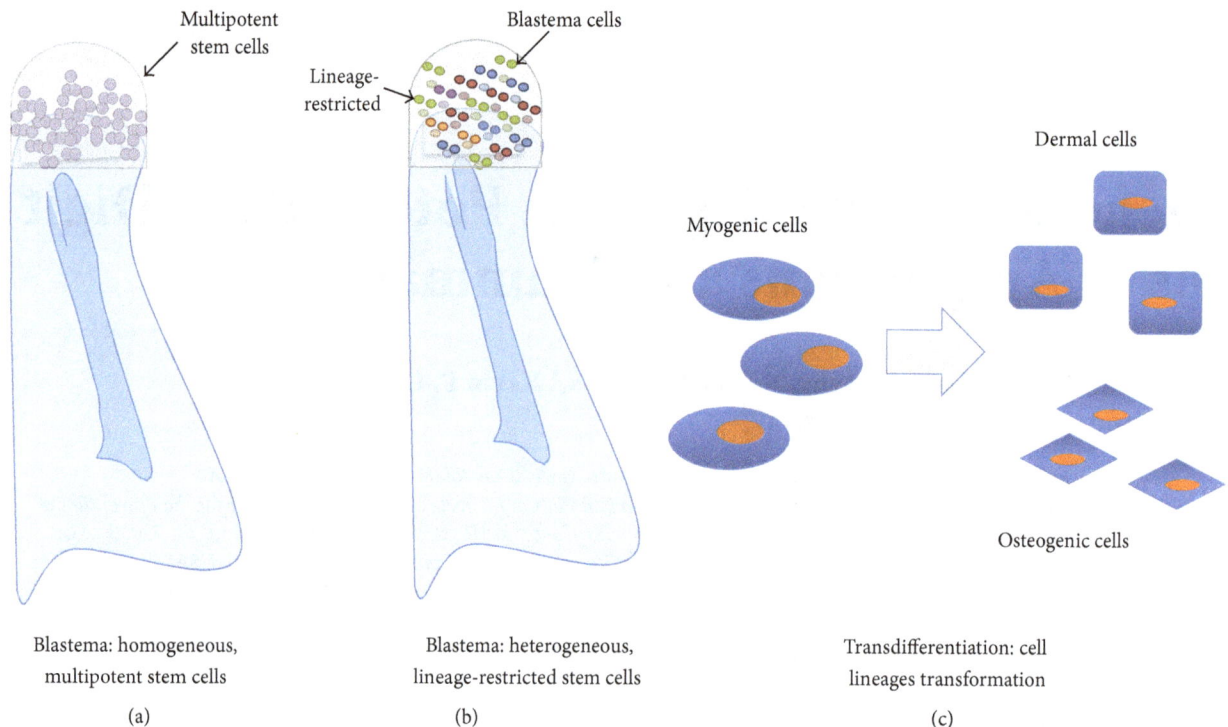

Blastema: homogeneous, multipotent stem cells

(a)

Blastema: heterogeneous, lineage-restricted stem cells

(b)

Transdifferentiation: cell lineages transformation

(c)

FIGURE 1: The blastema is a group of cells originating from the limb tissue local to the amputation site. (a) It was originally speculated that the blastema is a homogeneous structure of multipotent cells (purple dots), which would then form all the structures of the amputated digit tip or limb. (b) However, recent studies in the regeneration of both axolotl limb and zebrafish fin have demonstrated that the blastema cells are a heterogeneous assortment of lineage-restricted, unipotent progenitor cells (colorful dots). (c) Cell transdifferentiation might also play a role in the formation of blastema cells. Dermis and skeletal tissue, both of lateral plate mesodermal origin, have been shown to transdifferentiate. The blue area in (a, b) represents the remaining tissues following digit amputation.

nonregenerating wounds. Full-limb regeneration in adult urodele amphibians occurs in several overlapping stages including wound healing, dedifferentiation, and redevelopment, which is similar to natural embryonic limb development [20]. The first phase in wound healing involves the contraction of blood vessels and growth of the injured epidermis to cover the remaining limb stump. Blastema cells then accumulate underneath the healed epidermis, which forms a thickened structure at its apex, called the apical epithelial cap (AEC) [21, 22]. The proliferating blastema cells of newts consist of dedifferentiated cells derived from muscle, bone, skin, and other tissues, which serve as progenitors for regenerating the new limb. However, in axolotls stem cell activation in the form of satellite cells may also play a role in blastema formation [19]. Regeneration occurs by completely different mechanisms between these two different salamander species; thus care must be taken when interpreting results between newts and axolotls. Blastema and AEC formation are dependent on the activation of some unknown signals and several known signals such as ionic fluxes, nitric oxides, MARCKs protein, and trophic factors (e.g., the FGF, TGF, and BMP families) [22, 23] in the wound that consequently promote the formation of the blastema and the AEC. The growth and differentiation phase of the regenerative process includes many features recapitulating embryonic limb development but does exhibit some

differences compared to development de novo, for example, the size of the new limb, connection to the existing adult limb, and a nerve requirement [24].

2.1. Blastema Formation. The blastema is a group of cells originating from the limb tissue localized at the amputation site. The essential role of the blastema in limb regeneration has been investigated by Stocum and Cameron [25]. The cellular origin of blastemal cells, mechanisms of cellular release from mature tissue, dedifferentiation, accumulation of cells, blastema growth, and tissue patterning have all been the focus of extensive investigations.

2.1.1. Dedifferentiation. It was previously speculated that the blastema was comprised of a homogeneous population of multipotent cells (Figure 1(a)) that eventually form all the structures of the amputated digit tip or limb [26, 27]. An earlier study introduced fluorescent dextran-labeled myotubes into a regenerating limb stump and found the dye in the regenerated muscles and, in limited cases, the cartilage [28], suggesting the possibility that myofibers were capable of dedifferentiating into stem/progenitor cells and contributed to tissue regeneration. However, the possibility that the cells fused [29, 30] or that the dye leaked from the muscle into the cartilage cells when the myofibers dedifferentiated into single cells cannot be ruled out. Studies in both axolotl limb and zebrafish fin regeneration, using GFP- or transposon-based clonal analysis, have demonstrated that the cells are lineage-restricted, which suggests that the blastema is a heterogeneous assortment of lineage-restricted progenitor cells [31, 32] (Figure 1(b)). The cells may undergo dedifferentiation, but not completely to a multipotent state, as cell fates are limited to their developmental origin [33]. The dedifferentiation of Schwann cell precursors also releases paracrine factor to affect mammalian digit regenerations [17]. Muscle cells from presomitic mesoderm, Schwann cells from the neural fold, and epidermis from the lateral ectoderm are all derived from the same germ layer prior to maturity. In the past decades, many studies have presented evidence favoring the view that dedifferentiation with cell lineage switching occurs during newt limb regeneration, especially when the normal regenerative process is challenged (e.g., by irradiation or loss of a particular tissue). However, other studies in which axolotls were used suggested that stem cells are primarily involved (at least for muscle regeneration) and that lineage switching does not occur. More recently, a published study showed that, during limb regeneration, muscles were regenerated by completely different mechanisms in these two salamander species: (1) dedifferentiation, proliferation, and redifferentiation in newts and (2) satellite cells in axolotls [19, 33, 34]. Therefore, lineage switching may occur in newts under certain conditions, while this does not appear to occur in axolotls.

2.1.2. Resident Stem Cells. It is strongly believed that the cells in the blastema originate from dedifferentiated local tissue at the amputation site; however, adult stem cells (e.g., muscle satellite cells and possibly also the periosteum and dermis) [35–39] also contribute to the formation of the blastema, though the number of these endogenous cells may be insufficient to facilitate regeneration on their own. It has been shown that resident tissue stem/progenitor cells, rather than hematopoietic cells, contribute to the regeneration of amputated mouse digit tips [40, 41]. Adult stem cells in the nail bed are also thought to be involved in the regrowth of the amputated digit tip [42, 43].

2.1.3. Transdifferentiation. Transdifferentiation is a term typically used to describe a change in cell type from one mature cell type to another, also known as lineage reprograming [44]. Cellular transdifferentiation may also play a role in the tissue regeneration process. Dermis and skeletal tissue, both of lateral plate mesodermal origin, have been shown to transdifferentiate (Figure 1(c)) [45–47]. Transdifferentiation also appears to occur in lens and retina regeneration, where the pigmented epithelial cells dedifferentiate and then form lens epithelial cells or retinal neurons, respectively [1]. The term "transdifferentiation" has been used in the literature to refer to different, but related, processes, depending on whether the cells undergo a dedifferentiation process. Some use the general term "metaplasia" instead, which is independent of the mechanism used by cells to convert to a different cell type.

2.1.4. Extracellular Matrix (ECM) Involvement. Blastema cells may originate from host cells that are released from the tissue following injury-induced ECM breakdown. As cells are converted from a quiescent, fully differentiated state into a dedifferentiated state in the local surrounding matrix, many cellular changes occur. Actin cytoskeletal rearrangement, integrin disconnection from the matrix, and loss of cell polarity may induce the cells to suppress differentiation genes and upregulate genetic programs that allow the cells to reenter the cell cycle and consequently reacquire a state of "stemness" [25]. Alternatively, factors embedded in the ECM, such as cytokines, growth factors, and matrix cryptic peptides, are released upon the breakdown of the ECM and activate signaling pathways that trigger cellular dedifferentiation [48]. In addition to the stem/progenitor cell population, neural input/regrowth is also very important for the formation of the blastema [24]. In the absence of axons, the AEC forms but is not maintained, and the blastema never develops. If the nerves are removed after the blastema has formed, the limb will regenerate, but only to a limited extent, due to limited cell proliferation in the blastema. It is thought that the newly regenerated nerves stimulate the AEC to produce anterior gradient protein (AGP), which promotes the regeneration of denervated limbs [24, 49].

2.2. Blastema and the AEC. The AEC releases directional guidance signals to the blastema, allowing it to grow in the proper orientation. Two of the factors involved include transforming growth factor beta 1 (TGF-β1) and fibronectin, which are upregulated during blastema formation. Inhibition of TGF-β1 via SB-431542 decreases fibronectin expression and prevents blastema formation [50]. Conversely, signals from the blastema, such as the release of insulin-like growth factor (IGF), also trigger a response from the AEC [51]. Additionally, the cells in the blastema must proliferate to create enough progenitor cells to regrow the missing limb. The formation of blastema cells that accumulate under the AEC is not a recapitulation of embryonic limb development; it is a process that sets the urodeles apart from other tetrapod taxa [52, 53]. Various factors have been reported that promote blastema cell proliferation, including fibroblast growth factor- (FGF-) 1, 2, 8, and 10, transferrin, neuregulin, substance P, and AGP [54–57]. Although blastema cells proliferate rapidly, the cells of the AEC appear to be nonproliferative [49, 58], although migrating cells from the AEC do proliferate at later times [52, 53].

Patterning of the blastema cells into functional mature tissues has also been studied using various grafting experiments, which has demonstrated that the signals involved in reforming the tissues originate from the blastema [59]. A review by Tamura et al. describes several grafting experiments that demonstrate the positional memory of the blastema [60]. For example, a wrist-level blastema grafted to a more proximal stump did not grow until regeneration reached the wrist level, and the grafted blastema then grew into a supernumerary autopod (hand) [26]. Moreover, positional identity was found to be cell type-specific, such that cartilage-derived blastema, but not Schwann cell-derived cells, retained their positional identity [31, 61]. Some other experiments have demonstrated that blastemal cells have positional memory [31]. It is thought that fibroblastic cells may perform a similar function of maintaining positional identity in digit tip regeneration, because connective tissue fibroblasts from the terminal phalanx differ from those of the subterminal phalanx [62].

2.3. The AEC and AER. Limb regeneration partially recapitulates portions of embryonic limb development where the early developing embryo forms limb buds. The formation of the AEC is suggested to be a recapitulation of the apical ectodermal ridge (AER), a thickened epithelium at the distal end of the limb bud that functions as a signaling pathway to induce cell proliferation and maintains the mesenchymal cells in an undifferentiated state. The limb bud stops proliferating and begins to differentiate as the AER disappears [63]. The AER and AEC are considered to be functionally equivalent, with similar gene expression patterns, including the expression of *FGF-8* and *Sp-9* [64]. Proximal-distal patterning in the developing limbs is regulated by poorly understood interactions between FGFs secreted by the AER and Sonic hedgehog secreted from a posterior section of the limb bud, which in turn regulate the *Hox* genes [65, 66]. Retinoic acid regulates the *Meis homeobox* genes, which also affect proximal-distal patterning during both development and limb regeneration [67].

2.4. ECM Remodeling and MMP Activity. The ECM supports the architecture of the tissue during tissue regeneration. The activities of acid hydrolases and matrix metalloproteinases (MMPs), though traditionally known to play a role in mediating ECM turnover, have recently been demonstrated to actively participate in the regeneration process [68]. Regeneration failed in newts when amputated limbs were treated with the MMP inhibitor GM6001, demonstrating the essential involvement of MMPs in the regeneration process [69]. A comparison of normal, regenerating axolotls with regeneration-deficient short-toed axolotls revealed lower levels of MMP-8, MMP-9, and MMP-10 after amputation in the nonregenerating mutants, further highlighting the importance of MMPs in the regeneration process [70]. On the other hand, the participation of tissue inhibitors of metalloproteinases (TIMPs) is required to prevent excessive tissue hydrolysis and degradation induced by MMPs [71] such that dissociated cells at the amputation site begin to dedifferentiate into a more plastic stem cell phenotype [70].

Apart from the MMP activation in the early regenerative process, a transitional ECM develops that includes tenascin C, hyaluronic acid, and fibronectin, while the presence of collagens is reduced. Data suggest that tenascin C and hyaluronic acid can play instructive roles in the regenerative process [72, 73].

3. Mammalian Wound Healing

Many theories have been proposed to explain why successful regeneration occurs in urodele amphibians but not in mammals. First, the immune system has been shown to play a major role in the regeneration process of amputated limbs in newts [66, 74]. In mammals, fetal wounds can regenerate because they have an immature immune system; however, in adults, clearing pathogens appears to be evolutionarily favored compared to retaining the ability to regenerate a limb or digit [75]. Second, amphibians have retained limb regeneration-specific genes not found in mammals, which allow their cells to dedifferentiate [25]. A related theory is that mammals have evolved tumor suppression genes that inhibit regeneration [76]. The *Ink4a* locus is present in mammals but not amphibians; this region encodes the tumor suppression genes *p16ink4a* and Alternative Reading Frame *(ARF)*. Inactivation of both tumor suppressors retinoblastoma *(Rb)* and *ARF* allows terminally differentiated mammalian muscle cells to dedifferentiate [76]. An extension of this theory is that differentiated mammalian tissues can regenerate if the cells are induced to reenter the cell cycle, which occurs in the Murphy Roths Large (MRL) mouse and the p21-deficient mouse described below. Third, bioelectric signaling (e.g., membrane voltage polarity, ionic channels) may also play a role in the tissues' regeneration potential. Nonregenerating wounds display a positive polarity throughout the healing process, whereas in regenerating animals the polarity is initially positive but then quickly changes to negative polarity with the peak voltage occurring at the time of maximum cellular proliferation [77].

Wound healing is a complex process that is not yet fully understood. Mammalian wound healing of the skin and all organ systems has traditionally been divided into three major stages: inflammation, proliferation, and tissue remodeling [78]. Attempts have been made to correlate these three stages of mammalian wound healing with the three stages of amphibian regeneration (wound healing, dedifferentiation, and redevelopment) [79]. The phases of the regeneration processes in amphibians and mammals are summarized in Figure 2.

3.1. Inflammation. Immediately after injury, the body responds by stopping bleeding, which involves endothelial cell vasoconstriction and the activation of coagulation pathways. Platelets coagulate to form a fibrin clot comprised of collagen, fibronectin, and thrombin, while simultaneously releasing trophic factors and inflammation-associated cytokines. Neutrophils are the initial inflammatory cells that are recruited to the wound site. They release proteases and create reactive oxygen radicals to kill invading microbes and digest damaged tissue [80]. Monocytes are next recruited to the wound site and are converted into macrophages, while the neutrophils begin to undergo apoptosis. Macrophages remove bacteria, cellular debris, and dead neutrophils via phagocytosis and release signals that recruit more macrophages and fibroblasts to the wound site. It is unclear whether macrophages and/or neutrophils are absolutely required for wound healing, because a mutant mouse model that is deficient in macrophages and functional neutrophils is still capable of healing small wounds without creating an inflammatory response and heals without scar tissue formation [81].

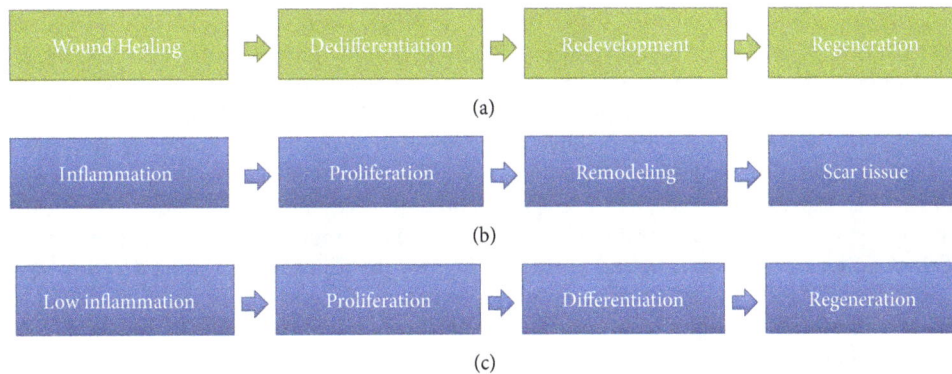

| Wound Healing | ⇒ | Dedifferentiation | ⇒ | Redevelopment | ⇒ | Regeneration |

(a)

| Inflammation | ⇒ | Proliferation | ⇒ | Remodeling | ⇒ | Scar tissue |

(b)

| Low inflammation | ⇒ | Proliferation | ⇒ | Differentiation | ⇒ | Regeneration |

(c)

FIGURE 2: Amphibian regeneration versus (a) attempted regeneration in mammals (b and c).

3.2. Proliferation. The proliferation stage begins approximately 4 days after injury and lasts for 10 days or more. During this period, epithelialization occurs via the expansion of skin keratinocytes. Some of inflammatory cytokines [interleukin-1 (IL-1) and tumor necrosis factor-α (TNF-α)] stimulate fibroblasts to synthesize and secrete keratinocyte growth factor-1 (KGF-1), KGF-2, and IL-6, which signal the keratinocytes to migrate and proliferate. Regulators of reepithelialization also include hepatocyte growth factor (HGF), FGFs, and epidermal growth factors (EGFs) released from injured tissues, which can stimulate receptor tyrosine kinases [82]. In contrast, TGFβ inhibits keratinocyte proliferation, and mice with mutations disrupting the TGFβ pathway have been observed to display faster wound healing [83, 84]. The provisional fibrin and fibronectin matrix formed during the inflammatory stage is reinforced by proteoglycans and other proteins synthesized by fibroblasts, which is then replaced by a stronger, more organized matrix composed of types I and III collagens. T lymphocytes migrate into the wound site after macrophage and fibroblast infiltration and are thought to influence the proliferative phase of wound healing [85]. Angiogenesis is stimulated by vascular endothelial growth factor A (VEGFA) and FGF-2, which stimulate endothelial cells to proliferate and form capillaries. Fibroblasts then transform into myofibroblasts to close the wound as a result of TGF-β1 and PDGF signaling [86].

3.3. Tissue Remodeling. The third phase of wound healing is tissue remodeling, which begins about a week after injury but can last for months or years after injury [87]. The remaining cells either migrate out of the wound or undergo apoptosis, leaving a scar consisting of mostly collagen and other ECM proteins and very few cells. During remodeling, type III collagen in the matrix is remodeled to the stronger type I collagen via MMPs, reducing the total type III collagen from 30% to approximately 10% [88]. Scar formation is the result of excess, unorganized collagen deposition [85] and is thought to be a mechanism to prevent the entry of microorganisms and to quickly provide mechanical support. Scars on the skin do not regrow hair follicles or sweat glands and are more sensitive to UV radiation [89].

After injury, basement membrane formation differs in the wound healing response between mammals and amphibians. In normal skin (both in mammals and in amphibians) the basement membrane lies between the epidermis and dermis and is comprised of collagen fibers, laminin, and other components. In mammals, a new basement membrane is formed between the new epidermis and dermis, which is then maintained during the wound healing process. This supports tissue integrity at the expense of scar formation. However, the basement membrane does not form during healing and only appears after regeneration is complete in amphibians [90]. If the basement membrane is induced in amphibians before regeneration is complete, scar formation occurs and regeneration ceases [91]. The basement membrane, however, may also play a beneficial role, as wound healing is impaired in mice lacking the basement membrane component nidogen 1 [92]. Nidogens 1 and 2 are basement membrane proteins that interact with laminin, collagen IV, and perlecan. The MRL mouse, a mouse model for systemic lupus erythematosus, was serendipitously found to regenerate multiple ear punches [93]. Unlike other mice, the MRL mouse forms a basement membrane during wound repair that is then removed during ear punch regeneration; this was found to be correlated to increased MMP activity and decreased TIMP activity [94].

3.4. MMP Activity and Wound Healing. MMPs are a family of zinc-dependent proteases and have been associated with wound healing, which involves extensive remodeling of the ECM [95, 96]. Wound sites express many MMPs, which facilitate various processes such as the infiltration of inflammatory cells, migration of fibroblasts, and angiogenesis. Although there is some substrate redundancy among MMPs, the interstitial collagenases are unique in their ability to degrade stromal collagens (types I, II, and III). These collagenases include MMP-1, MMP-8, MMP-13, and MMP-14 (MMP-14 is a membrane-bound MMP) [97]. An experiment in *Drosophila* demonstrated that a secreted MMP was required for basement membrane remodeling during wound healing [98], suggesting that MMP-14 does not play a major role in wound healing. MMP-13 is synthesized by cells in cartilage and bone and preferentially degrades type II collagen found in cartilage. MMP-8 is expressed primarily in neutrophils. MMP-1 (in humans) is expressed by most cells and can readily degrade all stromal collagens, but mainly types I and III. Human MMP-1 does not have an exact mouse homolog. MMP-1a (McolA) has only a 58% amino acid homology with human MMP-1 (and 74% nucleotide homology) [99]. This is in contrast with MMP-13, which shares >90% sequence homology with the mouse model. Murine MMP-13, unlike human MMP-13, has a broad expression profile, which is why it has served as a surrogate for MMP-1 in murine models [99, 100]. MMP-13-deficient mice exhibit normal wound healing [101]. Presumably, the loss of MMP-13 can be compensated for by other members of the MMP family, such as MMP-2 and MMP-14. MMP-8 was found to be upregulated in MMP-13-deficient wounds compared to controls; however, excess MMP-8 prevents proper tissue repair, as mice overexpressing MMP-8 demonstrate impaired wound healing [102].

Blocking the activity of MMPs with broad, nonspecific inhibitors results in delayed wound healing [103] and impaired stem cell migration and differentiation [104]. Mice with a mutation in collagen I that renders it insensitive to cleavage by MMP-1 demonstrate impaired tissue remodeling and severely delayed tissue healing [105, 106]. Many MMP-deficient mutants, however, do not demonstrate abnormalities in wound healing, with the exception of MMP3-deficient mice, which have a wound contraction defect [101], and MMP-8-deficient mice, which exhibit increased inflammation [107]. Therefore, the question remains as to whether an essential MMP has been found or if their contributions are due to multiple, overlapping MMPs. *MMP-9* and *MMP-13* double knockout mouse demonstrates delayed tissue healing, which is reversed upon topical treatment with recombinant MMP-9 and MMP-13 [108]. MMP-9 knockout mice displayed impaired cutaneous wound healing accompanied by defects in keratinocyte migration and collagen fibrillogenesis [109]; however, a lack of MMP-9 enhances the rate of wound closure in injured corneas [110]. Contrary to other MMPs, which are expressed at the front of advancing epithelial sheets and stimulate cell migration, MMP-9 acts to inhibit the rate of wound closure by inhibiting the replication of cells in the migrating epithelial sheet. Similarly, anti-MMP-9 treatments reduced fibrosis in soleus muscle regeneration [111]. Thus, although MMPs are essential for tissue regeneration, the specific role of each MMP is highly complex.

4. Digit/Appendage Regeneration in Humans

Children under the age of 10–15 have been documented to regenerate the tips of their fingers if the amputation is treated conservatively [9, 10]. Regeneration has been documented to restore finger shape, fingerprint, function, and the sense of touch. There were some cases where bone regrowth was documented; however, lengthening of the digit could have occurred via the distal growth of granulation tissue [11, 112]. Treatment of amputated digits with a skin flap prevents regeneration both in amphibians and in mammals [10, 113]. Similar conservative treatment of adult fingertip amputations has resulted in wound healing with no reported lengthening of the fingertip [114]; however, there has been a report of limited bone regrowth following surgical removal of the diaphysis of the 3rd phalanx in an adult [115]. Adult fingertip healing (in

individuals over 15 years old) with some documentation of bone regrowth was reported after treatment with a biological dressing based on chitin utilizing a "Hyphecan cap" (Hainan Kangda Marine Biomedical Corp., China) [116]; however, the amputated tip did not always grow to the full length. The Hyphecan occlusive dressing was also used in the treatment of other fingertip injuries [117]; however, the use of this material for the treatment of fingertip injuries outside of Hong Kong appears to be limited. A similar dressing was also used to treat burns in mice and was demonstrated to promote healing due to its modulation of TGFβ1 levels [118]. The use of a silver sulphadiazine dressing in 19 patients (aged 16 to 64 years) for the treatment of 21 distal fingertip amputations was reported with good to excellent results; however, documentation of bone regrowth (if any) was not presented [119]. Although regeneration is generally limited to the third phalanx in humans, there was a report of a child who suffered a crushing amputation at the proximal interphalangeal joints of her ring and little fingers and regenerated a distal phalanx with vestigial nail without the middle phalanx in her ring finger, though her little finger remained a stump [120].

4.1. Mouse Model for Digit/Appendage Regrowth. The newt and salamander regeneration models are useful for understanding the regeneration of an entire limb; however, as model systems, these are far removed from mammalian regeneration. The mouse is an ideal model to study digit tip regeneration, as the process in mice is similar to human fingertip regeneration. Both digit tips are similarly comprised of bones, tendons, muscles, skin, nerves, and blood vessels. Regeneration of the digit tip requires all these tissues to regrow in their proper locations and orientations to restore functionality. Several mouse models have been utilized to study digit regeneration [16]; however, there are differences between mammalian and axolotl regeneration besides their intrinsic regenerative abilities. For example, salamander limb regeneration is dependent on the presence of nerves; however, a denervated mouse digit tip can still regenerate, albeit at a reduced rate [121]. A recent study found that combinations of FGF8 and BMP7 gene therapy in neural cells in the dorsal root ganglia (DRG) were delivered to the limbs through the long axons of axolotls, suggesting major neural inputs of FGF and BMP in regulating blastema cell proliferation as well as controlling organ regeneration ability [22]. Denervation appears to affect the regenerative ability of the tissue by abrogating FGF signaling. FGF-2 is normally present in regenerating tissue but is not detectable after denervation [42]. Regenerating amphibians always form an AEC, which functionally mimics the AER during development; however, studies of the apical epithelium of regenerating digit tips are very limited, and there appears to be no AEC that forms in nonregenerating amputations [49].

4.2. Amputation Location. Studies of digit tip regeneration in mice have indicated a sharp transition between a tip that will regenerate and one that will not [40, 42, 112]. Regeneration is limited to the middle of the third phalanx. An amputation proximal to this region will result in

Phalanx

| 1 | 2 | 3 |

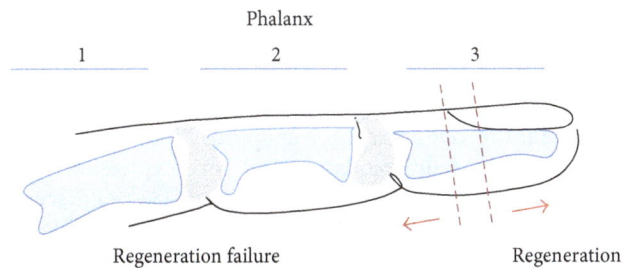

Regeneration failure   Regeneration

FIGURE 3: Amputation location affects the ability of digit/appendage to regenerate in mammals. Digit/appendage regrowth only occurs if the amputation site was distal to the middle to the 3rd phalanx, whereas digits amputated 2/3 through the 3rd phalanx do not regenerate.

a nonregenerative response, as presented in Figure 3. The ability of the amputated fingertip to regenerate is thought to be correlated with the presence of the nail bed, which grows continuously throughout life. The germinal matrix of the nail bed contains adult stem cells which are thought to be involved in the regrowth of the amputated digit tip [42]. Additionally, bone regrowth has been correlated to nail regrowth, and there is no bone regrowth in distal amputations when the nail is surgically ablated. Conversely, there is bone regrowth in proximal amputations where the bone is removed from the ventral surface of the digit but not the nail and matrix [12]. A nail transplantation study in the amputated proximal phalanges of rats showed limited bone regrowth when the nail was transplanted; however, no bone regrowth was seen without nail transplantation [122]. Wnt pathway activation of the nail stem cells appears to be required in order for blastema growth and digit tip regeneration to occur [42, 43]; however, the relationship between the nail and regeneration of the terminal phalanx is still unclear, as there are case studies of regenerative failure even when the nail root was present. Therefore, it was hypothesized that the nail bed is necessary but not sufficient for successful regeneration, perhaps aiding the scarless healing process [112].

4.3. The MRL Mouse Model.
The MRL/MpJ mouse strain has been commonly used as a model for autoimmune disease; it also has a unique capacity for wound healing and tissue regeneration without scar formation. Classically, this mouse strain displays an accelerated healing and tissue regeneration process after receiving an ear-hole punch. Moreover, 4-week-old MRL mice can regenerate their digits more quickly than control wild-type (WT) mice after having a distal digit amputated to the midpoint of the third phalanx [123]; however, when the digits from adult mice were amputated at the midpoint of the second phalanx, neither MRL mice nor controls could regenerate their digit tips [124, 125]. MRL mice (but not the WT controls) display blastema-like formation during the early stages after amputation; however, an apoptotic event eventually causes this structure to disappear. Altered ratios of collagens I and III, as well as differences in total collagen levels, have been demonstrated between MRL and WT mice, suggesting there would be differences in scar tissue formation, though not to the extent that there were

differences in the regeneration process [125]. In a recent study, we showed that the prevention of fibrosis formation with MMP-1 therapy resulted in better soft tissue regeneration within the amputated digits of adult mice [126]. Thus, the deposition of collagen occurs through an essential balance between ECM reconstruction and tissue regeneration.

4.4. Stem Cells and Blastema/Nonblastemal Dedifferentiation.
The regeneration of a newt or salamander limb is preceded by the formation of a proliferating blastema that is guided by the AEC. The mechanism of how this heterogeneous mass of dedifferentiated cells can then proceed to form a complete limb is still slowly being unraveled. Although there is no exact mammalian counterpart to the urodele blastema [127], digit tip regeneration in mice was shown to occur via the formation of a cluster of blastema-like mitotically active cells [128] that express BMP4 [112] as well as stem cell markers, including vimentin and Sca-1 [129]; however, the existence of a mammalian AEC during digit regeneration has not been demonstrated, which might explain the limited regeneration potential of the digit tip in mammals. Additionally, there is no evidence of dedifferentiation in the mammalian regenerating digit tip; however, this does not preclude the possibility that dedifferentiation may occur during mammalian digit tip regrowth.

Similar to lineage tracing studies in regenerating axolotl limbs and zebrafish fins, recent studies in mouse digit tip regeneration utilizing transgenic mice with Cre-mediated reporters corroborate the finding that the regenerated structures are lineage-dependent and derived from local tissues [40, 130]. Resident stem cells, which are already committed to become specific tissue types, are responsible for digit regeneration in mice; however, this does not rule out the possibility that terminally differentiated tissue can dedifferentiate into resident stem cells.

4.5. Msx1, Msx2, and BMP4.
The Msh homeobox (Msx) type 1 and type 2 transcriptional repressors are both expressed near the nail bed of neonatal mice and at the tips of developing digits [6]. It has been suggested that Msx-1 is required to maintain some cell types in an undifferentiated state and may be associated with urodele limb regeneration, and inactive msx genes also alter epithelial cell junction proteins during embryo implantation [131]. In amphibians, Msx-1 is initially upregulated and then downregulated during regeneration [132]. Fetal mice deficient in Msx-1, but not Msx-2, do not readily regenerate amputated digit tips; however, this can be restored in culture in a dose-dependent manner with the addition of bone morphogenetic protein 4 (BMP4) [6]. This study also demonstrated that blocking BMP4 signaling using Noggin (a BMP inhibitor) prevented fetal digit regeneration. Hence, mammalian digit regeneration was shown to be dependent on Msx1 and BMP2 modulation [133].

5. Future Directions

5.1. Promotion of Dedifferentiation.
Adult stem cells have been found to contribute to the regeneration of a number of

human tissues which are present in bone marrow, intestinal mucosa, superficial layers of the skin, liver, and the nail bed; however, regeneration of complex structures such as digits or limbs requires a greater number of progenitor cells than that naturally present in adult tissues. Urodele amphibians overcome this deficiency by producing more progenitor cells via dedifferentiation of terminally differentiated cells in the blastema; hence, regeneration could be enhanced in mammals by increasing mammalian dedifferentiation. Dedifferentiation refers to the ability of terminally differentiated somatic cells to revert to a more plastic progenitor cell state. The methods of somatic cell nuclear transfer, chromosome transfer, or fusion with ES cells have all been used to induce totipotency or pluripotency [134, 135]. Utilizing a combination of transcription factors, fibroblast cells can be converted into induced pluripotent stem cells (iPS cells) [136]. C2C12 myotubes, which are mature differentiated multinucleated muscle cells, have been shown to dedifferentiate when induced to express *Msx-1* [137], the microtubule-binding molecule myoseverin [138], and the small molecule, reversine [139], or when treated with extracts from regenerating newt limbs [140]. Muscle cells that were dedifferentiated upon treatment with reversine could be redifferentiated under the appropriate lineage-specific inducing conditions into cells of different lineages, including osteoblasts and adipocytes. Another method of dedifferentiating myotubes is cell cycle reentry by means of inhibiting the tumor suppression genes Rb and ARF [76], which indicates that dedifferentiation may be possible in mammals. We were able to label terminally differentiated, multinucleated myotubes with β-galactosidase via a Cre-Lox system [141]. Following muscle injury, we observed β-galactosidase-positive mononuclear cells, which were able to differentiate into different types of muscle cells, suggesting that these progenitor cells were the result of mammalian dedifferentiation during wound healing.

5.2. Pathway Activation. A number of novel signaling pathways that are involved in cell proliferation and tissue growth have been revealed recently, such as the Wnt and Hippo pathways. In particular, Wnt pathway activation has been shown to be involved in digit regeneration [42, 43]; moreover, genes, including *LRP6 and LRP5*, that are related to Wnt/beta-catenin signal transduction have been found to be differentially expressed (higher expression) in MRL mice, which can form blastema-like structures, compared to DBA and C57BL mice [123].

The Hippo signaling pathway has also been shown to regulate cell proliferation and stem cell function. While its downstream effector Yes-Associated Protein (YAP) contributes to cancer development, its activation also has beneficial roles in regenerative medicine applications. In particular, the Hippo pathway has direct regulating effects on stem cell proliferation and maintenance [142, 143] which may be important for inducing the accumulation of blastema-like cells. Thus, developing molecular tools that can activate the Wnt or Hippo pathways in the amputated digits of mammals might be capable of enhancing the regeneration process [144].

5.3. Electrical Stimulation. The effect of electric fields on regeneration was revealed when newt limbs were induced to dedifferentiate by only applying an electric field (i.e., no amputation) strong enough to induce electroporation with the absence of cell necrosis or apoptosis. The time courses for changes in dedifferentiation and gene expression were similar to that occurring after amputation [145]. Quiescent, terminally differentiated cells are electrically polarized; however, tumor cells and stem cells are generally depolarized. The application of an electric field could represent a novel approach to promote digit/appendage regeneration and could be used in combination with other approaches (e.g., pathway activation or growth factor delivery).

6. Summary

Urodele amphibians such as newts and salamanders can regenerate large portions of their bodies, including an entire limb. Limb regeneration in mammals is much more limited with only a portion of the terminal phalanx being capable of regenerating, and this is generally further limited to neonatal or young mammals. Proximal amputation usually results in incomplete wound healing and scar formation. Studying the molecular mechanisms of amphibian regeneration and mammalian wound healing could lead to novel therapeutic strategies to augment the regenerative response beyond the current natural limits of regeneration in mammals, including humans.

Conflicts of Interest

There are no conflicts of interest to declare.

Authors' Contributions

Yohan Choi and Fanwei Meng contributed equally to the manuscript.

Acknowledgments

The authors would like to acknowledge the financial support provided by the funding of the Department of Defense, the Texas Emerging Technology Foundation, and the Department of Pediatric Surgery, the University of Texas Medical School at Houston. The authors would also like to thank Mr. James H. Cummins and Dr. Lavanya Rajagopalan for editorial assistance.

References

[1] S. J. Odelberg, "Cellular plasticity in vertebrate regeneration," *Anatomical Record Part B: New Anatomist*, vol. 287, no. 1, pp. 25–35, 2005.

[2] A. L. Mescher, "The cellular basis of limb regeneration in urodeles," *International Journal of Developmental Biology*, vol. 40, no. 4, pp. 785–795, 1996.

[3] M. Han, X. Yang, G. Taylor, C. A. Burdsal, R. A. Anderson, and K. Muneoka, "Limb regeneration in higher vertebrates: developing a roadmap," *Anatomical Record Part B: New Anatomist*, vol. 287, no. 1, pp. 14–24, 2005.

[4] R. B. Borgens, "Mice regrow the tips of their foretoes," *Science*, vol. 217, no. 4561, pp. 747–750, 1982.

[5] A. D. Reginelli, Y. Q. Wang, D. Sassoon, and K. Muneoka, "Digit tip regeneration correlates with regions of Msx1 (Hox 7) expression in fetal and newborn mice," *Development*, vol. 121, no. 4, pp. 1065–1076, 1995.

[6] M. Han, X. Yang, J. E. Farrington, and K. Muneoka, "Digit regeneration is regulated by Msx1 and BMP4 in fetal mice," *Development*, vol. 130, no. 21, pp. 5123–5132, 2003.

[7] D. A. Neufeld and W. Zhao, "Phalangeal regrowth in rodents: postamputational bone regrowth depends upon the level of amputation," *Progress in Clinical and Biological Research*, vol. 383, pp. 243–252, 1993.

[8] H. Masaki and H. Ide, "Regeneration potency of mouse limbs," *Development Growth and Differentiation*, vol. 49, no. 2, pp. 89–98, 2007.

[9] B. S. Douglas, "Conservative management of guillotine amputation of the finger in children," *Australian Paediatric Journal*, vol. 8, no. 2, pp. 86–89, 1972.

[10] C. M. Illingworth, "Trapped fingers and amputated finger tips in children," *Journal of Pediatric Surgery*, vol. 9, no. 6, pp. 853–858, 1974.

[11] P. Vidal and M. G. Dickson, "Regeneration of the distal phalanx: a case report," *Journal of Hand Surgery*, vol. 18, no. 2, pp. 230–233, 1993.

[12] W. Zhao and D. A. Neufeld, "Bone regrowth in young mice stimulated by nail organ," *Journal of Experimental Zoology*, vol. 271, no. 2, pp. 155–159, 1995.

[13] B. A. Mast, R. F. Diegelmann, T. M. Krummel, and I. K. Cohen, "Scarless wound healing in the mammalian fetus," *Surgery Gynecology and Obstetrics*, vol. 174, no. 5, pp. 441–451, 1992.

[14] P. Samuels and A. K. W. Tan, "Fetal scarless wound healing," *Journal of Otolaryngology*, vol. 28, no. 5, pp. 296–302, 1999.

[15] K. M. Bullard, M. T. Longaker, and H. P. Lorenz, "Fetal wound healing: current biology," *World Journal of Surgery*, vol. 27, no. 1, pp. 54–61, 2003.

[16] Y. Choi, C. Cox, K. Lally, and Y. Li, "The strategy and method in modulating finger regeneration," *Regenerative Medicine*, vol. 9, no. 2, pp. 231–242, 2014.

[17] A. Johnston, S. Yuzwa, M. Carr et al., "Dedifferentiated schwann cell precursors secreting paracrine factors are required for regeneration of the mammalian digit tip," *Cell Stem Cell*, vol. 19, no. 4, pp. 433–448, 2016.

[18] L. Yu, M. Yan, J. Simkin et al., "Angiogenesis is inhibitory for mammalian digit regeneration," *Regeneration*, vol. 1, no. 3, pp. 33–46, 2014.

[19] T. Sandoval-Guzmán, H. Wang, S. Khattak et al., "Fundamental differences in dedifferentiation and stem cell recruitment during skeletal muscle regeneration in two salamander species," *Cell Stem Cell*, vol. 14, no. 2, pp. 174–187, 2014.

[20] S. V. Bryant, T. Endo, and D. M. Gardiner, "Vertebrate limb regeneration and the origin of limb stem cells," *International Journal of Developmental Biology*, vol. 46, no. 7, pp. 887–896, 2002.

[21] M. Suzuki, A. Satoh, H. Ide, and K. Tamura, "Nerve-dependent and -independent events in blastema formation during Xenopus froglet limb regeneration," *Developmental Biology*, vol. 286, no. 1, pp. 361–375, 2005.

[22] A. Satoh, A. Makanae, Y. Nishimoto, and K. Mitogawa, "FGF and BMP derived from dorsal root ganglia regulate blastema induction in limb regeneration in *Ambystoma mexicanum*," *Developmental Biology*, vol. 417, no. 1, pp. 114–125, 2016.

[23] A. F. Nogueira, C. M. Costa, J. Lorena et al., "Tetrapod limb and sarcopterygian fin regeneration share a core genetic programme," *Nature Communications*, vol. 7, Article ID 13364, 2016.

[24] D. L. Stocum, "The role of peripheral nerves in urodele limb regeneration," *European Journal of Neuroscience*, vol. 34, no. 6, pp. 908–916, 2011.

[25] D. L. Stocum and J. A. Cameron, "Looking proximally and distally: 100 years of limb regeneration and beyond," *Developmental Dynamics*, vol. 240, no. 5, pp. 943–968, 2011.

[26] K. Crawford and D. L. Stocum, "Retinoic acid coordinately proximalizes regenerate pattern and blastema differential affinity in axolotl limbs," *Development*, vol. 102, no. 4, pp. 687–698, 1988.

[27] L. E. Iten and S. V. Bryant, "The interaction between the blastema and stump in the establishment of the anterior-posterior and proximal-distal organization of the limb regenerate," *Developmental Biology*, vol. 44, no. 1, pp. 119–147, 1975.

[28] D. C. Lo, F. Allen, and J. P. Brockes, "Reversal of muscle differentiation during urodele limb regeneration," *Proceedings of the National Academy of Sciences of the United States of America*, vol. 90, no. 15, pp. 7230–7234, 1993.

[29] N. Terada, T. Hamazaki, M. Oka et al., "Bone marrow cells adopt the phenotype of other cells by spontaneous cell fusion," *Nature*, vol. 416, no. 6880, pp. 542–545, 2002.

[30] Z. Yang, Q. Liu, R. J. Mannix et al., "Mononuclear cells from dedifferentiation of mouse myotubes display remarkable regenerative capability," *Stem Cells*, vol. 32, no. 9, pp. 2492–2501, 2014.

[31] M. Kragl, D. Knapp, E. Nacu et al., "Cells keep a memory of their tissue origin during axolotl limb regeneration," *Nature*, vol. 460, no. 7251, pp. 60–65, 2009.

[32] S. Tu and S. L. Johnson, "Fate restriction in the growing and regenerating zebrafish fin," *Developmental Cell*, vol. 20, no. 5, pp. 725–732, 2011.

[33] S. Puri, A. E. Folias, and M. Hebrok, "Plasticity and dedifferentiation within the pancreas: development, homeostasis, and disease," *Cell Stem Cell*, vol. 16, no. 1, pp. 18–31, 2015.

[34] P. R. Tata, H. Mou, A. Pardo-Saganta et al., "Dedifferentiation of committed epithelial cells into stem cells in vivo," *Nature*, vol. 503, no. 7475, pp. 218–223, 2013.

[35] C. R. Kintner and J. P. Brockes, "Monoclonal antibodies identify blastemal cells derived from dedifferentiating muscle in newt limb regeneration," *Nature*, vol. 308, no. 5954, pp. 67–69, 1984.

[36] J. I. Morrison, S. Lööf, P. He, and A. Simon, "Salamander limb regeneration involves the activation of a multipotent skeletal muscle satellite cell population," *Journal of Cell Biology*, vol. 172, no. 3, pp. 433–440, 2006.

[37] K. Muneoka, W. F. Fox, and S. V. Bryant, "Cellular contribution from dermis and cartilage to the regenerating limb blastema in axolotls," *Developmental Biology*, vol. 116, no. 1, pp. 256–260, 1986.

[38] D. M. Gardiner, K. Muneoka, and S. V. Bryant, "The migration of dermal cells during blastema formation in axolotls," *Developmental Biology*, vol. 118, no. 2, pp. 488–493, 1986.

[39] C. Li, J. M. Suttie, and D. E. Clark, "Histological examination of antler regeneration in red deer (*Cervus elaphus*)," *The Anatomical Record. Part A, Discoveries in Molecular, Cellular, and Evolutionary Biology*, vol. 282, no. 2, pp. 163–174, 2005.

[40] Y. Rinkevich, P. Lindau, H. Ueno, M. T. Longaker, and I. L. Weissman, "Germ-layer and lineage-restricted stem/progenitors regenerate the mouse digit tip," *Nature*, vol. 476, no. 7361, pp. 409–414, 2011.

[41] Y. Rinkevich, Z. N. Maan, G. G. Walmsley, and S. K. Sen, "Injuries to appendage extremities and digit tips: a clinical and cellular update," *Developmental Dynamics*, vol. 244, no. 5, pp. 641–650, 2015.

[42] M. Takeo, W. C. Chou, Q. Sun et al., "Wnt activation in nail epithelium couples nail growth to digit regeneration," *Nature*, vol. 499, no. 7457, pp. 228–232, 2013.

[43] J. A. Lehoczky and C. J. Tabin, "Lgr6 marks nail stem cells and is required for digit tip regeneration," *Proceedings of the National Academy of Sciences of the United States of America*, vol. 112, no. 43, pp. 13249–13254, 2015.

[44] C. Jopling, S. Boue, and J. C. I. Belmonte, "Dedifferentiation, transdifferentiation and reprogramming: three routes to regeneration," *Nature Reviews Molecular Cell Biology*, vol. 12, no. 2, pp. 79–89, 2011.

[45] M. Bhaskaran, N. Kolliputi, Y. Wang, D. Gou, N. R. Chintagari, and L. Liu, "Trans-differentiation of alveolar epithelial type II cells to type I cells involves autocrine signaling by transforming growth factor β1 through the Smad pathway," *The Journal of Biological Chemistry*, vol. 282, no. 6, pp. 3968–3976, 2007.

[46] D. J. Pearton, Y. Yang, and D. Dhouailly, "Transdifferentiation of corneal epithelium into epidermis occurs by means of a multistep process triggered by dermal development signals," *Proceedings of the National Academy of Sciences of the United States of America*, vol. 102, no. 10, pp. 3714–3719, 2005.

[47] S. M. Boularaoui, K. M. Abdel-Raouf, N. S. Alwahab et al., "Efficient transdifferentiation of human dermal fibroblasts into skeletal muscle," *Journal of Tissue Engineering and Regenerative Medicine*, 2017.

[48] A. Q. Phan, J. Lee, M. Oei et al., "Positional information in axolotl and mouse limb extracellular matrix is mediated via heparan sulfate and fibroblast growth factor during limb regeneration in the axolotl (Ambystoma mexicanum)," *Regeneration*, vol. 2, no. 4, pp. 182–201, 2015.

[49] A. Satoh, S. V. Bryant, and D. M. Gardiner, "Nerve signaling regulates basal keratinocyte proliferation in the blastema apical epithelial cap in the axolotl (Ambystoma mexicanum)," *Developmental Biology*, vol. 366, no. 2, pp. 374–381, 2012.

[50] M. Lévesque, S. Gatien, K. Finnson et al., "Transforming growth factor: β signaling is essential for limb regeneration in axolotls," *PLoS ONE*, vol. 2, no. 11, Article ID e1227, 2007.

[51] F. Chablais and A. Jaźwińska, "IGF signaling between blastema and wound epidermis is required for fin regeneration," *Development*, vol. 137, no. 6, pp. 871–879, 2010.

[52] E. R. Zielins, R. C. Ransom, T. E. Leavitt, M. T. Longaker, and D. C. Wan, "The role of stem cells in limb regeneration," *Organogenesis*, vol. 12, no. 1, pp. 16–27, 2016.

[53] C. McCusker, S. V. Bryant, and D. M. Gardiner, "The axolotl limb blastema: cellular and molecular mechanisms driving blastema formation and limb regeneration in tetrapods," *Regeneration*, vol. 2, no. 2, pp. 54–71, 2015.

[54] A. Kumar, J. W. Godwin, P. B. Gates, A. A. Garza-Garcia, and J. P. Brockes, "Molecular basis for the nerve dependence of limb regeneration in an adult vertebrate," *Science*, vol. 318, no. 5851, pp. 772–777, 2007.

[55] B. Boilly, K. P. Cavanaugh, D. Thomas, H. Hondermarck, S. V. Bryant, and R. A. Bradshaw, "Acidic fibroblast growth factor is present in regenerating limb blastemas of axolotls and binds specifically to blastema tissues," *Developmental Biology*, vol. 145, no. 2, pp. 302–310, 1991.

[56] S. I. Munaim and A. L. Mescher, "Transferrin and the trophic effect of neural tissue on amphibian limb regeneration blastemas," *Developmental Biology*, vol. 116, no. 1, pp. 138–142, 1986.

[57] L. Wang, M. A. Marchionni, and R. A. Tassava, "Cloning and neuronal expression of a type III newt neuregulin and rescue of denervated, nerve-dependent newt limb blastemas by rhGGF2," *Journal of Neurobiology*, vol. 43, no. 2, pp. 150–158, 2000.

[58] E. D. Hay and D. A. Fischman, "Origin of the blastema in regenerating limbs of the newt *Triturus viridescens*. An autoradiographic study using tritiated thymidine to follow cell proliferation and migration," *Developmental Biology*, vol. 3, no. 1, pp. 26–59, 1961.

[59] J. Simkin, M. C. Sammarco, L. A. Dawson, P. P. Schanes, L. Yu, and K. Muneoka, "The mammalian blastema: regeneration at our fingertips," *Regeneration*, vol. 2, no. 3, pp. 93–105, 2015.

[60] K. Tamura, S. Ohgo, and H. Yokoyama, "Limb blastema cell: a stem cell for morphological regeneration," *Development Growth and Differentiation*, vol. 52, no. 1, pp. 89–99, 2010.

[61] M. Kragl and E. M. Tanaka, "Axolotl (Ambystoma mexicanum) limb and tail amputation," *Cold Spring Harbor Protocols*, vol. 4, no. 8, 2009.

[62] Y. Wu, K. Wang, A. Karapetyan et al., "Connective tissue fibroblast properties are position-dependent during mouse digit tip regeneration," *PLoS ONE*, vol. 8, no. 1, Article ID e54764, 2013.

[63] R. Zeller, L. Jackson-Grusby, and P. Leder, "The limb deformity gene is required for apical ectodermal ridge differentiation and anteroposterior limb pattern formation," *Genes & Development*, vol. 3, no. 10, pp. 1481–1492, 1989.

[64] A. Satoh, A. Makanae, and N. Wada, "The apical ectodermal ridge (AER) can be re-induced by wounding, wnt-2b, and fgf-10 in the chicken limb bud," *Developmental Biology*, vol. 342, no. 2, pp. 157–168, 2010.

[65] J. Zakany, G. Zacchetti, and D. Duboule, "Interactions between HOXD and Gli3 genes control the limb apical ectodermal ridge via Fgf10," *Developmental Biology*, vol. 306, no. 2, pp. 883–893, 2007.

[66] J. W. Godwin, A. R. Pinto, and N. A. Rosenthal, "Macrophages are required for adult salamander limb regeneration," *Proceedings of the National Academy of Sciences of the United States of America*, vol. 110, no. 23, pp. 9415–9420, 2013.

[67] N. Mercader, E. M. Tanaka, and M. Torres, "Proximodistal identity during vertebrate limb regeneration is regulated by Meis homeodomain proteins," *Development*, vol. 132, no. 18, pp. 4131–4142, 2005.

[68] T. A. Bhat, D. Nambiar, D. Tailor, A. Pal, R. Agarwal, and R. P. Singh, "Acacetin inhibits in vitro and in vivo angiogenesis and downregulates Stat signaling and VEGF expression," *Cancer Prevention Research*, vol. 6, no. 10, pp. 1128–1139, 2013.

[69] V. Vinarsky, D. L. Atkinson, T. J. Stevenson, M. T. Keating, and S. J. Odelberg, "Normal newt limb regeneration requires matrix metalloproteinase function," *Developmental Biology*, vol. 279, no. 1, pp. 86–98, 2005.

[70] N. Santosh, L. J. Windsor, B. S. Mahmoudi et al., "Matrix metalloproteinase expression during blastema formation in regeneration-competent versus regeneration-deficient amphibian limbs," *Developmental Dynamics*, vol. 240, no. 5, pp. 1127–1141, 2011.

[71] T. J. Stevenson, V. Vinarsky, D. L. Atkinson, M. T. Keating, and S. J. Odelberg, "Tissue inhibitor of metalloproteinase 1 regulates matrix metalloproteinase activity during newt limb regeneration," *Developmental Dynamics*, vol. 235, no. 3, pp. 606–616, 2006.

[72] S. Calve, S. J. Odelberg, and H.-G. Simon, "A transitional extracellular matrix instructs cell behavior during muscle regeneration," *Developmental Biology*, vol. 344, no. 1, pp. 259–271, 2010.

[73] E. Forsberg, E. Hirsch, L. Fröhlich et al., "Skin wounds and severed nerves heal normally in mice lacking tenascin-C," *Proceedings of the National Academy of Sciences of the United States of America*, vol. 93, no. 13, pp. 6594–6599, 1996.

[74] D. J. Milner and J. A. Cameron, "Muscle repair and regeneration: stem cells, scaffolds, and the contributions of skeletal muscle to amphibian limb regeneration," *Current Topics in Microbiology and Immunology*, vol. 367, pp. 133–159, 2013.

[75] B. J. Larson, M. T. Longaker, and H. P. Lorenz, "Scarless fetal wound healing: a basic science review," *Plastic and Reconstructive Surgery*, vol. 126, no. 4, pp. 1172–1180, 2010.

[76] K. V. Pajcini, S. Y. Corbel, J. Sage, J. H. Pomerantz, and H. M. Blau, "Transient inactivation of Rb and ARF yields regenerative cells from postmitotic mammalian muscle," *Cell Stem Cell*, vol. 7, no. 2, pp. 198–213, 2010.

[77] M. Levin, "Bioelectric mechanisms in regeneration: unique aspects and future perspectives," *Seminars in Cell and Developmental Biology*, vol. 20, no. 5, pp. 543–556, 2009.

[78] J. A. Schilling, "Wound healing," *Surgical Clinics of North America*, vol. 56, no. 4, pp. 859–874, 1976.

[79] H. Yokoyama, "Initiation of limb regeneration: the critical steps for regenerative capacity," *Development Growth and Differentiation*, vol. 50, no. 1, pp. 13–22, 2008.

[80] M. C. Sammarco, J. Simkin, A. J. Cammack et al., "Hyperbaric oxygen promotes proximal bone regeneration and organized collagen composition during digit regeneration," *PLoS ONE*, vol. 10, no. 10, Article ID e0140156, 2015.

[81] P. Martin, D. D'Souza, J. Martin et al., "Wound healing in the PU.1 null mouse—tissue repair is not dependent on inflammatory cells," *Current Biology*, vol. 13, no. 13, pp. 1122–1128, 2003.

[82] G. C. Gurtner, S. Werner, Y. Barrandon, and M. T. Longaker, "Wound repair and regeneration," *Nature*, vol. 453, no. 7193, pp. 314–321, 2008.

[83] C. Amendt, A. Mann, P. Schirmacher, and M. Blessing, "Resistance of keratinocytes to TGFβ-mediated growth restriction and apoptosis induction accelerates re-epitheliazation in skin wounds," *Journal of Cell Science*, vol. 115, no. 10, pp. 2189–2198, 2002.

[84] G. S. Ashcroft, X. Yang, A. B. Glick et al., "Mice lacking Smad3 show accelerated wound healing and an impaired local inflammatory response," *Nature Cell Biology*, vol. 1, no. 5, pp. 260–266, 1999.

[85] G. Broughton II, J. E. Janis, and C. E. Attinger, "The basic science of wound healing," *Plastic and Reconstructive Surgery*, vol. 117, no. 7, pp. 12S–34S, 2006.

[86] Y. Li, W. Foster, B. M. Deasy et al., "Transforming growth factor-β1 induces the differentiation of myogenic cells into fibrotic cells in injured skeletal muscle: a key event in muscle fibrogenesis," *American Journal of Pathology*, vol. 164, no. 3, pp. 1007–1019, 2004.

[87] M. B. Witte and A. Barbul, "General principles of wound healing," *Surgical Clinics of North America*, vol. 77, no. 3, pp. 509–528, 1997.

[88] H. N. Lovvorn III, D. T. Cheung, M. E. Nimni, N. Perelman, J. M. Estes, and N. S. Adzick, "Relative distribution and crosslinking of collagen distinguish fetal from adult sheep wound repair," *Journal of Pediatric Surgery*, vol. 34, no. 1, pp. 218–223, 1999.

[89] E. Due, K. Rossen, L. T. Sorensen, A. Kliem, T. Karlsmark, and M. Haedersdal, "Effect of UV irradiation on cutaneous cicatrices: a randomized, controlled trial with clinical, skin reflectance, histological, immunohistochemical and biochemical evaluations," *Acta Dermato-Venereologica*, vol. 87, no. 1, pp. 27–32, 2007.

[90] M. Globus, S. Vethamany-Globus, and Y. C. I. Lee, "Effect of apical epidermal cap on mitotic cycle and cartilage differentiation in regeneration blastemata in the newt, Notophthalmus viridescens," *Developmental Biology*, vol. 75, no. 2, pp. 358–372, 1980.

[91] D. L. Stocum and K. Crawford, "Use of retinoids to analyze the cellular basis of positional memory in regenerating amphibian limbs," *Biochemistry and Cell Biology*, vol. 65, no. 8, pp. 750–761, 1987.

[92] A. Baranowsky, S. Mokkapati, M. Bechtel et al., "Impaired wound healing in mice lacking the basement membrane protein nidogen 1," *Matrix Biology*, vol. 29, no. 1, pp. 15–21, 2010.

[93] L. D. Clark, R. K. Clark, and E. Heber-Katz, "A new murine model for mammalian wound repair and regeneration," *Clinical Immunology and Immunopathology*, vol. 88, no. 1, pp. 35–45, 1998.

[94] D. Gourevitch, L. Clark, P. Chen, A. Seitz, S. J. Samulewicz, and E. Heber-Katz, "Matrix metalloproteinase activity correlates with blastema formation in the regenerating MRL mouse ear hole model," *Developmental Dynamics*, vol. 226, no. 2, pp. 377–387, 2003.

[95] W. C. Parks, C. L. Wilson, and Y. S. López-Boado, "Matrix metalloproteinases as modulators of inflammation and innate immunity," *Nature Reviews Immunology*, vol. 4, no. 8, pp. 617–629, 2004.

[96] I. H. Bellayr, X. Mu, and Y. Li, "Biochemical insights into the role of matrix metalloproteinases in regeneration: challenges and recent developments," *Future Medicinal Chemistry*, vol. 1, no. 6, pp. 1095–1111, 2009.

[97] C. I. Coon, S. Fiering, J. Gaudet, C. A. Wyatt, and C. E. Brincker-hoff, "Site controlled transgenic mice validating increased expression from human matrix metalloproteinase (MMP-1) promoter due to a naturally occurring SNP," *Matrix Biology*, vol. 28, no. 7, pp. 425–431, 2009.

[98] L. J. Stevens and A. Page-McCaw, "A secreted MMP is required for reepithelialization during wound healing," *Molecular Biology of the Cell*, vol. 23, no. 6, pp. 1068–1079, 2012.

[99] M. Balbín, A. Fueyo, V. Knäuper et al., "Identification and enzymatic characterization of two diverging murine counterparts of human interstitial collagenase (MMP-1) expressed at sites of embryo implantation," *Journal of Biological Chemistry*, vol. 276, no. 13, pp. 10253–10262, 2001.

[100] C. E. Brinckerhoff and L. M. Matrisian, "Matrix metalloproteinases: a tail of a frog that became a prince," *Nature Reviews Molecular Cell Biology*, vol. 3, no. 3, pp. 207–214, 2002.

[101] B. Hartenstein, B. T. Dittrich, D. Stickens et al., "Epidermal development and wound healing in matrix metalloproteinase 13-deficient mice," *Journal of Investigative Dermatology*, vol. 126, no. 2, pp. 486–496, 2006.

[102] P. L. Danielsen, A. V. Holst, H. R. Maltesen et al., "Matrix metalloproteinase-8 overexpression prevents proper tissue repair," *Surgery*, vol. 150, no. 5, pp. 897–906, 2011.

[103] U. Mirastschijski, C. J. Haaksma, J. J. Tomasek, and M. S. Ågren, "Matrix metalloproteinase inhibitor GM 6001 attenuates keratinocyte migration, contraction and myofibroblast formation

in skin wounds," *Experimental Cell Research*, vol. 299, no. 2, pp. 465–475, 2004.

[104] I. Bellayr, K. Holden, X. Mu, H. Pan, and Y. Li, "Matrix metalloproteinase inhibition negatively affects muscle stem cell behavior," *International Journal of Clinical and Experimental Pathology*, vol. 6, no. 2, pp. 124–141, 2013.

[105] X. Liu, H. Wu, M. Byrne, J. Jeffrey, S. Krane, and R. Jaenisch, "A targeted mutation at the known collagenase cleavage site in mouse type I collagen impairs tissue remodeling," *Journal of Cell Biology*, vol. 130, no. 1, pp. 227–237, 1995.

[106] A. H. M. Beare, S. O'Kane, S. M. Krane, and M. W. J. Ferguson, "Severely impaired wound healing in the collagenase-resistant mouse," *Journal of Investigative Dermatology*, vol. 120, no. 1, pp. 153–163, 2003.

[107] A. Gutiérrez-Fernández, M. Inada, M. Balbín et al., "Increased inflammation delays wound healing in mice deficient in collagenase-2 (MMP-8)," *The FASEB Journal*, vol. 21, no. 10, pp. 2580–2591, 2007.

[108] N. Hattori, S. Mochizuki, K. Kishi et al., "MMP-13 plays a role in keratinocyte migration, angiogenesis, and contraction in mouse skin wound healing," *American Journal of Pathology*, vol. 175, no. 2, pp. 533–546, 2009.

[109] T. R. Kyriakides, D. Wulsin, E. A. Skokos et al., "Mice that lack matrix metalloproteinase-9 display delayed wound healing associated with delayed reepithelization and disordered collagen fibrillogenesis," *Matrix Biology*, vol. 28, no. 2, pp. 65–73, 2009.

[110] R. Mohan, S. K. Chintala, J. C. Jung et al., "Matrix metalloproteinase gelatinase B (MMP-9) coordinates and effects epithelial regeneration," *Journal of Biological Chemistry*, vol. 277, no. 3, pp. 2065–2072, 2002.

[111] M. Zimowska, K. H. Olszynski, M. Swierczynska, W. Streminska, and M. A. Ciemerych, "Decrease of MMP-9 activity improves soleus muscle regeneration," *Tissue Engineering. Part A*, vol. 18, no. 11-12, pp. 1183–1192, 2012.

[112] M. Han, X. Yang, J. Lee, C. H. Allan, and K. Muneoka, "Development and regeneration of the neonatal digit tip in mice," *Developmental Biology*, vol. 315, no. 1, pp. 125–135, 2008.

[113] A. L. Mescher, "Effects on adult newt limb regeneration of partial and complete skin flaps over the amputation surface," *Journal of Experimental Zoology*, vol. 195, no. 1, pp. 117–127, 1976.

[114] M. J. Allen, "Conservative management of finger tip injuries in adults," *Hand*, vol. 12, no. 3, pp. 257–265, 1980.

[115] L. H. McKim, "Regeneration of the distal phalanx," *Canadian Medical Association Journal*, vol. 26, no. 5, pp. 549–550, 1932.

[116] L. P. Lee, P. Y. Lau, and C. W. Chan, "A simple and efficient treatment for fingertip injuries," *Journal of Hand Surgery (British and European Volume)*, vol. 20, no. 1, pp. 63–71, 1995.

[117] A. S. Halim, C. A. Stone, and V. S. Devaraj, "The Hyphecan cap: a biological fingertip dressing," *Injury*, vol. 29, no. 4, pp. 261–263, 1998.

[118] R. M. Baxter, T. Dai, J. Kimball et al., "Chitosan dressing promotes healing in third degree burns in mice: gene expression analysis shows biphasic effects for rapid tissue regeneration and decreased fibrotic signaling," *Journal of Biomedical Materials Research. Part A*, vol. 101, no. 2, pp. 340–348, 2013.

[119] S. C. Buckley, S. Scott, and K. Das, "Late review of the use of silver sulphadiazine dressings for the treatment of fingertip injuries," *Injury*, vol. 31, no. 5, pp. 301–304, 2000.

[120] C. E. Cobiella, F. S. Haddad, and I. Bacarese-Hamilton, "Phalangeal metaplasia following amputation in a child's finger," *Injury*, vol. 28, no. 5-6, pp. 409–410, 1997.

[121] K. S. Mohammad and D. A. Neufeld, "Denervation retards but does not prevent toetip regeneration," *Wound Repair and Regeneration*, vol. 8, no. 4, pp. 277–281, 2000.

[122] K. S. Mohammad, F. A. Day, and D. A. Neufeld, "Bone growth is induced by nail transplantation in amputated proximal phalanges," *Calcified Tissue International*, vol. 65, no. 5, pp. 408–410, 1999.

[123] R. B. Chadwick, L. Bu, H. Yu et al., "Digit tip regrowth and differential gene expression in MRL/Mpj, DBA/2, and C57BL/6 mice," *Wound Repair and Regeneration*, vol. 15, no. 2, pp. 275–284, 2007.

[124] N. J. Turner, S. A. Johnson, and S. F. Badylak, "A histomorphologic study of the normal healing response following digit amputation in C57bl/6 and MRL/MpJ mice," *Archives of Histology and Cytology*, vol. 73, no. 2, pp. 103–111, 2011.

[125] D. L. Gourevitch, L. Clark, K. Bedelbaeva, J. Leferovich, and E. Heber-Katz, "Dynamic changes after murine digit amputation: the MRL mouse digit shows waves of tissue remodeling, growth, and apoptosis," *Wound Repair and Regeneration*, vol. 17, no. 3, pp. 447–455, 2009.

[126] X. Mu, I. Bellayr, H. Pan, Y. Choi, and Y. Li, "Regeneration of soft tissues is promoted by MMP1 treatment after digit amputation in mice," *PLoS ONE*, vol. 8, no. 3, Article ID e59105, 2013.

[127] K. Muneoka, C. H. Allan, X. Yang, J. Lee, and M. Han, "Mammalian regeneration and regenerative medicine," *Birth Defects Research Part C - Embryo Today: Reviews*, vol. 84, no. 4, pp. 265–280, 2008.

[128] D. A. Neufeld, "Partial blastema formation after amputation in adult mice," *Journal of Experimental Zoology*, vol. 212, no. 1, pp. 31–36, 1980.

[129] W. A. Fernando, E. Leininger, J. Simkin et al., "Wound healing and blastema formation in regenerating digit tips of adult mice," *Developmental Biology*, vol. 350, no. 2, pp. 301–310, 2011.

[130] J. A. Lehoczky, B. Robert, and C. J. Tabin, "Mouse digit tip regeneration is mediated by fate-restricted progenitor cells," *Proceedings of the National Academy of Sciences of the United States of America*, vol. 108, no. 51, pp. 20609–20614, 2011.

[131] X. Sun, C. B. Park, W. Deng, S. S. Potter, and S. K. Dey, "Uterine inactivation of muscle segment homeobox (Msx) genes alters epithelial cell junction proteins during embryo implantation," *FASEB Journal*, vol. 30, no. 4, pp. 1425–1435, 2016.

[132] L. Crews, P. B. Gates, R. Brown et al., "Expression and activity of the newt Msx-1 gene in relation to limb regeneration," *Proceedings of the Royal Society B: Biological Sciences*, vol. 259, no. 1355, pp. 161–171, 1995.

[133] L. Yu, M. Han, M. Yan, E.-C. Lee, J. Lee, and K. Muneoka, "BMP signaling induces digit regeneration in neonatal mice," *Development*, vol. 137, no. 4, pp. 551–559, 2010.

[134] K. Hochedlinger and R. Jaenisch, "Nuclear reprogramming and pluripotency," *Nature*, vol. 441, no. 7097, pp. 1061–1067, 2006.

[135] C. A. Cowan, J. Atienza, D. A. Melton, and K. Eggan, "Developmental biology: nuclear reprogramming of somatic cells after fusion with human embryonic stem cells," *Science*, vol. 309, no. 5739, pp. 1369–1373, 2005.

[136] K. Takahashi and S. Yamanaka, "Induction of pluripotent stem cells from mouse embryonic and adult fibroblast cultures by defined factors," *Cell*, vol. 126, no. 4, pp. 663–676, 2006.

[137] S. J. Odelberg, A. Kollhoff, and M. T. Keating, "Dedifferentiation of mammalian myotubes induced by msx1," *Cell*, vol. 103, no. 7, pp. 1099–1109, 2000.

[138] G. R. Rosania, Y.-T. Chang, O. Perez et al., "Myoseverin, a microtubule-binding molecule with novel cellular effects," *Nature Biotechnology*, vol. 18, no. 3, pp. 304–308, 2000.

[139] S. Chen, S. Takanashi, Q. Zhang et al., "Reversine increases the plasticity of lineage-committed mammalian cells," *Proceedings of the National Academy of Sciences of the United States of America*, vol. 104, no. 25, pp. 10482–10487, 2007.

[140] C. J. McGann, S. J. Odelberg, and M. T. Keating, "Mammalian myotube dedifferentiation induced by newt regeneration extract," *Proceedings of the National Academy of Sciences of the United States of America*, vol. 98, no. 24, pp. 13699–13704, 2001.

[141] X. Mu, H. Peng, H. Pan, J. Huard, and Y. Li, "Study of muscle cell dedifferentiation after skeletal muscle injury of mice with a Cre-Lox system," *PLoS ONE*, vol. 6, no. 2, Article ID e16699, 2011.

[142] L. Vermeulen, "Keeping stem cells in check: a Hippo balancing act," *Cell Stem Cell*, vol. 12, no. 1, pp. 3–5, 2013.

[143] A. C. Mullen, "Hippo tips the TGF-β scale in favor of pluripotency," *Cell Stem Cell*, vol. 14, no. 1, pp. 6–8, 2014.

[144] D. Basu, R. Lettan, K. Damodaran, S. Strellec, M. Reyes-Mugica, and A. Rebbaa, "Identification, mechanism of action, and antitumor activity of a small molecule inhibitor of Hippo, TGF-β, and Wnt signaling pathways," *Molecular Cancer Therapeutics*, vol. 13, no. 6, pp. 1457–1467, 2014.

[145] D. L. Atkinson, T. J. Stevenson, E. J. Park, M. D. Riedy, B. Milash, and S. J. Odelberg, "Cellular electroporation induces dedifferentiation in intact newt limbs," *Developmental Biology*, vol. 299, no. 1, pp. 257–271, 2006.

Localisation of Lactate Transporters in Rat and Rabbit Placentae

Nigel P. Moore,[1] Catherine A. Picut,[2] and Jeffrey H. Charlap[3]

[1]Ubrs GmbH, Postfach, 4058 Basel, Switzerland
[2]WIL Research, LLC, Hillsborough, NC 27278, USA
[3]WIL Research, LLC, Ashland, OH 44805, USA

Correspondence should be addressed to Nigel P. Moore; nigel.moore@ubrs.ch

Academic Editor: Carlo Pellicciari

The distribution of monocarboxylate transporter (MCT) isoforms 1 and 4, which mediate the plasmalemmal transport of L-lactic and pyruvic acids, has been identified in the placentae of rats and rabbits at different ages of gestation. Groups of three pregnant Sprague-Dawley rats and New Zealand White rabbits were sacrificed on gestation days (GD) 11, 14, 18, or 20 and on GD 13, 18, or 28, respectively. Placentae were removed and processed for immunohistochemical detection of MCT1 and MCT4. In the rat, staining for MCT1 was associated with lakes and blood vessels containing enucleated red blood cells (maternal vessels) while staining for MCT4 was associated with vessels containing nucleated red blood cells (embryofoetal vessels). In the rabbit, staining for MCT1 was associated with blood vessels containing nucleated red blood cells while staining for MCT4 was associated with vessels containing enucleated red blood cells. Strength of staining for MCT1 decreased during gestation in both species, but that for MCT4 was stronger than that for MCT1 and was consistent between gestation days. The results imply an opposite polarity of MCT1 and MCT4 across the trophoblast between rat and rabbit.

1. Introduction

Lactate transport is achieved by members of the monocarboxylate transporter (MCT) family, or SLC16 solute carrier family, a class of plasma membrane transport proteins. MCT isoforms 1–4 are symporters that mediate the proton-dependent transport of small monocarboxylic acids, particularly L-lactic acid; pyruvic acid; and the ketone bodies, acetoacetic (3-oxobutyric) acid and 3-hydroxybutyric acid [1, 2]. MCT1–4 have widely differing affinities for L-lactate and pyruvate. The different substrate affinities, as well as tissue distribution, between these four isoforms reflect their roles in energy metabolism. MCT1 and MCT2 are expressed in cells that use lactate as a respiratory fuel or for gluconeogenesis, while MCT3 and MCT4 are associated with lactate efflux from highly glycolytic cells, although MCT1 can also mediate lactate efflux under hypoxic conditions [3, 4].

Northern blot analysis of pooled human placenta (Clontech human tissue panels) has identified mRNA for MCT1 and MCT4, but not MCT2 [5, 6]; while early work reported the presence of MCT3 mRNA or protein expression [6, 7], this isoform was later reclassified as MCT4 [3, 6]. Subsequent Western blot analysis of individual human term placentae showed expression of both MCT1 and MCT4 [7–9], although the distributions of the two isoforms within the syncytiotrophoblast differ. While MCT1 is localised predominantly towards the basal plasma membrane opposed to the foetal blood, MCT4 is localised predominantly towards the maternal-facing microvillous plasma membrane. They are not strictly segregated, however, and there is some degree of coincidence [9–11]. In contrast, while both isoforms are found in the mouse placenta, their polarity is the opposite of that in human placenta. Specifically, MCT1 is localised predominantly on the apical side of the syncytiotrophoblast I, adjacent to the maternal blood, and MCT4 is found on the basal side of the syncytiotrophoblast II, adjacent to the foetal blood [10].

Beyond these findings in human and mouse placenta, nothing is known of the distribution of lactate transporters in the placenta. The purpose of the studies reported here was to investigate the localisation of MCT1 and MCT4 in the placentae of two species that are commonly used in reproduction toxicity studies, the rat and the rabbit.

2. Materials and Methods

2.1. Animals.
The animal facilities at WIL Research are fully accredited by AAALAC International, and all maintenance and experimental procedures were conducted in compliance with National Research Council guidelines [12]. All procedures were conducted according to Good Laboratory Practice.

Sexually mature, virgin female Crl:CD(SD) rats (Charles River Laboratories, Inc., Kingston, NY) and time-mated female New Zealand White Hra:(NZW)SPF rabbits (Covance Research Products, Inc., Greenfield, IN) were received in good health from the breeders. Rats were approximately eighty days old upon receipt; rabbits were approximately six months old and were received on GD 1. Each animal was examined on the day of receipt and uniquely identified by ear tag.

2.2. Animal Procedures.
All animals were kept in environmentally controlled rooms with a twelve-hour light/dark photoperiod, maintained at a temperature of $22\pm3°C$ (rats) or $19\pm3°C$ (rabbits), and relative humidity of $50\pm20\%$. Upon arrival and until pairing, all rats were housed individually in clean, stainless steel, wire-mesh cages suspended above cage-board that was changed at least thrice weekly. The rats were paired for mating in the home cage of the male. Following positive evidence of mating, designated GD 0, the females were returned to individual suspended wire-mesh cages. Rabbits were housed individually in clean, stainless steel cages suspended above ground corncob bedding which was changed twice weekly. For the duration of the study, animals were maintained on Certified Rodent LabDiet® 5002 or Certified Rabbit LabDiet® 5322 (PMI Nutrition International, St. Louis, MO); rabbits were also provided kale leaf.

Groups of three rats were euthanised by carbon dioxide inhalation on GD 11, 14, 18, or 20. Groups of three rabbits were euthanised by an intravenous injection of sodium pentobarbital via a marginal ear vein on GD 13, 18, or 28. The thoracic, abdominal, and pelvic cavities were opened by a ventral mid-line incision, and the contents were examined. The uterus was exposed, excised, and trimmed. Placentae were collected from three embryos or foetuses from each gestation day and retained in 10% neutral-buffered formalin. Viable foetuses were euthanised by a subcutaneous injection of sodium pentobarbital in the scapular region.

2.3. Immunohistochemistry.
Following fixation for 48–72 hours, placentae were transferred to 70% ethanol. The tissues were trimmed and processed into paraffin blocks, sectioned, and mounted on glass microscope slides.

MCT1 Staining. Slides were stained immunohistochemically using the Ventana Discovery XT automated slide staining system (Ventana Medical Systems Inc., Tucson, AZ). The Ventana DabMap detection system was used. Antigen retrieval was obtained using CC1 (cell conditioning solution, pH 8.0). The primary antibody was chicken anti-rat MCT1 (EMD Millipore, Billerica, MA; catalogue number AB1286-I), and the secondary antibody was donkey anti-chicken IgY (Jackson Immunoresearch, reference number 703-065-155).

MCT4 Staining. Antigen retrieval was obtained in a DIVA decloaker at full power for two 5-minute cycles. Background staining was blocked by two procedures, using hydrogen peroxide and the Stirrup blocking solution. Slides were stained immunohistochemically using the Ventana Discovery XT slide staining system with rabbit anti-human MCT4 (Biorbyt LLC, San Francisco, CA; catalogue number orb137272) as the primary antibody, and goat anti-rabbit as the secondary antibody (Jackson Immunoresearch, reference number 111-066-003). The avidin-biotin (ABC) detection method was used followed by diaminobenzidine (DAB) as the chromagen.

Slides were counterstained with haematoxylin. Qualitative microscopic examination, including determination of staining localisation and staining intensity on a scale of 1 to 4, was performed on the stained sections by a board-certified veterinary pathologist.

3. Results

3.1. Localisation of MCT in Rat Placenta.
Staining for both MCT1 and MCT4 in rat placentae on GD 11, 14, 18, and 20 was limited to the labyrinth zone. Within the labyrinth, MCT1 was present on the maternal side of the trophoblasts (Figures 1(a)–1(d)). The staining was moderate and formed a thick line outlining the lakes or vessels that were filled with enucleated red cells (i.e., the maternal vessels) and in many cases existed circumferentially around the trophoblasts forming a "chicken-wire" pattern. This moderately intense specific staining of the maternal side of trophoblasts, with nonstaining of the foetal side, was most apparent at GD 14 where there was a significant number of foetal vessels that were distinguished by foetal nucleated red cells and surrounded by unstained syncytiotrophoblasts. There was linear moderate staining of cytotrophoblasts at edge of maternal vessels (Figure 1(b)). At GD 11, there was a limited amount of labyrinth present, and foetal vessels were not readily identified in the deep labyrinth. There was moderate staining for MCT1 in cytotrophoblasts bordering maternal vessels, but no staining was present immediately around the foetal blood vessels (Figure 1(a)). Specific staining for MCT1 was less intense (mild) at the periphery of maternal vessels in the GD 18 and GD 20 placentae, when compared to GD 11 and GD 14, and there was nonspecific background staining of maternal, enucleated, red cells (Figures 1(c) and 1(d)). Few nucleated foetal red cells, surrounded by unstained trophoblasts, were identified in the GD 18 placentae and were rarely identified in the GD 20 placentae.

Staining for MCT4 in the labyrinth zone of the placenta was uniformly and very strongly positive (Table 1) and was limited to the foetal side of the syncytiotrophoblasts (Figures 1(e)–1(h)), forming a thick line outlining foetal blood vessels, which were characterised by nucleated foetal erythrocytes. Cells with larger nuclei were unstained cytotrophoblasts that border the maternal blood vessels which contain enucleated erythrocytes. The cellular location of the stain (i.e., whether basement membrane or cell membrane) could not be determined by light microscopy. Nucleated red cells were apparent within the lumina of foetal blood vessels at GD 11, 14, and 18 but were few at GD 18 and were exceedingly rare at GD 20.

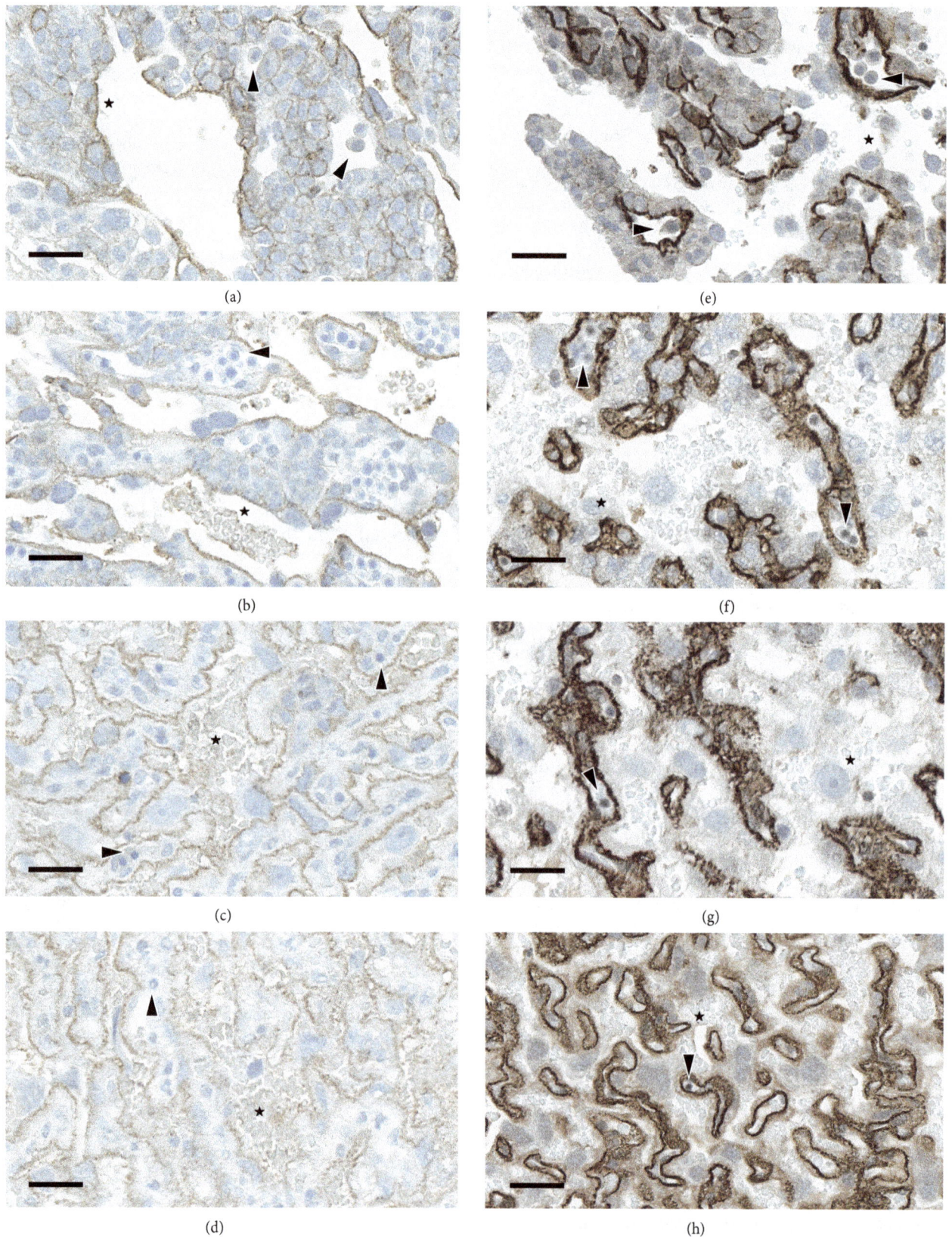

FIGURE 1: Localisation of MCT1 and MCT4 in rat placenta at four different ages during gestation. The staining of rat placenta with antibodies to (a–d) MCT1 and (e–h) MCT4 is shown for (a, e) GD 11; (b, f) GD 14; (c, g) GD 18; and (d, h) GD 20. Maternal blood vessels and cells are indicated with stars (★) and foetal blood vessels and cells are indicated with arrow heads (▶). Original objective 40x, bar 15 μm.

TABLE 1: Strength of staining for MCT isoforms 1 and 4 in rat and rabbit placentae at different days of gestation.

Species	GD	MCT1	MCT4
Rat	11	3	4+
	14	3	4+
	18	2	4+
	20	2	4+
Rabbit	13	2	4
	18	1	4
	28	0	4

Graded staining intensity: 0, none; 1, minimal; 2, mild; 3, moderate; 4, strong; 4+, very strong.

There was occasionally cell surface staining of glycogen cells of the giant cell layer and of the decidua; however, this staining was far less intense than that of syncytiotrophoblasts and was considered to represent nonspecific background staining.

3.2. Localisation of MCT in Rabbit Placenta.

There was mild staining of MCT1 along the lining of blood vessels containing nucleated foetal red cells at GD 13 (Figure 2(a)), which was reduced to minimal intensity at GD 18 (Figure 2(b)). These positive vessels were located at the periphery of the chorion and labyrinth. At GD 28, there was no specific positive staining of trophoblasts surrounding maternal red blood cells, and there were no vessels containing foetal red blood cells (Figure 2(c)). There was no specific positive staining for MCT1 within the body of the labyrinth at any stage of gestation. Both vessels and lakes filled with either maternal or foetal red cells were lined by syncytiotrophoblasts that had no surface staining for MCT1. There was slight nonspecific (i.e., background) staining of maternal red cells.

Staining for MCT4 was strong on GD 13, 18, and 28 and limited to the maternal side of the syncytiotrophoblasts (Figures 2(d)–2(f)). At GD 13 and 18, the positive staining formed a thick line outlining the maternal blood vessels throughout the entire labyrinth. At GD 28, there was more extensive staining of the cytoplasmic membrane of the syncytiotrophoblasts, where the staining was not limited to the side of the cell facing the lumen of the vessel. This more expansive staining resulted in a "chicken-wire" appearance. Nucleated foetal red cells are no longer present in the placenta at this stage of gestation. There was no staining along trophoblasts lining the foetal vessels at any stage.

There was nonspecific cell surface staining of glycogen cells of the giant cell layer and of the decidua and light background staining of maternal red blood cells.

4. Discussion

By using antibodies against the C-termini of MCT1 (chicken anti-rat) and MCT4 (rabbit anti-human), both isoforms have been identified in the placenta of rat and rabbit at different days of gestation. MCT1 was predominantly localised to the maternal side of the trophoblast in the rat, but to the foetal side in the rabbit. Conversely, MCT4 was localised towards the foetal side in the rat and the maternal side in the rabbit. The localisation of the two isoforms in the rabbit is similar to that previously reported in human placenta [9–11]; localisation in the rat is similar to that previously reported in mouse placenta [10].

The polarity remained constant throughout gestation in both species, as has also been reported for the mouse [10]. MCT1 is also localised at the basal membrane in the four-month old human placenta as well as the term placenta [10], which indicates that its distribution, at least in qualitative terms, also may not change during gestation.

The staining for both isoforms was generally stronger in the rat than in the rabbit at equivalent stages of gestation (Table 1). In both species, the expression of MCT1, as evidenced by strength of staining, appeared to decrease during gestation, while that of MCT4 remained consistent. The expression of both MCT1 and MCT4 mRNA is reported to diminish during later gestation, in contrast to that of the type 1 glucose transporter the expression of which was consistently intense [10].

Both MCT1 and MCT4 have a low degree of substrate-specificity, transferring a relatively wide range of substituted and unsubstituted monocarboxylic acids across the plasma membrane [13–16]. The major difference between the two isoforms is their substrate affinity. The affinity of MCT1 towards L-lactate and pyruvate is relatively high, within the normal physiological range for blood concentrations [13, 14], while that of MCT4 is relative low [15, 16]. These differences are postulated to underlie the function of the two isoforms, MCT1 sequestering lactate as a source for energy metabolism and growth and MCT4 releasing lactate during periods of high cellular production [1, 3, 4].

It follows, therefore, that human and rabbit trophoblasts have a high-affinity (MCT1) transporter on the foetal side and a low-affinity (MCT4) transporter on the maternal side, while murine trophoblasts have a low-affinity (MCT4) transporter on the foetal side and a high-affinity (MCT1) transporter on the maternal side. The MCT is a proton-sensitive symporter, and the rate and direction of plasmalemmal transfer will be determined not only by the concentration gradient of the monocarboxylate anion but also by the proton gradient, which may change during gestation as the foetus becomes net lactogenic and the foetal blood becomes acidic in comparison to that of the maternal blood towards the end of gestation.

In conclusion, there is a difference in the localisation of MCT1 and MCT4 between the murine placenta on the one hand and the rabbit and human placentae on the other. Essentially, the "polarity" of these isoforms across the trophoblast is reversed. The functional significance of this difference is unclear, although it has been postulated that it is relevant to the direction of substrate transport [10]. It may also reflect differences in the role of the trophoblast in lactate supply, either by transfer from the maternal blood or by metabolism of glucose or amino acids, and the disposition of maternally derived 2- and 3-hydroxyl- or carbonyl-substituted carboxylic acids that are also MCT substrates. Given the complexity of monocarboxylic acid transport across the trophoblast, further data from specifically designed, integrated studies are required to elucidate the functional significance

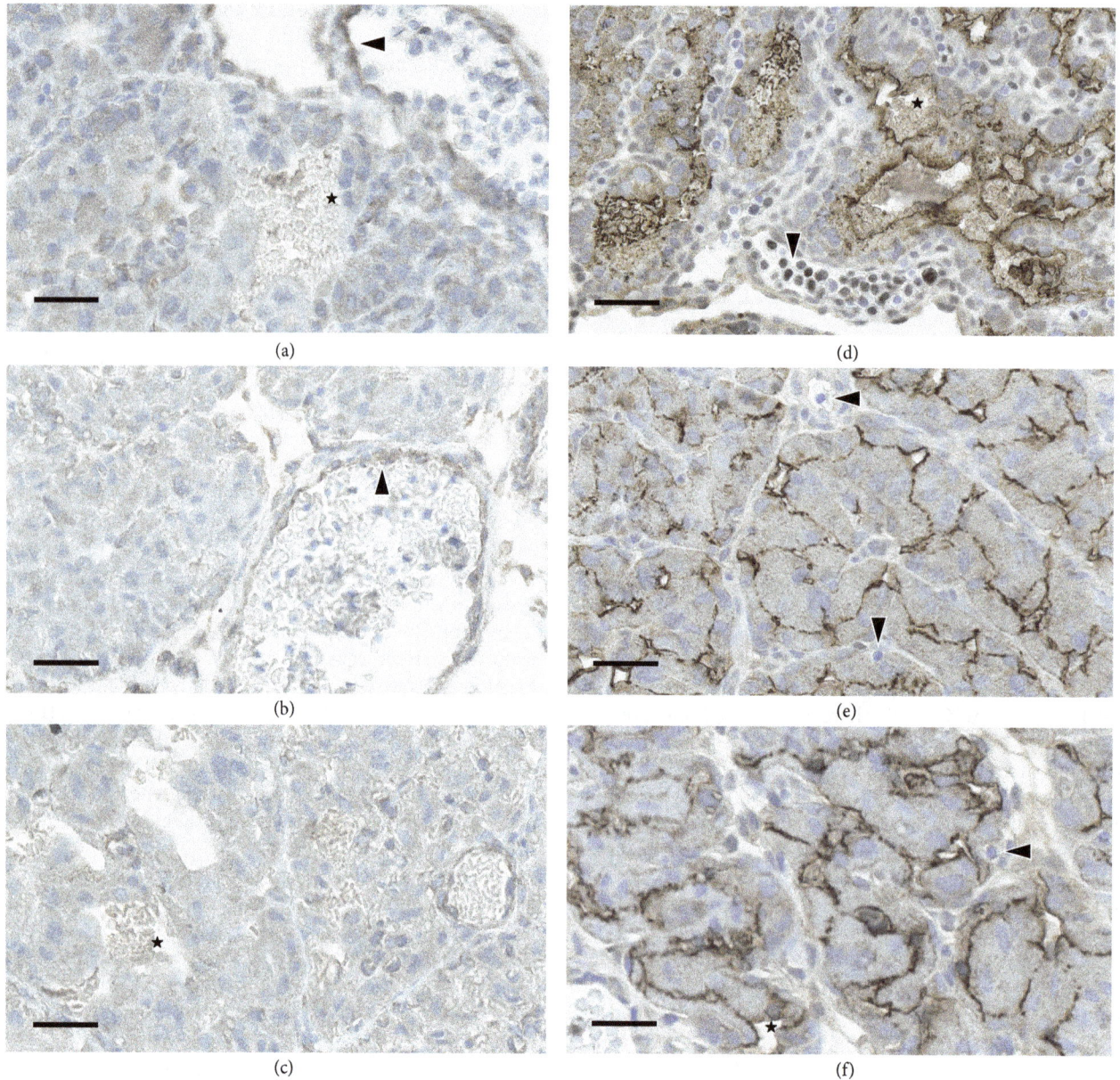

FIGURE 2: Localisation of MCT1 and MCT4 in rabbit placenta at three different ages during gestation. The staining of rabbit placenta with antibodies to (a–c) MCT1 and (d–f) MCT4 is shown for (a, d) GD 13; (b, e) GD 18; and (c, f) GD 28. Maternal blood vessels and cells are indicated with stars (★) and foetal blood vessels and cells are indicated with arrow heads (▶). Original objective 40x, bar 15 μm.

of the differences in MCT isoform localisation and changes in expression during gestation.

Abbreviations

GD: Gestation day(s)
MCT: Monocarboxylate transporter.

Disclosure

This work was conducted by Catherine A. Picut and Jeffrey H. Charlap under the conditions of their regular employment and by Nigel P. Moore as a paid consultant to the sponsors.

Current address of Jeffrey H. Charlap is Charles River Laboratories, Horsham, PA 19044, USA.

Competing Interests

There is no undisclosed conflict of interests regarding publication of this paper.

Acknowledgments

This work was funded by member companies of the Ethylene Oxide and Derivatives Producers' Association, Brussels, Belgium; a sector group of Cefic, the European chemical industry association. The authors gratefully acknowledge Sirena Hudgins (WIL Research) for technical expertise in the conduct of this work.

References

[1] A. P. Halestrap and N. T. Price, "The proton-linked monocarboxylate transporter (MCT) family: structure, function and regulation," *Biochemical Journal*, vol. 343, no. 2, pp. 281–299, 1999.

[2] V. N. Jackson and A. P. Halestrap, "The kinetics, substrate, and inhibitor specificity of the monocarboxylate (lactate) transporter of rat liver cells determined using the fluorescent intracellular pH indicator, $2',7'$-bis(carboxyethyl)-5(6)-carboxyfluorescein," *The Journal of Biological Chemistry*, vol. 271, no. 2, pp. 861–868, 1996.

[3] A. P. Halestrap, "The monocarboxylate transporter family–structure and functional characterization," *IUBMB Life*, vol. 64, no. 1, pp. 1–9, 2012.

[4] A. P. Halestrap and M. C. Wilson, "The monocarboxylate transporter family–role and regulation," *IUBMB Life*, vol. 64, no. 2, pp. 109–119, 2012.

[5] R.-Y. Lin, J. C. Vera, R. S. K. Chaganti, and D. W. Golde, "Human monocarboxylate transporter 2 (MCT2) is a high affinity pyruvate transporter," *The Journal of Biological Chemistry*, vol. 273, no. 44, pp. 28959–28965, 1998.

[6] N. T. Price, V. N. Jackson, and A. P. Halestrap, "Cloning and sequencing of four new mammalian monocarboxylate transporter (MCT) homologues confirms the existence of a transporter family with an ancient past," *Biochemical Journal*, vol. 329, no. 2, pp. 321–328, 1998.

[7] M. C. Wilson, V. N. Jackson, C. Heddle et al., "Lactic acid efflux from white skeletal muscle is catalyzed by the monocarboxylate transporter isoform MCT3," *The Journal of Biological Chemistry*, vol. 273, no. 26, pp. 15920–15926, 1998.

[8] P. Settle, C. P. Sibley, I. M. Doughty et al., "Placental lactate transporter activity and expression in intrauterine growth restriction," *Journal of the Society for Gynecologic Investigation*, vol. 13, no. 5, pp. 357–363, 2006.

[9] P. Settle, K. Mynett, P. Speake et al., "Polarized lactate transporter activity and expression in the syncytiotrophoblast of the term human placenta," *Placenta*, vol. 25, no. 6, pp. 496–504, 2004.

[10] A. Nagai, K. Takebe, J. Nio-Kobayashi, H. Takahashi-Iwanaga, and T. Iwanaga, "Cellular Expression of the Monocarboxylate Transporter (MCT) family in the placenta of mice," *Placenta*, vol. 31, no. 2, pp. 126–133, 2010.

[11] M. Willis, N. Zaidi, M. Li, A. Husain, and H. Kay, "765: defining the role of placental lactate transporters (MCT1 and MCT4) in preeclampsia," *American Journal of Obstetrics and Gynecology*, vol. 201, article S275, 2009.

[12] National Research Council, *Guide for the Care and Use of Laboratory Animals*, National Academy Press, Washington, DC, USA, 2011.

[13] L. Carpenter and A. P. Halestrap, "The kinetics, substrate and inhibitor specificity of the lactate transporter of Ehrlich-Lettre tumour cells studied with the intracellular pH indicator BCECF," *Biochemical Journal*, vol. 304, no. 3, pp. 751–760, 1994.

[14] R. C. Pools, S. L. Cranmer, A. P. Halestrap, and A. J. Levi, "Substrate and inhibitor specificity of monocarboxylate transport into heart cells and erythrocytes. Further evidence for the existence of two distinct carriers," *Biochemical Journal*, vol. 269, no. 3, pp. 827–829, 1990.

[15] K.-S. Dimmer, B. Friedrich, F. Lang, J. W. Deitmer, and S. Bröer, "The low-affinity monocarboxylate transporter MCT4 is adapted to the export of lactate in highly glycolytic cells," *Biochemical Journal*, vol. 350, no. 1, pp. 219–227, 2000.

[16] J. E. Manning Fox, D. Meredith, and A. P. Halestrap, "Characterisation of human monocarboxylate transporter 4 substantiates its role in lactic acid efflux from skeletal muscle," *The Journal of Physiology*, vol. 529, no. 2, pp. 285–293, 2000.

4

Effect of Antioxidant Water on the Bioactivities of Cells

Seong Gu Hwang,[1] Ho-Sung Lee,[2] Byung-Cheon Lee,[2] and GunWoong Bahng[3]

[1]Department of Animal Life and Environmental Science, Hankyong National University, Anseong 17579, Republic of Korea
[2]Institute for Information Technology Convergence, Division of Electrical Engineering, Korea Advanced Institute of Science and Technology, Daejeon 34138, Republic of Korea
[3]Department of Mechanical Engineering, The State University of New York Korea, Incheon 21985, Republic of Korea

Correspondence should be addressed to GunWoong Bahng; gwbahng3208@daum.net

Academic Editor: Richard Tucker

It has been reported that water at the interface of a hydrophilic thin film forms an exclusion zone, which has a higher density than ordinary water. A similar phenomenon was observed for a hydrated hydrophilic ceramic powder, and water turns into a three-dimensional cell-like structure composed of high density water and low density water. This structured water appears to have a stimulative effect on plant growth. This report outlines our study of antioxidant properties of this structured water and its effect on cell bioactivities. Culturing media which were prepared utilizing this antioxidant structured water promoted the viability of RAW 264.7 macrophage cells by up to three times. The same tendency was observed for other cells including IEC-6, C2Cl2, and 3T3-L1. Also, the cytokine expression of the splenocytes taken from a mouse spleen increased in the same manner. The water also appears to suppress the viability of cancer cell, MCF-7. These results strongly suggest that the structured water helps the activities of normal cells while suppressing those of malignant cells.

1. Introduction

There have been many arguments on the role of water in a cell and this debate still continues today [1]. Most of these debates are still primitive, focused on whether water is a simple solvent medium of diverse organic components in cytoplasm or it plays an essential role [2]. As the name suggests, most studies in molecular biology have been conducted focusing on the behavior of molecular components such as lipids, proteins, and other cellular molecules including various antioxidants. Despite the fact that water occupies the largest portion of a cell, it does not receive proper attention from academia in biology and medicine, likely because the structure of water and its effect on biological activities is still not yet well understood [3–5].

Recently it was claimed that there is a fourth phase of water similar to a liquid crystal, in addition to the three phases of water: solid, liquid, and gas [6]. This fourth phase takes place at the interface of a hydrophilic material such as Nafion® film as water molecules are arranged along the polarity of the surface. Since its density is higher than ordinary water,

microspheres in a suspension are excluded as the water is structured, and, based on this phenomenon, it was named as an exclusion zone [7–9]. Also, electric potential as high as −200 mV has been observed to develop across the boundary of the exclusion zone and outside of this region (exclusion zone negative). This potential is generated by the dissociation of water molecules into negative ions (OH-) and protons as it is structured [6]. This important finding implies that water itself can affect the growth and bioactivity of live beings.

Actually, it was reported that an exclusion zone is formed also around hydrophilic ceramic powder and can stimulate seed germination and early sapling growth in brown chickpea seeds. Root length and/or formation of shoots increased at least 2~3-fold [10]. In due course, structured water may also influence cell bioactivities. To confirm this possibility, an evaluation of the property of structured water composed of an exclusion zone and its effect on the bioactivities of cells are necessary. In this study, structured water formed by mixing with a hydrophilic ceramic powder was discovered to have an antioxidant property. In addition, to address any arguments regarding the effects of any possible dissolved

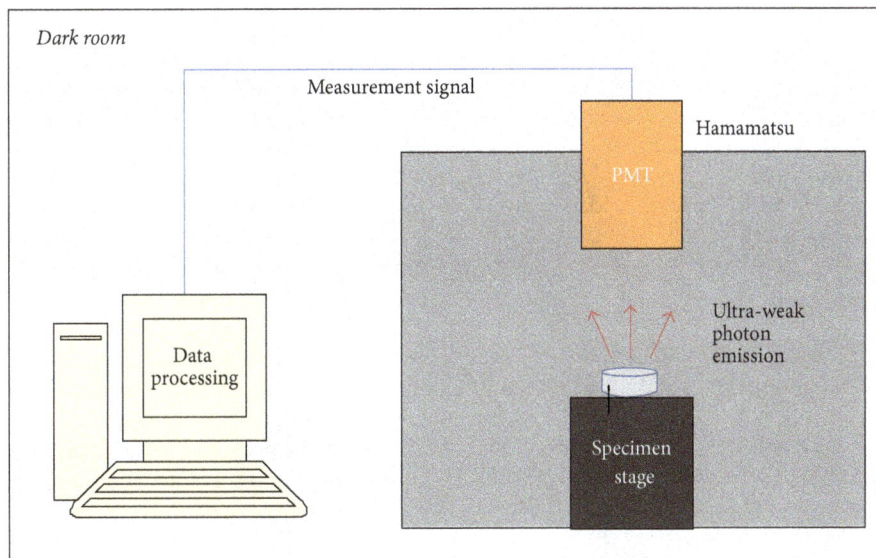

FIGURE 1: Schematic diagram of the ultra-weak photon emission measurement system.

mineral components, we have developed another method utilizing far-infrared ray waves from the ceramic powder to make structured water in a noncontact manner.

This structured water was utilized to prepare culture media for evaluating its effect on the bioactivities of cells, and this water appears to have a meaningful effect on the bioactivities of normal cells such as cell viability, phagocytic activity, and natural killer cell activity. Conversely, the viability of cancer cell was suppressed when it was incubated with the same medium made of structured water. These results imply that the role of water should be understood as a critical component rather than merely as a simple media in a cell.

2. Materials and Methods

2.1. Evaluation of Antioxidant Property of Water

2.1.1. Water Mixed with the Ceramic Powder. A hydrophilic ceramic powder, QELBY®, with particle size ranging from 40 nm to 1 μm, was obtained from Quantum Energy Co., Ltd. It was produced by finely grinding natural clay minerals of the feldspar family with silica as the main component (more than 60%).

To evaluate the antioxidant property, an ultra-weak photon emission was measured using a photomultiplier tube (Hamamatsu Photonics, E1341-01) under total darkness. Figure 1 shows the schematic diagram of the photon emission measurement system.

For the evaluation of the antioxidant property of water mixed with the powder, three kinds of samples were prepared. The same amount of bean sprout stems were mixed with deionized water only; deionized water mixed with natural silica powder; and deionized water mixed with QELBY powder, respectively. The concentration of the powder in each mixture was 0.01 g/ml. The number of photon emissions per one second was measured for 8 minutes and an average value was calculated. The measurement was repeated every day for

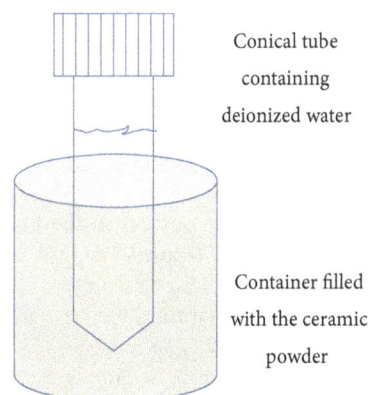

FIGURE 2: *Preparation of water in the noncontact method.* Deionized water in a 50 ml vial was put up into the ceramic powder for two days and used for the preparation of culturing media by mixing with the powder form DMEM.

seven days. In addition, the color change of the bean sprout stems was observed simultaneously.

2.1.2. Water Treated in the Noncontact Method. For the noncontact treatment of water, vials containing 50 ml of deionized water were put up into the ceramic powder halfway for 48 hours before it was used for the preparation of the culturing media. Deionized water was obtained by a water purification system that produced an ultra-pure water type I grade with electrical resistivity of 18.2 MΩcm and reduction of inorganics up to 99.99% (Young Lin Instrument, aqua MAX™ ultra 370 series). Figure 2 shows the schematic diagram of water treatment in a noncontact manner.

To evaluate the antioxidant property of water prepared in the noncontact method, the number of photons emitted per second was measured every day for 3 minutes at a time after adding 0.5 ml of 10 mM tert-Butyl hydroperoxide (TBHP) to 1 ml of deionized water (control) and noncontact treated deionized water (sample), respectively, for the duration of 4 days. TBHP, as an organic peroxide, is a strong oxidant and forms radical species of the form RO^* during the decomposition of its O-O bond [11]. It is more stable to compare with H_2O_2 and widely used as a source of free radical in cellular activity as well as an inducer of oxidative stress [12, 13].

2.2. Preparation of Culture Media

2.2.1. Culture Media Prepared by Mixing with Ceramic Powder. To avoid any possible contamination, the QELBY ceramic powder was heat treated using an electric muffle furnace for an hour at 800°C and then autoclaved just before mixing with Dulbecco's modified Eagle's medium (DMEM). DMEM with 10% of heat inactivated fetal bovine serum (FBS) as a supplementation and 1% penicillin-streptomycin (P/S) was prepared by mixing the powder form DMEM (Gibco®, 12800-017, USA) with deionized water. The liquid form DMEM was mixed with 1% in weight concentration of the powder and agitated for 5 minutes and then left for 2 days in a refrigerator maintained at 4°C. After 2 days, the suspended layer was filtered with Whatman® filter paper (grade 5) and the weight of the filtered paper was measured in order to estimate the total dissolved powder in the media. The filtered DMEM was diluted with a new fresh DMEM to certain concentrations (e.g., 25, 50, 100, 200, and 400 μg/ml) prior to preparing the culturing media.

The concentration of the powder in the filtered DMEM was calculated to be about 700 ppm~750 ppm. The final concentration of powder in the diluted media was estimated to be lower than 1 ppm. For the natural killer cell and splenocyte cell, RPMI 1640 medium was used instead of DMEM. Supplementary drawings SFig. 1~3, in Supplementary Material available online at https://doi.org/10.1155/2017/1917239, show the overall diagram of media preparation and cell viability measurement.

2.2.2. Culture Media with the Water Prepared in the Noncontact Method. Another set of culture media was prepared using a structured water prepared in a noncontact manner. A centrifuge conical tube containing 50 ml of deionized water was inserted into the ceramic powder for 48 hours or other specified durations at room temperature. Deionized water treated in this way was used to prepare the culture media by mixing with the DMEM powder. Supplementary drawing SFig. 4 shows the overall diagram of media preparation and cell viability measurement for the noncontact method.

2.3. Evaluation of Bioactivities of Cells

2.3.1. Viability of Macrophage Cell. Murine RAW 264.7 was obtained from the Korean Cell Line Bank (KCLB, Korea). To evaluate the viability of the macrophage cell, it was seeded in a 96-well plate at a density of 1×10^5 cells/ml in a final volume of 100 μl and allowed to adhere for 3 hours. It was then pretreated for 1 hour with 10 μg/ml lipopolysaccharide (LPS) for the stimulation before applying the prepared media.

After 1 hour, prepared media with increasing concentrations of dissolved QELBY ceramic powder was added to each well before incubation. Incubation was performed for 24 hours or 48 hours at 37°C in humidified 5% CO_2 atmosphere. After the incubation, media in the wells were suctioned and 100 μl of fresh media with 10 μl of CCK-8 solution were added per well and then incubated at 37°C for 2 hours. A cell counting kit (CCK-8, Dojindo, Japan) was used after 2 hours of treatment for the determination of cell viability using enzyme-linked immunosorbent assay (ELISA) microplate reader (TECAN, Switzerland) for absorbance at 450 nm. As a control, the viability of the cell cultured in DMEM without LPS stimulation was measured. The viability of the cell with LPS stimulation was measured also as a positive control. The viability of the treated cells was expressed as the percentage of control cells. A schematic diagram of the media preparation and cell viability evaluation is described in the supplementary drawing SFig. 1.

2.3.2. Phagocytic Activity. The phagocytic activity of a macrophage cell was evaluated by measuring the uptake of neutral red dye. Cells were cultured at 37°C with the prepared media in humidified 5% CO_2 for 24 hours and then 100 μl of 0.075% aseptic neutral red solution was added and cultured again for 1 hour. Following the end of culture, the plate was washed three times with phosphate-buffered saline, and a mixture of 100% ethanol and 99.9% acetic acid (1 : 1 v/v) was added to the 150 μl of cell lysate. The mixture was mixed fully and absorbance at 550 nm was measured using an ELISA reader. The schematic diagram of the media preparation and evaluation of the phagocytic activity is shown in the supplementary drawing SFig. 2.

2.3.3. Natural Killer Cell Activity. For the evaluation of the natural killer (NK) cell activity and viability of splenocyte, mice (C57BL/6) were sacrificed and their spleens were collected aseptically. To prepare the media, 0.1 g of QELBY ceramic powder was mixed with 10 ml of Roswell Park Memorial Institute (RPMI) 1640 medium containing 10% fetal bovine serum (FBS) and 1% penicillin-streptomycin and stood for 24 hours at 4°C. The suspended layer in differing amounts were used for the preparation of culture media, for example, 25, 50, 100, and 200 μg/ml by mixing with a new fresh RPMI 1640 medium.

Target cell (YAC-1) was seeded in a 96-well plate at a density of 5×10^3 cells/ml for 3 hours. An effector cell (isolated splenocyte) and target cells were mixed at effector : target ratio of 10 : 1 and incubated with the prepared media for 24 hours. NK cell activity was assessed after treatment with CCK-8 assay solution for 2 hours. The NK cell activity was calculated as follows based on the optical density (OD) measured by ELISA reader at 450 nm.

% NK cell activity

$$= \left[1 - \left\{ \frac{\{OD\,(effector + target) - OD\,effector\}}{OD\,target} \right\} \right] \quad (1)$$

$$\times\, 100.$$

TABLE 1: *Average number of ultra-weak photon emissions per one second.* Measurements were taken for eight minutes, from bean sprout stems stored in deionized water only, water mixed with the silica powder, and water mixed with the QELBY ceramic powder.

Time of measurement	16 hours later	6 days later	7 days later
(1) Control, deionized water only	27.86	27.97	35.65
(2) Deionized water + silica (0.1 g/ml)	30.31	39.49	39.20
(3) Deionized water + QELBY (0.1 g/ml)	27.28	18.16	23.78

Supplementary drawing SFig. 3 shows the overall cell culture process and natural killer cell activity measurement.

2.3.4. Splenocyte and Cytokine Expression.

Mice spleens were collected aseptically to prepare spleen cell suspension. The evaluation of splenocyte viability was carried out according to the CCK-8 method. Splenocytes were seeded in a 96-well plate at 1×10^5 cells/ml for 1 hour. The prepared media with different amounts of the suspended layer, 0, 25, 50, 100, and 200 μg/ml, from the RPMI 1640 mixed with the 1% of QELBY powder stood for 48 hours, were added to the cells and incubated for 24 and 48 hours. CCK-8 reagent was added to the cell suspension and the optical density was measured 2 hours later at 450 nm using a microplate reader.

Changes in cytokine expression of murine splenocytes were measured also. Splenocytes were seeded in a 6-well plate at 5×10^5 cells/ml for 3 hours and then incubated for 24 hours with the prepared culture media. Total RNA was extracted using TRIzol® reagent (Takara Korea, Korea) according to the manufacturer's instructions. One μg of RNA was used to acquire the complementary DNA (cDNA) using the protocol provided by M-MuLV reverse transcriptase (Fermentas, Lithuania) for reverse transcription polymerase chain reaction (RT-PCR). Specific primers were used to amplify different genes. PCR products were then separated by electrophoresis using 1.5% agarose stained with ethidium bromide, followed by a UV transillumination. A PCR analysis was done three times and a densitometry analysis was carried out using Lane® 1D software. Beta actin (β-actin), interleukin-12 (IL-12), and interferon-gamma (INF-γ) expressions were analyzed.

2.3.5. Breast Cancer Cell.

The viability of the human breast cancer cell, MCF-7, was measured in the same way using both kinds of culture media. The suspended layer was taken in different amounts from the premixed DMEM with 1% of QELBY ceramic powder after 48 hours of storage in a refrigerator maintained at 4°C. Additionally, to see the effect of water preparation process, DMEM mixed with the 1% of QELBY powder was agitated for 3 minutes and then centrifuged at 3,000 rpm for 5 minutes. Different amounts of the supernatant were collected and used for the preparation of culture media.

Also, the DMEM culture medium which was prepared using only the water that was treated for 48 hours in a noncontact method was used for the evaluation of cell viability.

2.4. Statistical Analysis.

All experiments were performed for the specified number of n indicated in the figure captions, from $n = 4$ to $n = 7$, and results are presented as mean ± SD. The cellular activities in intact cultures were taken for 100% of the cultures. Statistical significance was determined using a one-way analysis of variance (ANOVA) followed by Duncan's Multiple Range Test. A value of $p < 0.05$ was considered statistically significant.

3. Results

3.1. Antioxidant Property of Water

3.1.1. Water Mixed with the Powder.

The QELBY® powder showed a good hydrophilic property and colloid formation was observed when mixed with water. For the bean sprout stems mixed with water prepared in a contact method, the numbers of photons emitted are not much different among the three specimens measured 16 hours later as shown in Figure 3(a). However, the number of photon emissions from the deionized water mixed with silica powder (red line) increased after 6 days followed by the deionized water only (blue line) as shown in Figure 3(b). After seven days, it was almost the same as shown in Figure 3(c). In comparison, the deionized water mixed with the QELBY® ceramic powder (green line) clearly showed a lower photon emission even after seven days as shown in Figures 3(b) and 3(c). Table 1 shows the average number of ultra-weak photon emission per one second measured for eight minutes from the bean sprout stems mixed with different kinds of water prepared in the contact method.

In the deionized water or deionized water mixed with the silica powder as shown in Figures 4(a), 4(b), and 4(c), the color of the bean sprout stems turned dark brown or brown. However, the color of the bean sprout stems stored in the deionized water mixed with the QELBY ceramic powder did not change significantly and was light brown after seven days as shown in Figure 4(c). These results indicate that bean sprout stems in the deionized water and the deionized water mixed with silica powder oxidized faster than those stored in the deionized water mixed with the QELBY powder.

In order to indicate the change of color quantitatively, an arbitrary unit from 1 (white, no change) to 5 (dark brown, severely changed) has been adopted. Figure 5 shows the variation of color with the incubation time, 16 hours, 6 days, and 7 days, for the bean sprouts shown in Figure 4. Until 6 days of incubation, there was no difference in color between deionized water only and in deionized water mixed with QELBY ceramic powder. Next day, on the 7th day of incubation, the color of the bean sprouts in the deionized water mixed with QELBY® powder hardly changed but that of the bean sprouts in deionized water began to change.

(a) Sixteen hours later

(b) Six days later

(c) Seven days later

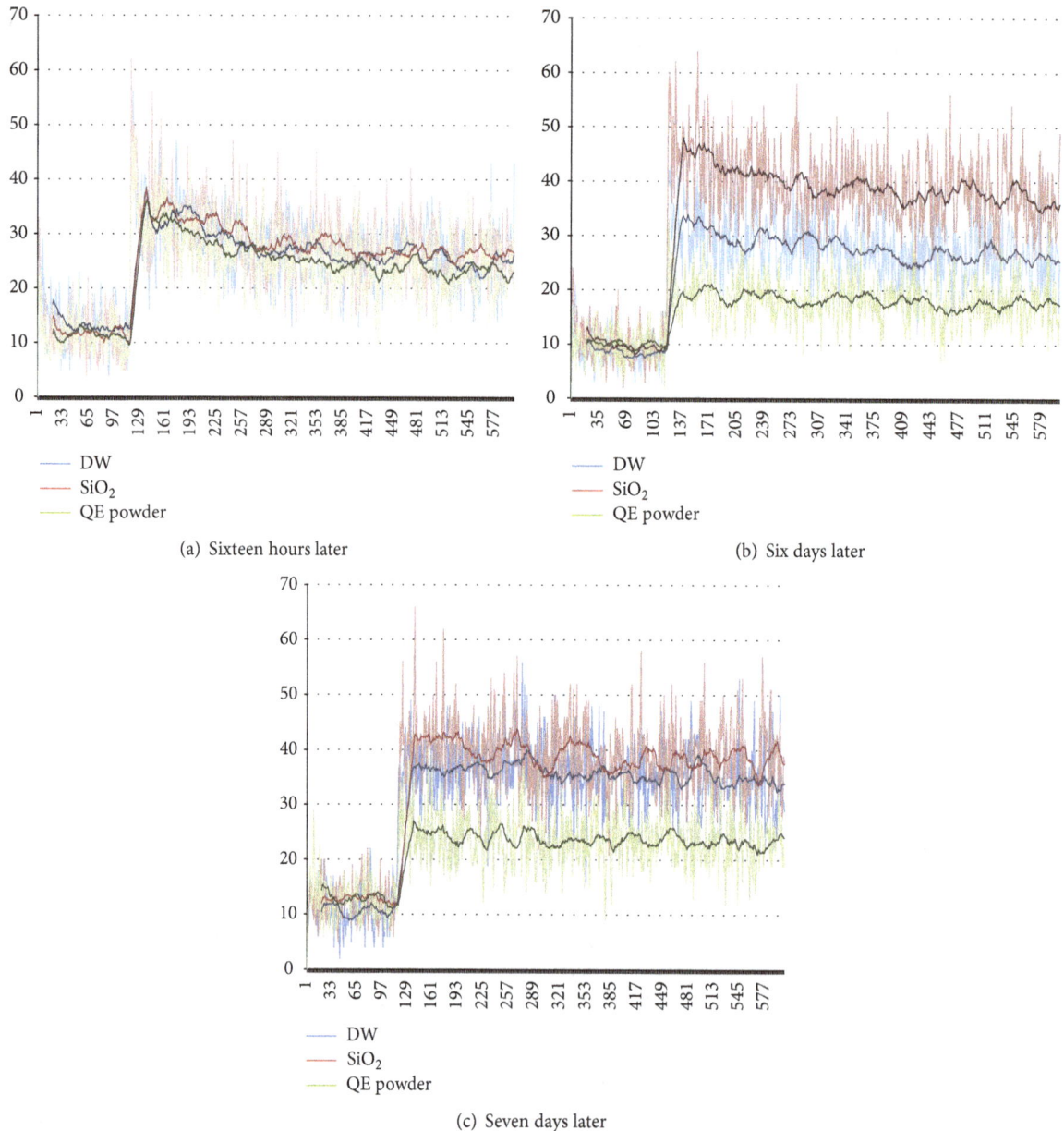

FIGURE 3: *Measurement of ultra-weak photon emission with a photomultiplier tube.* The blue line (DW), red line (SiO_2), and the green line (QE powder) indicate the number of photons from deionized water only, deionized water mixed with silica powder, and deionized water mixed with the QELBY powder, respectively. The number of photons per one second was measured for eight minutes. The vertical axis depicts the number of photons, and the horizontal axis represents seconds elapsed.

The color of bean sprouts in deionized water mixed with silica powder rapidly changed continuously from white to dark brown from the beginning all the way to the last 7th day.

3.1.2. Water Treated in the Noncontact Method. Experimental measurements show that the photon emission from the deionized water increased on the second day of measurement. On days 3 and 4, the number of photon emissions was lower than those from day 2. However, they were still higher than that of day 1 as shown in Figure 6(a). In comparison, the number of photon emissions from the deionized water treated in a noncontact manner showed a dramatic decrease on day 4, reaching one-fourth of its initial level after 93 h of exposure, as shown in Figure 6(b).

Figure 7 is redrawn from the data shown in Figure 6 for direct comparison to the control data. After 16 hours of the noncontact treatment, the treated water already shows a lower number of photon emissions compared to the control as shown in Figure 7(a). This difference increased after 93 hours of exposure and its level was about one-fourth of the control (Figure 7(b)). This indicates that the antioxidant property of the water treated in the noncontact manner increased as exposure time increased.

(a) Sixteen hours later

(b) Six days later

(c) Seven days later

FIGURE 4: *Observation of the color change of bean sprout stems.* (1) Deionized water; (2) deionized water mixed with silica powder; (3) deionized water mixed with QELBY ceramic powder, respectively.

3.2. Effect of Culturing Media on the Cell Viability

3.2.1. Culturing Media Prepared by Mixing with Powder. The viability of RAW 264.7 macrophage cell increased in a dose dependent manner in the amount of the added suspended layer as shown in Figure 8. Compared to the control, it increased 1.5-fold in a media of $200 \, \mu g/ml$ concentration. There was no significant difference in the viability of cells regardless of incubation time length (24 hours or 48 hours). Note that all of the samples incubated with the prepared culture media after LPS stimulation showed a higher viability compared to the samples treated with LPS stimulation only or incubated without LPS stimulation.

The prepared culture media increased the phagocytic activity also as shown in Figure 9. The activity almost doubled

for the concentration of $5 \, \mu g/ml$ and almost the same level of activity was maintained up to the $80 \, \mu g/ml$. Unlike the viability, the dose dependent manner was not observed. For reference, the activity of RAW 264.5 macrophage cell stimulated with LPS only increased about 1.5-fold compared to the control.

A measurement of NK cell activity shows that the prepared media has a positive effect in a dose dependent manner as shown in Figure 10, resulting in a 3-fold increase for the medium mixed with $200 \, \mu g/ml$ of suspended layer.

The incubation of mouse splenocytes with the prepared media showed a significant dose dependent effect on the viability as shown in Figure 11 although it is not as dramatic as in the case of phagocytic activity. The reverse

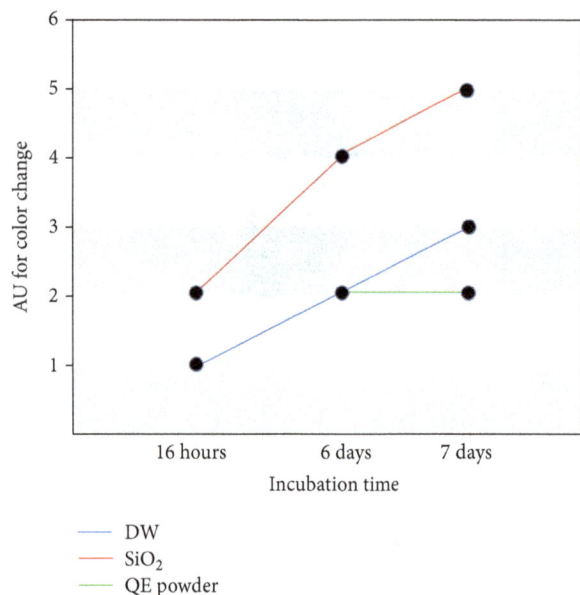

FIGURE 5: *Variation of color of the bean sprout stems with incubation time.* Arbitrary unit (AU) in 5-point scale is adopted to indicate the color change, that is, from AU1, white, and no change in color, to AU5, dark brown, and severely changed.

transcription polymerase chain reaction (RT-PCR) shows that both interleukin-12 (IL-12) and interferon-gamma (INF-γ) expressions were upregulated dose dependently by incubating with the prepared media as shown in Figure 12. This result confirms the increased activity of the splenocyte cells shown in Figure 11.

3.2.2. Culturing Media Prepared with Water Treated in the Noncontact Method. For the culture medium prepared by mixing the DMEM powder with only deionized water treated in a noncontact manner, the viability of the macrophage cell was almost doubled or tripled depending on the length of incubation time as shown in Figure 13. Since the macrophage cell is sensitive to stimulation from outside as an immune cell, other kinds of normal cells besides macrophage cell, including Rattus norvegicus (IEC-6), mouse myoblast (C2C12), and murine adipocyte (3T3-L1), were incubated in the same way to confirm whether the same tendency is observed or not. Although there were some differences in the level of viability, all of them showed the same tendency, that is, increased cell viability as shown in Figure 13. The effect of incubation time was not so large compared to that of the macrophage cell. The unexposed deionized water was used for the preparation of control for all kinds of cells.

To observe the effect of the difference in exposure time on the viability of macrophage cell, culture media was prepared using deionized water which was treated in the noncontact manner for 6, 12, 24, and 48 hours, respectively. All of them showed an almost 3-fold increase in viability with a slight difference among them as shown in Figure 14. This means that even a short period of exposure time shorter than 6 hours may be enough to increase the viability of macrophage cell.

3.2.3. Effect of Culturing Media on a Breast Cancer Cell. We performed the same experiment to observe its effect on the viability of an abnormal cell. A breast cancer cell, MCF-7, was incubated in the same manner and the viability was suppressed in a concentration-dependent manner as shown in Figure 15. It was inhibited to a level below 80% compared to the control for a concentration of 200 μg/ml. The same trend was observed in the media prepared with a water treated in a noncontact manner for 48 hours as indicated by the bar of EQ in Figure 15.

To see whether there is any influence on the cell viability of preparation process, the QELBY ceramic powder was mixed with the DMEM and centrifuged for 5 minutes right after mixing with the powder rather than waiting for 48 hours of standing time. The supernatant after the centrifuge was used for the preparation of culturing media in a different amount. Results shown in Figure 16 indicate that the effect is slightly lower for the same concentration, 100 and 200 μg/ml each, compared to those of the data shown in Figure 15. At the concentration of 400 μg/ml, the viability was suppressed almost to 70% of the control. This result implies that 48 hours of standing time may not be necessary to see the effect of the culturing media prepared in a contact method.

4. Discussions

There has been an argument regarding the physicochemical properties of interfacial water and the possibility that it may influence biological activities depending on the hydrophobic or hydrophilic properties of the cell membrane and other organic components of cells. It was claimed that water in contact with a hydrophobic surface will have a lower density and a better adsorption of various surfactants and proteins from water [14]. This approach is based only on the variation of the density of interfacial water.

It has been reported that the increase of the ultra-weak photon emission is associated with an oxidation reaction. The number of photon emissions is reduced when an antioxidant component such as catechin is added to water [15] while it increases under an oxidative atmosphere [16, 17]. Therefore, the number of photon emissions is expected to decrease for water with an antioxidant property. In this study, it was clearly demonstrated that the structured water prepared through a mixture with the hydrophilic ceramic powder exhibits an antioxidant property. Antioxidants inhibit the oxidation of lipids, proteins, or other molecular components in cells [18], and the generation of an ultra-weak photon emission is also associated with the oxidation of proteins and amino acids in bean sprout stems [19]. This indicates not only that there is a variation in the density of water, but also that an electric property can be a critical factor in the role of water. Actually, it has been found that an electric potential as high as −200 mV exists across the boundary between the exclusion zone and outside of this region (exclusion zone negative), in addition to the difference in density [6].

Interestingly, the water treated in the noncontact manner also demonstrates an antioxidant property as shown in Figures 6 and 7. The level of antioxidant property increases with longer exposure time. The energy source for the formation

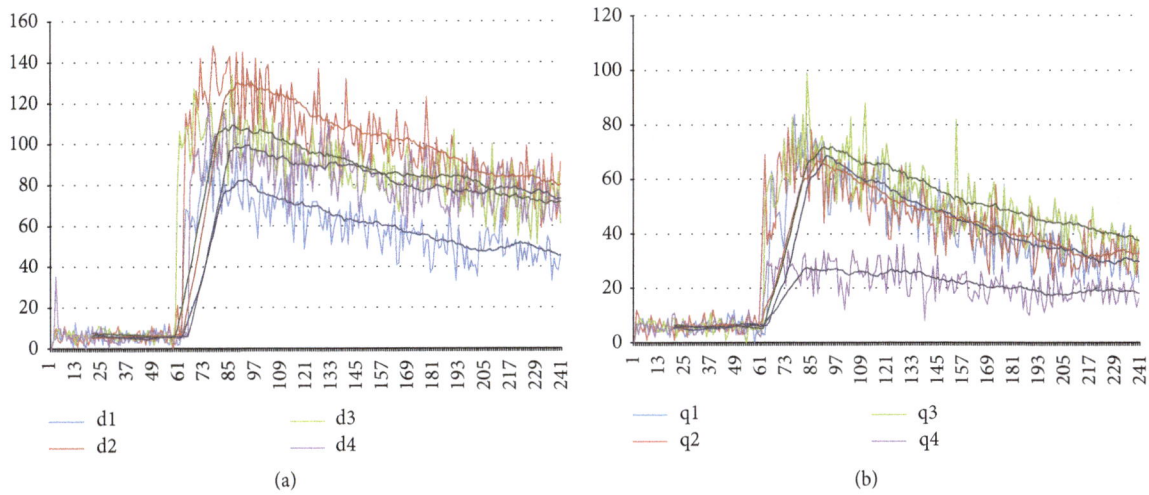

FIGURE 6: *Results of ultra-weak photon emission measurements after mixing 0.5 ml of 10 mM tert-Butyl hydroperoxide with 1 ml of deionized water.* The duration of measurement is 3 min. The vertical axis represents the number of photons per second and the horizontal axis represents seconds elapsed. (a) Deionized water. The number of photon emissions increased compared to day 1 (d1, d2, d3, and d4 indicate the day of measurement, resp.). (b) Deionized water treated in the noncontact manner. The number of photon emissions decreased drastically at day 4 (q1, q2, q3, and q4 indicate the day of measurement, resp.).

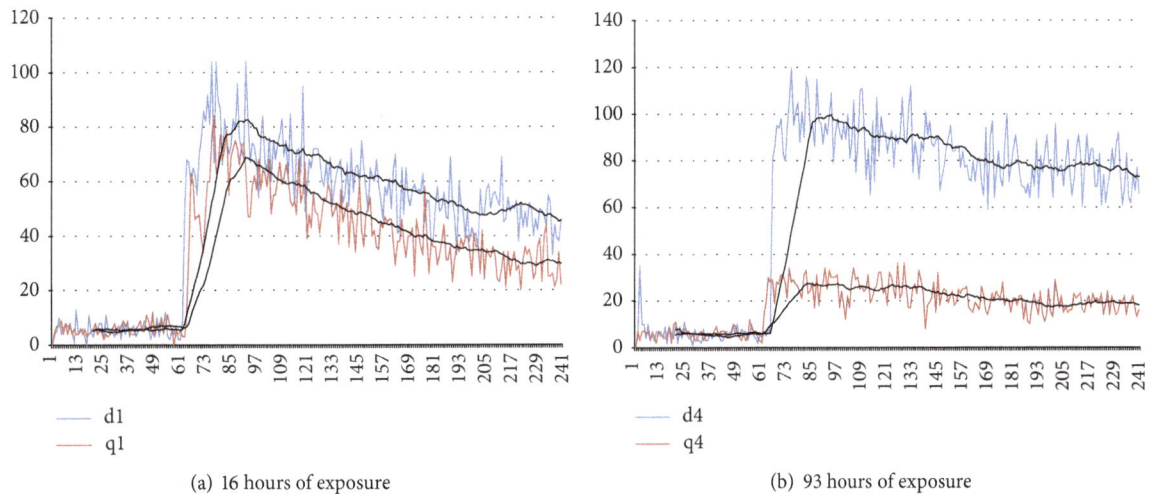

FIGURE 7: *Results of the ultra-weak photon emission measurement of deionized water (blue line) and noncontact treated deionized water (red line), respectively.* (a) 16 hours later. (b) 93 hours later. The antioxidant property levels of the noncontact treated water is about one-fourth of the control.

of structured water in the noncontact manner is possibly the infrared rays radiating from the ceramic powder [20–22]. This technique is quite important in the analysis of experimental results since only the structure of water was modified without any addition of material components to the water.

The modification of water structure seems to be a sort of continuous and gradual phenomenon. According to the experimental results of the ultra-weak photon emission (Figures 6 and 7) which was carried out for 93 hours in the noncontact manner, the water can be structured for quite a long time continuously through exposure to the powder for a longer time. However, the effects of exposure time on cell viability show that less than 6 hours is enough to obtain

the increased viability of the RAW-264.7 cell (Figure 14). This implies that it may not be necessary to make a highly structured water by treating the water for a longer time for biological purposes. In other words, cells are very sensitive to the structure of water, which is supported by the fact that even a very small amount of premixed structured water, 0.5% (Figure 9), is effective in increasing phagocytic activity. Additionally, the structured water is stable enough to show these effects, even when mixed with a new fresh culture medium and shaken well. Only a small amount of structured water is necessary to increase bioactivities of cells as shown in Figure 8.

The culturing media mixed with QELBY powder or prepared with the antioxidant structured water exhibit an

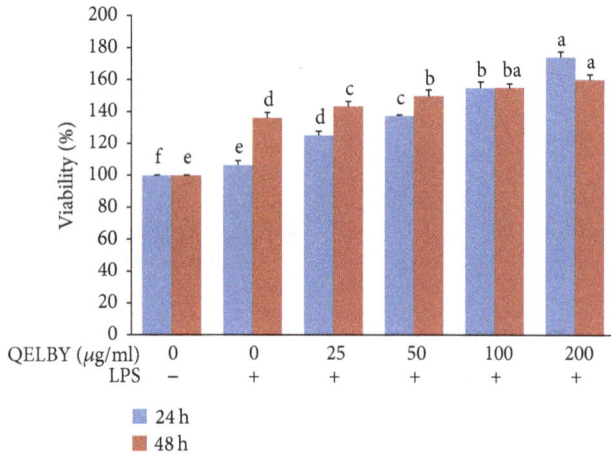

FIGURE 8: *Evaluation of the viability of RAW 264.7 macrophage cell.* Culture media were prepared by mixing with different amount of suspended layer from the DMEM mixed with 1% of QELBY ceramic powder and left for 48 hours. The blue and red bars indicate the length of incubating time, 24 hours and 48 hours, respectively, with the prepared media. Data are mean ± SD (*n* = 4). Bars with different superscript for each standing time are significantly different at *p* < 0.05.

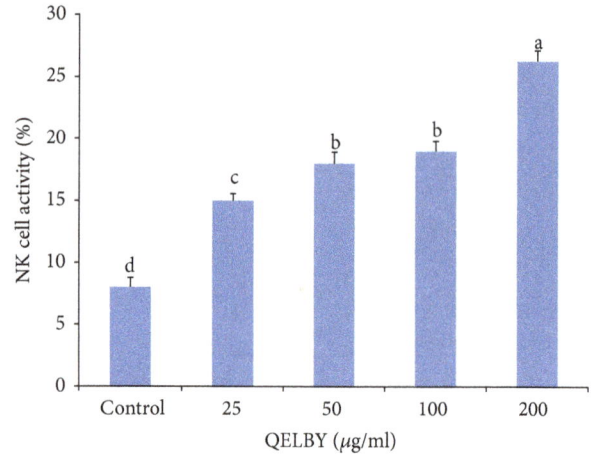

FIGURE 10: *Evaluation of natural killer cell activity.* Effector cell (isolated spleen cell) and target cells (YAC-I) were cultured at an effector : target ratio of 10 : 1 in RPMI 1640 media. The culturing media were prepared by mixing with different amount of suspended layer from the RPMI 1640 media mixed with 1% of QELBY powder and left for 24 hours. NK cell activity was assessed using CCK-8 assay to measure cell cytotoxicity against YAC-I cells. The results are expressed as mean ± SD (*n* = 5). Bars with different superscript are significantly different at *p* < 0.05.

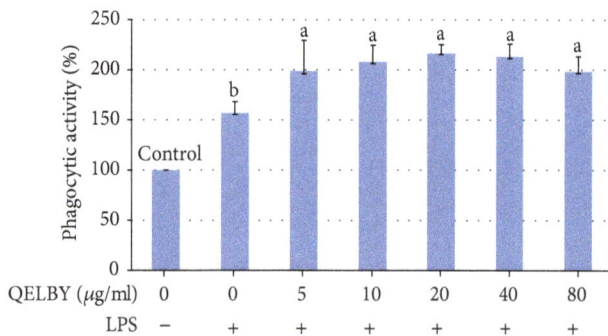

FIGURE 9: *Evaluation of phagocytic activity.* Phagocytic activity was measured by the uptake of neutral red dye. Culture media were prepared by mixing different amounts of suspended layers from the DMEM mixed with 1% of QELBY powder and left for 48 hours. Data are mean ± SD (*n* = 5). Bars with different superscript are significantly different at *p* < 0.05.

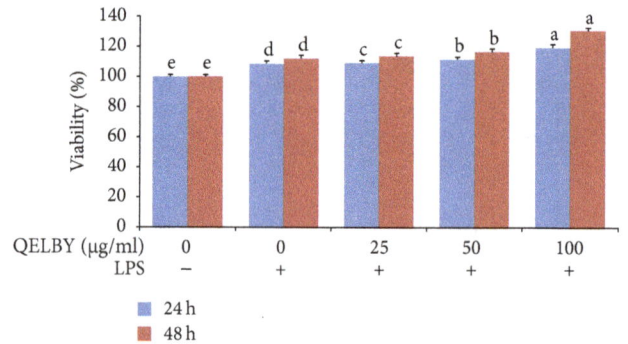

FIGURE 11: *Evaluation of viability of mouse splenocytes incubated for 24 and 48 hours.* The data shows a significant dose dependent effect on the viability of mouse splenocytes. The results are expressed as mean ± SD (*n* = 5). Bars with different superscript are significantly different at *p* < 0.05.

increase in the viability of normal cells. Recognizing that the same tendency was observed for both kinds of media prepared either by mixing with the QELBY powder or by mixing with the deionized water treated in the noncontact manner, the experimental results reported here must be interpreted as it relates to the antioxidant property of water generated by the modification of water structure, rather than the material components.

It is expected that a reactive oxygen species scavenging effect may occur through this media as other nutritional antioxidants [23–26]. As a result, the electron transport chain in the mitochondrion membrane may be activated, resulting in the increased viability for normal cells. Considering the principle of the CCK-8 assay kit, this explanation is more convincing since the formazan dye is formed as tetrazolium salt

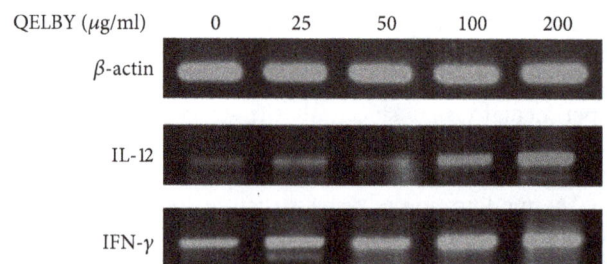

FIGURE 12: *Changes in cytokine expression of murine splenocytes incubated with the prepared media for 24 hours.* Reverse transcription polymerase chain reaction (RT-PCR) shows that both L-12 and INF-γ expressions were upregulated dose dependently by the incubation with the prepared media.

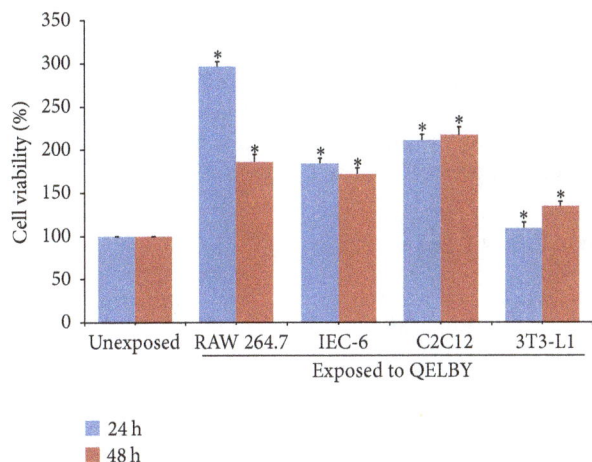

FIGURE 13: *The effects of the media prepared with the deionized water treated in a noncontact manner on the viability of various cells. The deionized water was treated for 48 hours. Blue and red bars indicate the length of incubating time, 24 hours and 48 hours, respectively. Data are mean ± SD ($n = 6$). Bars with (∗) are significantly different compared to their unexposed counterpart group ($p < 0.05$).*

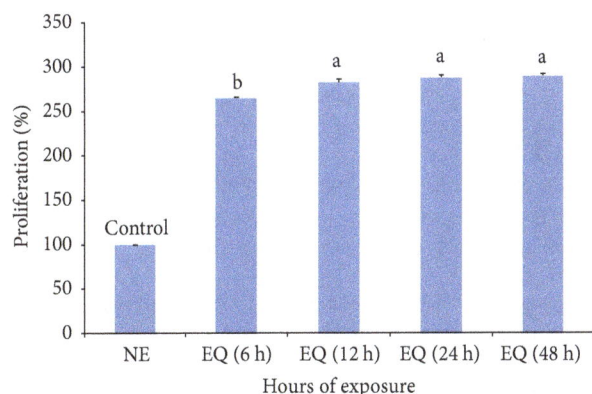

FIGURE 14: *The effects of the exposure time on the viability of RAW 264.7 macrophage cell. Deionized water was exposed to the QELBY ceramic powder for a specified time at room temperature before preparation of DMEM. Cells were seeded in different plates for each exposure time (NE: DMEM was prepared with the water, which was not exposed; EQ: DMEM prepared with the water, which was exposed to QELBY powder for specified times). Data are mean ± SD ($n = 7$). Bars with different superscript are significantly different at $p < 0.05$.*

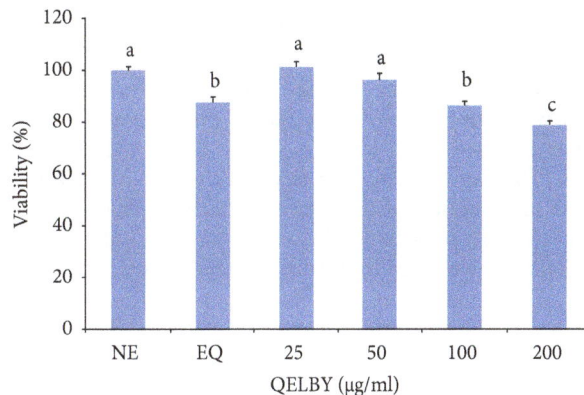

FIGURE 15: *The effect of the culturing media on the viability of a breast cancer cell, MCF-7. NE in the diagram indicates that unexposed deionized water was used to prepare the medium as a control. EQ means that deionized water treated in the noncontact manner for 48 hours was used for the preparation of medium. The number indicates the amount of the suspended layer which was taken from the premixed DMEM with 1% of QELBY® ceramic powder and left for 48 hours to prepare the media. Data are mean ± SD ($n = 6$). Means with different superscript are significantly different at $p < 0.05$.*

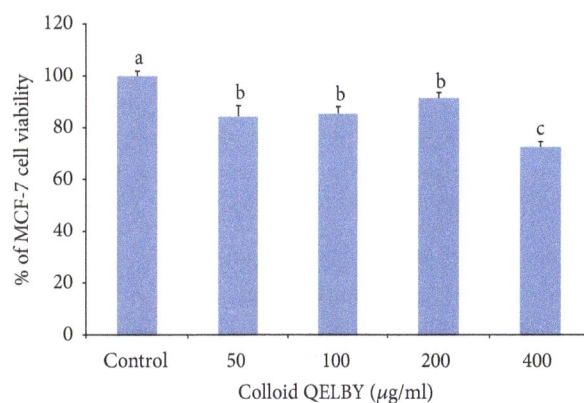

FIGURE 16: *The effect of the QELBY colloid on the viability of MCF-7 breast cancer cells. The QELBY colloid was prepared by mixing with 1% of QELBY power and DMEM and centrifuged at 3,000 rpm. A different amount of the supernatant was used for the preparation of DMEM media. Data are mean ± SD ($n = 4$). Means with different superscript are significantly different at $p < 0.05$.*

in the CCK-8 solution which is reduced by the dehydrogenase in the electron transport chain.

Another possibility is in the increased proton concentration. When the electric potential is generated between the exclusion zone and outside of it, a large amount of protons is generated in the outside region [6]. This means that the supply of proton can be enhanced when the water is structured. Considering the fact that the mitochondrial membrane potential, Ψm, is composed of potentials from ion and proton gradients, the increased amount of proton may contribute to the increased Ψm. Actually, it was reported that Ψm was increased up to 1.36-fold for RAW 264.7 and

HD11 cells after exposing them to the QELBY powder in the noncontact manner for 48 or 72 hours [27]. The increased Ψm may contribute to the increased viability and increased ATP production. However, this was not observed for these cells.

On the other hand, the same culturing media show the opposite trend for malignant cancer cells. This is quite an interesting phenomenon since the same stimulator results in the opposite reactions depend on the cells, normal or malignant. At this time, the mechanism is unclear, but the increased Ψm may provide clues for further research. According to the previous report, the Ψm was observed to increase also for the HeLa cell by more than 1.36-fold, while the intracellular ATP level decreased after 72 hours of exposure [27]. It is well known that cancer cells have a more hyperpolarised Ψm than normal cells [28–32] and can be larger than 50% than normal cells [33]. It has been reported that 10% value alterations from its optimum in Ψm can cause serious variability on ATP synthesis, by as large as a 90% decrease [34]. Thus, the hyperpolarity in Ψm of a cancer cell may be related to the malfunction of the electron transport chain, that is, oxidative phosphorylation. Warburg reported that cancer cells' metabolism occurs through aerobic glycolysis unlike normal cells, which rely on aerobic respiration [35]. The increase in Ψm and decrease in ATP production in the HeLa cell cultured in the noncontact manner with the QELBY powder imply that the variation in Ψm is one of the causes for the decreased viability of the cancer cell, MCF-7. Definitely, experiments on the effect of structured water on the potential Ψm is required for a detailed mechanism.

It has been claimed that continued oxidative stress can lead to chronic inflammation, which in turn could mediate most chronic diseases including cancer [36, 37]. In this report, the antioxidant property of the structured water, increased viability and cytokine expression, and suppression of cancer cell were described. More detailed anti-inflammatory effects of QELBY powder were reported elsewhere [38]. It is presumed that all of these results, antioxidant, anti-inflammation, and anticancer effects of QELBY ceramic powder, are strongly correlated to each other. Additionally, it was observed that DNA exposed to QELBY was protected from the damage caused by free radicals [27]. These results suggest that a future study on the effect of QELBY should focus on its overall holistic influence on a living body rather than focusing on molecular level cell cycle regulatory system.

Again, these results strongly suggest that water plays a critical role in cellular metabolism, more than previously understood. Water is not merely a simple liquid, but instead plays an important role in cellular and molecular mechanisms by stimulating the enhanced bioactivities of normal cells, while suppressing those of malignant cancer cells. All of these results strongly point to the positive role of structured water in cellular bioactivities.

5. Conclusion

The experimental results reported here imply that both the materialistic nutritional components of culture media and water structure can be critical factors in the analysis of bioactivities of cells. Structured water, either prepared through mixture with a hydrophilic ceramic powder, or treated in the noncontact manner, exhibited antioxidant property. The increase in cellular bioactivities is strongly related to the antioxidant property of structured water. Structured water appears to exhibit the same properties as an antioxidant. Since no antioxidant components were added in the water treated in the noncontact method, these properties are solely related to the structure of water. These experimental results strongly suggest that water can play an essential role in bioactivities of cells, depending on its antioxidant effects, based on its structure. Water itself can be an active constituent in cell biology, as are many other cellular molecules.

Disclosure

The funder did not have any additional roles in the study design, data collection and analysis, decision to publish, or preparation of the manuscript. The funder only provided financial support and research materials.

Conflicts of Interest

The authors declare that there are no conflicts of interest regarding the publication of this paper.

Acknowledgments

The authors appreciate Mr. Min Young Ahn for carrying out experiments on the cell viability measurement. The financial support provided by the Quantum Energy Co. Ltd. to SUNY Korea, the Korea Advanced Institute of Science and Technology, and Hankyung National University is greatly appreciated.

References

[1] P. Ball, "Water as an active constituent in cell biology," *Chemical Reviews*, vol. 108, no. 1, pp. 74–108, 2008.

[2] P. Jungwirth, "Biological water or rather water in biology?" *Journal of Physical Chemistry Letters*, vol. 6, no. 13, pp. 2449–2451, 2015.

[3] X. F. Pang, *Water*, World Scientific Publishing, Singapore, 2014.

[4] B. Bagchi, *Water in Biological and Chemical Processes*, Cambridge University Press, Cambridge, UK, 2013.

[5] G. H. Pollack, "Cell electrical properties: reconsidering the origin of the electrical potential," *Cell Biology International*, vol. 39, no. 3, pp. 237–242, 2015.

[6] G. Pollack, *The Fourth Phase of Water*, Ebner & Sons Publishers, Seattle, Washington, USA, 2013.

[7] J.-M. Zheng, W.-C. Chin, E. Khijniak, E. Khijniak Jr., and G. H. Pollack, "Surfaces and interfacial water: evidence that hydrophilic surfaces have long-range impact," *Advances in Colloid and Interface Science*, vol. 127, no. 1, pp. 19–27, 2006.

[8] J. Zheng and G. H. Pollack, "Long-range forces extending from polymer-gel surfaces," *Physical Review E: Statistical, Nonlinear, and Soft Matter Physics*, vol. 68, Article ID 031408, 2003.

[9] C.-S. Chen, W.-J. Chung, I. C. Hsu, C.-M. Wu, and W.-C. Chin, "Force field measurements within the exclusion zone of water," *Journal of Biological Physics*, vol. 38, no. 1, pp. 113–120, 2012.

[10] A. Sharma, D. Toso, K. Kung, G. Bahng, and G. H. Pollack, "QELBY®-Induced Enhancement of Exclusion Zone Buildup and Seed Germination," *Advances in Materials Science and Engineering*, vol. 2017, pp. 1–10, 2017.

[11] G. Franz and R. A. Sheldon, *Ullmann's Encyclopedia of Industrial Chemistry*, Wiley-VCH, Weinheim, Germany, 2005.

[12] C. H. Kennedy, D. F. Church, G. W. Winston, and W. A. Pryor, "Tert-butyl hydroperoxide-induced radical production in rat liver mitochondria," *Free Radical Biology and Medicine*, vol. 12, no. 5, pp. 381–387, 1992.

[13] O. Kučera, R. Endlicher, T. Roušar et al., "The effect of *tert*-butyl hydroperoxide-induced oxidative stress on lean and steatotic rat hepatocytes *in vitro*," *Oxidative Medicine and Cellular Longevity*, vol. 2014, Article ID 752506, 12 pages, 2014.

[14] E. A. Vogler, "Structure and reactivity of water at biomaterial surfaces," *Advances in Colloid and Interface Science*, vol. 74, no. 1–3, pp. 69–117, 1998.

[15] J. A. Ives, E. P. A. V. Wijk, N. Bat et al., "Ultraweak photon emission as a non-invasive health assessment: a systematic review," *PLoS ONE*, vol. 9, no. 2, Article ID e87401, 2014.

[16] K. Nakamura and M. Hiramatsu, "Ultra-weak photon emission from human hand: influence of temperature and oxygen concentration on emission," *Journal of Photochemistry and Photobiology B: Biology*, vol. 80, no. 2, pp. 156–160, 2005.

[17] A. Rastogi and P. Pospíšl, "Spontaneous ultraweak photon emission imaging of oxidative metabolic processes in human skin: effect of molecular oxygen and antioxidant defense system," *Journal of Biomedical Optics*, vol. 16, no. 9, Article ID 096005, 2011.

[18] Y. S. Velioglu, G. Mazza, L. Gao, and B. D. Oomah, "Antioxidant activity and total phenolics in selected fruits, vegetables, and grain products," *Journal of Agricultural and Food Chemistry*, vol. 46, no. 10, pp. 4113–4117, 1998.

[19] F. Khabiri, R. Hagens, C. Smuda et al., "Non-invasive monitoring of oxidative skin stress by ultraweak photon emission (UPE)-measurement. I: Mechanisms of UPE of biological materials," *Skin Research and Technology*, vol. 14, no. 1, pp. 103–111, 2008.

[20] A. P. Sommer, M. K. Haddad, and H. J. Fecht, "Light effect on water viscosity: implications for ATP biosynthesis," *Scientific Reports*, vol. 5, Article ID 12029, 2015.

[21] B. Chai, H. Yoo, and G. H. Pollack, "Effect of radiant energy on near-surface water," *Journal of Physical Chemistry B*, vol. 113, no. 42, pp. 13953–13958, 2009.

[22] T.-K. Leung, C.-F. Chan, P.-S. Lai, C.-H. Yang, C.-Y. Hsu, and Y.-S. Lin, "Inhibitory effects of far-infrared irradiation generated by ceramic material on murine melanoma cell growth," *International Journal of Photoenergy*, vol. 2012, Article ID 646845, 8 pages, 2012.

[23] P. Duh, "Antioxidant activity of burdock (Arctium lappa linné): Its scavenging effect on free-radical and active oxygen," *Journal of the American Oil Chemists' Society*, vol. 75, no. 4, pp. 455–461, 1998.

[24] T. N. Selvam, V. Venkatakrishnan, S. Damodar Kumar, and P. Elumalai, "Antioxidant and tumor cell suppression potential of Premna serratifolia Linn leaf," *Toxicology International*, vol. 19, no. 1, pp. 31–34, 2012.

[25] Y. Saito, Y. Yoshida, T. Akazawa, K. Takahashi, and E. Niki, "Cell death caused by selenium deficiency and protective effect of antioxidants," *Journal of Biological Chemistry*, vol. 278, no. 41, pp. 39428–39434, 2003.

[26] J.-H. Jang and Y.-J. Surh, "Protective effect of resveratrol on β-amyloid-induced oxidative PC12 cell death," *Free Radical Biology and Medicine*, vol. 34, no. 8, pp. 1100–1110, 2003.

[27] H. T. Lee, D. Han, J. B. Lee, G. W. Bahng, J. D. Lee, and J. W. Yoon, "Biological effects of indirect contact with QELBY® powder on nonmacrophagic and macrophage-derived cell lines," *Journal of the Preventive Veterinary Medicine*, vol. 40, no. 1, pp. 1–6, 2016.

[28] S. D. Bernal, T. J. Lampidis, I. C. Summerhayes, and L. B. Chen, "Rhodamine-123 selectively reduces clonal growth of carcinoma cells in vitro," *Science*, vol. 218, no. 4577, pp. 1117–1119, 1982.

[29] K. K. Nadakavukaren, J. J. Nadakavukaren, and L. B. Chen, "Increased rhodamine 123 uptake by carcinoma cells," *Cancer Research*, vol. 45, pp. 6093–6099, 1985.

[30] S. Davis, M. J. Weiss, J. R. Wong, T. J. Lampidis, and L. B. Chen, "Mitochondrial and plasma membrane potentials cause unusual accumulation and retention of rhodamine 123 by human breast adenocarcinoma-derived MCF-7 cells," *Journal of Biological Chemistry*, vol. 260, no. 25, pp. 13844–13850, 1985.

[31] L. B. Chen, "Mitochondrial membrane potential in living cells," *Annual Review of Cell Biology*, vol. 4, pp. 155–181, 1988.

[32] B. G. Heerdt, M. A. Houston, and L. H. Augenlicht, "Growth properties of colonic tumor cells are a function of the intrinsic mitochondrial membrane potential," *Cancer Research*, vol. 66, no. 3, pp. 1591–1596, 2006.

[33] B. G. Heerdt, M. A. Houston, and L. H. Augenlicht, "The intrinsic mitochondrial membrane potential of colonic carcinoma cells is linked to the probability of tumor progression," *Cancer Research*, vol. 65, no. 21, pp. 9861–9867, 2005.

[34] G. Bagkos, K. Koufopoulos, and C. Piperi, "ATP synthesis revisited: New avenues for the management of mitochondrial diseases," *Current Pharmaceutical Design*, vol. 20, no. 28, pp. 4570–4579, 2014.

[35] O. Warburg, "On the origin of cancer cells," *Science*, vol. 123, no. 3191, pp. 309–314, 1956.

[36] S. Reuter, S. C. Gupta, M. M. Chaturvedi, and B. B. Aggarwal, "Oxidative stress, inflammation, and cancer: how are they linked?" *Free Radical Biology and Medicine*, vol. 49, no. 11, pp. 1603–1616, 2010.

[37] Y.-W. Kim, X. Z. West, and T. V. Byzova, "Inflammation and oxidative stress in angiogenesis and vascular disease," *Journal of Molecular Medicine*, vol. 91, no. 3, pp. 323–328, 2013.

[38] J. D. Lee, E. J. S. Vergara, S. H. Choi, S. G. Hwang, and G. W. Bahng, "Anti-inflammatory activity of quantum energy living body on lipopolysaccharide-induced murine RAW 264.7 macrophage cell line," *Bioceramics Development and Applications*, vol. 6, Article ID 089, 2016.

5

Epigenetic Contribution of High-Mobility Group A Proteins to Stem Cell Properties

Vincenzo Giancotti ⑩,[1,2] **Natascha Bergamin,**[3] **Palmina Cataldi,**[3] **and Claudio Rizzi**[3]

[1]*Department of Life Science, University of Trieste, Trieste, Italy*
[2]*Trieste Proteine Ricerche, Palmanova, Udine, Italy*
[3]*Division of Pathology, Azienda Ospedaliero-Universitaria, Udine, Italy*

Correspondence should be addressed to Vincenzo Giancotti; giancotti@alice.it

Academic Editor: Richard Tucker

High-mobility group A (HMGA) proteins have been examined to understand their participation as structural epigenetic chromatin factors that confer stem-like properties to embryonic stem cells (ESCs), induced pluripotent stem cells (iPSCs), and cancer stem cells (CSCs). The function of HMGA was evaluated in conjunction with that of other epigenetic factors such as histones and microRNAs (miRs), taking into consideration the posttranscriptional modifications (PTMs) of histones (acetylation and methylation) and DNA methylation. HMGA proteins were coordinated or associated with histone and DNA modification and the expression of the factors related to pluripotency. CSCs showed remarkable differences compared with ESCs and iPSCs.

1. Introduction

Three polypeptides HMGA1a, HMGA1b (together HMGA1), and HMGA2 are high-mobility group A nuclear phosphoproteins that are highly expressed in undifferentiated and cancer cells, but that are noticeably absent in adult differentiated cells. Using previous nomenclature, these proteins were identified as HMGI, HMGY, and HMGI-C, respectively. The high levels of expression in embryos, which is followed by a gradual decrease and the need for these genes to remain unaltered, suggest that HMGA proteins play fundamental roles in normal development [1–3].

Why are HMGA proteins considered epigenetic factors?

If epigenetics comprises processes and molecular factors that modify the three-dimensional structure of chromatin without altering the primary sequence of DNA, then HMGA proteins should be considered epigenetic factors because they are architectural elements that modify the global structure of chromatin as well as organizing specific sites of expression in cooperation/competition with histones and in cooperation with other factors involved in epigenetic gene expression processes. If so, HMGA proteins should accompany embryonic stem cells (ESCs) through the various differentiating

lineages. ESCs are blastocyst-derived stem cells that show self-renewal and invasion as natural properties, together with pluripotency, that is, the capability to differentiate and give rise to many progressive specific lineages to build a complete organism. ESCs constitute then the logical reference system to interpret two other types of stem cell: induced pluripotent stem cells (iPSCs) and cancer stem cells (CSCs).

iPSCs were artificially produced for the first time by Takahashi and Yamanaka through ectopic expression of Oct4, Sox2, Klf4, and cMyc (together OSKM) in murine somatic cells [4] and by Thompson's group in human cells by replacing Klf4 and cMyc with factors LIN28 and NANOG [5]. LIN28 expression leads directly to the expression of HMGA proteins and the induced cells show properties similar to ESCs, with self-renewal capacity, invasion, and pluripotency of yielding cells useful for regenerative medicine. Since these breakthroughs, many studies have found that induced pluripotency is also feasible by using other methodologies and molecules including HMGA proteins [6–11]. We focused on HMGA proteins in iPSCs because HMGA proteins are as highly expressed in these cells as in ESCs [1–3].

Tumours and cancer cell lines express at least one type of HMGA proteins (HMGA1 or HMGA2) and show a high level of oncogenic transformation [12]. CSCs are a subpopulation of cancer cells that have some characteristics similar to ESCs and iPSCs including self-renewal and invasiveness. Moreover, they exhibit resistance to eradication by therapy; however, currently, their pattern of differentiating into normal cell lineages remains unknown. Although the properties of CSCs are well understood, their origin is controversial; in heterogeneous tumour masses, they represent a small fraction of cells, whose origin is uncertain and which are likely cancer type dependent. In any case, CSCs have been reported to express epithelial-mesenchymal-transition (EMT) factors as well as HMGA proteins, and they should be considered a high oncogenically transformed system [13].

In our previous review [12], we discussed the expression of HMGA proteins and pathways involved in seven types of cancer. We examined, in detail, results obtained by six different research groups that worked on the same breast cancer cell line, MDA-MB-231, which shows a triple-negative phenotype. All the authors agreed on reporting high levels of expression of both HMGA1 and HMGA2 in MDA-MB-231 cells, which have some properties of stem cells (self-renewal and invasion), while the property of metastasis is a specific characteristic of tumour cells. From the analysis of the results from published studies on seven cancers (breast, colorectal, prostate, lung, thyroid, ovarian, and brain), HMGA proteins were found to be derived from many active pathways such as Wnt/β-catenin, RAS/RAF, TGF-β, PI3K/Akt, and IL-6/Stat3, and, at same time, they induced these pathways, establishing an interconnected and self-stimulating process that drives cells towards high level of oncogenic transformation. These cells, likely CSCs, express high levels of both HMGA1 and HMGA2; this might constitute an essential element of resistant cancer cells characterized by well-defined self-renewal and invasion factors.

Here, we extend the analysis that we carried out on cancer cell lines and tumours to ESCs and iPSCs, because the three types of cells share original factors that constitute an early starting point of ESCs and iPSCs in development and, conversely, a rather stable positioning for CSCs. To this end, we examined only some ESCs varieties among differentiating lineages (because of limitations in the length of the review) and discuss cancer properties and iPSCs.

2. Chromatin Epigenetic Network

The main properties of ESCs are self-renewal and pluripotency which allow an increase in the number of stem cells necessary to build the whole organism and differentiation of these cells into all opportune lineages that give rise to all tissues. HMGA proteins are highly expressed in such cells [2].

Self-renewal and pluripotency of ESCs are assured by the presence of a few specific factors such as OCT4, SOX2, KLF4, cMYC, NANOG, and LIN28 whose expression is due to a precise chromatin structure derived from epigenetic modifying events that regulate chromatin organization and, consequently, gene expression in all type of cells. These events include the following:

(a) DNA methylation (m)/demethylation;

(b) histone acetylation (Ac)/deacetylation;

(c) histone methylation (me)/demethylation;

(d) alteration of the nucleosomal structure;

(e) regulation of gene expression by microRNAs (miRs) and long noncoding RNAs (lncRNAs).

These events do not occur independently of each other; rather they are connected to confer a precise functional structure to large or small parts of the chromatin.

DNA can be methylated at the cytosine 5-position of CpGs by DNA methyl-transferases (DNMTs). Unmethylated or hypomethylated DNA participates to the formation of an open (or active or unrepressed) euchromatin structure which allows the high levels of gene expression needed for ESCs to differentiate in various lineages. DNA modification (as well as histone modifications) is so determinative that a different degree of methylation can promote differentiation into an alternative lineage. DNA methylation exerts a repressive effect on pluripotency and initiates differentiation. In contrast, demethylated DNA allows iPSCs to acquire pluripotency similar to that of ESCs. Repression of DNA methylation by inhibiting DNMTs preserves the pluripotency of ESCs, while active DNMTs (such as DNMT1) induce the transition from pluripotency to multipotency [14, 15].

Another aspect of the polyhedral regulation of chromatin is posttranscriptional modifications (PTMs) of histones, which mainly consists of acetylation and methylation, particularly in lysines (K) of histone H3. Acetylation of lysines eliminates the positive charges that enable interactions with negatively charged DNA phosphates to compact the chromatin. Therefore, acetylation promotes open or unrepressed chromatin as unmethylated DNA. H3K9Ac, if present in promoters, activates transcription [16]. This modification is associated with the self-renewing capacity and pluripotency of ESCs and iPSCs. Activation of chromatin by H3K9Ac is coupled in the same action by H3K4me2/3: H3K9 hyperacetylation and H3K4 methylation induce the expression of pluripotent genes such as Oct4 and NANOG to maintain self-renewal [17].

Changing acetylated H3K9Ac into methylated H3K9 (H3K9me2/3) results in a closed or repressed heterochromatin structure to which H3K27me3 strongly contributes. Indeed, H3K9me3 is considered a barrier to efficient induction of somatic cells into iPSCs [18]. In ESCs, NANOG and lysine demethylase 1 together repress the genes involved in development, and NANOG shortens the cell cycle length by positively regulating the CDK6 kinase gene in the G1/S transition. The proliferation capability of MSCs before differentiation is guaranteed by the pair of self-renewal factors NANOG/OCT4 [19, 20].

Lysine methylation results from the action of the catalytic subunit EZH2 (enhancer of zest 2) of the polycomb complex 2 (PcG2) [21–27]. Through H3K27me3, EZH2 represses genes involved in both differentiation and cancer. During differentiation, EZH2 allows the transition from pluripotency to multipotency and progressively decreases self-renewal and proliferation up to mature differentiated cells. In cancer, EZH2 does not repress self-renewal that is retained. Although

EZH2 is in any case a repressor, it can act differently on the basis of the other factors accompanying it. For example, tumour suppressors such as p16^{INK4a} are repressed in cancer, but activated in differentiated cells [23, 24]. Indeed, Song et al. [28] defined EZH2 as a candidate oncogenic driver in a study on MDA-MB-231 and 4T1 triple-negative breast cancer cells. EZH2 overexpression in triple-negative breast cancer cells was shown to be related to self-renewal, migration, invasion, and tumour suppressor silencing. Consequently, the use of agents such as ZLD 1039, which inhibits EZH2 activity, stops metastasis. Moreover, it was reported that inhibition of the histone deacetylases (HDACs) also shows inhibition similar to that of EZH2 in an anticancer treatment. Here we must mention the striking difference in cancer cells compared to ESCs and iPSCs, in which EZH2 expression and histone deacetylation are associated with differentiation, that is, with a decrease of the proliferation.

As mentioned above, H3K4me3 and H3K27me3 regulate an open or closed (unrepressed or repressed, resp.) chromatin structure. However, these two different modifications of histone H3, which have opposite functions, may be present in the same promoter, referred to as bivalence [29]. There is a functional dualism in which the preponderance of one modification or the other allows the activation or repression of a gene.

The addition or removal of modifications in both DNA and histones needs an alteration of the compact nucleosomal structure that is achieved by the specific remodeling ATP-dependent enzymes SWI/SNF, ISWI, and CHD. These chromatin remodeling agents are also able to change the position of nucleosomes along the DNA sequence modifying then the length of the linker DNA where histone H1 is bound [30–33]. Remodeling factors and PTMs are related. For example, the repression of the remodeling factor Snf5 upregulates H3K27m3 and increases p16^{INK4a} repression in cancer [23, 24, 34].

3. Searching for the Location of the HMGA Proteins in the Chromatin Epigenetic Network

3.1. Relationships between HMGA Proteins, EZH2, and Proliferation Factors. The possible effects of either overexpression or repression of EZH2 in cancers such as breast, bladder, gastric, hepatocellular, lung, thyroid, and tongue are shown in Figure 1. HMGA proteins show actions consistent with those of EZH2 in promoting tumours and proliferation [12, 28, 35–41]. Active EZH2 induces and activates, in conjunction with HMGA, tumour-promoting factors and proliferation, repressing differentiating factors such as runt-domain transcription 3 factor (RUNX3), p57 cyclin-CDK inhibitor 1C (CDKN1C), and cadherin 1 (CDH1) [28, 42–49]. Inhibition of EZH2 by ZLD1039 or miR-26a no longer induces tumour invading factors such as metalloprotease 2 and 9 (MMP2/9) and those related to epithelial-mesenchymal-transition (EMT) and, in contrast, induces differentiating factors. Further support to the connected action of EZH2 and HMGA proteins derives from studies on other cancers

in which EZH2 and HMGA (frequently HMGA2) converge towards the oncogenic achievement. In prostate cancer [50–52], EZH2 overexpression correlates with high levels of oncogenic transformation and is due to the loss of miR-let-7, the miRs' family known as the main repressor of HMGA proteins. In breast cancer and non-small cell lung cancer (NSCLC), in which HMGA proteins are overexpressed [12], the protein MUC1-C activates EZH2 promoter through induction of the pRb-E2F pathway [52]. These relationships will be more extensively discussed in Figures 2 and 3. In bladder cancer [38–40], HMGA2 is upregulated and EMT established, while E-cadherin is repressed. Notably, EZH2 also induces EMT and represses E-cadherin promoting metastasis [53, 54]. In conclusion, EZH2, HMGA proteins, and miR-let-7 family are strictly linked in determining the cellular state in which other factors participate, such as the LIN 28 proteins (partners of miR-let-7), as we are going to illustrate in the following paragraphs.

3.2. HMGA Proteins and Factors of Pluripotency and Proliferation. The miR-let-7 family and LIN28 proteins have been described as tumour suppressors and tumour inducers, respectively [41, 55–60]. In normal development, the opposing actions of LIN28 and let-7 axis assure proper timing for development, proliferation, and differentiation. In this axis, HMGA2 participates [61–65]. As shown in Figure 2(a), the predominance of let-7 can result from both reduced expression of LIN28 and decreased activity of EZH2, that, in contrast, can be activated by let-7 repression [51, 66]. Increased activity of let-7 allows negative regulation of cancer factors such as RAS, MYC, HMGA1, and HMGA2; in other words, the origin and maintenance of CSCs are impeded [67]. Disturbance of the double-negative feedback loop causes severe effects as shown in Figure 2(b) [66, 68]. Many possible actions can decrease let-7 expression. LIN28 can be overexpressed by oncogenic factors such as MUC-1 or through a feedback loop with MYC [69]. let-7, initially increased by chemotherapeutic treatments, can decrease de novo because of tumour acquired resistance following, for example, irradiation or cisplatin therapy that likely increases EZH2 in human non-small cancer lung cells (NSCLCs) [68]. Consistently, in pancreatic cancer cells, EZH2 depletion decreases resistance to doxorubicin and gemcitabine, allowing p27 expression and apoptosis induction [70, 71]. Moreover, long noncoding RNA can downregulate let-7 and consequently increase LIN28, which is then in a position to induce oncogenesis and establish CSCs [72]. HMGA1 and HMGA2 are deeply implicated in the triangulation of the factors and events shown in Figure 2 because they belong to groups of factors that grant self-renewal capacity to cells. However, it should be noted that Figure 2 presents an incomplete view of the complex relationships that link other factors such as Sox2, [67, 73, 74].

The couple proteins Rb and E2F are a well-known complex involved in proliferation, because E2F induces the expression of target genes that are proliferation factors in cancer. An unphosphorylated (or hypophosphorylated) form of Rb participates in an E2F complex; this status prevents the transcription of E2F-dependent tumour-promoting factors.

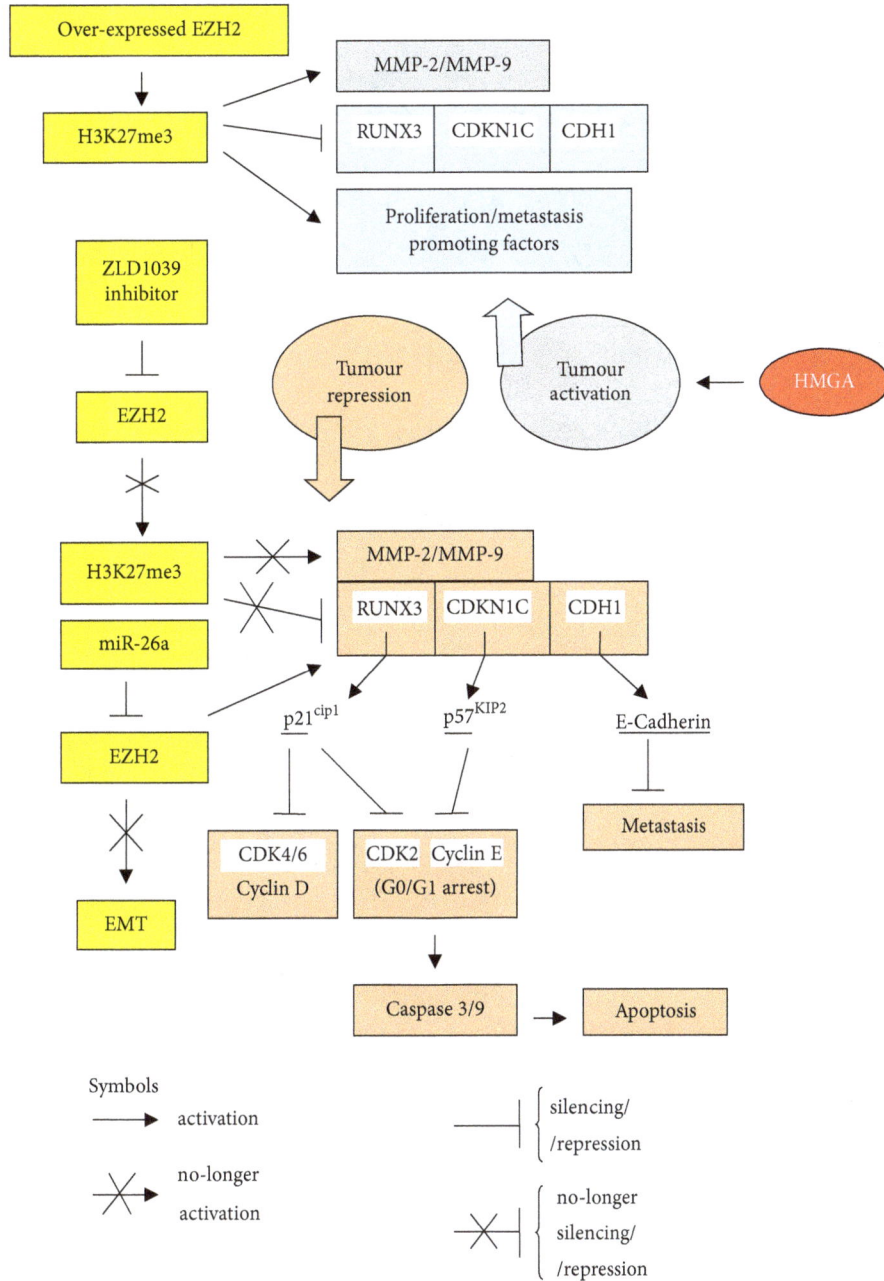

FIGURE 1: Alternative cascades of events following EZH2 overexpression/repression in cancer cells. CDH: cadherin; CDK: cyclin-dependent kinase; CDKN1C: cyclin-CDK inhibitor p57; EMT: epithelial-mesenchymal-transition; EZH2: enhancer of zeste 2; MMP: metalloprotease; p21^{cip1}: p21 CDK-interacting protein 1; p57^{KIP2}: p57 kinase inhibitory protein 2; RUNX3: runt-domain transcription factor 3.

pRb hyperphosphorylated by cyclin D1/CDK4/6 dissociates from the complex, inducing E2F factors, cell cycle progression, and proliferation (see Figure 3(a)) [75, 76]. Inactivation of the free E2F can result from repression by INK4A family repressors (p15, p16, p18, and p19) and CIP/KIP family repressors (p21, p27, and p57) that prevent the G1/S transition [77–79]. Proteins such as p16 are active in normal tissues but absent in cancer tissues or highly proliferating stem cells. It is worthwhile to mention that enzymes, such as HDACs, which modify histones, are able to modify other proteins. A parallel action (Figure 3(b)) is carried out by HDAC

inhibitors, because HDACs are proliferation promoters. For example, HDAC2 is highly expressed in tumours and related to p16; by inhibiting HDAC2, p16 activity is promoted and cells are arrested in G1/S [80]. Similarly, in pituitary tumourigenesis, HMGA2 displaces HDAC1 from the complex Rb/E2F1, leaving the latter in an active acetylated form [81]. We should note that HMGA and HDACs are consistent in inducing proliferation. The schemes in Figures 3(a) and 3(b) may no longer be valid if there are upstream events that activate pathways or modify gene structures (such as mutation and amplification of DNA and histone

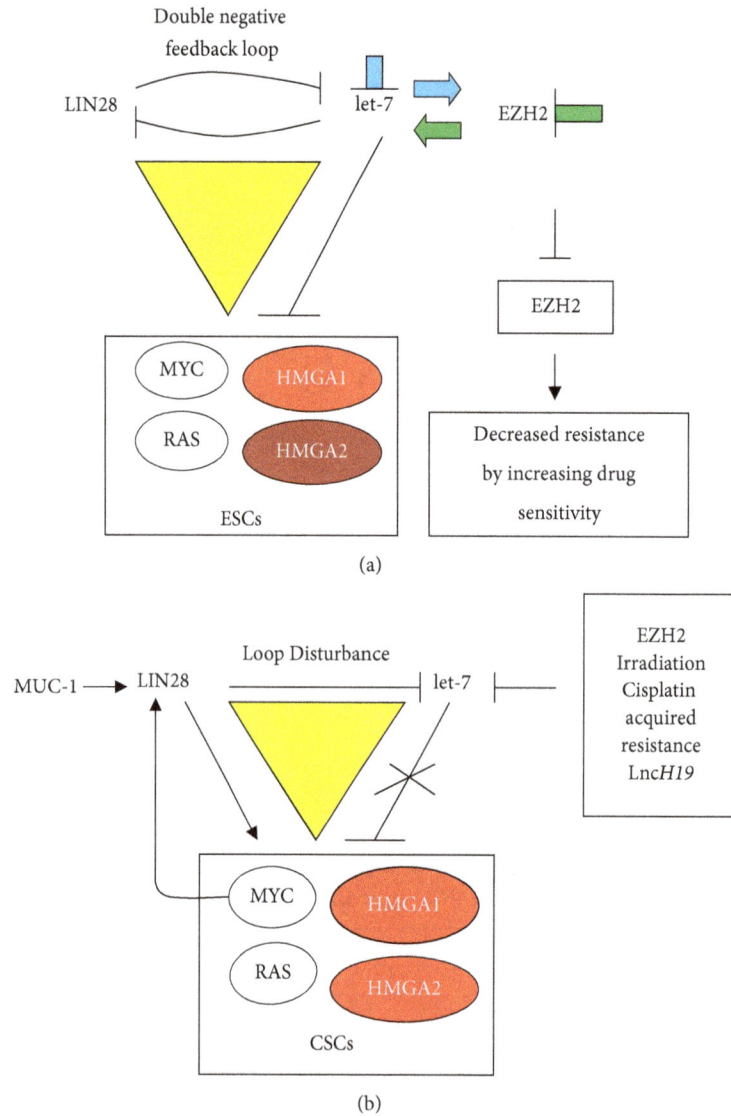

FIGURE 2: LIN28-miR-let-7 feedback loop and its regulation in ESCs and CSCs. Symbols are the same as shown in Figure 1.

epigenetic modifications) which could influence the factors under discussion (Figure 3(c)). First, the upstream expression of HMGA1 and HMGA2 could change the effect of linked factors. Indeed, expression of the tumour-promoting HMGA proteins (expressed also in ESCs) represses factors such as p16, whose repressing effect is reversed and, consequently, that of Rb [82–85]. Indeed, the active *Ink4a/Arf* locus, which expresses p16 and p19, is blocked by HMGA2, which in turn is repressed by miR-let-7b; elevated expression of miR-let-7b reduces the self-renewal capacity of neuronal stem cells (NSCs) [86]. Finally, a parallel action, shown in Figure 3(c), is exerted in mice by EZH2 according to the paper by He et al. [87]. EZH2 represses p16^{Ink4a} through H3K27me3 resulting in CDK4/6 upregulation and cardiomyocyte proliferation.

3.3. HMGA Proteins in Adipogenesis and Osteogenesis. In MSC differentiation, the canonical Wnt/β-catenin pathway works with DNA and histone modifications and factors

specific for MSC differentiation [88]. Interestingly, alternative differentiation of MSCs into adipocytes or osteocytes occurs, depending on different combinations of histone H3 modifications with active or repressed Wnt/β-catenin that involves gene activation/repression leading to adipogenesis or osteogenesis [89]. The predifferentiation stage of MSCs is characterized by an active canonical Wnt/β-catenin pathway with H3K4me and without H3K27me, which gives rise to open chromatin expressing c-Myc and cyclin D1. The process of differentiation can follow two paths, depending on the Wnt/β-catenin status [90–93]. If the Wnt/β-catenin pathway remains active, β-catenin translocated to the nucleus activates the expression of TCF/LEF dependent genes specific for osteogenic differentiation such as Runx2, Dlx5, and Osterix, and at the same time the adipogenic genes C/EBPα and PPARγ are repressed [94–100]. In contrast, if high levels of EZH2 are present, Wnt/β-catenin is repressed by H3K27me, while the expression of C/EBPα and PPARγ is

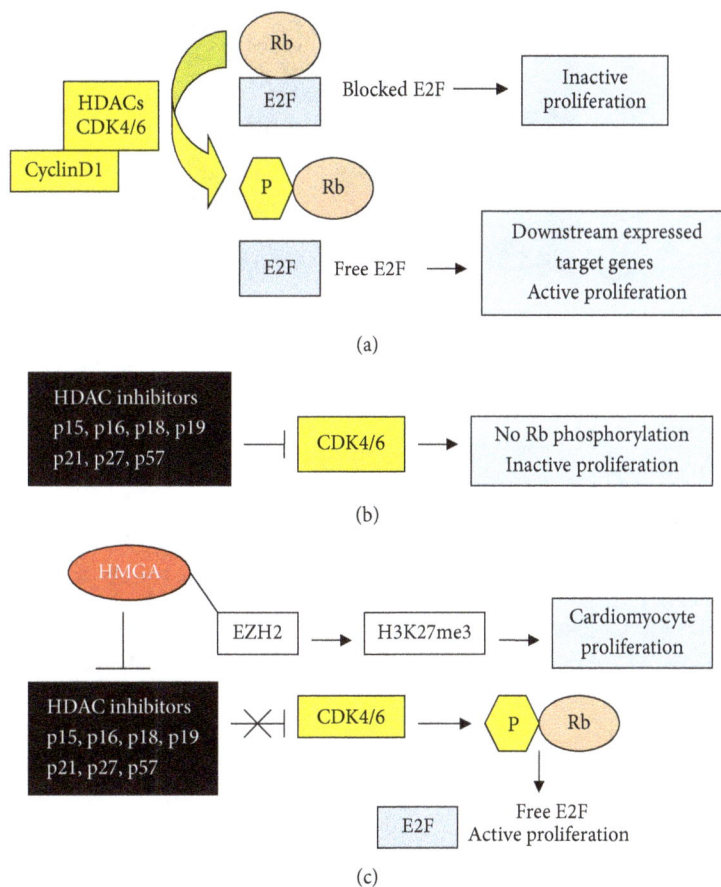

FIGURE 3: Rb and E2F in CSCs. (a) Phosphorylated Rb dissociates from E2F that induces proliferation factors; (b) blocking of Rb CDK4/6 kinases prevents Rb phosphorylation and inactivates proliferation; (c) HMGA proteins inhibit CDK4/6 Rb kinases and block cycle repressing-factors: proliferation is reactivated. Symbols are the same as shown in Figure 1.

activated [85, 93]. Figure 4 summarizes the progress of the two differentiating lineages. HMGA2 protein induces adipogenesis rather than osteogenesis [101]. However, it is interesting to note that the data in Figure 4 reflects normal development, whereas, in cancer, the Wnt/β-catenin pathway and HMGA expression are always consistent; that is, they serve as tumour promoters. In other words, the processes underlying both differentiation and cancer show the presence of repressive epigenetic factors that are apparently contradictory, considering the enormous differences between the two phenotypes. It is evident that repression in the two systems does not follow the same repressive gene pattern. In adipogenesis, HMGA2 is involved in two functions [101, 102]. On the one hand, it guarantees that undifferentiated preadipocytes from MSCs have an open chromatin structure that is needed to initiate differentiation. To this end, HMGA2 activates factors in the C/EBP family and PPAR-γ and, at the same time, EZH2 induces H3K27me3, repressing Wnt/β-catenin, which is needed because this pathway is an osteogenic promoter rather than an adipogenic one. On the other hand, HMGA2 in conjunction with the STAT3 pathway allows proliferation to produce fat masses [102].

Both adipogenesis and osteogenesis are strongly miR-dependent; however, in Figure 4, we indicate only the miR-30

family among a myriad of miRs discussed elsewhere [103]. Members of the let-7 family of miRs are strong repressors of HMGA proteins, as in cancer [12]. The repression of HMGA2 by let-7 (and other miRs) strongly promotes osteogenesis and inhibits adipogenesis [104, 105] and is linked to both Wnt/β-catenin and EZH2 as shown in Figure 4. HMGA2, present in both preadipocytes and preosteocytes, guarantees an open chromatin structure that initiates the two lineages through the factors introduced above. HMGA2 disappears soon after this in osteogenesis, whereas, in adipogenesis, it gradually decreases over time. It is absent in mature differentiated cells, but still present in stem cells that constitute the reserve for replacing dead cells. However, HMGA2 repression by let-7 allows osteogenesis to proceed, while adipogenesis is repressed because C/EBP and PPAR-γ are not activated.

We have focused our discussion on the differentiation of MSCs from the mesoderm based on the factors introduced above. Table 1 summarizes in a concise form the relationships between the Wnt pathway, miR-let-7, HMGA2, and EZH2 which are involved in MSCs differentiation, beginning from mesoderm and progressing to four mature differentiated cells: adipocytes, osteocytes, myocytes, and cardiocytes. The marks (+) and (−), indicating a positive or a negative contribution, respectively, to the differentiating process, are rather

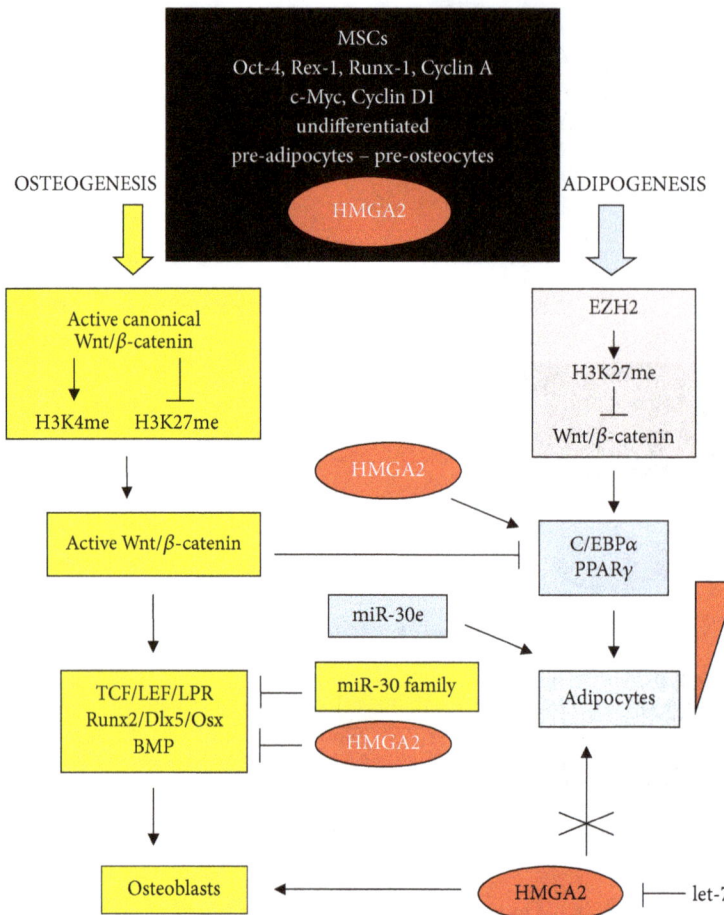

FIGURE 4: Alternative adipocyte and osteocyte differentiating pathways of MSCs based on EZH2 and Wnt/β-catenin actions. BMP: bone morphogenetic protein; C/EBP: CCAAT/enhancer binding protein; Dlx: distal-less homeobox; LPR: low-density lipoprotein receptor related protein; Osx: Osterix; PPARγ: peroxisome proliferator-activated receptor γ; Runx: runt-related transcription factor. Symbols are the same as shown in Figure 1.

TABLE 1: Wnt/β-catenin, miR-let-7, HMGA2, and EZH2 action in mesenchymal stem cells (MSCs) differentiation.

Differentiation lineage	Canonical Wnt	miR let-7	HMGA2	EZH2
Adipogenesis	−	−	+	+
Osteogenesis	+	+	−	−
Myogenesis				
Early	+	+	−	−
Late	−			
Cardiogenesis	−	+	−	−

(+): positive contribution; (−): negative contribution.

simplistic and incomplete in showing a complex program that is characterized by progressive changes, with factors from each stage still expressed in subsequent stages. In other words, differentiation and development are frequently used in a generic manner, although they refer to different and overlapping processes: from pluripotency to multipotency and monopotency (of stem cells as the reserves to regenerate

tissues); from proliferation and invasion to the maturation of nonproliferating cells; and consequently from factors and pathways of pluripotency and invasion to molecules that are characteristic of differentiated cells. As shown in Table 1, it can be difficult to determine the precise point of action of these factors.

3.4. HMGA Proteins in Myogenesis. Adipogenesis and osteogenesis, initiated by MSCs, are discussed above. Myogenesis and osteogenesis deserve additional comments.

Pluripotency and proliferation of ESCs are assured by factors such as IMP2, cMyc, NRAS, and HMGA2. MyoD is a factor of myogenic differentiation that represses proliferation through long noncoding MyoD RNA (LncMyoD) [106] once a proper number of cells to be terminally differentiated are produced. It is conceivable that HMGA2 is repressed because its expression is strictly associated with the above factors (Figure 5, yellow). The repression of pluripotency factors indicates the end of myogenic proliferation. In contrast, the EMT factors Snail/2 repress myogenic differentiation because they are associated with the invasion and lack of differentiation of cells [107, 108] (Figure 5, blue). MyoD repression by

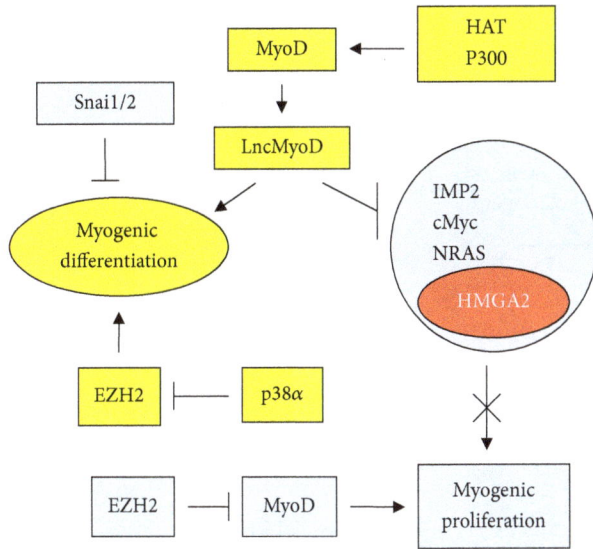

FIGURE 5: Relationship between MyoD, pluripotency factors, and EZH2 in myogenic differentiation. Symbols are the same as shown in Figure 1.

EZH2 stimulates proliferation [109], while EZH2 degradation in response to the phosphorylation of p38α kinase arrests proliferation, allowing differentiation to prevail [110] (Figure 5). Inhibition of EZH2 decreases H3K27me3 modifications, and the transition from MSCs to differentiated cartilage is increased [111]. Moreover, in differentiation, MyoD factor is acetylated; HAT p300 acetylates MyoD during myogenic differentiation and increases its transcriptional activity [112, 113]. It is interesting to note that EZH2 repression (and consequently H3K27me3 repression) induces osteogenic and myogenic differentiation and suppresses tumour formation. Wnt3a is one of the ligands that can induce the canonical β-catenin pathway [114]. The expression of HMGA proteins and proliferation are induced through the association of β-catenin, TCF, and LEF [115–117]. In myogenesis, this occurs early for later differentiation of cells. Wnt3a action (early stage) overlaps the initiation of MyoD expression, when Wnt3a activity should be ending [118–120]. In Table 1, these two states are shown.

A large number of miRs involved in myogenic differentiation have been reported. Horak et al. [121] introduced a list of miRs involved in skeletal muscle development. Among these miRs, we show the action of miR-1, miR-133, and miR-206 in Figure 6. These miRs, also reported by Chen et al. [122], are also involved in myogenesis together with miR-34b [123], miR-16 [124], and miR-195/497 [125]. The middle of Figure 6 shows the contribution of various miRs to the promotion of myogenic differentiation (left side, yellow) which results in the inhibition of myogenic proliferation (right side, blue). The activating/repressing events are rather complex. miR-1 and miR-206 downregulate histone deacetylase 4 (HDAC4). Inhibition of deacetylases reduces proliferation of cancer cells. Consequently, as shown in Figure 6, HDAC4 inhibition promotes myogenic differentiation. According to the study by Chen et al. [122] proliferation and differentiation are

mutually exclusive in skeletal muscle formation in which miR-1 and miR-206 are inducers of differentiation while miR-133 is an inducer of proliferation, assuming it is not blocked by HDAC1/2 (Figure 6). In this context, HMGA2 protein is involved in tissue regeneration because its expression expands muscle proliferating myoblast progenitors [126]. Moreover, HMGA2 targets IGF2BP2 (also named IMP2), which in turn induces many genes that promote cell growth, including cMyc and SP1. For example, IMP2 and its homolog IMP1 are involved in neuronal precursor cell proliferation, along with HMGA2; in adult neuronal stem cells, let-7 downregulates IMP proteins and HMGA2 [83, 127–129]. Finally, the inhibition of HMGA1 (a self-renewal factor) by miRs 195/497 and that of cyclin D1 (a cell cycle promoter) by miR-206 induce differentiation (left) and downregulation of proliferation (right) (Figure 6).

Many transcription factors allow stem cells to be either normal or cancerous. Snail1 and Slug (also named Snail2), are some of these transcription molecules. Indeed, Snail1 and Slug, by repressing the membrane protein claudin-1, activate EMT in both normal canine kidney cells (MDCK) and MDA-MB-231 breast cancer cells [130]. In Figure 6, we show that cells engaged in differentiation should lose their invasion capability, which is a property of self-renewing cells. miR-30a and miR-206 downregulate, in myogenic differentiation, Snail1/2 which are associated with stemness as above discussed. Finally, in the upper part of Figure 6, the action of Bcl-2 is illustrated. This is an antiapoptotic agent and a general inducer of proliferation. There is a four-side relationship that links Bcl-2, HMGA2, p53, and miR-34a in which p53 is a positive inducer of miR-34a which, in turn, inhibits Bcl-2. In contrast, HMGA2 is an inducer of Bcl-2 and, consequently, proliferation [131–135]. In myogenesis, the downregulation of Bcl-2 by miR-16 and miR-34b [123, 124] results in the inhibition of myogenic proliferation (Figure 6).

To better understand the location of HMGA proteins in myogenesis, we examined their action in satellite stem cells, which are postnatal stem cell stock for muscle regeneration. If this function is not required, satellite stem cells remain in an nonproliferating quiescent state (Figure 7) that is characterized by Pax7 [136], a known satellite stem cell marker, the repression of the growing factor HMGA2, and the absence of both the proliferating index Ki67 and the differentiation-related factor MyoD [137]. Chromatin is in an open and permissive state because of histone H3 modifications and is ready to receive environmental information to activate development [136–138]. Once factors in the microenvironment are produced in response to a request for regeneration, satellite stem cells are activated by the proliferation-inducing HMGA2/IGF2BP2, the cell cycle inducer cyclin D1, and an increase in H3K27me3. Many specific myogenic factors such as MyoD and Myf5 are expressed and growth starts [125, 139, 140]. Once an appropriate number of cells have been produced and the action of HMGA proteins is no longer necessary, their expression is repressed as specifically reported for HMGA1 [141, 142], and differentiation proceeds to completion. In conclusion, the data shown in Figures 5–7 suggest the involvement of HMGA proteins, in conjunction with many other factors, to produce proliferating cells that

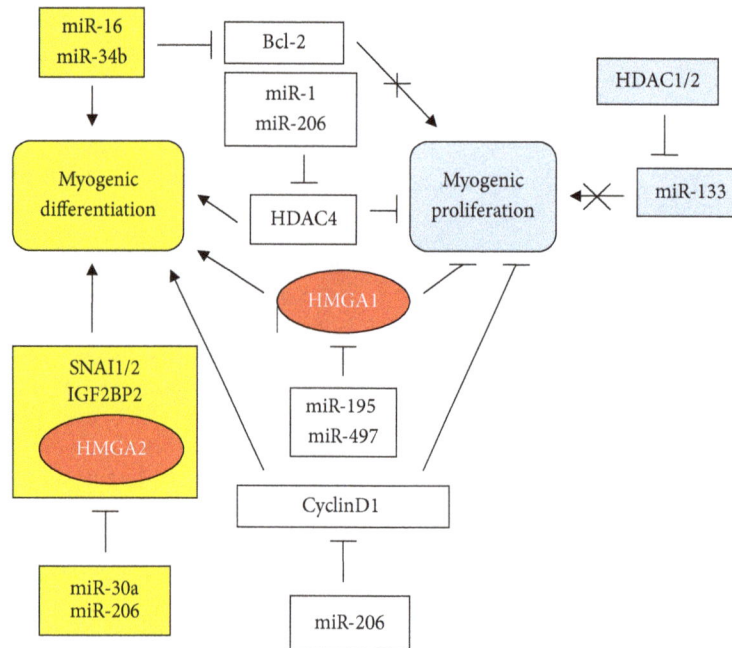

FIGURE 6: miRs, HDACs, and HMGA proteins in myogenic differentiation. Left side (yellow): induction of differentiation; right side (blue): suppression of proliferation. Symbols are the same as shown in Figure 1.

can be differentiated from myogenic stem cells. Because members of the let-7 family of miRs are important repressors of HMGA protein (and other self-renewal factors) expression, Table 1 shows a positive (+) contribution for let-7 that indicates differentiation, while the contribution is negative (−) for proliferation.

3.5. HMGA Proteins in Cardiomyogenesis. miR-let-7 overexpression is required for human embryonic stem cell-derived cardiomyocytes (hESCs, CM) [143]. To this end, as shown in Figure 8, self-renewal/proliferation factors such as Lin28, NANOG, Oct4, and HMGA2 that characterize hESCs should be repressed, for example, by miR-125b. miR-125b overexpression, by downregulating self-renewal factors, allows unrepressed let-7 to induce differentiation to cardiac muscle cells [144].

As in myogenesis, in cardiogenesis, Wnt3a activates proliferation if the β-catenin pentadegradating complex is inactive because of the modification of one or more of its component such as CK1. β-catenin then accumulates in the nucleus, where, in association with TCF/LEF, it induces specific gene expression for proliferation. If the pentacomplex is active in degrading β-catenin, then proliferation is hampered and cardiomyocytes are activated for differentiation [145]. miR-1 is the main regulator of vertebrate cardiomyogenesis [146]; its overexpression promotes differentiation of cardiomyocytes from multipotent MSCs by downregulation of Wnt3a, which is a canonical inducer of proliferation (Figure 8). Lu et al. [146] also report that the expression of let-7b in cardiomyocytes (CM) is similar to that of miR-1. The conclusion is that, considering only the differentiation stage, there is a positive (+) contribution by let-7 and a negative (−) one by Wnt, as in myogenesis (Table 1). Notably, some ligands,

such as Wnt-5a and Wnt-11, act as repressors of canonical β-catenin signaling promoting the differentiation of cardiac progenitors [147–149].

The transition from hESCs or iPSCs could occur, for example, as a result of exposure to bone morphologic protein-4 (BMP-4) [150], which induces an early mesodermal differentiation stage by repressing SOX2 and promoting SLUG, MSX2, and EMT. At this stage, cells still show proliferation properties; however, these are specifically directed towards cardiomyocyte production (Figure 9). Pluripotent stem cells factors such as SOX2 are repressed and the canonical Wnt pathway is responsible for proliferation. Indeed, an active Akt signal (because its repressor PTEN has been deleted) induces the β-catenin pathway which promotes proliferation of cardiac progenitor cells [151]. To activate late cardiomyocyte differentiation of already proliferating cells, it is necessary to block Wnt signaling. To this end, there are many choices: protein factors such as secreted frizzled related protein 2 (Sfrp2), dickkopf protein 1 (DKK1), or synthetic chemical compounds (such as IWR-1 and IWP-1) can inhibit Wnt [152–156].

Figure 9 indicates Gata4 and NKx2.5, as two factors that characterize cardiogenic differentiation. GATA4 in an acetylated form (as MyoD in myogenesis) that promotes cardiogenic differentiation. Nucleosomal remodeling and deacetylase (NuRD) is able to deacetylase GATA4 that, in this form, does not induce cardiomyocyte differentiation. Indeed, deacetylases support proliferation rather than differentiation [157]. NKx2.5 is positively regulated by HMGA2 through Smad1/4 of the TGF-β pathway and it is a crucial factor for cardiogenesis [158]. Phosphorylated NKx2.5 by p38γ kinase is translocated into the nucleus and, together with GATA4, forms a protein complex that is critical for cardiomyocyte differentiation because it maintains the cardiac progenitor

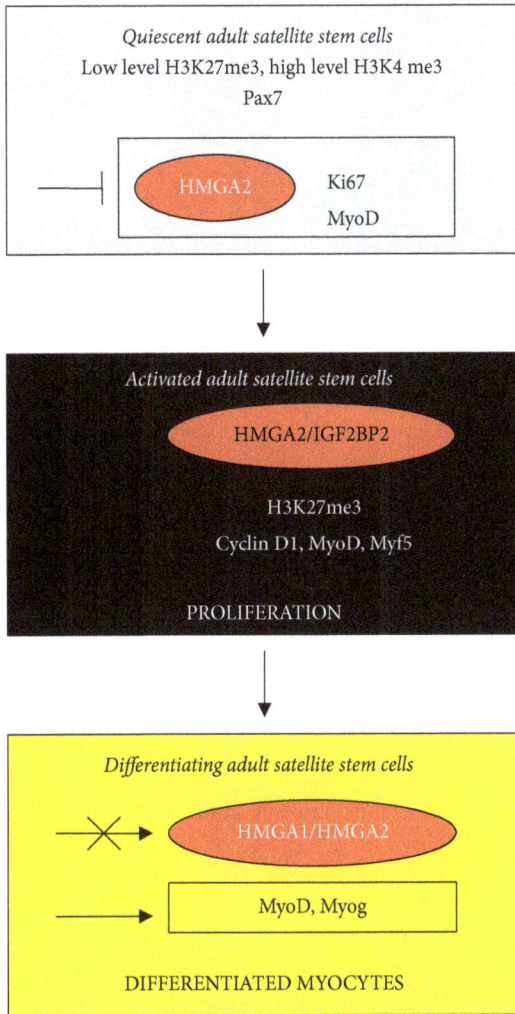

FIGURE 7: From quiescent adult satellite stem cells to differentiated myocytes. Symbols are the same as shown in Figure 1.

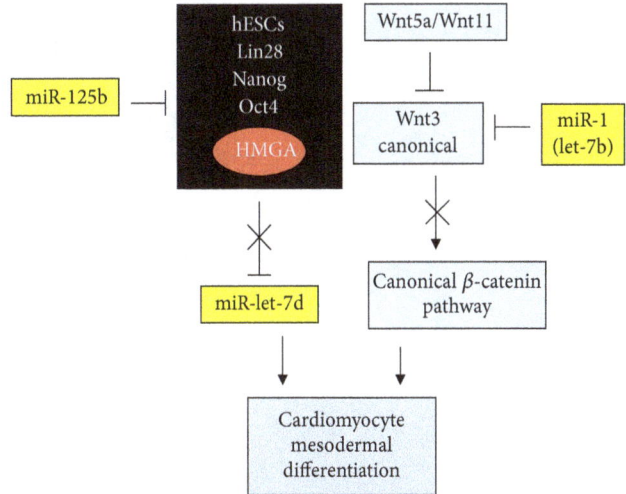

FIGURE 8: From hESCs to cardiomyocyte differentiation through miRs and Wnt activities. Symbols are the same as shown in Figure 1.

cells (CPCs) state. In this context, NKx2.5 enhancer contains modified H3 forms: H3K9Ac, H3K27Ac, and H3K4me3 [159–161]. Notably, NKx2.5 is not expressed in undifferentiated hiPSCs that, in contrast, express Oct4, one of the canonical factors of pluripotency. The induction of NKx2.5 and GATA4 expression requires chromatin modifications of the enhancers by SWI/SNF, whose ATPase Brg1 is a main component of the modifying complex, by HMGA1, and by modified forms of histone H3 (see above) [160, 162, 163]. The SWI/SNF machinery is constantly modifying the chromatin from ESCs until the cells are differentiated, and in the proliferative state they are accompanied by HMGA proteins. However, during development, SWI/SNF activity is progressively modulated by different SWI/SNF subunits, DNA modification, post-transnational protein modifications, and miR action. For example, the change from H3K27me3 to H3K27Ac regulates the change from an inactive gene to an active one in CPCs.

3.6. Direct Involvement of HMGA Proteins in Stem Cell Induction and Maintenance. iPSCs develop because of the ectopic expression of pluripotent ESC factors; it is therefore consistent that HMGA proteins are expressed in iPSCs as well as in ESCs. Accordingly, both HMGA1 and HMGA2 have been shown to be highly expressed in iPSCs and to contribute to reprogramming efficiency [10, 164, 165].

OCT4, SOX2, and NANOG maintain the undifferentiated state of ESCs and, if expressed in these cells, of iPSCs and CSCs [166]. The three factors also guarantee pluripotency of ESCs and possibly of iPSCs, but pluripotency of CSCs is questionable if it means these cells are capable of differentiating into normal cells. The three factors are DNA-binding proteins similar to HMGA and histones; however, these factors show secondary and tertiary structures that are different from those of HMGA proteins which are considered unstructured/disordered polypeptides [167, 168]. OCT4 with a helix-turn-helix (H-T-H) containing domain, SOX2 with an HMGB-box containing domain, and NANOG with a homeodomain (HD) containing domain cooperatively bind to the DNA, altering the bending, kinking, looping, and unwinding that allow the action of other factors on the chromatin. The three factors interact with the DNA at the major groove (OCT4), at the minor groove (SOX2), and at both grooves and the DNA backbone (NANOG) to recognize specific DNA sequences and AT-rich regions [169–171]. Interestingly, HMGA proteins recognize as well AT-rich DNA sequences but not in a specific way. Because HMGA proteins are the most abundant proteins bound to DNA after histones and they occupy a large portion of the chromatin, it is conceivable that HMGA proteins predispose the chromatin to receive OCT4, SOX2, and NANOG by sliding along the DNA to find their specific binding sequences. This is much more than a hypothesis because Shah et al. in hESCs showed that HMGA1 binds directly to cMYC, Sox2, and Lin28 promoters and induces the expression of these proteins [10].

Previously, we hypothesized that the expression of HMGA is related to the high level of cell transformation and resistance, in other words to CSCs [12]. Indeed, we found that the expression of HMGA2, although absent in some samples

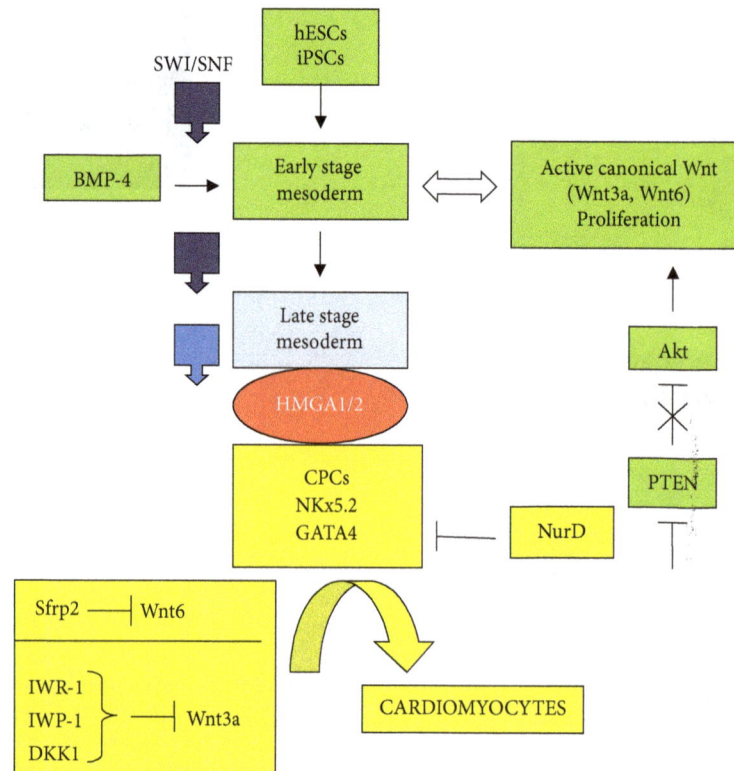

FIGURE 9: Wnt pathway repression in cardiomyocyte differentiation: from ESCs to cardiomyocytes through cardiac progenitor cells (CPCs). Symbols are the same as shown in Figure 1.

of colorectal cancer (CRC), correlates with cell budding and vascular invasion [172]. Moreover, Kaur et al. [8] showed that HMGA2 is expressed in primary glioblastoma tumours and that its expression strongly correlates with CD133+ expression, a marker of stemness. The authors concluded that HMGA2 should be considered as a stem-like factor of glioblastoma cells, guaranteeing clonogenicity, invasion, and malignant properties. Further support is found in a recent paper by Sun et al. [11] in which it was reported that "HMGA2 increased the expression of the stem cell markers CD44, ALDH1, Sox2, and Oct4 and the EMT-related factors Snail and β-catenin" in gastric cancer cells.

As mentioned briefly above, many studies have shown the possibility of obtaining iPSCs using molecules other than Yamanaka OSKM [4] or Thomson OS-LIN28-NANOG factors [5]. Among these molecules are HMGA proteins, whose expression is strongly associated with that of LIN28. Therefore, it should be possible to obtain iPSCs through HMGA. Indeed, Shah et al. [10] first demonstrated this possibility in an exhaustive study on HMGA1. The report showed the following:

(1) The expression of HMGA1 induces reprogramming in adult somatic cells to an undifferentiated phenotype with pluripotency characteristic of iPSCs.

(2) Differentiating hESCs show decreased expression of HMGA1, along with a decrease in pluripotency factors OCT4, SOX2, and NANOG, which suggests

that HMGA1 maintains the undifferentiated state of hESCs.

(3) Hyperexpressing HMGA1 in hESCs (that already express HMGA) not only blocks differentiation further, but also increases the levels of pluripotent gene OCT4, SOX2, NANOG, and cMYC expression. The same type of experiment in MSCs showed higher expression of LIN28 (among other factors), demonstrating a feedback loop between HMGA1 and LIN28, which suggests that the loop between cMYC and LIN28 shown in Figure 2 may be valid for all master reprogramming factors.

(4) If HMGA1 is overexpressed in somatic cells already transinduced with OSKM factors, the reprogramming rate is increased, stem cells survive, and proliferation is observed, whereas, following HMGA1 knockdown, OSKM factors are repressed.

A subsequent study on HMGA1 [173] in glioblastoma (GBM) stem cells (SCs) confirmed the above relationship between HMGA1 and pluripotency factors and added evidence of the epigenetic contribution of HMGA1 in SCs. This study focused on the axis between HMGA1/pluripotency factors and miR-296-5p; the results are summarized in Figure 10. This miR is a repressor of the stem cell phenotype in GBM; however, its action is abolished by the repression of its promoter by DNMT methylation. Repression of DNMT by 5-azacytidine reactivates miR-296-5p. On the other hand, HMGA1, which

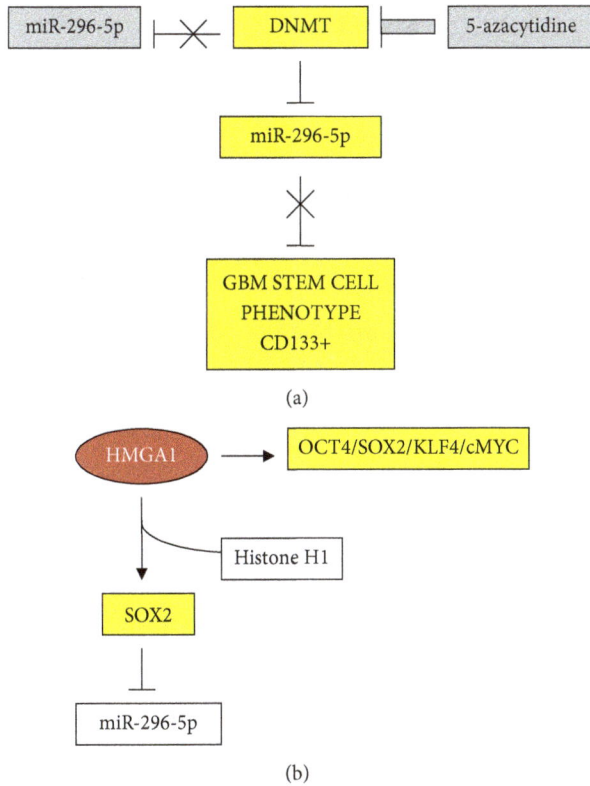

FIGURE 10: HMGA1 participation in stem cell establishment. (a) DNMTs repress miR-296-5p which can no longer repress the GBM stem cell phenotype. The inhibition of DNMTs by 5-azacytidine no longer blocks miR-296-5p and stem cells are not induced. (b) HMGA1 stimulates the expression of four master pluripotent factors and, with particular efficiency, SOX2 which inhibits miR-296-5p and allows stem cell establishment. Symbols are the same as shown in Figure 1.

displaces histone H1 from promoters, induces pluripotency factors and particularly SOX2, which represses miR-296-5p. Interestingly, HMGA2 is not involved in miR-296-5p regulation, while the members of miR-let7 family are associated with both proteins.

Two papers from the same research group [174, 175] indicated that it is possible to directly and efficiently reprogram various somatic cells into human induced neural stem cells (hiNSCs) by coexpressing SOX2 and HMGA2. Reprogramming is hampered by miR-let-7b, a member of the well-known HMGA repressor family. Interestingly, if the reprogramming is carried out in the umbilical cord blood derived MSCs that already express HMGA2, reprogramming occurs more easily than in somatic cells.

4. Conclusion

The present review is not focused on ESCs, iPSCs, and CSCs for which there are exhaustive reports, some of which we referred to here. Rather, we aimed to elucidate stem cell systems to examine the contribution of HMGA proteins. HMGA proteins are highly expressed in all three systems and related to the structure of chromatin (as already known), as well as

global organization and specific gene expression/repression that regulate the development, self-renewal, proliferation, invasion, and EMT of normal and cancer cells. Because of the above discussed properties, it is logical that they have been considered for targeting in cancer therapy. The link between HMGA proteins, CSCs, and drug resistance was further established and HMGA identified as target for sensitizing cancer cells to drug treatments [176, 177]. Cordycepin reduced the expression of the EMT factors in 72 melanoma patient samples comprising HMGA2, Twist1, and ZEB1, and the inhibition of HMGA2 sensitized gastric cancer cells to chemotherapy treatment [178, 179]. A codelivery therapy that inhibits HMGA2 by siRNA and acts on DNA by doxorubicin showed efficacy in CRC and the same dual treatment inhibits cell growth, vimentin (stemness marker), and MMP9 (invasion marker) in breast cancer cells [180, 181]. More studies should be carried out using combination of more drugs focusing on the finding that inhibition of HMGA proteins reduces the resistance of cancer cells to the treatment.

We took into consideration only two histone PTMs, that is, acetylation and methylation, omitting other modifications such as phosphorylation. Similarly, we did not discuss the PTMs by HMGA (although existing), because papers we reviewed did not discuss their modifications.

Though HMGA proteins are present in all three stem systems, their biological contributions to the maintenance and development of stemness of ESCs, iPSC, and CSCs are quite different, as we discussed at many points above. Established ESCs and iPSCs are considered similar and are characterized by the expression of the same factors, including HMGA, even though they are formed differently. Indeed, whereas ESCs are the natural product of the blastocyst system and do not undergo a preceding differentiation state, iPSCs are derived from somatic cells or from cells having a lower degree of stemness and with the so-called *memory* of the starting state that could reemerge in certain conditions. HMGA proteins are reexpressed in iPSCs but are absent in differentiated cells. However, as shown in Figure 11, the same cooperating set of epigenetic factors are utilized by ESCs to differentiate and, in opposite way, by somatic cells to form iPSCs. Considering these factors, it should be evident that each of them does not work alone; rather, they operate in concert with the other shown factors in a reciprocally modulated fashion, to which other factors not reported in Figure 11 also contribute. For example, the enzyme EZH2 is a component of a complex containing many other molecules that ensure its enzymatic activity. Consequently, an open chromatin structure does not imply the absence of repressed genes and, in contrast, closed chromatin does not imply their presence. Moreover, context-dependent differences, that can modulate or even reverse a preceding result are well established. In other words, HMGA proteins expressed in different stem cells do not necessarily have the same function. HMGA proteins are expressed at every stage in ESC development, throughout which the capacity to proliferate is still needed, including in reactivated quiescent stem cells in reservoir niches. The expression of HMGA proteins is silent once the cells differentiate and mature. Presumably, iPSCs follow a similar pathway of differentiation, but no data are available. Histone acetylation is one among the main modes of epigenetic modification and is associated with HMGA. Acetylated histones accompany HMGA expression in both ESCs and iPSCs and likely in chromatin regions with active gene expression. The deacetylation of histones by HDACs restores the positive charge of lysines, strengthens interactions with the DNA, and increases the compactness of chromatin where HDACs act. In these regions, the expression of reprogramming factors is more difficult (as in somatic cells), and, in order to obtain iPSCs, histones must be acetylated as in ESCs and have consistently high levels of HMGA proteins (Figure 11). Histone deacetylase inhibitors (HDACi) leave the chromatin in an open state as needed for the expression of pluripotent/self-renewal factors, then HMGA and HDACi are able to function in conjunction; consistently, H3K9Ac is present in both ESCs and iPSCc. However, in cancers HDACi are used as antiproliferative agents; that is, an anticancer action requires acetylated histones that are associated with an open state chromatin and the expression of HMGA, which should decrease with anticancer treatments. The difference between ESCs/iPSCs and CSCs is evident and likely involves modes of action with other epigenetic factors and modifications such as methylation. Differentiation of ESCs results from the repressive action of H3K27me3, modified by the enzyme EZH2. However, methylation cannot be considered a repressive modification itself, because H3K4me3 contributes to an open state of the chromatin (Figure 11). On the one hand, there are HMGA, H3K9Ac, and H3K4me3, and, on the other hand, there are H3K27me3, no HMGA, and H3K9me3. An aberrant situation was found in cancer, where H3K27me3 and HMGA are both present; in this case, EZH2/H3K27me3 inhibitors have been suggested as anticancer therapeutic agents. The difference in expression of EZH2/H3K27me3 in ESCs and iPSCs (low level) in comparison with CSCs (high level) extends even to cancer systems. Two examples are illustrative. Luo et al. [182] reported that EZH2/H3K27me3 promotes invasion, metastasis, and EMT of laryngeal squamous carcinoma cells and, at the same time, represses E-cadherin. This is consistent with the above information. However, Cardenas et al. [183] reported that EZH2 inhibition (not expression) promotes EMT in ovarian cancer cells, whereas its expression represses ZEB2, which is a main EMT promoting factor [12]. Both tumours express HMGA proteins [184–186]. In addition, Yi et al. [187] reported that EZH2 promotes ovarian cancer migration and invasion by inhibiting a repressor of MMP2/9, which are tumour promoters (as shown in Figure 1).

DNA methylation by DNMTs has a repressive action on chromatin, similar to that of miR-let-7 family members. In contrast, demethylated DNA is present in ESCs and promotes iPSCs (open chromatin). Demethylated DNA has an important role in the reactivation of genes that induce pluripotency. Indeed, the cooccurrence of H3K9me3 and methylated DNA results in incomplete reprogramming. In contrast, the methylation of DNA by an active DNMT1 is crucial for the transition from pluripotency to multipotency. Demethylation can result from the presence of the enzyme TET1, which catalyses the transformation of 5mC into 5hmC (5-hydroxymethylcytosine) and subsequent formation of unmethylated cytosine [188]. The loss of

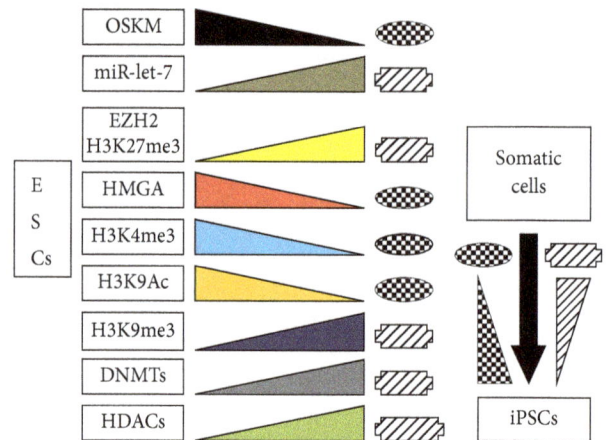

FIGURE 11: Changes in the action of factors from ESCs to somatic cell (coloured triangles) and reversed action (black arrow) from somatic cells to iPSCs.

TET1 leaves methylated DNA, that is, repressed chromatin, which induces cell migration and E-cadherin repression via EZH2/H3K27me3 in colon cancer cells [189]. Once again, we note an aberrant association between epigenetic factors in comparison with that normally found in ESCs and iPSCs.

In conclusion, we examined HMGA proteins, which are well-known cancer promoters, and compared their functions in ESCs and iPSCs as epigenetic chromatin-modifying factors associated with self-renewal, proliferation, invasion, and EMT stemness properties. HMGA proteins are involved at every stage of epigenetic stem cell regulation, up to the last moment when proliferation is required. Their levels of expression can remain the same or differ with the coexpression/modification of other factors, which are linked through a multicomponent molecular machinery that manages chromatin accessibility, first in the formation/maintenance of ESCs, iPSCs, and CSCs and subsequently in the use of these systems based on the specific biological context. We examined how DNA accessibility and gene expression are dependent on a multifactorial machinery whose composition could explain the contradictory results deriving from considering the action of only one factor. This may be expressed in the saying: *one swallow does not make a summer*.

Abbreviations

Akt: Serine/threonine kinase 1
BMP: Bone morphogenetic protein
CDH: Cadherin
CDK: Cyclin-dependent kinase
CDKN1C: p57 Cyclin-CDK inhibitor
C/EBP: CCAAT/enhancer binding protein
CM: Cardiomyocytes
CSCs: Cancer stem cells
Dkk-1:   Wnt inhibitor dickkopf homolog 1
Dlx: Distal-less homeobox
DNMT: DNA methyl transferase
DNMTi: DNA methyl transferase inhibitor
E2F: E2 transcription factor

EMT:	Epithelial-mesenchymal-transition
ESCs:	Embryonic stem cells
EZH2:	Enhancer of zest 2
GATA4:	Protein transcription factor 4 binding to GATA DNA sequence
HAT:	Histone acetyl transferase
HDAC:	Histone deacetylase
HDACi:	Histone deacetylase inhibitor
H3K9Ac:	Histone H3 acetylated at lysine 9
H3K9me2/3:	Histone H3 bi- or trimethylated at lysine 9
H3K4me3:	Histone H3 trimethylated at lysine 4
H3K27me3:	Histone H3 trimethylated at lysine 27
5hmC:	5-hydroxymethylcytosine
HMGA:	High-mobility group A
iPSCs:	Induced pluripotent stem cells
IWP-1:	Inhibitor of Wnt production 1
IWR-1:	Inhibitor of Wnt response 1
LEF:	Lymphoid enhancer binding factor
LncRNA:	Long-noncoding RNA
LPR:	Low-density lipoprotein receptor related
miR:	microRNA
5mC:	5-Methylcytosine
MMP:	Metalloprotease
MSCs:	Mesenchymal stem cells
NKx2.5:	NK2 homeobox protein 5
NuRD:	Nucleosomal remodeling and deacetylase
OSKM:	Oct4, Sox2, Klf4, cMyc
Osx:	Osterix
P:	Phosphate group
p21^{cip1}:	p21 CDK-interacting protein 1
p57^{kip2}:	p57 kinase inhibitor protein 2
PPAR:	Peroxisome proliferator-activated receptor
PTEN:	Phosphatase and tension homolog
PTMs:	Posttranslational modifications
Rb:	Retinoblastoma protein
RUNX:	Runt-domain transcription factor
Sfrp2:	Secreted frizzled related protein 2
SWI/SNF:	Switch/sucrose nonfermentable
TCF:	T-cell binding factor
Wnt:	Wingless-related integration site.

Conflicts of Interest

The authors declare that they have no conflicts of interest.

Acknowledgments

Thanks are due to Trieste Proteine Ricerche for administrative and graphical assistance and to the University of Trieste for library support.

References

[1] X. Zhou, K. F. Benson, H. R. Ashar, and K. Chada, "Mutation responsible for the mouse pygmy phenotype in the developmentally regulated factor HMGI-C," *Nature*, vol. 376, no. 6543, pp. 771–774, 1995.

[2] G. Chiappetta, V. Avantaggiato, R. Visconti et al., "High level expression of the HMGI (Y) gene during embryonic development," *Oncogene*, vol. 13, no. 11, pp. 2439–2446, 1996.

[3] U. Hirning-Folz, M. Wilda, V. Rippe, J. Bullerdiek, and H. Hameister, "The expression pattern of the Hmgic gene during development," *Genes, Chromosomes and Cancer*, vol. 23, no. 4, pp. 350–357, 1998.

[4] K. Takahashi and S. Yamanaka, "Induction of pluripotent stem cells from mouse embryonic and adult fibroblast cultures by defined factors," *Cell*, vol. 126, no. 4, pp. 663–676, 2006.

[5] J. Yu, M. A. Vodyanik, K. Smuga-Otto et al., "Induced pluripotent stem cell lines derived from human somatic cells," *Science*, vol. 318, no. 5858, pp. 1917–1920, 2007.

[6] K. Hawkins, "Cell signalling pathways underlying induced pluripotent stem cell reprogramming," *World Journal of Stem Cells*, vol. 6, no. 5, pp. 620–628, 2014.

[7] K. Izgi, H. Canatan, and B. Iskender, "Current status in cancer cell reprogramming and its clinical implications," *Journal of Cancer Research and Clinical Oncology*, vol. 143, no. 3, pp. 371–383, 2017.

[8] H. Kaur, S. Z. Ali, L. Huey et al., "The transcriptional modulator HMGA2 promotes stemness and tumorigenicity in glioblastoma," *Cancer Letters*, vol. 377, no. 1, pp. 55–64, 2016.

[9] K. Liu, Y. Song, H. Yu, and T. Zhao, "Understanding the roadmaps to induced pluripotency," *Cell Death & Disease*, vol. 5, no. 5, Article ID e1233, 2014.

[10] S. N. Shah, C. Kerr, L. Cope et al., "HMGA1 reprograms somatic cells into pluripotent stem cells by inducing stem cell transcriptional networks," *PLoS ONE*, vol. 7, no. 11, Article ID e48533, 2012.

[11] J. Sun, B. Sun, D. Zhu et al., "HMGA2 regulates CD44 expression to promote gastric cancer cell motility and sphere formation," *American Journal of Cancer Research*, vol. 7, no. 2, pp. 260–274, 2017.

[12] V. Giancotti, P. Cataldi, and C. Rizzi, "Roles of HMGA proteins in cancer: Expression, pathways, and redundancies," *Journal of Modern Human Pathology*, vol. 1, no. 6, pp. 44–62, 2016.

[13] T. Kawaguchi, *Cancer Metastasis and Cancer Stem Cell/Niche*, Bentham Science Publishers, 2016.

[14] M. Farlik, F. Halbritter, F. Müller et al., "DNA Methylation Dynamics of Human Hematopoietic Stem Cell Differentiation," *Cell Stem Cell*, vol. 19, no. 6, pp. 808–822, 2016.

[15] J. J. Trowbridge, J. W. Snow, J. Kim, and S. H. Orkin, "DNA methyltransferase 1 is essential for and uniquely regulates hematopoietic stem and progenitor cells," *Cell Stem Cell*, vol. 5, no. 4, pp. 442–449, 2009.

[16] V. Sapountzi, I. R. Logan, and C. N. Robson, "Cellular functions of TIP60," *The International Journal of Biochemistry & Cell Biology*, vol. 38, no. 9, pp. 1496–1509, 2006.

[17] L. Chen and G. Q. Daley, "Molecular basis of pluripotency," *Human Molecular Genetics*, vol. 17, no. 1, pp. R23–R27, 2008.

[18] S. Matoba, Y. Liu, F. Lu et al., "Embryonic development following somatic cell nuclear transfer impeded by persisting histone methylation," *Cell*, vol. 159, no. 4, pp. 884–895, 2014.

[19] C.-C. Tsai, P.-F. Su, Y.-F. Huang, T.-L. Yew, and S.-C. Hung, "Oct4 and Nanog directly regulate Dnmt1 to maintain self-renewal and undifferentiated state in mesenchymal stem cells," *Molecular Cell*, vol. 47, no. 2, pp. 169–182, 2012.

[20] X. Zhang, I. Neganova, S. Przyborski et al., "A role for NANOG in G1 to S transition in human embryonic stem cells through direct binding of CDK6 and CDC25A," *The Journal of Cell Biology*, vol. 184, no. 1, pp. 67–82, 2009.

[21] R. Cao, L. Wang, H. Wang et al., "Role of histone H3 lysine 27 methylation in polycomb-group silencing," *Science*, vol. 298, no. 5595, pp. 1039–1043, 2002.

[22] A. P. Bracken, D. Pasini, M. Capra, E. Prosperini, E. Colli, and K. Helin, "EZH2 is downstream of the pRB-E2F pathway, essential for proliferation and amplified in cancer," *EMBO Journal*, vol. 22, no. 20, pp. 5323–5335, 2003.

[23] S. K. Kia, M. M. Gorski, S. Giannakopoulos, and C. P. Verrijzer, "SWI/SNF mediates polycomb eviction and epigenetic reprogramming of the INK4b-ARF-INK4a locus," *Molecular and Cellular Biology*, vol. 28, no. 10, pp. 3457–3464, 2008.

[24] I. Oruetxebarria, F. Venturini, T. Kekarainen et al., "p16INK4a Is Required for hSNF5 Chromatin Remodeler-induced Cellular Senescence in Malignant Rhabdoid Tumor Cells," *The Journal of Biological Chemistry*, vol. 279, no. 5, pp. 3807–3816, 2004.

[25] B. G. Wilson, X. Wang, X. Shen et al., "Epigenetic antagonism between polycomb and SWI/SNF complexes during oncogenic transformation," *Cancer Cell*, vol. 18, no. 4, pp. 316–328, 2010.

[26] J. A. Simon and C. A. Lange, "Roles of the EZH2 histone methyltransferase in cancer epigenetics," *Mutation Research - Fundamental and Molecular Mechanisms of Mutagenesis*, vol. 647, no. 1-2, pp. 21–29, 2008.

[27] D. Zeng, M. Liu, and J. Pan, "Blocking EZH2 methylation transferase activity by GSK126 decreases stem cell-like myeloma cells," *Oncotarget*, vol. 8, no. 2, pp. 3396–3411, 2017.

[28] X. Song, T. Gao, N. Wang et al., "Selective inhibition of EZH2 by ZLD1039 blocks H3K27 methylation and leads to potent antitumor activity in breast cancer," *Scientific Reports*, vol. 6, Article ID 20864, 2016.

[29] A. Harikumar and E. Meshorer, "Chromatin remodeling and bivalent histone modifications in embryonic stem cells," *EMBO Reports*, vol. 16, no. 12, pp. 1609–1619, 2015.

[30] R. Dubey, A. M. Lebensohn, Z. Bahrami-Nejad et al., "Chromatin-remodeling complex SWI/SNF controls multidrug resistance by transcriptionally regulating the drug efflux pump ABCB1," *Cancer Research*, vol. 76, no. 19, pp. 5810–5821, 2016.

[31] N. E. Wiest, S. Houghtaling, J. C. Sanchez, A. E. Tomkinson, and M. A. Osley, "The SWI/SNF ATP-dependent nucleosome remodeler promotes resection initiation at a DNA double-strand break in yeast," *Nucleic Acids Research*, vol. 45, no. 10, pp. 5887–5900, 2017.

[32] S. Deindl, W. L. Hwang, S. K. Hota et al., "ISWI remodelers slide nucleosomes with coordinated multi-base-pair entry steps and single-base-pair exit steps," *Cell*, vol. 152, no. 3, pp. 442–452, 2013.

[33] B. J. Manning and T. Yusufzai, "The ATP-dependent chromatin remodeling enzymes CHD6, CHD7, and CHD8 exhibit distinct nucleosome binding and remodeling activities," *The Journal of Biological Chemistry*, vol. 292, no. 28, pp. 11927–11936, 2017.

[34] M. S. Isakoff, C. G. Sansam, P. Tamayo et al., "Inactivation of the Snf5 tumor suppressor stimulates cell cycle progression and cooperates with p53 loss in oncogenic transformation," *Proceedings of the National Acadamy of Sciences of the United States of America*, vol. 102, no. 49, pp. 17745–17750, 2005.

[35] D. C.-C. Voon, H. Wang, J. K. W. Koo et al., "EMT-induced stemness and tumorigenicity are fueled by the EGFR/Ras pathway," *PLoS ONE*, vol. 8, no. 8, Article ID e70427, 2013.

[36] Z. Y. Wu, S. M. Wang, Z. H. Chen et al., "MiR-204 regulates HMGA2 expression and inhibits cell proliferation in human thyroid cancer," *Cancer Biomarkers*, vol. 15, no. 5, pp. 535–542, 2015.

[37] X.-P. Zhao, H. Zhang, J.-Y. Jiao, D.-X. Tang, Y.-L. Wu, and C.-B. Pan, "Overexpression of HMGA2 promotes tongue cancer metastasis through EMT pathway," *Journal of Translational Medicine*, vol. 14, no. 1, article no. 26, 2016.

[38] G. L. Yang, L. H. Zhang, J. J. Bo et al., "Overexpression of HMGA2 in bladder cancer and its association with clinicopathologic features and prognosis: HMGA2 as a prognostic marker of bladder cancer," *European Journal of Surgical Oncology*, vol. 37, no. 3, pp. 265–271, 2011.

[39] X. Ding, Y. Wang, X. Ma et al., "Expression of HMGA2 in bladder cancer and its association with epithelial-to-mesenchymal transition," *Cell Proliferation*, vol. 47, no. 2, pp. 146–151, 2014.

[40] Z. Shi, X. Li, D. Wu et al., "Silencing of HMGA2 suppresses cellular proliferation, migration, invasion, and epithelial–mesenchymal transition in bladder cancer," *Tumor Biology*, vol. 37, no. 6, pp. 7515–7523, 2016.

[41] K.-J. Chen, Y. Hou, K. Wang et al., "Reexpression of let-7g microRNA inhibits the proliferation and migration via K-Ras/HMGA2/snail axis in hepatocellular carcinoma," *BioMed Research International*, vol. 2014, Article ID 742417, 2014.

[42] C. G. Kleer, Q. Cao, S. Varambally et al., "EZH2 is a marker of aggressive breast cancer and promotes neoplastic transformation of breast epithelial cells," *Proceedings of the National Acadamy of Sciences of the United States of America*, vol. 100, no. 20, pp. 11606–11611, 2003.

[43] S.-B. Gao, Q.-F. Zheng, B. Xu et al., "EZH2 represses target genes through H3K27-dependent and H3K27-independent mechanisms in hepatocellular carcinoma," *Molecular Cancer Research*, vol. 12, no. 10, pp. 1388–1397, 2014.

[44] R. Hubaux, K. L. Thu, B. P. Coe, C. Macaulay, S. Lam, and W. L. Lam, "EZH2 promotes E2F-driven SCLC tumorigenesis through modulation of apoptosis and cell-cycle regulation," *Journal of Thoracic Oncology*, vol. 8, no. 8, pp. 1102–1106, 2013.

[45] X. Dang, A. Ma, L. Yang et al., "MicroRNA-26a regulates tumorigenic properties of EZH2 in human lung carcinoma cells," *Cancer Genetics*, vol. 205, no. 3, pp. 113–123, 2012.

[46] Q. I.-C. Song, Z.-B. Shi, Y.-T. Zhang et al., "Downregulation of microRNA-26a is associated with metastatic potential and the poor prognosis of osteosarcoma patients," *Oncology Reports*, vol. 31, no. 3, pp. 1263–1270, 2014.

[47] X.-Z. Chi, J.-O. Yang, K.-Y. Lee et al., "RUNX3 suppresses gastric epithelial cell growth by inducing p21 WAF1/CiP1 expression in cooperation with transforming growth factor β-activated SMAD," *Molecular and Cellular Biology*, vol. 25, no. 18, pp. 8097–8107, 2005.

[48] K. Ito, "RUNX3 in oncogenic and anti-oncogenic signaling in gastrointestinal cancers," *Journal of Cellular Biochemistry*, vol. 112, no. 5, pp. 1243–1249, 2011.

[49] A. H. Juan, S. Wang, K. D. Ko et al., "Roles of H3K27me2 and H3K27me3 Examined during Fate Specification of Embryonic Stem Cells," *Cell Reports*, vol. 17, no. 5, pp. 1369–1382, 2016.

[50] A. Matsika, B. Srinivasan, C. Day et al., "Cancer stem cell markers in prostate cancer: An immunohistochemical study of ALDH1, SOX2 and EZH2," *Pathology*, vol. 47, no. 7, pp. 622–628, 2015.

[51] D. Kong, E. Heath, W. Chen et al., "Loss of let-7 up-regulates EZH2 in prostate cancer consistent with the acquisition of cancer stem cell signatures that are attenuated by BR-DIM," *PLoS ONE*, vol. 7, no. 3, Article ID e33729, 2012.

[52] H. Rajabi, M. Hiraki, A. Tagde et al., "MUC1-C activates EZH2 expression and function in human cancer cells," *Scientific Reports*, vol. 7, no. 1, article no. 7481, 2017.

[53] Z. L. Gan, M. Xu, R. Hua, C. Tan, J. Zhang, Y. Gong et al., "The polycomb group protein EZH2 induces epithelial-mesenchymal transition and pluripotent phenotype of gastric cancer

by binding to PTEN promoter," *Journal of Hematology & Oncology*, vol. 11, no. 1, p. 9, 2018.

[54] C. Wang, X. Liu, Z. Chen et al., "Polycomb group protein EZH2-mediated E-cadherin repression promotes metastasis of oral tongue squamous cell carcinoma," *Molecular Carcinogenesis*, vol. 52, no. 3, pp. 229–236, 2013.

[55] N. M. Alajez, W. Shi, D. Wong et al., "Lin28b Promotes Head and Neck Cancer Progression via Modulation of the Insulin-Like Growth Factor Survival Pathway," *Oncotarget*, vol. 3, no. 12, pp. 1641–1652, 2012.

[56] Y. Y. Wang, T. Ren, Y. Y. Cai, and X. Y. He, "MicroRNA let-7a inhibits the proliferation and invasion of nonsmall cell lung cancer cell line 95D by regulating K-RAS and HMGA2 gene expression," *Cancer Biotherapy and Radiopharmaceuticals*, vol. 28, no. 2, pp. 131–137, 2013.

[57] T. Wu, J. Jia, X. Xiong et al., "Increased expression of Lin28B associates with poor prognosis in patients with oral squamous cell carcinoma," *PLoS ONE*, vol. 8, no. 12, Article ID e83869, 2013.

[58] X.-R. Wang, H. Luo, H.-L. Li et al., "Overexpressed let-7a inhibits glioma cell malignancy by directly targeting k-ras, independently of pten," *Neuro-Oncology*, vol. 15, no. 11, pp. 1491–1501, 2013.

[59] X.-G. Mao, M. Hütt-Cabezas, B. A. Orr et al., "LIN28A facilitates the transformation of human neural stem cells and promotes glioblastoma tumorigenesis through a pro-invasive genetic program," *Oncotarget*, vol. 4, no. 7, pp. 1050–1064, 2013.

[60] E. Piskounova, C. Polytarchou, J. E. Thornton et al., "Lin28A and Lin28B inhibit let-7 MicroRNA biogenesis by distinct mechanisms," *Cell*, vol. 147, no. 5, pp. 1066–1079, 2011.

[61] J. Tsialikas and J. Romer-Seibert, "LIN28: Roles and regulation in development and beyond," *Development*, vol. 142, no. 14, pp. 2397–2404, 2015.

[62] F. Cimadamore, A. Amador-Arjona, C. Chen, C.-T. Huang, and A. V. Terskikh, "SOX2-LIN28/let-7 pathway regulates proliferation and neurogenesis in neural precursors," *Proceedings of the National Acadamy of Sciences of the United States of America*, vol. 110, no. 32, pp. E3017–E3026, 2013.

[63] A. L. Morgado, C. M. P. Rodrigues, and S. Solá, "MicroRNA-145 Regulates Neural Stem Cell Differentiation Through the Sox2-Lin28/let-7 Signaling Pathway," *Stem Cells*, vol. 34, no. 5, pp. 1386–1395, 2016.

[64] M. R. Copley, S. Babovic, C. Benz et al., "The Lin28b-let-7-Hmga2 axis determines the higher self-renewal potential of fetal haematopoietic stem cells," *Nature Cell Biology*, vol. 15, no. 8, pp. 916–925, 2013.

[65] R. Grant Rowe, L. D. Wang, S. Coma et al., "Developmental regulation of myeloerythroid progenitor function by the Lin28b-let-7-Hmga2 axis," *The Journal of Experimental Medicine*, vol. 213, no. 8, pp. 1497–1512, 2016.

[66] F. Peng, T. T. Li, K. L. Wang, G. Q. Xiao, J. H. Wang, H. D. Zhao et al., "H19/let-7/LIN28 reciprocal negative regulatory circuit promotes breast cancer stem cell maintenance," *Cell Death Disease*, vol. 8, no. 1, p. e2569, 2017.

[67] T.-C. Chang, L. R. Zeitels, H.-W. Hwang et al., "Lin-28B transactivation is necessary for Myc-mediated let-7 repression and proliferation," *Proceedings of the National Acadamy of Sciences of the United States of America*, vol. 106, no. 9, pp. 3384–3389, 2009.

[68] J. Yin, J. Zhao, W. Hu et al., "Disturbance of the let-7/LIN28 doublenegative feedback loop is associated with radio- and

[69] M. Alam, R. Ahmad, H. Rajabi, and D. Kufe, "MUC1-C induces the LIN28B!LET-7!HMGA2 axis to regulate self-renewal in NSCLC," *Molecular Cancer Research*, vol. 13, no. 3, pp. 449–460, 2015.

[70] A. V. Ougolkov, V. N. Bilim, and D. D. Billadeau, "Regulation of pancreatic tumor cell proliferation and chemoresistance by the histone methyltransferase enhancer of zeste homologue 2," *Clinical Cancer Research*, vol. 14, no. 21, pp. 6790–6796, 2008.

[71] J. Bai, M. Ma, M. Cai et al., "Inhibition enhancer of zeste homologue 2 promotes senescence and apoptosis induced by doxorubicin in p53 mutant gastric cancer cells," *Cell Proliferation*, vol. 47, no. 3, pp. 211–218, 2014.

[72] Y. Wei, Z. Liu, and J. Fang, "H19 functions as a competing endogenous RNA toregulate human epidermal growth factor receptor expression by sequestering let-7c in gastric cancer," *Molecular Medicine Reports*, vol. 17, no. 2, pp. 2600–2606, 2018.

[73] C.-S. Chien, M.-L. Wang, P.-Y. Chu et al., "Lin28B/Let-7 regulates expression of Oct4 and Sox2 and reprograms oral squamous cell carcinoma cells to a stem-like state," *Cancer Research*, vol. 75, no. 12, pp. 2553–2565, 2015.

[74] N. Yang, L. Hui, Y. Wang, H. Yang, and X. Jiang, "Overexpression of SOX2 promotes migration, invasion, and epithelial-mesenchymal transition through the Wnt/β-catenin pathway in laryngeal cancer Hep-2 cells," *Tumor Biology*, vol. 35, no. 8, pp. 7965–7973, 2014.

[75] J. Kato, H. Matsushime, S. W. Hiebert, M. E. Ewen, and C. J. Sherr, "Direct binding of cyclin D to the retinoblastoma gene product (pRb) and pRb phosphorylation by the cyclin D-dependent kinase CDK4," *Genes & Development*, vol. 7, no. 3, pp. 331–342, 1993.

[76] S. A. Ezhevsky, H. Nagahara, A. M. Vocero-Akbani, D. R. Gius, M. C. Wei, and S. F. Dowdy, "Hypo-phosphorylation of the retinoblastoma protein (pRb) by cyclin D:Cdk4/6 complexes results in active pRb," *Proceedings of the National Acadamy of Sciences of the United States of America*, vol. 94, no. 20, pp. 10699–10704, 1997.

[77] N. P. Pavletich, "Mechanisms of cyclin-dependent kinase regulation: Structures of Cdks, their cyclin activators, and Cip and INK4 inhibitors," *Journal of Molecular Biology*, vol. 287, no. 5, pp. 821–828, 1999.

[78] P. D. Jeffrey, L. Tong, and N. P. Pavletich, "Structural basis of inhibition of CDK-cyclin complexes by INK4 inhibitors," *Genes & Development*, vol. 14, no. 24, pp. 3115–3125, 2000.

[79] A. A. Russo, L. Tong, J.-O. Lee, P. D. Jeffrey, and N. P. Pavletich, "Structural basis for inhibition of the cyclin-dependent kinase Cdk6 by the tumour suppressor p16(INK4a)," *Nature*, vol. 395, no. 6699, pp. 237–243, 1998.

[80] J. K. Kim, J. H. Noh, J. W. Eun et al., "Targeted inactivation of HDAC2 restores p16^{INK4a} activity and exerts antitumor effects on human gastric cancer," *Molecular Cancer Research*, vol. 11, no. 1, pp. 62–73, 2013.

[81] M. Fedele, R. Visone, I. De Martino et al., "HMGA2 induces pituitary tumorigenesis by enhancing E2F1 activity," *Cancer Cell*, vol. 9, no. 6, pp. 459–471, 2006.

[82] V. Muthusamy, C. Hobbs, C. Nogueira et al., "Amplification of CDK4 and MDM2 in malignant melanoma," *Genes, Chromosomes and Cancer*, vol. 45, no. 5, pp. 447–454, 2006.

[83] J. Nishino, I. Kim, K. Chada, and S. J. Morrison, "Hmga2 promotes neural stem cell self-renewal in young but not old mice

by reducing p16Ink4a and p19Arf Expression," *Cell*, vol. 135, no. 2, pp. 227–239, 2008.

[84] A. Schuldenfrei, A. Belton, J. Kowalski et al., "HMGA1 drives stem cell, inflammatory pathway, and cell cycle progression genes during lymphoid tumorigenesis," *BMC Genomics*, vol. 12, article no. 549, 2011.

[85] K.-R. Yu, S.-B. Park, J.-W. Jung et al., "HMGA2 regulates the in vitro aging and proliferation of human umbilical cord blood-derived stromal cells through the mTOR/p70S6K signaling pathway," *Stem Cell Research*, vol. 10, no. 2, pp. 156–165, 2013.

[86] A. Tzatsos and N. Bardeesy, "Ink4a/Arf Regulation by let-7b and Hmga2: A Genetic Pathway Governing Stem Cell Aging," *Cell Stem Cell*, vol. 3, no. 5, pp. 469–470, 2008.

[87] A. He, Q. Ma, J. Cao et al., "Polycomb repressive complex 2 regulates normal development of the mouse heart," *Circulation Research*, vol. 110, no. 3, pp. 406–415, 2012.

[88] S. Siddiqi, J. Mills, and I. Matushansky, "Epigenetic remodeling of chromatin architecture: Exploring tumor differentiation therapies in mesenchymal stem cells and sarcomas," *Current Stem Cell Research & Therapy*, vol. 5, no. 1, pp. 63–73, 2010.

[89] M. Visweswaran, S. Pohl, F. Arfuso et al., "Multi-lineage differentiation of mesenchymal stem cells—to Wnt, or not Wnt," *International Journal of Biochemistry & Cell Biology*, vol. 68, pp. 139–147, 2015.

[90] D. A. Glass II and G. Karsenty, "Canonical Wnt signaling in osteoblasts is required for osteoclast differentiation," *Annals of the New York Academy of Sciences*, vol. 1068, no. 1, pp. 117–130, 2006.

[91] S. Kang, C. N. Bennett, I. Gerin, L. A. Rapp, K. D. Hankenson, and O. A. MacDougald, "Wnt signaling stimulates osteoblastogenesis of mesenchymal precursors by suppressing CCAAT/enhancer-binding protein α and peroxisome proliferator-activated receptor γ," *The Journal of Biological Chemistry*, vol. 282, no. 19, pp. 14515–14524, 2007.

[92] H.-X. Li, X. Luo, R.-X. Liu, Y.-J. Yang, and G.-S. Yang, "Roles of Wnt/β-catenin signaling in adipogenic differentiation potential of adipose-derived mesenchymal stem cells," *Molecular and Cellular Endocrinology*, vol. 291, no. 1-2, pp. 116–124, 2008.

[93] L. Wang, Q. Jin, J.-E. Lee, I.-H. Su, and K. Ge, "Histone H3K27 methyltransferase Ezh2 represses Wnt genes to facilitate adipogenesis," *Proceedings of the National Acadamy of Sciences of the United States of America*, vol. 107, no. 16, pp. 7317–7322, 2010.

[94] M. Visweswaran, L. Schiefer, F. Arfuso, R. J. Dilley, P. Newsholme, and A. Dharmarajan, "Wnt antagonist secreted frizzled-related protein 4 upregulates adipogenic differentiation in human adipose tissue-derived mesenchymal stem cells," *PLoS ONE*, vol. 10, no. 2, Article ID e0118005, 2015.

[95] C. Haxaire, E. Haÿ, and V. Geoffroy, "Runx2 Controls Bone Resorption through the Down-Regulation of the Wnt Pathway in Osteoblasts," *The American Journal of Pathology*, vol. 186, no. 6, pp. 1598–1609, 2016.

[96] J. S. Heo, S. G. Lee, and H. O. Kim, "Distal-less homeobox 5 is a master regulator of the osteogenesis of human mesenchymal stem cells," *International Journal of Molecular Medicine*, vol. 40, no. 5, pp. 1486–1494, 2017.

[97] K. Felber, P. M. Elks, M. Lecca, and H. H. Roehl, "Expression of osterix is regulated by FGF and Wnt/β-catenin signalling during osteoblast differentiation," *PLoS ONE*, vol. 10, no. 12, Article ID e0144982, 2015.

[98] Y. Han, C. Y. Kim, H. Cheong, and K. Y. Lee, "Osterix represses adipogenesis by negatively regulating PPARγ transcriptional activity," *Scientific Reports*, vol. 6, Article ID 35655, 2016.

[99] X.-Y. Zhao, X.-Y. Chen, Z.-J. Zhang et al., "Expression patterns of transcription factor PPARγ and C/EBP family members during in vitro adipogenesis of human bone marrow mesenchymal stem cells," *Cell Biology International*, vol. 39, no. 4, pp. 457–465, 2015.

[100] B. Henriquez, M. Hepp, P. Merino et al., "C/EBPβ binds the P1 promoter of the Runx2 gene and up-regulates Runx2 transcription in osteoblastic cells," *Journal of Cellular Physiology*, vol. 226, no. 11, pp. 3043–3052, 2011.

[101] Y. Xi, W. Shen, L. Ma et al., "HMGA2 promotes adipogenesis by activating C/EBPβ-mediated expression of PPARγ," *Biochemical and Biophysical Research Communications*, vol. 472, no. 4, pp. 617–623, 2016.

[102] Y. Yuan, Y. Xi, J. Chen et al., "STAT3 stimulates adipogenic stem cell proliferation and cooperates with HMGA2 during the early stage of differentiation to promote adipogenesis," *Biochemical and Biophysical Research Communications*, vol. 482, no. 4, pp. 1360–1366, 2017.

[103] H. Kang and A. Hata, "The role of microRNAs in cell fate determination of mesenchymal stem cells: Balancing adipogenesis and osteogenesis," *BMB Reports*, vol. 48, no. 6, pp. 319–323, 2015.

[104] H. Wang, Z. Sun, Y. Wang et al., "miR-33-5p, a novel mechano-sensitive microRNA promotes osteoblast differentiation by targeting Hmga2," *Scientific Reports*, vol. 6, Article ID 23170, 2016.

[105] J. Wei, H. Li, S. Wang et al., "Let-7 enhances osteogenesis and bone formation while repressing adipogenesis of human stromal/mesenchymal stem cells by regulating HMGA2," *Stem Cells and Development*, vol. 23, no. 13, pp. 1452–1463, 2014.

[106] C. Gong, Z. Li, K. Ramanujan et al., "A long non-coding RNA, LncMyoD, regulates skeletal muscle differentiation by blocking IMP2-mediated mRNA translation," *Developmental Cell*, vol. 34, no. 2, pp. 181–191, 2015.

[107] V. D. Soleimani, H. Yin, A. Jahani-Asl et al., "Snail Regulates MyoD Binding-Site Occupancy to Direct Enhancer Switching and Differentiation-Specific Transcription in Myogenesis," *Molecular Cell*, vol. 47, no. 3, pp. 457–468, 2012.

[108] Y. Tang and S. J. Weiss, "Snail/Slug-YAP/TAZ complexes cooperatively regulate mesenchymal stem cell function and bone formation," *Cell Cycle*, vol. 16, no. 5, pp. 399–405, 2017.

[109] I. Marchesi, F. P. Fiorentino, F. Rizzolio, A. Giordano, and L. Bagella, "The ablation of EZH2 uncovers its crucial role in rhabdomyosarcoma formation," *Cell Cycle*, vol. 11, no. 20, pp. 3828–3836, 2012.

[110] S. Consalvi, A. Brancaccio, A. Dall'agnese, P. L. Puri, and D. Palacios, "Praja1 E3 ubiquitin ligase promotes skeletal myogenesis through degradation of EZH2 upon p38α activation," *Nature Communications*, vol. 8, Article ID 13956, 2017.

[111] G.-I. Im and K.-J. Shin, "Epigenetic approaches to regeneration of bone and cartilage from stem cells," *Expert Opinion on Biological Therapy*, vol. 15, no. 2, pp. 181–193, 2015.

[112] M. Hamed, S. Khilji, J. Chen, and Q. Li, "Stepwise acetyltransferase association and histone acetylation at the Myod1 locus during myogenic differentiation," *Scientific Reports*, vol. 3, article no. 2390, 2013.

[113] L. Wei, N. Jamonnak, J. Choy, Z. Wang, and W. Zheng, "Differential binding modes of the bromodomains of CREB-binding protein (CBP) and p300 with acetylated MyoD," *Biochemical*

and *Biophysical Research Communications*, vol. 368, no. 2, pp. 279–284, 2008.

[114] J. K. Van Camp, S. Beckers, D. Zegers, and W. Van Hul, "Wnt Signaling and the Control of Human Stem Cell Fate," *Stem Cell Reviews and Reports*, vol. 10, no. 2, pp. 207–229, 2014.

[115] P. Wend, S. Runke, and K. Wend, "WNT10B/β-catenin signalling induces HMGA2 and proliferation in metastatic triple-negative breast cancer," *EMBO Molecular Medicine*, vol. 5, no. 2, pp. 264–279, 2013.

[116] W.-Y. Cai, T.-Z. Wei, Q.-C. Luo et al., "The wnt-β-catenin pathway represses let-7 microrna expression through transactivation of Lin28 to augment breast cancer stem cell expansion," *Journal of Cell Science*, vol. 126, no. 13, pp. 2877–2889, 2013.

[117] A. Morishita, M. R. Zaidi, A. Mitoro et al., "HMGA2 is a driver of tumor metastasis," *Cancer Research*, vol. 73, no. 14, pp. 4289–4299, 2013.

[118] A. Abou-Elhamd, A. F. Alrefaei, G. F. Mok et al., "Klhl31 attenuates β-catenin dependent Wnt signaling and regulates embryo myogenesis," *Developmental Biology*, vol. 402, no. 1, pp. 61–71, 2015.

[119] Y. C. Pan, X. W. Wang, H. F. Teng, Y. J. Wu, H. C. Chang, and S. L. Chen, "Wnt3a signal pathways activate MyoD expression by targeting cis-elements inside and outside its distal enhancer," *Bioscience Reports*, vol. 35, Article ID e00180, 2015.

[120] L. Zhuang, J.-A. Hulin, A. Gromova et al., "Barx2 and Pax7 have antagonistic functions in regulation of Wnt signaling and satellite cell differentiation," *Stem Cells*, vol. 32, no. 6, pp. 1661–1673, 2014.

[121] M. Horak, J. Novak, and J. Bienertova-Vasku, "Muscle-specific microRNAs in skeletal muscle development," *Developmental Biology*, vol. 410, no. 1, pp. 1–13, 2016.

[122] J. Chen, T. E. Callis, and D. Wang, "microRNAs and muscle disorders," *Journal of Cell Science*, vol. 122, no. 1, pp. 13–20, 2009.

[123] Z. Tang, H. Qiu, L. Luo et al., "miR-34b Modulates Skeletal Muscle Cell Proliferation and Differentiation," *Journal of Cellular Biochemistry*, vol. 118, no. 12, pp. 4285–4295, 2017.

[124] X. Jia, H. Ouyang, B. A. Abdalla, H. Xu, Q. Nie, and X. Zhang, "miR-16 controls myoblast proliferation and apoptosis through directly suppressing Bcl2 and FOXO1 activities," *Biochimica et Biophysica Acta - Gene Regulatory Mechanisms*, vol. 1860, no. 6, pp. 674–684, 2017.

[125] H. Qiu, J. Zhong, L. Luo et al., "Regulatory axis of miR-195/497 and HMGA1-Id3 governs muscle cell proliferation and differentiation," *International Journal of Biological Sciences*, vol. 13, no. 2, pp. 157–166, 2017.

[126] Z. Li, J. A. Gilbert, Y. Zhang et al., "An HMGA2-IGF2BP2 Axis Regulates Myoblast Proliferation and Myogenesis," *Developmental Cell*, vol. 23, no. 6, pp. 1176–1188, 2012.

[127] J. Nishino, S. Kim, Y. Zhu, H. Zhu, and S. J. Morrison, "A network of heterochronic genes including Imp1 regulates temporal changes in stem cell properties," *eLife*, vol. 2, 2013.

[128] Y. Fujii, Y. Kishi, and Y. Gotoh, "IMP2 regulates differentiation potentials of mouse neocortical neural precursor cells," *Genes to Cells*, vol. 18, no. 2, pp. 79–89, 2013.

[129] M. Janiszewska, M. L. Suvà, N. Riggi et al., "Imp2 controls oxidative phosphorylation and is crucial for preservin glioblastoma cancer stem cells," *Genes & Development*, vol. 26, no. 17, pp. 1926–1944, 2012.

[130] O. M. Martínez-Estrada, A. Cullerés, F. X. Soriano et al., "The transcription factors Slug and Snail act as repressors of Claudin-1 expression in epithelial cells," *Biochemical Journal*, vol. 394, no. 2, pp. 449–457, 2006.

[131] F. Frasca, A. Rustighi, R. Malaguarnera et al., "HMGA1 inhibits the function of p53 family members in thyroid cancer cells," *Cancer Research*, vol. 66, no. 6, pp. 2980–2989, 2006.

[132] V. Tarasov, P. Jung, B. Verdoodt et al., "Differential regulation of microRNAs by p53 revealed by massively parallel sequencing: *miR-34a* is a p53 target that induces apoptosis and G_1-arrest," *Cell Cycle*, vol. 6, no. 13, pp. 1586–1593, 2007.

[133] F. Esposito, M. Tornincasa, P. Chieffi, I. De Martino, G. M. Pierantoni, and A. Fusco, "High-mobility group A1 proteins regulate p53-mediated transcription of Bcl-2 gene," *Cancer Research*, vol. 70, no. 13, pp. 5379–5388, 2010.

[134] T.-C. Chang, E. A. Wentzel, O. A. Kent et al., "Transactivation of miR-34a by p53 broadly influences gene expression and promotes apoptosis," *Molecular Cell*, vol. 26, no. 5, pp. 745–752, 2007.

[135] B. Mansoori, A. Mohammadi, S. Shirjang, and B. Baradaran, "HMGI-C suppressing induces P53/caspase9 axis to regulate apoptosis in breast adenocarcinoma cells," *Cell Cycle*, vol. 15, no. 19, pp. 2585–2592, 2016.

[136] K. C. Lilja, N. Zhang, A. Magli et al., "Pax7 remodels the chromatin landscape in skeletal muscle stem cells," *PLoS ONE*, vol. 12, no. 4, Article ID e0176190, 2017.

[137] T. Laumonier, F. Bermont, P. Hoffmeyer, V. Kindler, and J. Menetrey, "Human myogenic reserve cells are quiescent stem cells that contribute to muscle regeneration after intramuscular transplantation in immunodeficient mice," *Scientific Reports*, vol. 7, no. 1, article no. 3462, 2017.

[138] M. A. Rudnicki, F. le Grand, I. McKinnell, and S. Kuang, "The molecular regulation of muscle stem cell function," *Cold Spring Harbor Symposium on Quantitative Biology*, vol. 73, pp. 323–331, 2008.

[139] S. Alonso-Martin, A. Rochat, D. Mademtzoglou et al., "Gene Expression Profiling of Muscle Stem Cells Identifies Novel Regulators of Postnatal Myogenesis," *Frontiers in Cell and Developmental Biology*, vol. 4, 2016.

[140] A. C. Panda, K. Abdelmohsen, J. L. Martindale et al., "Novel RNA-binding activity of MYF5 enhances Ccnd1/Cyclin D1 mRNA translation during myogenesis," *Nucleic Acids Research*, vol. 44, no. 5, pp. 2393–2408, 2016.

[141] J. Brocher, B. Vogel, and R. Hock, "HMGA1 down-regulation is crucial for chromatin composition and a gene expression profile permitting myogenic differentiation," *BMC Cell Biology*, vol. 11, article no. 64, 2010.

[142] D. Di Marcantonio, D. Galli, C. Carubbi et al., "PKCε as a novel promoter of skeletal muscle differentiation and regeneration," *Experimental Cell Research*, vol. 339, no. 1, pp. 10–19, 2015.

[143] K. T. Kuppusamy, D. C. Jones, H. Sperber et al., "Let-7 family of microRNA is required for maturation and adult-like metabolism in stem cell-derived cardiomyocytes," *Proceedings of the National Acadamy of Sciences of the United States of America*, vol. 112, no. 21, pp. E2785–E2794, 2015.

[144] S. S. Y. Wong, C. Ritner, S. Ramachandran et al., "MiR-125b promotes early germ layer specification through lin28/let-7d and preferential differentiation of mesoderm in human embryonic stem cells," *PLoS ONE*, vol. 7, no. 4, Article ID e36121, 2012.

[145] F. Laco, J.-L. Low, J. Seow et al., "Cardiomyocyte differentiation of pluripotent stem cells with SB203580 analogues correlates with Wnt pathway CK1 inhibition independent of p38 MAPK signaling," *Journal of Molecular and Cellular Cardiology*, vol. 80, pp. 56–70, 2015.

[146] T.-Y. Lu, B. Lin, Y. Li et al., "Overexpression of microRNA-1 promotes cardiomyocyte commitment from human cardiovascular

progenitors via suppressing WNT and FGF signaling pathways," *Journal of Molecular and Cellular Cardiology*, vol. 63, pp. 146–154, 2013.

[147] J. A. Bisson, B. Mills, J.-C. P. Helt, T. P. Zwaka, and E. D. Cohen, "Wnt5a and Wnt11 inhibit the canonical Wnt pathway and promote cardiac progenitor development via the Caspase-dependent degradation of AKT," *Developmental Biology*, vol. 398, no. 1, pp. 80–96, 2015.

[148] E. D. Cohen, M. F. Miller, Z. Wang, R. T. Moon, and E. E. Morrisey, "Wnt5a and wnt11 are essential for second heart field progenitor development," *Development*, vol. 139, no. 11, pp. 1931–1940, 2012.

[149] M. P. Flaherty, T. J. Kamerzell, and B. Dawn, "Wnt signaling and cardiac differentiation," *Progress in Molecular Biology and Translational Science*, vol. 111, pp. 153–174, 2012.

[150] A. Richter, L. Valdimarsdottir, H. E. Hrafnkelsdottir et al., "BMP4 promotes EMT and mesodermal commitment in human embryonic stem cells via SLUG and MSX2," *Stem Cells*, vol. 32, no. 3, pp. 636–648, 2014.

[151] W. Luo, X. Zhao, H. Jin et al., "Akt1 signaling coordinates bmp signaling and β-catenin activity to regulate second heart field progenitor development," *Development*, vol. 142, no. 4, pp. 732–742, 2015.

[152] J. Schmeckpeper, A. Verma, L. Yin et al., "Inhibition of Wnt6 by Sfrp2 regulates adult cardiac progenitor cell differentiation by differential modulation of Wnt pathways," *Journal of Molecular and Cellular Cardiology*, vol. 85, pp. 215–225, 2015.

[153] Y. Ren, M. Y. Lee, S. Schliffke et al., "Small molecule Wnt inhibitors enhance the efficiency of BMP-4-directed cardiac differentiation of human pluripotent stem cells," *Journal of Molecular and Cellular Cardiology*, vol. 51, no. 3, pp. 280–287, 2011.

[154] J. Weng, H. Zhang, C. Wang et al., "MIR-373-3p Targets DKK1 to Promote EMT-Induced Metastasis via the Wnt/ β-Catenin Pathway in Tongue Squamous Cell Carcinoma," *BioMed Research International*, vol. 2017, Article ID 6010926, 2017.

[155] Y. Yang, X.-X. Chen, W.-X. Li et al., "EZH2-mediated repression of Dkk1 promotes hepatic stellate cell activation and hepatic fibrosis," *Journal of Cellular and Molecular Medicine*, vol. 21, no. 10, pp. 2317–2328, 2017.

[156] J. E. Hudson and W.-H. Zimmermann, "Tuning Wnt-signaling to enhance cardiomyogenesis in human embryonic and induced pluripotent stem cells," *Journal of Molecular and Cellular Cardiology*, vol. 51, no. 3, pp. 277–279, 2011.

[157] A. S. Garnatz, Z. Gao, M. Broman, S. Martens, J. U. Earley, and E. C. Svensson, "FOG-2 mediated recruitment of the NuRD complex regulates cardiomyocyte proliferation during heart development," *Developmental Biology*, vol. 395, no. 1, pp. 50–61, 2014.

[158] K. Monzen, Y. Ito, A. T. Naito et al., "A crucial role of a high mobility group protein HMGA2 in cardiogenesis," *Nature Cell Biology*, vol. 10, no. 5, pp. 567–574, 2008.

[159] S. Gregoire, G. Li, A. C. Sturzu, R. J. Schwartz, and S. M. Wu, "YY1 Expression Is Sufficient for the Maintenance of Cardiac Progenitor Cell State," *Stem Cells*, vol. 35, no. 8, pp. 1913–1923, 2017.

[160] I. Lei, L. Liu, M. H. Sham, and Z. Wang, "SWI/SNF in cardiac progenitor cell differentiation," *Journal of Cellular Biochemistry*, vol. 114, no. 11, pp. 2437–2445, 2013.

[161] C. J. A. Ramachandra, A. Mehta, P. Wong, and W. Shim, "ErbB4 Activated p38γ MAPK Isoform Mediates Early Cardiogenesis Through NKx2.5 in Human Pluripotent Stem Cells," *Stem Cells*, vol. 34, no. 2, pp. 288–298, 2016.

[162] B. Duncan and K. Zhao, "HMGA1 mediates the activation of the CRYAB promoter by BRG1," *DNA and Cell Biology*, vol. 26, no. 10, pp. 745–752, 2007.

[163] S. He, M. K. Pirity, W.-L. Wang et al., "Chromatin remodeling enzyme Brg1 is required for mouse lens fiber cell terminal differentiation and its denucleation," *Epigenetics & Chromatin*, vol. 3, no. 1, article no. 21, 2010.

[164] A. C. Planello, J. Ji, V. Sharma et al., "Aberrant DNA methylation reprogramming during induced pluripotent stem cell generation is dependent on the choice of reprogramming factors," *Cell Regeneration*, vol. 3, no. 1, article no. 4, 2014.

[165] C.-S. Yang, C. G. Lopez, and T. M. Rana, "Discovery of non-steroidal anti-inflammatory drug and anticancer drug enhancing reprogramming and induced pluripotent stem cell generation," *Stem Cells*, vol. 29, no. 10, pp. 1528–1536, 2011.

[166] D. Yesudhas, M. Batoo, M. A. Anwar, S. Panneerselvam, and S. Choi, "Proteins recognizing DNA: Structural uniqueness and versatility of DNA-binding domains in stem cell transcription factors," *Gene*, vol. 8, no. 8, article no. 192, 2017.

[167] E. Maurizio, L. Cravello, L. Brady et al., "Conformational role for the C-terminal tail of the intrinsically disordered high mobility group A (HMGA) chromatin factors," *Journal of Proteome Research*, vol. 10, no. 7, pp. 3283–3291, 2011.

[168] R. Sgarra, S. Zammitti, A. Lo Sardo et al., "HMGA molecular network: from transcriptional regulation to chromatin remodeling," *Biochimica et Biophysica Acta—Gene Regulatory Mechanisms*, vol. 1799, no. 1-2, pp. 37–47, 2010.

[169] A. Soufi, G. Donahue, and K. S. Zaret, "Facilitators and impediments of the pluripotency reprogramming factors' initial engagement with the genome," *Cell*, vol. 151, no. 5, pp. 994–1004, 2012.

[170] M. Wegner, "From head to toes: the multiple facets of Sox proteins," *Nucleic Acids Research*, vol. 27, no. 6, pp. 1409–1420, 1999.

[171] Y. Hayashi, L. Caboni, D. Das et al., "Structure-based discovery of NANOG variant with enhanced properties to promote self-renewal and reprogramming of pluripotent stem cells," *Proceedings of the National Acadamy of Sciences of the United States of America*, vol. 112, no. 15, pp. 4666–4671, 2015.

[172] C. Rizzi, P. Cataldi, A. Iop et al., "The expression of the high-mobility group A2 protein in colorectal cancer and surrounding fibroblasts is linked to tumor invasiveness," *Human Pathology*, vol. 44, no. 1, pp. 122–132, 2013.

[173] H. Lopez-Bertoni, B. Lal, N. Michelson et al., "Epigenetic modulation of a miR-296-5p:HMGA1 axis regulates Sox2 expression and glioblastoma stem cells," *Oncogene*, vol. 35, no. 37, pp. 4903–4913, 2016.

[174] K.-R. Yu, J.-H. Shin, J.-J. Kim et al., "Rapid and Efficient Direct Conversion of Human Adult Somatic Cells into Neural Stem Cells by HMGA2/let-7b," *Cell Reports*, vol. 10, no. 3, pp. 441–452, 2015.

[175] J.-J. Kim, J.-H. Shin, K.-R. Yu et al., "Direct conversion of human umbilical cord blood into induced neural stem cells with SOX2 and HMGA2," *International Journal of Stem Cells*, vol. 10, no. 2, pp. 227–234, 2017.

[176] D. D'Angelo, P. Mussnich, R. Rosa, R. Bianco, G. Tortora, and A. Fusco, "High mobility group A1 protein expression reduces the sensitivity of colon and thyroid cancer cells to antineoplastic drugs," *BMC Cancer*, vol. 14, article 851, 2014.

[177] M. Z. Akhter and M. R. Rajeswari, "Interaction of doxorubicin with a regulatory element of hmga1 and its in vitro anti-cancer

activity associated with decreased HMGA1 expression," *Journal of Photochemistry and Photobiology B: Biology*, vol. 141, pp. 36–46, 2014.

[178] P. Zhang, C. Huang, C. Fu et al., "Cordycepin (3'-deoxyadenosine) suppressed HMGA2, Twist1 and ZEB1-dependent melanoma invasion and metastasis by targeting miR-33b," *Oncotarget*, vol. 6, no. 12, pp. 9834–9853, 2015.

[179] X. Yang, Q. Zhao, H. Yin, X. Lei, and R. Gan, "MiR-33b-5p sensitizes gastric cancer cells to chemotherapy drugs via inhibiting HMGA2 expression," *Journal of Drug Targeting*, vol. 25, no. 7, pp. 653–660, 2017.

[180] H. Siahmansouri, M. H. Somi, Z. Babaloo et al., "Effects of HMGA2 siRNA and doxorubicin dual delivery by chitosan nanoparticles on cytotoxicity and gene expression of HT-29 colorectal cancer cell line," *Journal of Pharmacy and Pharmacology*, pp. 1119–1130, 2016.

[181] P. Eivazy, F. Atyabi, F. Jadidi-Niaragh et al., "The impact of the codelivery of drug-siRNA by trimethyl chitosan nanoparticles on the efficacy of chemotherapy for metastatic breast cancer cell line (MDA-MB-231)," *Artificial Cells, Nanomedicine and Biotechnology*, vol. 45, no. 5, pp. 889–896, 2017.

[182] H. Luo, Y. Jiang, S. Ma et al., "EZH2 promotes invasion and metastasis of laryngeal squamous cells carcinoma via epithelial-mesenchymal transition through H3K27me3," *Biochemical and Biophysical Research Communications*, vol. 479, no. 2, pp. 253–259, 2016.

[183] H. Cardenas, J. Zhao, E. Vieth, K. P. Nephew, and D. Matei, "EZH2 inhibition promotes epithelial-to-mesenchymal transition in ovarian cancer cells," *Oncotarget*, vol. 7, no. 51, pp. 84453–84467, 2016.

[184] O. K. Kim, E. J. Seo, E. J. Choi, S. I. Lee, Y. W. Kwon, J. H. Jang et al., "Crucial role of HMGA1 in the self-renewal and drug resistance of ovarian cancer stem cells," *Experimental & Molecular Medicine*, vol. 48, no. 8, p. e255, 2016.

[185] Y. Liu, Y. Wang, Y. Zhang, J. Fu, and G. Zhang, "Knockdown of HMGA1 expression by short/small hairpin RNA inhibits growth of ovarian carcinoma cells," *Biotechnology and Applied Biochemistry*, vol. 59, no. 1, pp. 1–5, 2012.

[186] K. Zhuang, Q. Wu, S. Jiang, H. Yuan, S. Huang, and H. Li, "CCAT1 promotes laryngeal squamous cell carcinoma cell proliferation and invasion," *American Journal of Translational Research*, vol. 8, no. 10, pp. 4338–4345, 2016.

[187] X. Yi, J. Guo, J. Guo et al., "EZH2-mediated epigenetic silencing of TIMP2 promotes ovarian cancer migration and invasion," *Scientific Reports*, vol. 7, no. 1, article no. 3568, 2017.

[188] A. Tsagaratou, C.-W. J. Lio, X. Yue, and A. Rao, "TET methylcytosine oxidases in T cell and B cell development and function," *Frontiers in Immunology*, vol. 8, article no. 220, 2017.

[189] Z. Zhou, H.-S. Zhang, Y. Liu et al., "Loss of TET1 facilitates DLD1 colon cancer cell migration via H3K27me3-mediated down-regulation of E-cadherin," *Journal of Cellular Physiology*, vol. 233, no. 2, pp. 1359–1369, 2018.

6

Class-Specific Histone Deacetylase Inhibitors Promote 11-Beta Hydroxysteroid Dehydrogenase Type 2 Expression in JEG-3 Cells

Katie L. Togher,[1,2,3] **Louise C. Kenny,**[1,3] **and Gerard W. O'Keeffe**[2,3,4]

[1]*Department of Obstetrics and Gynaecology, Cork University Maternity Hospital, University College Cork, Cork, Ireland*
[2]*APC Microbiome Institute, Biosciences Institute, University College Cork, Cork, Ireland*
[3]*INFANT Centre, Cork University Maternity Hospital, University College Cork, Cork, Ireland*
[4]*Department of Anatomy and Neuroscience, University College Cork, Cork, Ireland*

Correspondence should be addressed to Gerard W. O'Keeffe; g.okeeffe@ucc.ie

Academic Editor: Wiljan J. A. J. Hendriks

Exposure to maternal cortisol plays a crucial role in fetal organogenesis. However, fetal overexposure to cortisol has been linked to a range of short- and long-term adverse outcomes. Normally, this is prevented by the expression of an enzyme in the placenta called 11-beta hydroxysteroid dehydrogenase type 2 (11β-HSD2) which converts active cortisol to its inactive metabolite cortisone. Placental 11β-HSD2 is known to be reduced in a number of adverse pregnancy complications, possibly through an epigenetic mechanism. As a result, a number of pan-HDAC inhibitors have been examined for their ability to promote 11β-HSD2 expression. However, it is not known if the effects of pan-HDAC inhibition are a general phenomenon or if the effects are dependent upon a specific class of HDACs. Here, we examined the ability of pan- and class-specific HDAC inhibitors to regulate 11β-HSD2 expression in JEG3 cells. We find that pan-, class I, or class IIa HDAC inhibition promoted 11β-HSD2 expression and prevented cortisol or interleukin-1β-induced decrease in its expression. These results demonstrate that targeting a specific class of HDACs can promote 11β-HSD2 expression in JEG3 cells. This adds to the growing body of evidence suggesting that HDACs may be crucial in maintaining normal fetal development.

1. Introduction

The glucocorticoid hypothesis proposes that overexposure of the fetus to glucocorticoids may produce long lasting effects on fetal development that subsequently increase disease risk later in life [1]. The glucocorticoid hypothesis is affirmed by studies that have shown that elevated maternal cortisol is associated with heightened HPA activity [2] and alterations in brain structure [3] in affected offspring. At the core of this process is the placental enzyme 11β-hydroxysteroid dehydrogenase type 2 (11β-HSD2), an enzyme that is expressed primarily within the syncytiotrophoblast of the placenta where it catalyses the conversion of active cortisol into its inactive product cortisone, thereby controlling the levels of cortisol that reach the fetus [4]. A number of preclinical and clinical studies have demonstrated a reduction in the placental expression of 11β-HSD2 following exposure to prenatal stress [5], anxiety [6], and following maternal infection

[7]. In addition to this, placental *HSD11B2* mRNA levels are reduced in pregnancy complications such as preeclampsia [8], intrauterine growth restriction (IUGR) [9], preterm birth (PTB) [10], and low birth weight (LBW) [11].

A complex repertoire of molecular pathways have been shown to be involved in regulating placental *HSD11B2* expression. Inhibition of the mitogen-activated protein kinases (MAPK) ERK1/2 increases *HSD11B2* expression [12], whilst suppressing p38 reduces 11β-HSD2 activity [13]. *HSD11B2* is increased by activation of peroxisome proliferator-activated receptor delta (PPARδ) [14] through recruitment of the SP1 transcription factor (TF) [15]. Similarly, activation of the hedgehog signalling [16] and forskolin-induced activation of the cyclic AMP (cAMP) pathway increases *HSD11B2* expression [17]. More recently, epigenetic mechanisms have been linked to 11β-HSD2 regulation. The most widely studied epigenetic mechanisms are DNA methylation and histone acetylation. Histone acetylation is regulated by histone acetyl transferase (HATs) and histone deacetylase (HDACs) enzymes. HATs add acetyl groups onto the N-terminal tail of histone proteins which increases gene expression [18]. HDACs remove them, thereby repressing transcription [19]. In humans, 18 HDACs have been discovered and they are classed into four main families: class I (HDACs 1, 2, 3, and 8), class II (HDACs 4, 5, 6, 7, 9, and 10), class III (SIRT1, SIRT2, SIRT3, SIRT4, SIRT5, SIRT6, and SIRT7), and class IV (HDAC 11) [20].

Recently, a significant emphasis has been placed on in vitro studies to tease apart the precise epigenetic mechanisms involved in regulating placental 11β-HSD2 expression. Global knock down of DNA methylation using the demethylating agent 5-aza-2′-deoxycytidine (5-aza) in JEG-3 cells has been shown to increase the expression of a number steroidogenic genes including *HSD11B2*, indicating a direct link for regulation of *HSD11B2* expression by methylation [21]. Despite advancements being made in understanding the role of methylation in 11β-HSD2 expression, little focus has been placed on examining the role that HDACs play in regulating 11β-HSD2. The present study aimed to investigate the role of histone acetylation in regulating basal and stressor-induced changes in 11β-HSD2 protein expression in an in vitro placenta model using small molecule pharmacological inhibitors.

2. Methods

2.1. Cell Culture and Treatment.
JEG-3 cells were grown in Dulbecco's modified Eagle's medium (DMEM): F12 (Sigma), with 10% fetal calf serum (FCS), 100 nM L-Glutamine, 100 U/ml penicillin, and 10 μg/ml streptomycin (Sigma). Cells were maintained at 37°C in a humidified atmosphere of 5% CO_2. 50,000 cells per well were plated on a 24-well plate and were treated with 1, 5, or 10 μM of MC1568, MS275, or SAHA (Selleckchem). Where indicated, 10 ng/ml interleukin-1β (IL-1β; Promokine) or 2 μM cortisol (Cort; Santa Cruz) was added for 24 h before HDAC inhibitor (HDI).

2.2. MTT Assay.
To assess cell viability, a thiazolyl blue tetrazolium bromide (MTT) solution was added to the cells at a concentration of 1 mg/ml in HBSS (Sigma). Following a 2-hour (h) incubation at 37°C, the cells were lysed in DMSO (Sigma). Absorbance was measured at a wavelength of 540 nm with a reference wavelength of 630 nm.

2.3. Immunocytochemistry.
At the experimental end point, cultures were fixed in 4% paraformaldehyde (PFA) in PBS for 10 min. Following 3 × 5 min washes in 10 mM PBS containing 0.02% Triton X-100 (PBS-T), cultures were incubated in blocking solution (5% BSA in PBS-T) for 1 h at room temperature. Where indicated, cultures were incubated in the following primary antibodies: 11β-HSD2 (1:250; Santa Cruz), AcH3 (1:250; Santa Cruz), GR (1:250; Santa Cruz), or IL1R1 (1:250; Invitrogen) diluted in 1% BSA in 10 mM PBS at 4°C for 16 h. Following 3 × 5 min washes in PBS-T, cells were incubated in the appropriate Alexa Fluor 488-conjugated or 594-conjugated secondary antibodies (1:1000; Invitrogen) diluted in 1% BSA in 10 mM PBS at room temperature

for 2 h. Cultures were counterstained with DAPI (1:3000; Sigma). Cells were imaged under an Olympus IX70 inverted microscope with Olympus DP70 camera and AnalysisD™ software.

2.4. RNA Extraction and Real-Time PCR.
RNA was extracted from JEG-3 cells 24 hours after seeding and term human placental tissue using Trizol Reagent (Life Technologies). Placental tissue was homogenised with a pestle and mortar and JEG-3 cells were removed from flasks by scraping and incubated in Trizol for 10 min and RNA extraction proceeded according to the manufacturer's instructions. 500 ng of RNA was reverse-transcribed using a high capacity cDNA Reverse Transcription Kit (Applied Biosystems) in a 20 μl reaction mixture consisting of 2.0 μl 10x RT Buffer, 0.8 μl 25x dNTP mix (100 mM), 2.0 μl 10x RT Random Primers, 1.0 μl Reverse Transcriptase, and 4.2 μl Nuclease-free H_2O, using the following parameters: 25°C for 10 min; 37°C for 120 min; 85°C for 5 min; 4°C for at least 10 min. The cDNA was stored at −80°C prior to use. For real-time PCR, samples were run in duplicate using TaqMan® Gene Expression Assay (Applied Biosystems) for *HSD11B2* using *18S* as a reference gene under the following parameters: 50°C for 2 min; 95°C for 10 min; 40 repetitions of 95°C for 15 s; and annealing/elongating at 60°C for 1 min.

2.5. Immunohistochemistry.
Histological placental sections (6 μM) were incubated in blocking solution (5% bovine serum albumin (BSA)) for 1 h at room temperature. Sections were treated with 10% H_2O_2 for 5 min, washed in 10 mM Phosphate Buffered Saline (PBS), and blocked for 1 h in 10% normal goat serum in 10 mM PBS with 0.4% Triton X. Sections were incubated in primary antibody to 11β-HSD2 (1:250; Santa Cruz) in 1% normal goat serum in 10 mM PBS with 0.4% Triton X overnight at 4°C. Following a 3 × 10 min wash in 10 mM PBS, sections were incubated with a biotinylated secondary antibody (1:200; Vector Labs) for 2 h at room temperature. Following another 3 × 10 min wash in 10 mM PBS, sections were incubated in ABC solution (1:200; Vector Labs) for 45 min at room temperature followed by immersion in diaminobenzidine substrate/chromogen reagent for 2-3 min at room temperature. Sections were dehydrated, cleared, mounted, and imaged using an Olympus AX70 Provis upright microscope.

2.6. Statistical Analysis.
For real-time PCR, expression levels were calculated using the 2-delta-Ct threshold method [22]. For immunocytochemistry, the fluorescence intensity of individual cells that were immunopositive for 11β-HSD2 or AcH3 was measured by densitometry using Image J analysis software (Rasband, WJ, http://rsb.info.nih.gov/ij/). The relative fluorescence intensity of 11β-HSD2 or AcH3 was calculated as the average fluorescence intensity after subtraction of the background noise. Data was analysed using GraphPad Prism v 5 (GraphPad Software Inc., San Diego, California). Where indicated, data was analysed (as per Section 2.3) with unpaired Student's *t*-test or one-way ANOVA with Tukey's post hoc testing. Values of $p < 0.05$ were considered statistically significant.

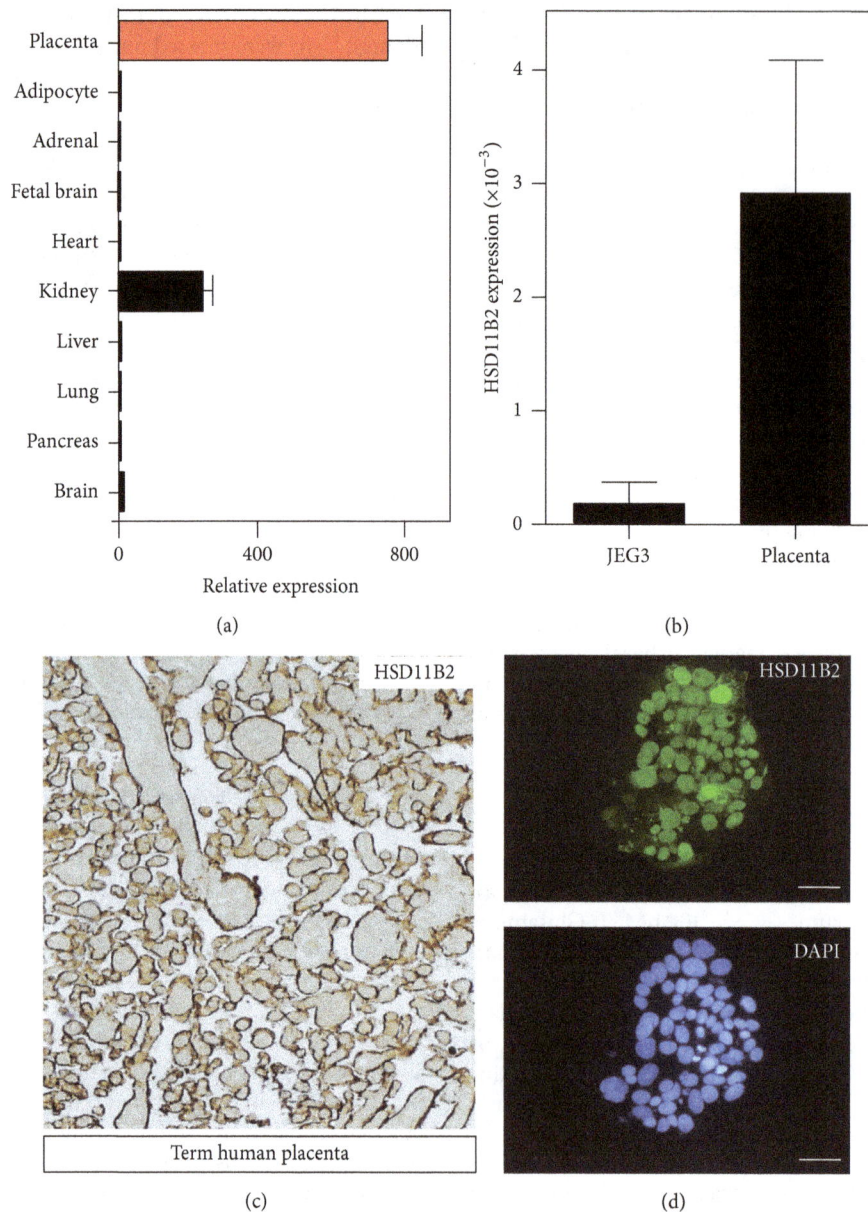

FIGURE 1: (a) Expression of data derived from the BioGPS database showing relative *HSD11B2* expression across multiple human tissues. (b) Real-time PCR showing *HSD11B2* expression in the term human placenta and in JEG-3 cells using the 2-delta-Ct method ($N = 3$, $p > 0.05$, unpaired Student's t-test; housekeeping gene 18S). Representative photomicrographs of (c) a term human placenta and (d) JEG-3 cells immunocytochemically stained for 11β-HSD2. Scale bar = 50 μm.

3. Results

3.1. Distribution of HSD11B2 in the Human Placenta and JEG-3 Cells. We utilized the BioGPS database, an online platform that enables the examination of relative levels of gene expression across multiple human tissues [23]. Using this directory, we confirmed the highest levels of *HSD11B2* in the placenta, followed by the kidneys, with very little expression seen in other tissues (Figure 1(a)), which was confirmed by immunohistochemistry on human term placental samples (Figure 1(c)). We next aimed to validate the use of the human choriocarcinoma cell line, JEG-3 cells. JEG-3 cells are a widely

used in vitro model of placental trophoblast cells and have previously been demonstrated to be an abundant source of endogenous 11β-HSD2 [24, 25]. In agreement with this, real-time PCR confirmed the expression of *HSD11B2* mRNA in JEG-3 cells, with placental RNA used as positive control (Figure 1(b)). Immunohistochemical staining preformed 24 hours after seeding also confirmed abundant expression of expression of 11β-HSD2 protein in JEG-3 cells (Figure 1(d)).

3.2. Pan-HDAC Inhibition Increases 11β-HSD2 Expression in JEG-3 Cells. HDACs can be divided into four distinct families, of particular interest are class I (HDAC1, HDAC2,

(a)

(b)

(c)

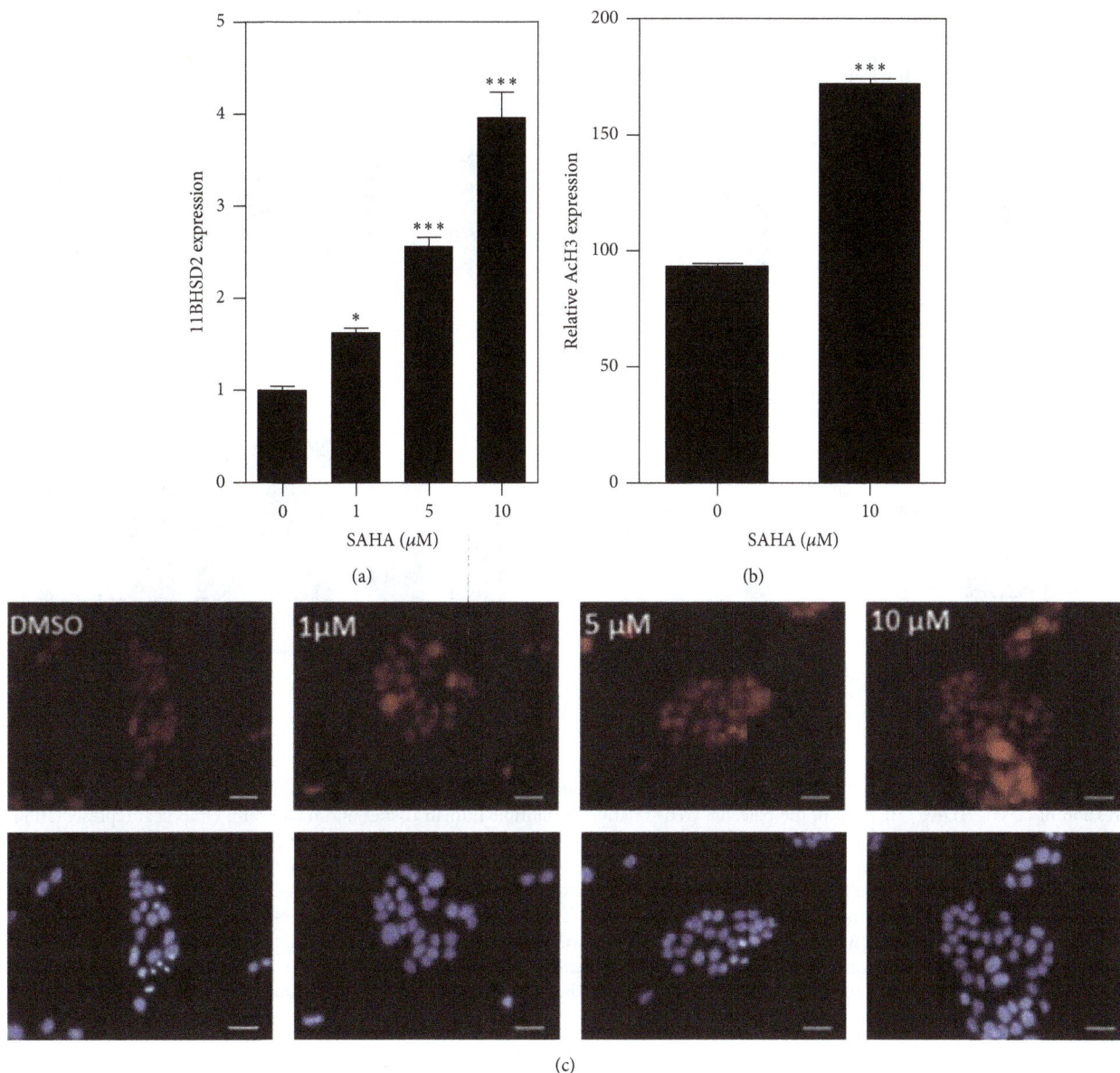

Figure 2: Epigenetic regulation of 11β-HSD2 expression. Graphical representation of (a) 11β-HSD2 and (b) AcH3 expression in JEG-3 cells treated with 0–10 μM of SAHA for 24 h. Data are expressed as mean ± SEM. (c) Representative photomicrographs of JEG-3 cells immunocytochemically stained for 11β-HSD2 ($^*p < 0.05$, $^{***}p < 0.001$ compared to 0 μM; (a) one-way ANOVA with post hoc Tukey's and (b) unpaired Student's t-test; 25 cells per group per experiment; $N = 3$). Scale bar = 50 μm.

HDAC5, and HDAC8) and class II (HDAC5, HDAC6, HDAC7, HDAC9, and HDAC10) HDACs [26]. We used the BioGPS database to examine the relative expression levels of these different HDACs in the human placenta. Class I and class II HDACs were widely expressed in the placenta (see Supplementary Figure 1 in Supplementary Material available online at https://doi.org/10.1155/2017/6169310); however, HDAC1 (class I) and HDAC5 (class IIa) had the highest relative levels of expression in the placenta compared to other tissues (Figures 3(a) and 3(e)). Given the widespread expression of HDACs, we next sought to determine the effect of global HDAC inhibition on placenta 11β-HSD2 protein expression. We treated JEG-3 cells with SAHA, a competitive

inhibitor of both class I and class II HDACs [27]. An initial dose response experiment was carried out 24 hours after seeding where JEG-3 cells were treated with concentrations of SAHA ranging within 1–10 μM for 24 h, followed by immunocytochemical staining for 11β-HSD2. The relative expression of 11β-HSD2 protein was quantified using densitometry. A one-way ANOVA revealed a significant overall effect of SAHA treatment on 11β-HSD2 expression ($F_{(3,8)}$ = 5.5, p = 0.02). Tukey's post hoc test revealed a significance difference between the vehicle and 10 μM SAHA group ($p <$ 0.05) (Figure 2(a)). As the effects of SAHA were significant at 10 μM, we also immunocytochemically stained for p-Ac-histone H3 (S11/K15) (pAcH3) in this group and found a

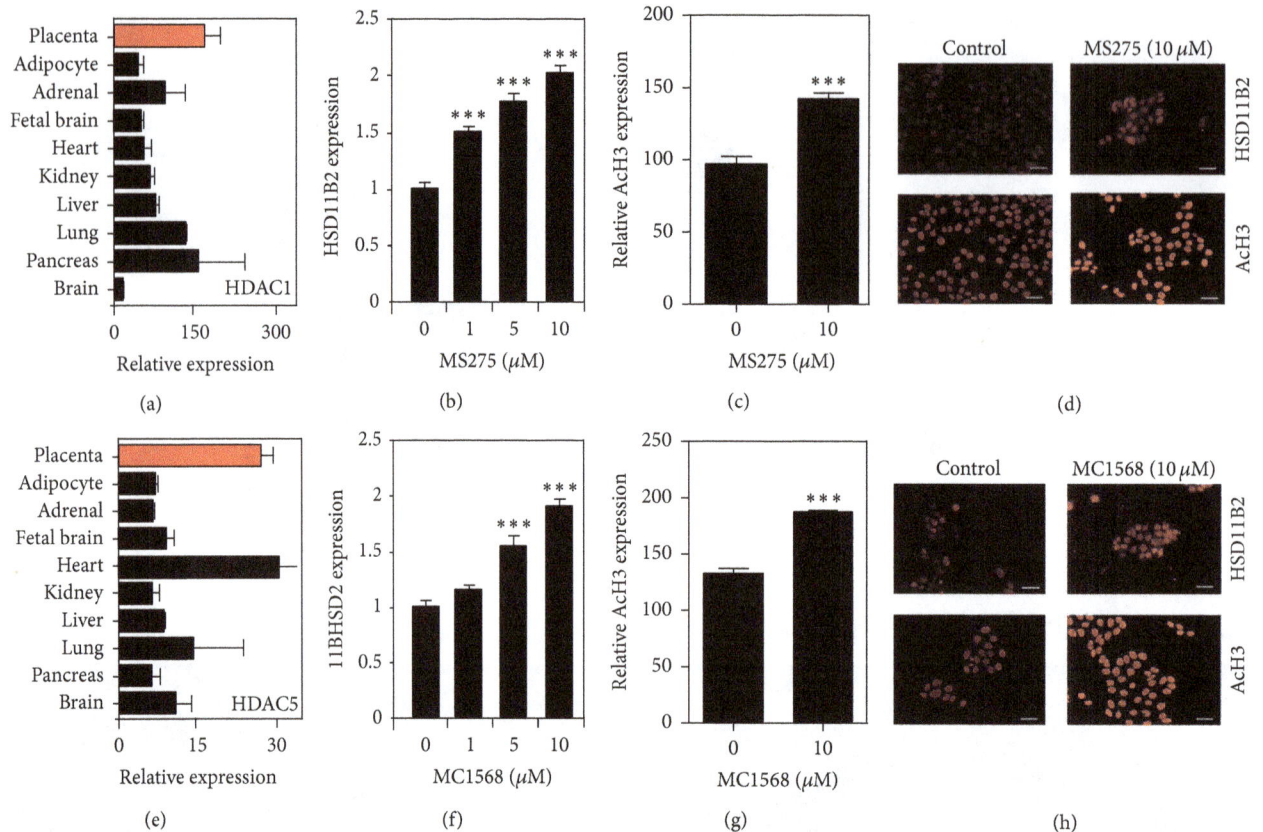

(a)

(b)

(c)

(d)

(e)

(f)

(g)

(h)

FIGURE 3: Class-specific HDACs on 11β-HSD2 regulation in the placenta. (a) Expression data from the BioGPS database showing the relative expression of class I HDAC, HDAC 1 in the placenta (red) relative to multiple human tissues and fetal brain. Graphical representation of (b) 11β-HSD2 and (c) AcH3 expression in JEG-3 cells treated with 0–10 μM of class I HDAC inhibitor MS275 for 24 h. (d) Representative photomicrographs of JEG-3 cells immunocytochemically stained for 11β-HSD2 and AcH3 after treatment with (0–10 μM) MS275 for 24 h. (e) Expression data from the BioGPS database showing the relative expression of class II HDAC, HDAC 5 in the placenta (red) relative to multiple human tissues and fetal brain. Graphical representation of (f) 11β-HSD2 and (g) AcH3 expression in JEG-3 cells treated with 0–10 mM of class IIa HDAC inhibitor MC1568 for 24 h. Data are expressed as mean ± SEM. (h) Representative photomicrographs of JEG-3 cells immunocytochemically stained for 11β-HSD2 and AcH3 after treatment with (0–10 μM) MC1568 for 24 h. Data are expressed as mean ± SEM ($^{***}p < 0.001$ compared to 0 μM; (b, f) one-way ANOVA with post hoc Tukey's and (c, g) unpaired Student's t-test; 25 cells per group per experiment; $N = 3$). Scale bar = 50 μm.

significant increase in the levels of pAcH3 in cells treated with 10 μM SAHA for 24 h ($p < 0.001$) (Figure 2(b)). Overall, these data indicate that pan-HDAC inhibition increases the levels of pAcH3 (which has been shown to correlate with gene expression) and 11β-HSD2 expression in JEG-3 cells.

3.3. Class-Specific HDAC Inhibitors (HDI) Promote 11β-HSD2 Expression in JEG-3 Cells.

We next investigated if the effects of pan-HDAC inhibition on 11β-HSD2 expression were class-specific using a class I-specific HDI (MS275) [28] and a class IIa-specific HDI (MC1568) [29]. JEG-3 cells were treated with increasing concentrations (0–10 μM) of MS275 or MC1568 for 24 h before being immunocytochemically stained for 11β-HSD2 and quantified using densitometry. A one-way ANOVA revealed a significant overall effect of both MS275 ($F_{(3,8)} = 95.89$, $p < 0.0001$) and MC1568 ($F_{(3,8)} = 53.69$, $p < 0.0001$) treatment. Tukey's post hoc test showed that MS275 or MC1568 promoted a significant increase in 11β-HSD2 protein expression with a significant difference

observed between the control and HDI-treated groups at concentrations of 1 μM ($p < 0.05$), 5 μM ($p < 0.0001$), and 10 μM ($p < 0.0001$) (Figures 3(b) and 3(f)). We also examined pAcH3 levels using densitometry and found a significant increase in the levels of pAcH3 in cells treated with 10 μM MC1568 or MS275 for 24 h ($p < 0.001$) (Figures 3(c) and 3(g)). These data show that class I and class IIa inhibition can promote 11β-HSD2 protein expression in JEG-3 cells.

3.4. Cortisol and IL-1β Decrease 11β-HSD2 Expression Which Is Prevented by MC1568.

Given that alterations in placental HSD11B2 expression are seen in pregnancies complicated with stress or infection [5, 7], we next sought to determine if the biological mediators of stress (Cort) and infection (IL-1β) altered 11β-HSD2 protein expression at the cellular level. Having confirmed using immunocytochemistry that the glucocorticoid receptor (GR) and interleukin 1 receptor, type I (IL1R1), were expressed in JEG-3 cells (Figure 4(a)), we carried out an MTT assay to establish a concentration of

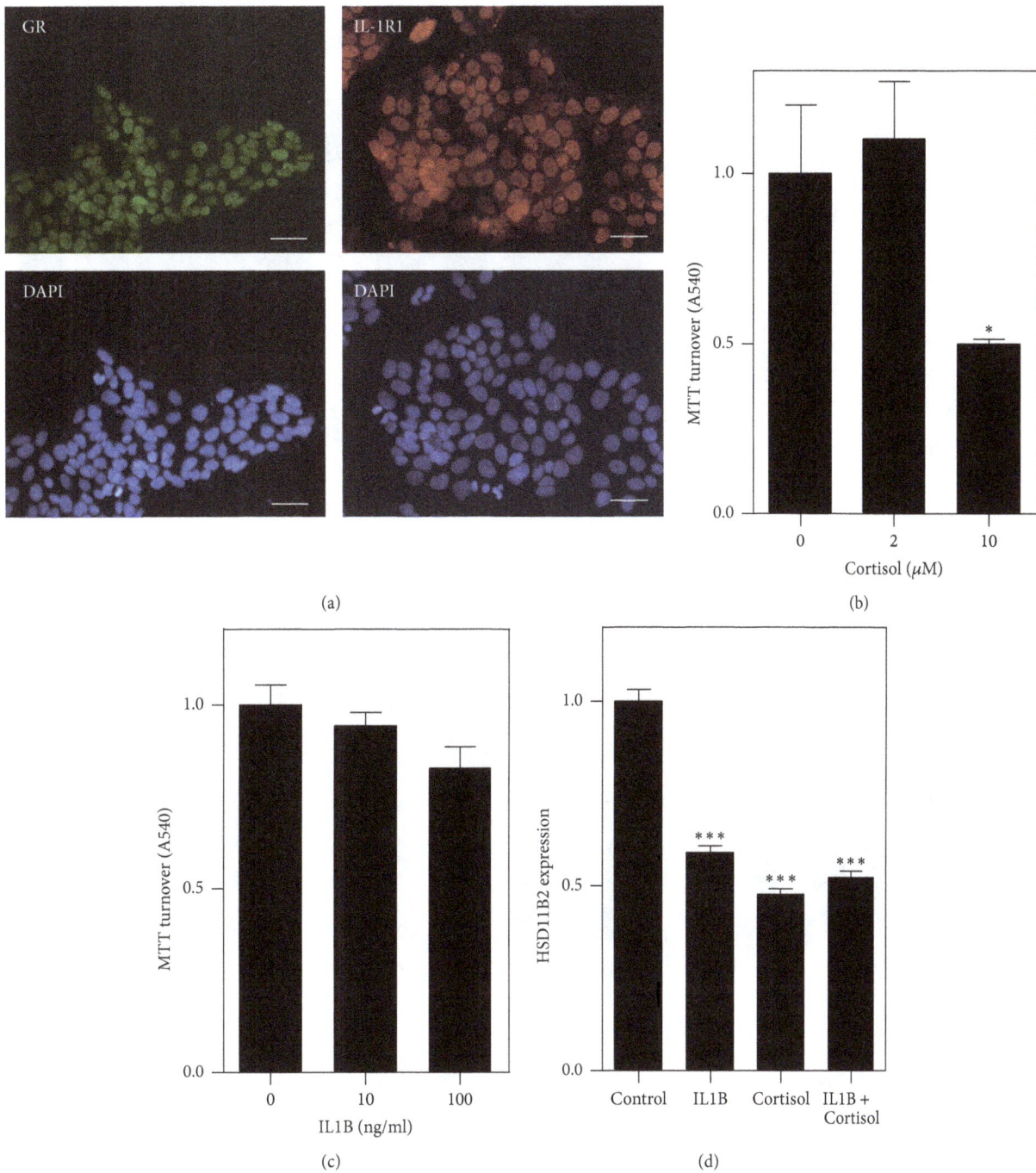

FIGURE 4: Cortisol and IL-1β response in JEG-3 cells. (a) Representative photomicrographs of JEG-3 cells immunocytochemically stained for the glucocorticoid receptor (GR; green) and (c) the interleukin 1 receptor, type I (IL-1R1; red). The second panel shows the corresponding DAPI stained image. (b, c) MTT assay examining the viability of JEG3 cells treated with either 0–10 μM cortisol (b) or 0–100 ng/ml IL-1β (c) for 24 h in vitro. (d) Graphical representation showing the levels 11β-HSD2 in JEG-3 cells exposed to a vehicle (control), 10 ng/ml IL-1β or 2 μM cortisol for 24 h. Data are expressed as mean ± SEM ($^*p < 0.05$ compared to 0 μM, $^{***}p < 0.001$ compared to control; one-way ANOVA with post hoc Tukey's; (d) 100 cells per group per experiment; $N = 3$). Scale bar = 50 μm.

Cort and IL-1β that did not affect cell viability. JEG-3 cells were treated with Cort (0–10 μM) or IL-1β (0–100 ng/ml) for 24 h and MTT assays were performed. An ANOVA showed an overall effect of Cort and IL-1β treatment on cell viability, with a difference observed with 10 μM Cort (Figure 4(b))

and 100 ng/ml IL-1β (Figure 4(c)) groups ($p < 0.05$). JEG-3 cells were then treated with 2 μM of Cort or 10 ng/ml IL-1β (concentrations that did not affect cell viability) for 24 h before being fixed and immunocytochemically stained for 11β-HSD2. Using densitometry, we observed a reduction in

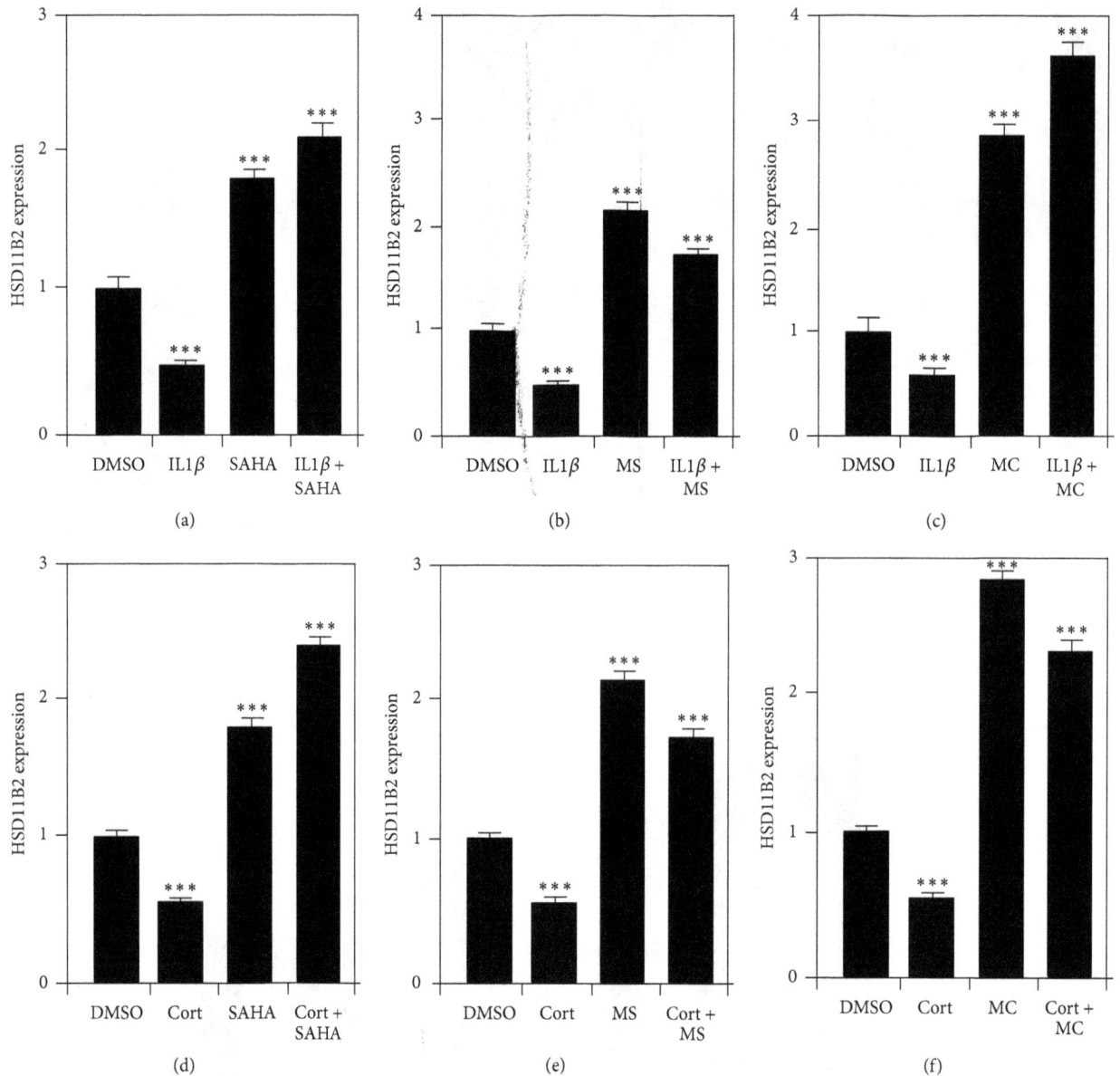

FIGURE 5: SAHA, MS275, and MC1568 prevent cortisol and IL1β-induced decreases in 11β-HSD2 expression. Graphical representation and 11β-HSD2 expression in JEG-3 cells treated with 2 μM Cort or 10 ng/ml IL-1β in the presence or absence of 10 μM SAHA (a, d), MS275 (b, e), or MC1568 (c, f) for 24 h. Data are expressed as mean ± SEM. (***$p < 0.001$ compared to DMSO; one-way ANOVA with post hoc Tukey's test; 25 cells per group per experiment; $N = 3$.)

11β-HSD2 protein expression following exposure to Cort and IL-1β (Figure 4(d)).

3.5. HDIs Can Restore 11β-HSD2 Expression in an Environment of Stress and Inflammation.

After identifying Cort and IL-1β as potential biological mediators causing a decrease in 11β-HSD2 protein expression, we next aimed to determine if HDIs could counteract these effects of cortisol and IL-1β on 11β-HSD2 protein expression. After plating for 24 hours, JEG-3 cells were treated with 10 μM of SAHA, MC1568, or MS275 followed by cortisol or IL-1β before being fixed and immunocytochemically stained for 11β-HSD2 protein. Densitometry revealed that pretreatment of JEG-3 cells with

nonspecific inhibitor SAHA attenuated the effect of IL-1β and Cort (SAHA: 2.8±0.15; SAHA + IL-1β: 3.0±0.2; SAHA + Cort: 3.350±0.19) (Figures 5(a) and 5(d)) on 11β-HSD2 expression. Similarly, treatment of JEG-3 cells with either class I-specific HDI, MS275 (MS275: 2.2 ± 0.06; MS275 + IL-1β: 3.18 ± 0.06; MS275 + Cort: 2.2 ±0.09) (Figures 5(b) and 5(e)), or class IIa-specific HDI MC1568 (MC1568: 1.8 ± 0.08; MC1568 + IL-1β: 2.3 ± 0.1; MC1568 + Cort: 1.4 ± 0.07) (Figures 5(c) and 5(f)) was sufficient to attenuate the effect of both Cort and IL-1β on 11β-HSD2 expression. These data show that exposure to heightened levels of Cort and IL-1β can reduce the levels of 11β-HSD2 protein in JEG-3 cells and that this effect that can be prevented by HDAC inhibition.

4. Discussion

The aim of this study was to examine the role of epigenetic regulators in the control of 11β-HSD2 protein expression in placental cells. We used the in vitro placental model JEG-3 cells, as, despite their limitations, they are a well-established cell line commonly used to mimic placental trophoblast cells [30]. We employed pharmacological inhibitors of HDACs to modulate histone acetylation and examined the impact of this on 11β-HSD2 protein expression. Finally, to assess the potential of these compounds to regulate 11β-HSD2 expression under conditions of stress and inflammation, cells were exposed to biological mediators of these conditions, namely, exogenous cortisol and IL-1β.

HSD11B2 has previously been shown to localise in trophoblast cells, with highest expression observed in the syncytiotrophoblast [31, 32]. In line with these studies, we demonstrated that 11β-HSD2 protein is strongly expressed in the term human placenta. To model trophoblast cells in vitro, we used the human choriocarcinoma cell line, JEG-3 cells. We found that these cells express *HSD11B2* mRNA making them a useful and convenient model to examine the molecular mechanisms that regulate 11β-HSD2 expression.

Using the BioGPS database, we demonstrated high expression of class 1 HDACs 1, 2, 3, and 8 and class 2 HDACs 5, 4, 7, and 9, suggesting a role for HDAC proteins in the placenta. Based on these findings, we used a SAHA, a pan-HDAC inhibitor, and demonstrated a dose-dependent increase in 11β-HSD2 protein expression. To confirm that the increase in 11β-HSD2 protein expression was paralleled by an increase in histone acetylation, we immunocytochemically stained the cells for AcH3 and showed a similar dose-dependent increase AcH3. This is in contradiction to previous studies, where *HSD11B2* expression was reported to be unchanged in JEG-3 cells following treatment with broad-spectrum class I and class II inhibitor trichostatin A [24]. However, the dose of TSA (300 nm) used in these studies was much smaller than the dose at which we observed an effect (10 uM) and we have identified that the effect of HDAC inhibition on 11β-HSD2 expression is dose-dependent.

HDACs play a diverse role during fetal development [26]. Global knockdown of HDAC3 [33] HDAC1 [34] and HDAC7 [34] results in fetal lethality; however, mice lacking HDAC6 develop normally [35]. HDACs have also been shown to be important regulators of placental development as inhibition of class II HDACs has been shown to impair trophoblast differentiation through interactions with Hypoxia-inducible factor [36]. Additionally, interaction of HDACs with the STAT-1 TF may contribute to inhibition of IFN-γ-inducible gene expression in trophoblast cells, thereby protecting the placenta cells from maternal immune rejection and contributing to a successful pregnancy [37]. This broad range of functions of HDACs suggests that global inhibition could result in detrimental effects; therefore, a more specific inhibition could represent an optimal method for modifying 11β-HSD2 expression. To determine if HDAC regulation of 11β-HSD2 protein expression is class-specific, we used class-specific pharmacological HDAC inhibitors. We observed a similar increase in 11β-HSD2 protein expression

with class-specific inhibition of either class I or class IIa HDACs, suggesting that many HDACS are likely involved in regulating 11β-HSD2 protein expression. Whilst this is the first study to examine the effects of class-specific inhibitors on HSD11B2 expression, it is interesting to note that previous studies have demonstrated a class-specific effect of HDACs on the regulation of other placental genes. Specifically, matrix metalloproteinase 9 has been shown to be regulated by class II but not class I HDACs [38].

Placental *HSD11B2* has been shown to be reduced in a number of adverse pregnancy conditions including anxiety, stress, and infection [5–7]. As elevations in proinflammatory cytokines and steroids are observed in these conditions [39, 40], we used cortisol and IL-1β to mimic an environment of stress and inflammation. We have previously demonstrated a reduction in 11β-HSD2 protein expression in JEG-3 cells following administration of IL-1β [7]. In this study, we also report a decrease in 11β-HSD2 expression following cortisol administration. In contrast, Ni and colleagues have previously shown an increase in 11β-HSD2 expression in primary human trophoblast cells exposed to Cort [17]. However, this study used primary cells which highlights the need for further study of these questions in primary trophoblast cells. Additionally, the maximum dose of cortisol used was 1 μM, whereby we observed a decrease at 2 μM. It is possible that cortisol may act in an adaptive way to induce 11β-HSD2, thereby protecting the fetus from high maternal glucocorticoids but, at a certain threshold cortisol, may begin to negatively impact 11β-HSD2 expression. Interestingly, broad or either class-specific HDAC inhibitors were sufficient to prevent the cortisol and IL-1β-induced decreases in 11β-HSD2 expression. This raises the possibility of targeting key epigenetics modulators to protect the fetal glucocorticoid barrier and untimely fetal glucocorticoid overexposure. However, given the critical role of epigenetic marks in fetal development, nonspecific inhibition of HDACs, even at class level, could produce detrimental effects on fetal development; therefore, identifying more specifically the precise epigenetic mechanism mediating HSD11B2 regulation using knockdown or overexpression of individual HDACs would allow the development of a more targeted approach. The advancement of targeted nanoparticles to deliver chemotherapeutic agents directly to the placenta represents an exciting new avenue to alter placental epigenetic mediators without interfering with the fetus [41]. Notably, we also observe potentiation of the effects of SAHA on HSD11B2 expression when administered with cortisol. Once activated, the GR can bind to many coactivator proteins with known HAT activity [42]. The combined inhibition of HDACs by SAHA with the potential increase in HAT activity caused by GR activation from exogenous cortisol may explain this enhanced 11β-HSD2 protein expression. This relationship further highlights the complexity of 11β-HSD2 regulation and the epigenetic landscape and confirms the need for more studies examining how placental 11β-HSD2 protein is controlled under both basal and pathological conditions.

Here, we provide evidence of a role for histone acetylation in the regulation of 11β-HSD2 in the placenta; a limitation is that the present study used JEG-3 cells. Although we confirmed 11β-HSD2 to be abundantly expressed in this cell line and that *HSD11B2* levels are comparable between JEG-3 cells and the human placenta, there are potential caveats associated with using JEG-3 cells [43]. As such replicating the current study in primary trophoblasts will help to clarify the functional role of HDACs in the regulation of 11β-HSD2 protein expression in the placenta. However, the present study demonstrates a role for HDACs in the regulation of a key enzyme that maintains the fetal glucocorticoid barrier under basal and pathological conditions. It is likely that a combination of different epigenetic modifiers including HDACs are involved in regulating 11β-HSD2 expression. As HDACs have a broad role in regulating fetal development, inhibition of all HDACs could be detrimental to the developing fetus. Therefore, unravelling the role of individual HDACs in 11β-HSD2 regulation, using more specific pharmacological inhibitors or targeted knockdown of HDACs, will be crucial to understanding the epigenetic mechanisms that regulate 11β-HSD2 expression and for developing novel protective pharmacotherapies for the human placenta.

Competing Interests

The authors declare no conflict of interests regarding the publication of this paper.

Acknowledgments

The authors acknowledge grant support in the form of Research Centres Grant (Louise C. Kenny) (Grant no. INFANT-12/RC/2272) and Research Frontiers Program Grant (Gerard W. O'Keeffe) (Grant no. 10/RFP/NES2786), from Science Foundation Ireland, and from a Translational Research Access Program (TRAP) Award from the School of Medicine UCC (Gerard W. O'Keeffe/Louise C. Kenny).

References

[1] R. M. Reynolds, "Glucocorticoid excess and the developmental origins of disease: two decades of testing the hypothesis—2012 curt richter award winner," *Psychoneuroendocrinology*, vol. 38, no. 1, pp. 1–11, 2013.

[2] E. P. Davis, L. M. Glynn, F. Waffarn, and C. A. Sandman, "Prenatal maternal stress programs infant stress regulation," *Journal of Child Psychology and Psychiatry and Allied Disciplines*, vol. 52, no. 2, pp. 119–129, 2011.

[3] C. Buss, E. P. Davis, B. Shahbaba, J. C. Pruessner, K. Head, and C. A. Sandman, "Maternal cortisol over the course of pregnancy and subsequent child amygdala and hippocampus volumes and affective problems," *Proceedings of the National Academy of Sciences of the United States of America*, vol. 109, no. 20, pp. E1312–E1319, 2012.

[4] K. Chapman, M. Holmes, and J. Seckl, "11β-hydroxysteroid dehydrogenases: intracellular gate-keepers of tissue glucocorticoid action," *Physiological Reviews*, vol. 93, no. 3, pp. 1139–1206, 2013.

[5] C. J. Peña, C. Monk, and F. A. Champagne, "Epigenetic effects of prenatal stress on 11β-hydroxysteroid dehydrogenase-2 in the placenta and fetal brain," *PLoS ONE*, vol. 7, no. 6, Article ID e39791, 2012.

[6] E. Conradt, B. M. Lester, A. A. Appleton, D. A. Armstrong, and C. J. Marsit, "The roles of DNA methylation of *NR3C1* and *11β-HSD2* and exposure to maternal mood disorder in utero on newborn neurobehavior," *Epigenetics*, vol. 8, no. 12, pp. 1321–1329, 2013.

[7] M. E. Straley, K. L. Togher, A. M. Nolan, L. C. Kenny, and G. W. O'Keeffe, "LPS alters placental inflammatory and endocrine mediators and inhibits fetal neurite growth in affected offspring during late gestation," *Placenta*, vol. 35, no. 8, pp. 533–538, 2014.

[8] W. Hu, X. Weng, M. Dong, Y. Liu, W. Li, and H. Huang, "Alteration in methylation level at 11β-hydroxysteroid dehydrogenase type 2 gene promoter in infants born to preeclamptic women," *BMC Genetics*, vol. 15, article 96, 2014.

[9] J. Dy, H. Guan, R. Sampath-Kumar, B. S. Richardson, and K. Yang, "Placental 11β-hydroxysteroid dehydrogenase type 2 is reduced in pregnancies complicated with idiopathic intrauterine growth restriction: evidence that this is associated with an attenuated ratio of cortisone to cortisol in the umbilical artery," *Placenta*, vol. 29, no. 2, pp. 193–200, 2008.

[10] E. Kajantie, L. Dunkel, U. Turpeinen et al., "Placental 11β-hydroxysteroid dehydrogenase-2 and fetal cortisol/cortisone shuttle in small preterm infants," *Journal of Clinical Endocrinology and Metabolism*, vol. 88, no. 1, pp. 493–500, 2003.

[11] V. Mericq, P. Medina, E. Kakarieka, L. Márquez, M. C. Johnson, and G. Iñiguez, "Differences in expression and activity of 11β-hydroxysteroid dehydrogenase type 1 and 2 in human placentas of term pregnancies according to birth weight and gender," *European Journal of Endocrinology*, vol. 161, no. 3, pp. 419–425, 2009.

[12] H. Guan, K. Sun, and K. Yang, "The ERK1/2 signaling pathway regulates 11beta-hydroxysteroid dehydrogenase type 2 expression in human trophoblast cells through a transcriptional mechanism," *Biology of Reproduction*, vol. 89, no. 4, article 92, 2013.

[13] A. Sharma, H. Guan, and K. Yang, "The p38 mitogen-activated protein kinase regulates 11β-hydroxysteroid dehydrogenase type 2 (11β-HSD2) expression in human trophoblast cells through modulation of 11β-HSD2 messenger ribonucleic acid stability," *Endocrinology*, vol. 150, no. 9, pp. 4278–4286, 2009.

[14] L. Julan, H. Guan, J. P. Van Beek, and K. Yang, "Peroxisome proliferator-activated receptor δ suppresses 11β-hydroxysteroid dehydrogenase type 2 gene expression in human placental trophoblast cells," *Endocrinology*, vol. 146, no. 3, pp. 1482–1490, 2005.

[15] P. He, Z. Chen, Q. Sun, Y. Li, H. Gu, and X. Ni, "Reduced expression of 11β-hydroxysteroid dehydrogenase type 2 in preeclamptic placentas is associated with decreased PPARγ but increased PPARα expression," *Endocrinology*, vol. 155, no. 1, pp. 299–309, 2014.

[16] H. Zhu, C. Zou, X. Fan et al., "Upregulation of 11beta-hydroxysteroid dehydrogenase type 2 expression by Hedgehog ligand contributes to the conversion of cortisol into cortisone," *Endocrinology*, vol. 157, no. 9, pp. 3529–3539, 2016.

[17] X. T. Ni, T. Duan, Z. Yang, C. M. Guo, J. N. Li, and K. Sun, "Role of human chorionic gonadotropin in maintaining 11β-hydroxysteroid dehydrogenase Type 2 expression in human placental syncytiotrophoblasts," *Placenta*, vol. 30, no. 12, pp. 1023–1028, 2009.

[18] X.-J. Yang, "Lysine acetylation and the bromodomain: a new partnership for signaling," *BioEssays*, vol. 26, no. 10, pp. 1076–1087, 2004.

[19] Y. Murakami, "Histone deacetylases govern heterochromatin in every phase," *EMBO Journal*, vol. 32, no. 17, pp. 2301–2303, 2013.

[20] A. Brandl, T. Heinzel, and O. H. Krämer, "Histone deacetylases: salesmen and customers in the post-translational modification market," *Biology of the Cell*, vol. 101, no. 4, pp. 193–205, 2009.

[21] K. Hogg, W. P. Robinson, and A. G. Beristain, "Activation of endocrine-related gene expression in placental choriocarcinoma cell lines following DNA methylation knock-down," *Molecular Human Reproduction*, vol. 20, no. 7, pp. 677–689, 2014.

[22] T. D. Schmittgen and K. J. Livak, "Analyzing real-time PCR data by the comparative CT method," *Nature Protocols*, vol. 3, no. 6, pp. 1101–1108, 2008.

[23] C. Wu, C. Orozco, J. Boyer et al., "BioGPS: an extensible and customizable portal for querying and organizing gene annotation resources," *Genome Biology*, vol. 10, no. 11, article R130, 2009.

[24] R. Alikhani-Koopaei, F. Fouladkou, F. J. Frey, and B. M. Frey, "Epigentic regulation of 11β-hydroxysteroid dehydrogenase type 2 expression," *Journal of Clinical Investigation*, vol. 114, no. 8, pp. 1146–1157, 2004.

[25] J. Tremblay, D. B. Hardy, I. E. Pereira, and K. Yang, "Retinoic acid stimulates the expression of 11β-hydroxysteroid dehydrogenase type 2 in human choriocarcinoma JEG-3 cells," *Biology of Reproduction*, vol. 60, no. 3, pp. 541–545, 1999.

[26] M. Haberland, R. L. Montgomery, and E. N. Olson, "The many roles of histone deacetylases in development and physiology: implications for disease and therapy," *Nature Reviews Genetics*, vol. 10, no. 1, pp. 32–42, 2009.

[27] W. S. Xu, R. B. Parmigiani, and P. A. Marks, "Histone deacetylase inhibitors: molecular mechanisms of action," *Oncogene*, vol. 26, no. 37, pp. 5541–5552, 2007.

[28] T. U. Bracker, A. Sommer, I. Fichtner, H. Faus, B. Haendler, and H. Hess-Stumpp, "Efficacy of MS-275, a selective inhibitor of class I histone deacetylases, in human colon cancer models," *International Journal of Oncology*, vol. 35, no. 4, pp. 909–920, 2009.

[29] L. M. Collins, L. J. Adriaanse, S. D. Theratile, S. V. Hegarty, A. M. Sullivan, and G. W. O'Keeffe, "Class-IIa histone deacetylase inhibition promotes the growth of neural processes and protects them against neurotoxic insult," *Molecular Neurobiology*, vol. 51, no. 3, pp. 1432–1442, 2015.

[30] K. Orendi, V. Kivity, M. Sammar et al., "Placental and trophoblastic in vitro models to study preventive and therapeutic agents for preeclampsia," *Placenta*, vol. 32, no. 1, pp. S49–S54, 2011.

[31] Z. S. Krozowski, S. E. Rundle, C. Wallace et al., "Immunolocalization of renal mineralocorticoid receptors with an antiserum against a peptide deduced from the complementary deoxyribonucleic acid sequence," *Endocrinology*, vol. 125, no. 1, pp. 192–198, 1989.

[32] H. P. Chen, Y. T. Zhao, and T. C. Zhao, "Histone deacetylases and mechanisms of regulation of gene expression," *Critical Reviews in Oncogenesis*, vol. 20, no. 1-2, pp. 35–47, 2015.

[33] R. L. Montgomery, M. J. Potthoff, M. Haberland et al., "Maintenance of cardiac energy metabolism by histone deacetylase 3 in mice," *Journal of Clinical Investigation*, vol. 118, no. 11, pp. 3588–3597, 2008.

[34] R. L. Montgomery, C. A. Davis, M. J. Potthoff et al., "Histone deacetylases 1 and 2 redundantly regulate cardiac morphogenesis, growth, and contractility," *Genes and Development*, vol. 21, no. 14, pp. 1790–1802, 2007.

[35] Y. Zhang, S. Kwon, T. Yamaguchi et al., "Mice lacking histone deacetylase 6 have hyperacetylated tubulin but are viable and develop normally," *Molecular and Cellular Biology*, vol. 28, no. 5, pp. 1688–1701, 2008.

[36] E. Maltepe, G. W. Krampitz, K. M. Okazaki et al., "Hypoxia-inducible factor-dependent histone deacetylase activity determines stem cell fate in the placenta," *Development*, vol. 132, no. 15, pp. 3393–3403, 2005.

[37] J. C. Choi, R. Holtz, and S. P. Murphy, "Histone deacetylases inhibit IFN-γ-inducible gene expression in mouse trophoblast cells," *Journal of Immunology*, vol. 182, no. 10, pp. 6307–6315, 2009.

[38] M. Poljak, R. Lim, G. Barker, and M. Lappas, "Class i to III histone deacetylases differentially regulate inflammation-induced matrix metalloproteinase 9 expression in primary amnion cells," *Reproductive Sciences*, vol. 21, no. 6, pp. 804–813, 2014.

[39] M. E. Coussons-Read, M. L. Okun, and C. D. Nettles, "Psychosocial stress increases inflammatory markers and alters cytokine production across pregnancy," *Brain, Behavior, and Immunity*, vol. 21, no. 3, pp. 343–350, 2007.

[40] E. Baibazarova, C. Van De Beek, P. T. Cohen-Kettenis, J. Buitelaar, K. H. Shelton, and S. H. M. Van Goozen, "Influence of prenatal maternal stress, maternal plasma cortisol and cortisol in the amniotic fluid on birth outcomes and child temperament at 3 months," *Psychoneuroendocrinology*, vol. 38, no. 6, pp. 907–915, 2013.

[41] T. J. Kaitu'u-Lino, S. Pattison, L. Ye et al., "Targeted nanoparticle delivery of doxorubicin into placental tissues to treat ectopic pregnancies," *Endocrinology*, vol. 154, no. 2, pp. 911–919, 2013.

[42] P. J. Barnes, I. M. Adcock, and K. Ito, "Histone acetylation and deacetylation: importance in inflammatory lung diseases," *The European Respiratory Journal*, vol. 25, no. 3, pp. 552–563, 2005.

[43] D. I. Sokolov, K. N. Furaeva, O. I. Stepanova et al., "Changes in functional activity of JEG-3 trophoblast cell line in the presence of factors secreted by placenta," *Archives of Medical Research*, vol. 46, no. 4, pp. 245–256, 2015.

7

Exploring Seipin: From Biochemistry to Bioinformatics Predictions

Aquiles Sales Craveiro Sarmento ⓘ,[1] Lázaro Batista de Azevedo Medeiros ⓘ,[1] Lucymara Fassarella Agnez-Lima ⓘ,[1] Josivan Gomes Lima ⓘ,[2] and Julliane Tamara Araújo de Melo Campos ⓘ[1]

[1]Laboratório de Biologia Molecular e Genômica, Departamento de Biologia Celular e Genética, Centro de Biociências, Universidade Federal do Rio Grande do Norte, Natal, RN, Brazil
[2]Departamento de Medicina Clínica, Hospital Universitário Onofre Lopes, Universidade Federal do Rio Grande do Norte, Natal, RN, Brazil

Correspondence should be addressed to Julliane Tamara Araújo de Melo Campos; tamara_bio@yahoo.com.br

Academic Editor: Michael Peter Sarras

Seipin is a nonenzymatic protein encoded by the *BSCL2* gene. It is involved in lipodystrophy and seipinopathy diseases. Named in 2001, all seipin functions are still far from being understood. Therefore, we reviewed much of the research, trying to find a pattern that could explain commonly observed features of seipin expression disorders. Likewise, this review shows how this protein seems to have tissue-specific functions. In an integrative view, we conclude by proposing a theoretical model to explain how seipin might be involved in the triacylglycerol synthesis pathway.

1. Introduction

Lipodystrophies are rare diseases related to adipose tissue commitment [1]. For years, researchers pursued several candidate genes to associate and explain the biochemical mechanisms that underlie the clinical manifestations. The two pioneers of congenital lipodystrophy studies were Waldemar Berardinelli in 1954 [2], followed by Martin Seipin in 1963 [3]. Described as autosomal recessive diseases, different genes were tested as a candidate related to the physiopathology of lipodystrophies. In 2001, Magré et al. associated mutations in a specific locus of the 11q13 chromosome with type 2 Berardinelli-Seip congenital lipodystrophy (BSCL type 2) [4]. Therefore, they named the protein encoded from the *BSCL2* gene as "seipin" as a tribute to Martin Seip. Because the molecular function of seipin was unknown, several investigators started to study its role in the biology of adipogenesis. Seventeen years later, seipin is still far away from being fully understood.

Seipin is a protein located in the endoplasmic reticulum (ER) membrane [5]. The *BSCL2* gene is highly expressed in the testis and some regions of the human brain, such as the spinal cord, frontal lobe cortex, and regions related to the regulation of energy balance, such as the hypothalamus and brainstem [4, 6, 7]. In mice, its expression is high in the motor and somatosensory cortex, mesencephalic nucleus, cranial motor nuclei, thalamic and hypothalamic nuclei, reticular formation of brainstem, and vestibular complex [8]. The human protein atlas databank (https://www.proteinatlas.org/ENSG00000168000-BSCL2/tissue) [9, 10] also confirms that the *BSCL2* gene is highly transcribed in the human brain. Besides, seipin is upregulated during *in vitro* hormone-induced adipogenesis [4, 11] and high expression in fully differentiated adipose tissue isolated from mice was also observed [12].

Two of the most famous primary bioinformatics databanks, NCBI [13] and UniProt [14], reveal three *BSCL2* transcription variants that produce three seipin isoforms. Seipin isoform 1 has 398 amino acids, while isoforms 2 and 3 have 287 and 462, respectively (Figure 1). Isoform 3 has a larger N terminus sequence, while isoform 2 is the most different from the other two, mainly at the C terminus site. After amino acids TGLR, only isoforms 1 and 3 are similar, even if the alignment tool tried to group isoform 2. Isoform 1 seems to be the most important seipin isoform and is considered the canonical one for the UniProt Site. Its sequence provides the nomenclature for more than 30 seipin mutations [4, 15–31]. Isoforms 1 and 3 are switched between these two databanks and some confusion might occur. However, we chose the nomenclature based on UniProt.

2. Seipin as an Oligomeric Transmembrane Protein

In 2006, Lundin et al. predicted and experimentally confirmed that seipin is a transmembrane protein with two hydrophobic helices. They demonstrated its C and N terminus facing cytosol and concluded that seipin has a core/looping region inside the ER lumen. At the time, Lundin considered seipin with 462 amino acids and made some bioinformatics predictions with that isoform [33]. Many authors confirmed seipin regions localized through the ER membrane, although there are some differences in amino acids positions among the predictions (Table 1 and Figure 2) [4, 6, 33, 34].

In 2010, Binns et al. found that *Saccharomyces cerevisiae* seipin is a large protein complex. They suggested a stable homooligomer model of about nine subunits with a radially symmetric shape. For the authors, that structure resembled a toroid and appeared to be involved in the lipid droplet (LD) assembly organization [49]. Sim et al. confirmed seipin homooligomerization in human cell culture but found 12 subunits in a circular configuration [50]. The topology of seipin lacks evidence for any enzymatic domains or activity and some authors suggest that it may act as a scaffold for other proteins or play a structural role in membranes [51] (Figure 3).

Seipin, as an ER protein, affects the homeostasis of this organelle directly or indirectly in a tissue-specific way. The ER is a tubular organization specialized in the synthesis, mobility, and transport of proteins in eukaryotic cells. The tertiary structure of these macromolecules is essential for cell survival, and their accumulation out of the native conformations can compromise proteostasis [52]. As we will see in this review, seipin might be one of the proteins that can have a compromised folding and elicit an intracellular phenomenon called ER stress. Poorly folded proteins are able to bind chaperones and elicit the unfolded protein response (UPR). UPR is a marker of ER stress characterized by sequential reactions that may culminate in stress adaptation with protein ubiquitination and proteasome degradation. However, in unsolved stress situations, autophagy or apoptosis may arise [52–55].

3. Lipid Metabolism, Adipogenesis, and Lipid Droplets

Lipid anabolism consists of some reactions including the ones which synthesize fatty acids (FA) and triacylglycerol (TG). TG synthesis depends on FA availability and occurs in most cells, but mainly in adipocytes and hepatocytes (Figure 3).

Cells usually receive FA from lipoproteins of blood in a fed situation and can use that lipid for energy production or TG synthesis for storage into LDs [56, 57]. Lipid droplets are born from ER and are present in almost all eukaryotic cell types. LDs work in lipid metabolism and energy production as the intracellular "house" of some neutral lipids, such as TG and cholesterol esters [58]. Small amounts of lipids may exist in the aqueous ambient of ER but, as their number increases during TG synthesis, LDs may bud on a monolayer surface. When they sufficiently grow towards the cytosol, they might become independent organelles [59], as shown in Figure 3.

Essentially, all cells have the potential to store TGs into LDs. Nevertheless, around 90% of these lipids are inside LDs in white adipose tissue (WAT). One of the first steps of TG catabolism is the lipolysis: the reactions that turn TG into glycerol and FA and that mostly happen in WAT. Indeed, this is the unique tissue able to supply FA to other tissues. Glucagon is a positive regulator of WAT lipolysis, allowing release and transport of FA to muscles, liver, and other tissues. There, oxidative steps of lipid catabolism can occur to produce energy [60].

While lipolysis and TG synthesis should happen mostly in fully differentiated WAT, the former is usually "silenced" during adipogenesis [45, 61]. Adipocyte differentiation starts with mesenchymal stem cells (MSCs) that produce the preadipocytes in the first step. Next, the preadipocytes turn into fully differentiated adipocytes in the second step. During adipogenesis, the protein peroxisome proliferator-activated receptor gamma (PPARγ) is "the master regulator," acting as a transcription factor. PPARγ has its activity regulated in a tissue-specific manner, through coactivators and corepressors. Secondly, some transcription factors, such as CCAAT/enhancer-binding protein beta (C/EBPβ) and cyclic AMP-responsive element-binding protein 1 (CREB1), are also important [62, 63].

4. Seipin-Related Diseases: Lipodystrophy and Seipinopathy

All disorders involving seipin are characterized by some nervous tissue commitment. It is interesting to note how the same protein is involved in distinct clinical manifestations: Berardinelli-Seip congenital lipodystrophy type 2 (BSCL type 2) and seipinopathies. BSCL type 2, classified as one of the most dangerous of lipodystrophies, is a recessive disease caused by loss-of-function mutations, characterized by a severe adipose tissue disorder that might affect cognition-related nervous tissue regions [20, 64, 65]. "Seipinopathies," a term created to refer to specific motor neuropathies, are dominant diseases caused by gain-of-function mutations, mostly related to nervous tissue disorders: Silver syndrome (SS) and distal hereditary motor neuropathy (dHMN) [6].

Seipin study not only is crucial for rare conditions such as BSCL type 2 and seipinopathies but also might be important in obesity pathogenesis or treatment, because of the relationship of that protein with adipose tissue homeostasis. Next, we will discuss some seipin studies to understand the biochemistry of how that protein can affect lipid metabolism and be the central player of these diseases. We chose to divide

```
Iso1_398    ------------------------------------------------------------MVNDPPVP
Iso2_287    ------------------------------------------------------------MVNDPPVP
Iso3_462    MSTEKVDQKEEAGEKEVCGDQIKGPDKEEEPPAAASHGQGWRPGGRAARNARPEPGARHPALPAMVNDPPVP

cons                                                                    ********


Iso1_398    ALLWAQEVGQVLAGRARRLLLQFGVLFCTILLLLWVSVFLYGSFYYSYMPTVSHLSPVHFYYRTDCDSSTTS
Iso2_287    ALLWAQEVGQVLAGRARRLLLQFGVLFCTILLLLWVSVFLYGSFYYSYMPTVSHLSPVHFYYRTDCDSSTTS
Iso3_462    ALLWAQEVGQVLAGRARRLLLQFGVLFCTILLLLWVSVFLYGSFYYSYMPTVSHLSPVHFYYRTDCDSSTTS

cons        ***********************************************************************


Iso1_398    LCSFPVANVSLTKGGRDRVLMYGQPYRVTLELELPESPVNQDLGMFLVTISCYTRGGRIISTSSRSVMLHYR
Iso2_287    LCSFPVANVSLTKGGRDRVLMYGQPYRVTLELELPESPVNQDLGMFLVTISCYTRGGRIISTSSRSVMLHYR
Iso3_462    LCSFPVANVSLTKGGRDRVLMYGQPYRVTLELELPESPVNQDLGMFLVTISCYTRGGRIISTSSRSVMLHYR

cons        ***********************************************************************


Iso1_398    SDLLQMLDTLVFSSLLLFGFAEQKQLLEVELYADYRENSYVPTTGAIIEIHSKRIQLYGAYLRIHAHFTGLR
Iso2_287    SDLLQMLDTLVFSSLLLFGFAEQKQLLEVELYADYRENSYVPTTGAIIEIHSKRIQLYGAYLRIHAHFTGLR
Iso3_462    SDLLQMLDTLVFSSLLLFGFAEQKQLLEVELYADYRENSYVPTTGAIIEIHSKRIQLYGAYLRIHAHFTGLR

cons        ***********************************************************************


Iso1_398    YLLYNFPMTCAFIGVASNFTFLSVIVLFSYMQWVWGGIWPRHRFSLQVNIRKRDNSRKEVQRRISAHQPGPE
Iso2_287    LTSEKE----TIPGRKSNEGSLLI--------------------SQGLKA-RRSQLRNQMLQRMVRALKIPQ
Iso3_462    YLLYNFPMTCAFIGVASNFTFLSVIVLFSYMQWVWGGIWPRHRFSLQVNIRKRDNSRKEVQRRISAHQPGPE

cons            :       :: *   **   * :           *   ::  :*.: *::: :*:        *:


Iso1_398    GQEESTPQSDVTEDGESPEDPSGTEGQLSEEEKPDQQPLSGEEELEPEASDGSGSWEDAALLTEANLPAPAP
Iso2_287    GQRVSCPRRR------------------------------------------------------------
Iso3_462    GQEESTPQSDVTEDGESPEDPSGTEGQLSEEEKPDQQPLSGEEELEPEASDGSGSWEDAALLTEANLPAPAP

cons        **. * *:


Iso1_398    ASASAPVLETLGSSEPAGGALRQRPTCSSS
Iso2_287    ------------------------NQISSP
Iso3_462    ASASAPVLETLGSSEPAGGALRQRPTCSSS

cons                                **.
```

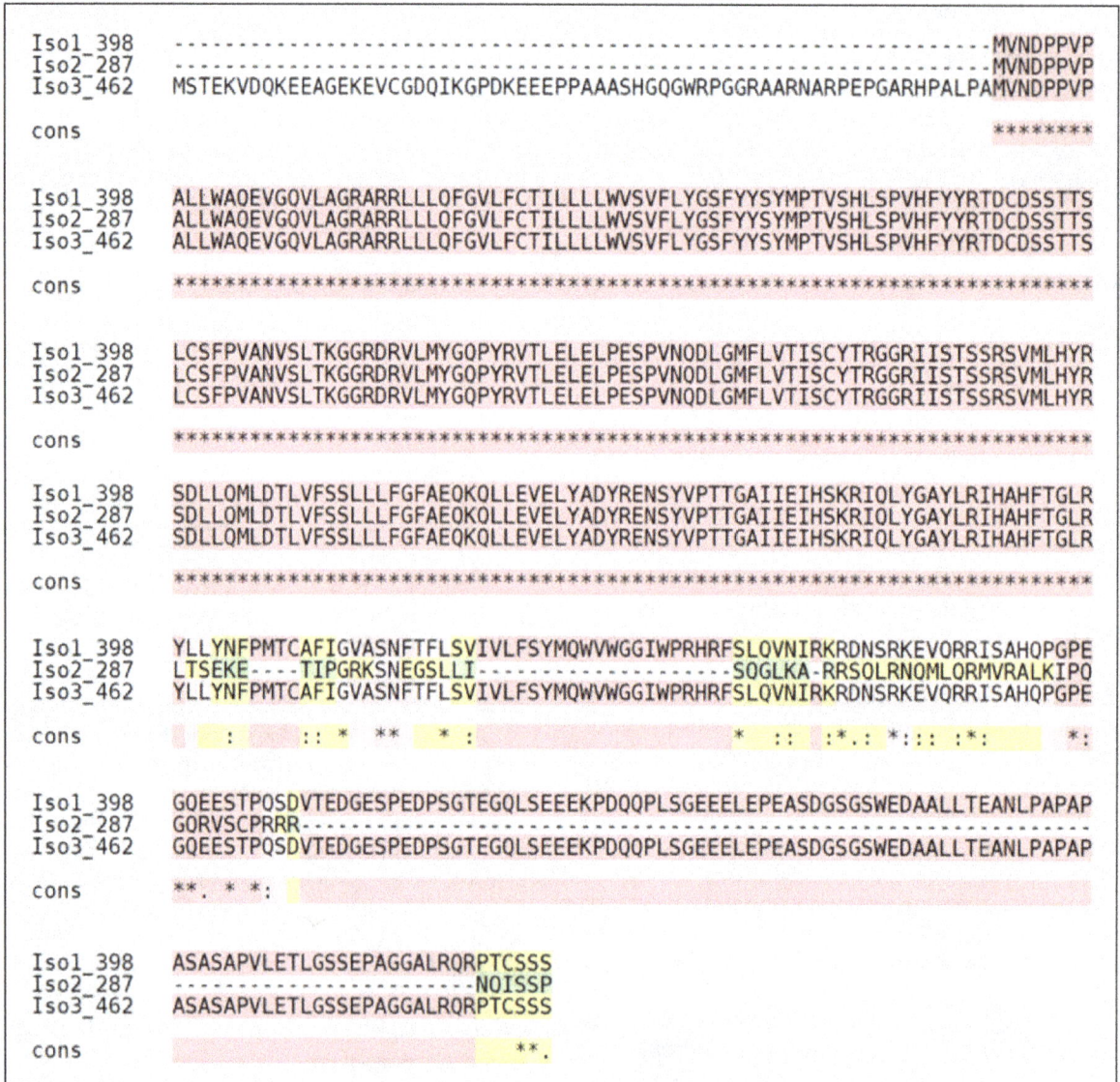

FIGURE 1: **Human seipin isoforms.** Multiple alignments of seipin isoforms were performed through T-Coffee [32]. Isoform 3 is the biggest with 462 amino acids, followed by 1 and 2 with 398 and 287 amino acids, respectively. UniProt [14] considers seipin isoform 1 as the canonical one. Pink color represents identical alignments; yellow corresponds to similar alignments; and green regions show different alignments. * corresponds to an equal match and the differences are highlighted by . and : symbols. Cons: consensus sequence; Iso: isoform.

TABLE 1: Transmembrane regions of seipin.

Isoform	Year	Transmembrane regions (amino acid position)		Reference
1 (398 aa)	2001	28-49	237-258	[4]
1 (398 aa)	2004	28-49	237-258	[34]
1 (398 aa)	UniProt (accessed in 2018)	27-47	243-263	[14]
3 (462 aa)	2006	95-117	294-316	[33]
3 (462 aa)	2008	95-117	273-336	[6]

Source: reviewed papers. AA: amino acids.

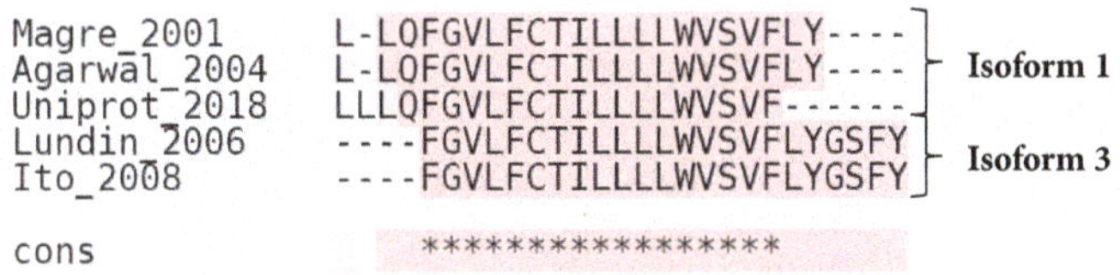

```
Magre_2001     L-LQFGVLFCTILLLLWVSVFLY----  ⎤
Agarwal_2004   L-LQFGVLFCTILLLLWVSVFLY----  ⎬ Isoform 1
Uniprot_2018   LLLQFGVLFCTILLLLWVSVF------  ⎦
Lundin_2006    ----FGVLFCTILLLLWVSVFLYGSFY  ⎤
Ito_2008       ----FGVLFCTILLLLWVSVFLYGSFY  ⎦ Isoform 3

cons                   *****************
```

(a) First seipin transmembrane region

```
Magre_2001     ----------------------------IGVASNFTFLSVIVLFSYMQWV--------------  ⎤
Agarwal_2004   ----------------------------IGVASNFTFLSVIVLFSYMQWV--------------  ⎬ Isoform 1
Uniprot_2018   --------------------------------FTFLSVIVLFSYMQWVWGGIW----------  ⎦
Lundin_2006    ------------------FPMTCAFIGVASNFTFLSVIVLF---------------------  ⎤
Ito_2008       LYGAYLRIHAHFTGLRYLLYNFPMTCAFIGVASNFTFLSVIVLFSYMQWVWGGIWPRHRFSLQV  ⎦ Isoform 3

cons                                       **********
```

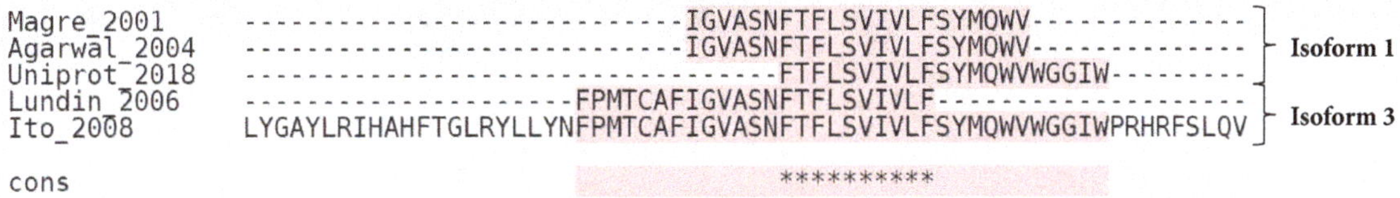

(b) Second seipin transmembrane region

FIGURE 2: **Alignment of transmembrane regions of seipin.** Multiple alignments of seipin isoforms were performed through T-Coffee [32]. Many authors predicted the transmembrane regions of seipin and the amino acid positions are reviewed in Table 1 [4, 6, 14, 33, 34]. It is possible to observe that, even with differences, some regions are conserved in the prediction for the same isoform or between different isoforms. Isoform 2 was omitted due to the low number of works with it. Pink color represents identical alignments; yellow corresponds to similar alignments; and green regions show different alignments. * corresponds to an equal match. Cons: consensus sequence.

the sections into adipogenic and nonadipogenic cells inspired by the paper of Yang et al., who proposed and proved different seipin functions for these two models [61].

5. Seipin Loss-of-Function

5.1. Seipin Loss-of-Function through Silencing or Knockout in Adipogenic Models. Situations that impair *BSCL2* gene expression from a quantitative point of view are associated with problems in adipocyte maturation. This is strictly related to TG synthesis reduction and lipolysis stimulation (Figure 4(a)). 3T3-L1 mouse preadipocyte stem cells stimulated to differentiation in a *BSCL2* knockdown condition were associated with downregulation of genes responsible for TG accumulation. Additionally, the same group performed TG quantification by a direct measurement technique and showed lower TG content when compared with controls [11]. Other independent researches confirmed the association of these cells with reduced total lipid accumulation and content [61]. In other models, such as *BSCL2* knockdown murine embryonic fibroblasts (MEFs) stimulated to differentiation, decreased TG accumulation, and adipocyte maturation impairment, were found. Interestingly, that same group observed an increase in protein kinase A- (PKA-) activated lipolysis. Likewise, adipogenesis was restored with lipolysis inhibition [45]. TG synthesis impairment was also associated with downregulation of two important enzymes of this pathway: 1-acyl-sn-glycerol-3-phosphate acyltransferase 2 (*AGPAT2*) and the phosphatidate phosphatase *LPIN1* (lipin1) in *BSCL2* knockdown during adipogenesis [11, 45].

Many studies have tried to explain why adipogenesis impairment is related to seipin loss-of-function. Researchers observed that induction of *BSCL2* transcription is an important event for the adipocyte differentiation and expression of some important adipogenic factors [12]. Chen et al. determined that *BSCL2* expression is only essential

during the second phase of adipogenesis: differentiation of preadipocytes to fully differentiated adipocytes. *BSCL2* mRNA interference during that phase caused lower *PPARG* gene expression and adipogenesis impairment. However, treatment with a PPARγ agonist rescued the process, proving that seipin is important for adipogenesis cascade at an upstream point compared with PPARγ [11]. Another independent study also observed *PPARG* downregulation during mouse adipogenesis in a *BSCL2* knockdown situation [61]. The ER stress is dangerous to adipogenesis because of the suppression of PPARγ expression [66]. Nonetheless, the research failed to associate *BSCL2* gene knockdown, adipogenesis, and ER stress [45].

5.2. Seipin Loss-of-Function through Mutations in Adipogenic Models. Situations that impair *BSCL2* gene expression from a qualitative point of view are also related to problems in adipocyte maturation. In the same way, this is strictly related to a decrease in TG synthesis (Figure 4(b)). Researchers observed a reduction in LD formation, associated with impaired lipid accumulation capacity during adipogenic differentiation of fibroblasts in patients with mutations E189X and R275X in seipin [67]. Moreover, other studies showed a reduction of almost 50% in TG content during differentiation of 3T3-L1 mouse preadipocyte stem cells carrying the A212P mutation. That group also observed downregulation of lipogenic genes and PPARγ. As expected, adipogenesis was partially restored by agonists of that protein [68].

The most discussed clinical disease associated with seipin loss-of-function disorders is type 2 Berardinelli-Seip congenital lipodystrophy (BSCL type 2). The main characteristics of this lipodystrophy are the almost complete absence of adipose tissue together with mild to severe intellectual impairment. This condition is one of the most dangerous among human lipodystrophies and is associated with some cases of hypertrophic cardiomyopathy and secondary mitochondrial

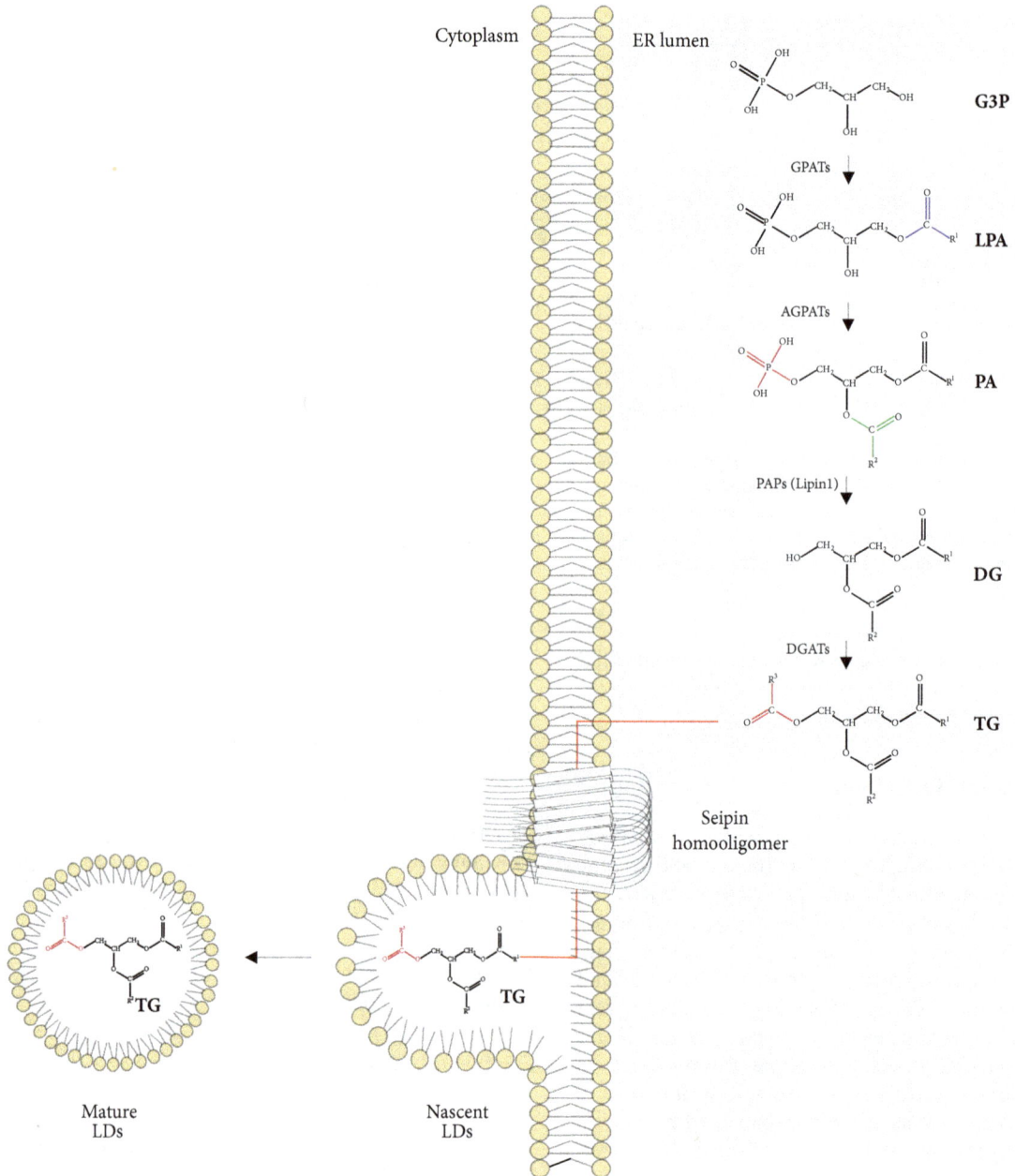

FIGURE 3: **Triacylglycerol synthesis and usual seipin localization.** During triacylglycerol synthesis, glycerol-3-phosphate acyltransferases (GPATs) catalyze the acylation at sn-1 position of glycerol-3-phosphate (G3P) and origin lysophosphatidic acid (LPA). Then, 1-acyl-sn-glycerol-3-phosphate acyltransferases (AGPATs) catalyze the acylation at sn-2 of LPA and give rise to phosphatidic acid (PA). Later, phosphatidate phosphatases (PAPs), as lipin1, can remove the phosphate group from PA and produce diacylglycerol (DG). Finally, diacylglycerol o-acyltransferases (DGATs) catalyze the acylation at the sn-3 position and give rise to triacylglycerol (TG) [35, 36]. In the same context, seipin comes as an oligomeric endoplasmic reticulum (ER) transmembrane protein that acts in lipid droplet (LD) assembly. ER and LDs were found to be neighbors, and seipin is concentrated in the communication regions between them, enabling the transfer of lipids recently synthetized to nascent LDs [37–39]. Pieces of the illustrations are from the SMART website [40].

dysfunction [20, 25, 69]. The loss of body fat is also one of the most significant features, affecting both mechanically or metabolically active adipose tissue [70].

Regarding ER stress, 3T3-L1 mouse preadipocyte stem cells presenting the A212P mutation in the seipin gene and stimulated to adipocyte differentiation showed an increase in UPR [68]. Interestingly, this mutation was shown to change

seipin localization in more than one study [12, 50], indicating that seipin mutations might activate UPR even if the protein is not in ER. Regarding adipocytes, independent authors observed that ER stress attenuates adipogenesis through repression of PPARγ [66]. However, other research showed that seipin mutations that increase ER stress without compromising seipin function (N88S) are not enough to impair

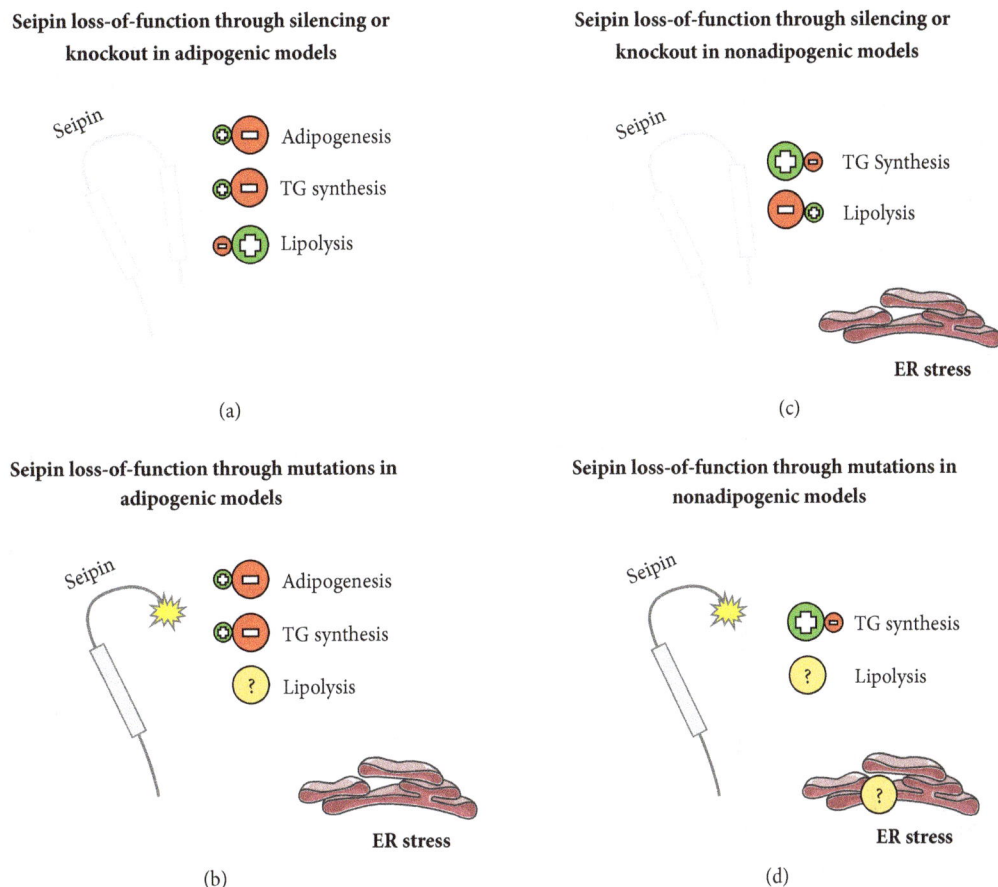

FIGURE 4: **Seipin loss-of-function.** We are proposing 4 different general models that usually happen with frequency under seipin loss-of-function. (a) Seipin loss-of-function through silencing or knockout in adipogenic models. Adipocyte maturation and TG synthesis are impaired, and ER stress was not found. (b) Seipin loss-of-function through mutations in adipogenic models. Adipocyte maturation and TG synthesis are impaired, ER stress was positively found, and there is a lack of information about the lipolysis situation. (c) Seipin loss-of-function through silencing or knockout in nonadipogenic models. TG synthesis was increased, lipolysis is impaired, and ER stress was positively found; (d) seipin loss-of-function through mutations in nonadipogenic models. TG synthesis was increased, and there is a lack of information about ER stress and lipolysis. (b) and (d) are the most representative situations of Berardinelli-Seip congenital lipodystrophy (BSCL) type 2. Positive symbols represent a process that is usually increased, while the negative symbols represent the opposite. The interrogation symbol represents a process that needs to be studied further. Pieces of the illustrations are from the SMART website [40].

adipogenesis [68]. In such a way, seipin loss-of-function seems to be more important to adipogenesis commitment than possible ER stress.

5.3. Seipin Loss-of-Function through Silencing or Knockout in Nonadipogenic Models. Situations that impair *BSCL2* gene expression from a quantitative point of view were also studied in other cells without adipogenic stimulation. This is strictly related to stimulation of TG synthesis and a decrease in lipolysis (Figure 4(c)). Yeast model with deletion of Fld1, a human seipin homologue, shows an increase in TG synthesis and formation of supersized (giant) LDs [37, 38]. Human HeLa cells with *BSCL2* mRNA silenced also showed similar results, and overexpression of wild-type seipin reversed this phenotype. Additionally, the same group showed that *BSCL2* mRNA silencing in 3T3-L1 mouse preadipocyte stem cells positively regulates TG synthesis. However, they also observed small and clustered LDs, different from the yeast

model [71]. In the same way, a *Drosophila* salivary gland model with *BSCL2* gene deletion presented ectopic LD formation and increased TG synthesis [72].

Mouse *BSCL2* gene deletion in hepatocytes showed LDs increased in number and size, as well as expression of genes implicated in their formation and stability. Stearoyl-CoA desaturase-1 (SCD1) acts in FA synthesis and seems to be negatively regulated in the presence of wild-type *BSCL2* in hepatocytes. Hence, SCD1 knockdown reversed the phenotype associated with seipin deficiency. For the authors, *BSCL2* knockdown induces SCD1 expression and activity, leading to TG synthesis stimulation, lipolysis reduction, and LD expansion, at least in hepatocytes [73].

Liu et al. studied the effect of *BSCL2* gene deletion in fully differentiated mouse adipose tissue and found a progressive lipodystrophy. As expected, signals of TG accumulation were observed together with lipolysis impairment [74]. Similarly, Chen et al. observed increased lipid uptake gene expression, stimulation of TG synthesis, and FA synthesis in mouse residual epidermal WAT with *BSCL2* gene deletion [75], which means that fully differentiated WAT seems to manifest characteristics of nonadipogenic cells during seipin expression disorders. The inclusion of fully differentiated adipocytes together with nonadipogenic cells might be unusual, but in terms of lipolysis and TG synthesis, these cells share similarities with nonadipogenic models. Indeed, both are not going through adipogenesis.

Specifically in mice adipocyte, *BSCL2* deletion was also associated with p38-mitogen-activated protein kinase- (MAPK-) dependent apoptosis increase. The authors observed that fibroblast growth factor 21 (FGF21) improved the consequences of seipin loss-of-function, through inhibition of p38-MAPK activity. This led to increased adiponectin plasma levels and metabolic homeostasis improvement [76]. Interestingly, mouse brown adipose tissue (BAT) functions do not seem to be affected much by *BSCL2* gene deletion when compared with WAT [77]. Indeed, deletion of seipin during mouse BAT adipogenesis is not sufficient to impair the whole process [78]. In spite of this, mouse fully differentiated BAT with *BSCL2* gene deletion displayed altered thermogenic capacity and some insulin resistance [79].

Seipin function is also important for assembly and homeostasis of LDs. In the human epidermoid carcinoma human cell line model (A431), *BSCL2* knockout triggered an aberrant LD morphology. The authors observed that seipin is enriched at the point of contact between ER and LDs, as shown in Figure 3, and the mutant seipin is not in that location anymore. For them, ER-LD contacts are morphologically abnormal in cells without seipin and this protein is also important for the stabilization of the contact between these two organelles, facilitating LD growth [80]. In the same year, another group showed similar results, proposing that seipin is required at the moment of transition between nascent to mature LDs, enabling lipid transfer to the nascent LDs in ER-LD contact sites [39].

Seipin is also highly expressed in testis. Testicular tissue from mice with *BSCL2* gene deletion also showed signals of TG synthesis stimulation such as a phosphatidic acid (PA) increase compared with controls. The authors discussed that lipid metabolism might be essential for testicular homeostasis and showed that mutations in seipin are also related to altered spermatozoid morphology, a phenomenon called teratozoospermia [81]. Indeed, further works with mouse *BSCL2* gene deletion observed male infertility with the absence of spermatozoids (azoospermia) and concluded that seipin presence is important for the late phase of spermatogenesis [77].

In neuron-specific seipin-knockout mice, Zhou et al. also observed downregulation of *PPARG*. It means that the PPARγ-seipin relation is not only restricted to adipogenesis or adipose tissue [82]. Further work showed that *BSCL2* knockout mice could impair proliferation and differentiation of hippocampal cells through reduced PPARγ levels. Indeed, this was a phenomenon reversed by PPARγ agonists [83]. Similar results were recently obtained by a group who showed that seipin deficiency is related to lower expression of α-amino-3-hydroxy-5-methyl-4-isoxazolepropionic acid (AMPA) receptor through reduced ERK-CREB activities. Together, this phenomenon was reversed by PPARγ activation [84]. Other studies also showed a reduced PPARγ level in mouse neuronal *BSCL2* knockout. These mice had impaired adaptation to amyloid neurotoxins when compared with wild-type seipin mice. Together, the authors observed that neuroinflammation could be intensified in absence of seipin, through PPARγ reduction [85]. Recently, Wang et al. showed that seipin deficiency in dopaminergic neurons enhances phosphorylation of α-synuclein and induces inflammation in an ER stress-independent manner. This phenomenon can trigger loss of dopaminergic neurons and is related to a decrease and increase in PPARγ and glycogen synthase kinase-3 beta (GSK3β) activities, respectively [86].

Zhou et al. and Wang et al. did not find evidence of ER stress in neuron-specific seipin-knockout mice, respectively [82, 86]. In contrast, Liu et al. observed ER stress markers positively induced in adipose-specific seipin-knockout mice [74]. Zowalaty et al. also showed ER stress and increased apoptosis in mammary gland alveolar epithelial cells with *BSCL2* gene deletion and revealed the importance of seipin for lactation [87]. *BSCL2* knockout was also associated with intensified cerebral ER stress after a transient middle cerebral artery occlusion situation in another study [88]. Moreover, *BSCL2* gene deletion in hepatocytes in mice also showed an increase of ER stress markers [73]. Interestingly, in a yeast model, ER stress was shown to stimulate LD formation and TG synthesis [89]. Therefore, this is a phenomenon that deserves more study in situations of seipin loss-of-function, as not all of the mutant seipin are able to accumulate in the ER and activate UPR [6].

5.4. Seipin Loss-of-Function through Mutations in Nonadipogenic Models. Situations that impair *BSCL2* gene expression from a qualitative perspective were also studied in other cells without adipogenic conditions. This is strictly related to TG synthesis stimulation (Figure 4(d)). Szymanski et al. studied a human fibroblast line obtained from a patient with a *BSCL2* nonsense mutation. Compared with the control, these cells presented many smaller and aberrant LDs [38]. In another study using *in vivo* seipin E253X mutation, the researchers also observed teratozoospermia and enlarged ectopic LDs in testis [81].

A study with some seipin mutations, such as A212P + S64R, Y106C, Y225L, and V99L, in human Epstein Barr virus-transformed lymphocytes showed more numerous and smaller LDs when compared with controls. Surprisingly, this might not be associated with increased TG synthesis, as the authors found TG content reduction in these cells when compared to the controls. However, the authors quantified TG after an overnight deprivation of serum in culture medium [90]. We hypothesize that this is a different situation compared with the *in vivo* constant hypertriglyceridemia, commonly found in BSCL type 2 patients [1]. Yet, these mutations do not significantly reduce seipin function as an interference mRNA (Figure 4(c)). A212P mutation, for example, does not seem to affect seipin quantitative expression but alters its localization [12, 50]. That substitution also does not affect the binding of seipin and lipin1 [50], a phenomenon that will be discussed further. Besides, that mutation could not inhibit LD formation in 3T3-L1 mouse preadipocyte stem cells without stimulation to differentiation [61].

Researchers found that hepatic steatosis is a common characteristic of people who inherited seipin loss-of-function mutations [1]. As we reviewed, previous works showed an increase in liver TG content in mice with *BSCL2* gene deletion when compared with controls [45, 91–93]. Nonetheless, lipodystrophy secondary effects have some importance for these results. Some authors observed that the specific seipin deficiency in mouse adipose tissue is mainly responsible for dyslipidemia, lipodystrophy, hepatic steatosis, and insulin resistance [94]. In such a way, metabolic dysfunction should not be expected in nonadipocyte-specific seipin deficiency. This was exactly the same result observed by Chen et al. in a mouse liver-specific deletion of the *BSCL2* gene, as these animals did not develop hepatic steatosis, proving that *BSCL2* is not autonomous to liver lipid homeostasis [93]. Similar results were also achieved by Wang et al. [95]. Researchers also observed that specific seipin deficiency in developing adipose tissue from mice is mainly responsible for lipodystrophy but not for severe hepatic steatosis, glucose intolerance, and insulin resistance [96]. Hence, we believe that BSCL type 2 metabolic features in nonadipogenic cells are caused by either secondary lipodystrophy-associated dysfunctions or some tissue-specific seipin loss-of-function. It is not correct to think that a phenotype, such as hepatic steatosis, is exclusively caused by seipin loss-of-function, without considering that patients' hepatocytes are constantly exposed to a high triacylglycerol blood content, for example.

6. Seipin Gain-of-Function

6.1. Seipin Gain-of-Function through Overexpression in Adipogenic Models. To our knowledge, seipin wild-type quantitative overexpression during adipogenesis was studied only in a few works (Figure 5(a)), and Yang et al. observed that the preadipocyte reaches the differentiation in these conditions [61].

6.2. Seipin Gain-of-Function through Mutations in Adipogenic Models. Mutations that modify *BSCL2* gene expression from a qualitative perspective and give seipin a gain-of-function are not associated with problems in adipocyte maturation (Figure 5(b)). As expected, 3T3-L1 mouse preadipocyte stem cells stimulated to differentiate in the presence of N88S and S90L seipin gain-of-function mutations did not show defective adipogenesis. Besides, the authors observed that the presence of ER stress secondary to these mutations is not enough to cause adipogenesis impairment [68].

6.3. Seipin Gain-of-Function through Overexpression in Nonadipogenic Models. Situations that modify *BSCL2* gene expression from a qualitative perspective and give seipin

a gain-of-function were studied in other cells without adipogenesis stimulation. This was strictly related to a decrease in TG synthesis (Figure 5(c)). Overexpression of wild-type seipin reduced TG and LD biosynthesis in both human HeLa and NIH3T3 cells [71]. Overexpression of seipin lacking its C terminus sequence or wild-type seipin dramatically reduced LD formation in hepatocytes AML-12 and 3T3-L1 mouse preadipocyte stem cells without differentiation stimulus [61].

The fully differentiated adipocytes have similar behaviors compared to other cells that are not in adipogenesis. Cui et al. overexpressed the human seipin exclusively in the mature adipocytes of mice. They observed a decrease in adipose tissue mass and lipolysis increase. Together, they concluded that seipin inhibits the lipid storage in mature adipocytes [97]. Here, we would like to invite you to compare Figure 4(c) with Figure 5(c). Even in opposite models, they point to a common finding that was concluded by Cui et al. [97]: in nonadipogenic cells (including mature adipocytes), seipin acts to limit lipid storage through TG synthesis impairment and to favor the lipolysis.

6.4. Seipin Gain-of-Function through Mutations in Nonadipogenic Models. Situations that modify *BSCL2* gene expression from a qualitative perspective and give seipin a gain-of-function are not associated with problems in adipocyte maturation and lipodystrophy. However, they give rise to the following seipinopathies (Figure 5(d)). SS and dHMN are two motor neuropathies that affect distal limb muscles. Individuals with SS have atrophy of the hands as the most marked manifestation and have mild to severe spasticity of the lower limbs, indicating involvement of upper motor neurons as well. Dominant mutations that compromise the N-glycosylation motif of seipin, such as N88S and S90L, are one of the genetic causes of SS and dHMN. These substitutions can form protein aggregates, a mechanism shared with some neurodegenerative disorders [5, 98, 99]. In a *BSCL2* knockout *Drosophila* model, the S90L mutation rescued the fat body, proving that it is gain-of-function. [72].

Ito et al. studied mouse brain neuroblastoma (N2a) cells expressing N88S and S90L seipin mutations. In these cases, they observed a strong interaction between seipin and the chaperone calnexin (*CANX*), responsible for the correct folding of new ER-synthetized proteins, compared to controls. They suggested that seipin mutations were able to strongly activate the UPR and suggested that the unfolded seipin would physically attract these kinds of chaperones. In addition, ER stress seemed to be so intense that it triggered apoptosis. For them, neurodegeneration observed in motor neuropathies could be explained by the strong ER stress consequences [100]. Moreover, an *in vivo* model with transgenic mice expressing N88S seipin also confirmed ER stress as a significant phenomenon associated with seipinopathy [101]. That mutant seipin also localized in the ER, suggesting evidence of the presence of an unfolded mutated protein as a cause of ER stress [6, 100]. The mechanism illustrating ER stress in seipinopathy is shown in Figure 6.

Seipin gain-of-function through overexpression in
adipogenic models

(a)

Seipin gain-of-function through mutations in
adipogenic models

(b)

Seipin gain-of-function through overexpression in
nonadipogenic models

(c)

Seipin gain-of-function through mutations in
nonadipogenic models

(d)

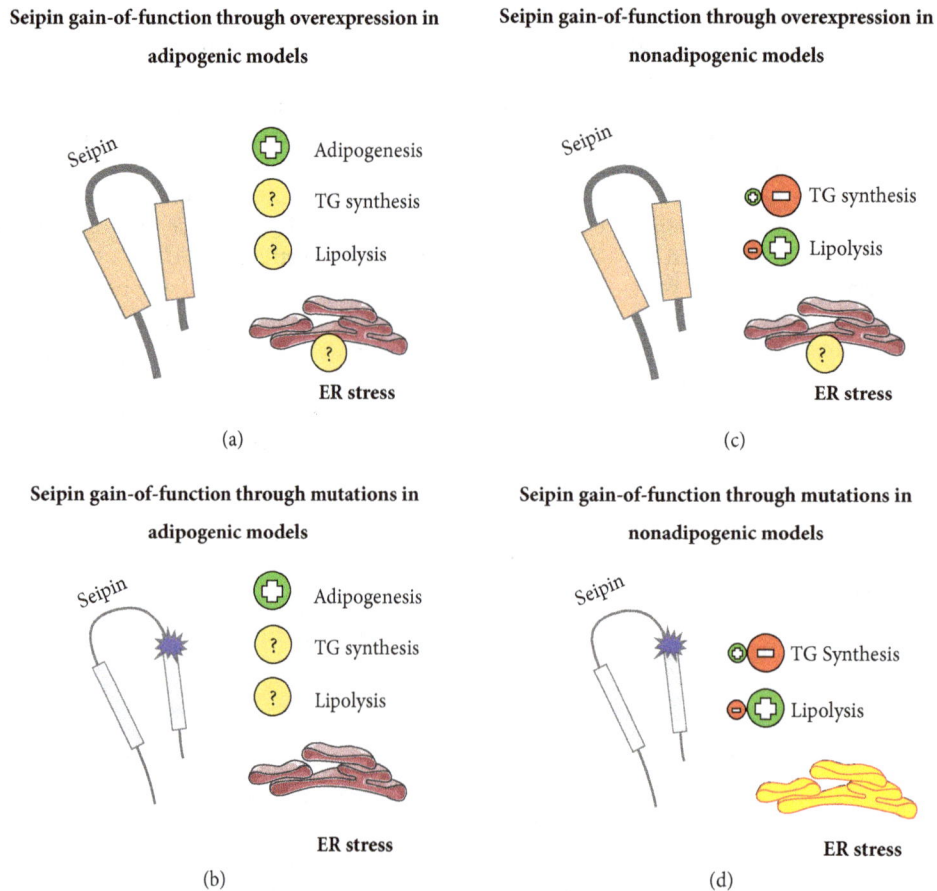

FIGURE 5: **Seipin gain-of-function.** We are proposing 4 different general models that usually happen with frequency under seipin gain-of-function. (a) Seipin gain-of-function through overexpression in adipogenic models. Adipocyte maturation was not impaired, but other parameters are still not clear. (b) Seipin gain-of-function through mutations in adipogenic models. Adipocyte maturation was not impaired even with ER stress. (c) Seipin gain-of-function through overexpression in nonadipogenic models. There was TG synthesis decrease, with lipolysis increase. (d) Seipin gain-of-function through mutations in nonadipogenic models. TG synthesis impairment was observed together with lipolysis increase and strong and significant ER stress. (b) and (d) models are related to seipinopathy. Positive symbols represent a process that is usually increased, while the negative symbols represent the opposite. The interrogation symbol represents a process that needs to be studied further. Pieces of the illustrations are from the SMART website [40].

Autophagy can be one of the many ER stress consequences. Indeed, mutant seipin in seipinopathy is also associated with the activation of that pathway in a mouse model [102], human embryonic kidney 293 cells (HEK293), and human non-small cell lung carcinoma cell line (H1299) [48]. Likewise, the expression of mutated seipin caused small diffuse LDs to fuse into larger ones and autophagy inhibition could reverse it. Not surprisingly, when the ER stress stimulator tunicamycin was added to the normal cells, it mimicked mutated seipin behavior. For them, autophagy is activated as an adaptive response to engulf and break down the LDs [48]. Other researchers manipulated astrocytoma and motor neurons to express N88S seipin and observed positive markers for ER stress but also a TG content diminution. As expected, supplementation with oleic acid and lipolysis inhibition were able to reverse these phenotypes. The authors proposed that lipolysis inhibition and new LD formation are associated with good prognosis of seipinopathy [103].

7. Merging of Genetic Concepts

In the previous sections, we divided the text to better understand the specific cell situations in seipin dysfunction. However, lipodystrophy can be found in animal models with *BSCL2* knockout and in human subjects with missense or nonsense mutations. More than 30 recessive *BSCL2* loss-of-function mutations are the cause of this lipodystrophy [4, 15–31] and it is still not clear if there is significant clinical difference among these kinds of mutations. However, we do know that it is a recessive condition and that only subjects who carry mutations in both alleles manifest BSCL type 2 (Figure 7(b)), even if the mutations are in different regions of the same gene (Figure 7(c)). The lipodystrophy clinical features are out of the scope of our review and can be found elsewhere [1, 20, 104]. It is also possible to find subjects with gain-of-function mutations in seipinopathies, a dominant condition that needs at least one mutated allele to

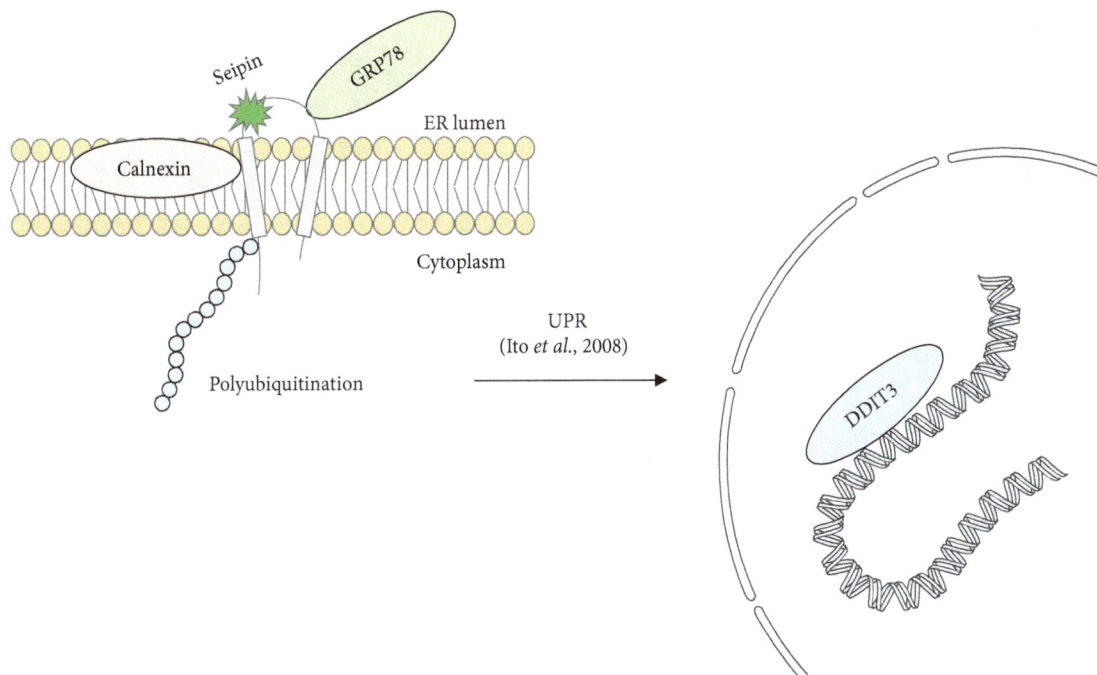

FIGURE 6: **Gain-of-function mutations in seipin elicit ER stress in seipinopathies.** Ito et al. proposed a model in which N88S and S90L mutations are able to disturb the seipin glycosylation site and generate ER stress and the unfolded protein response (UPR) [6]. They observed increased apoptosis as a result of the process. DDIT3 (also called CHOP) is a transcription factor responsible for the positive regulation of proapoptotic genes in response to ER stress. Calnexin (*CANX*) and GRP78 (also called BIP) are chaperones that work in UPR. As we reviewed, ER stress is not a phenomenon exclusively related to seipinopathy and might also be important for lipodystrophies. Pieces of the illustrations are from the SMART website [40].

manifest the phenotypical condition (Figures 7(d) and 7(e)). To our knowledge, there is no case of a subject carrying both mutations of lipodystrophy and seipinopathy (N88S or S80L) in the same or in different alleles of the *BSCL2* gene. The clinical features of seipinopathies are also out of the scope of our review and can be found elsewhere [5, 105–107].

Interestingly, different from seipinopathy or lipodystrophy, a rare fatal seipin neurodegenerative syndrome was also described. Referred to as Celia syndrome or progressive encephalopathy with or without lipodystrophy (PELD), this disease is caused by the R329X mutation in seipin. This is related to exon 7 gene skipping and was initially supposed to be a loss-of-function mutation. Indeed, heterozygous patients do not develop any symptoms (Figure 7(f)). However, some homozygous patients also do not develop lipodystrophy, suggesting that adipogenic functions might still be preserved (Figure 7(g)). These patients develop a fatal neurodegeneration, associated with child death by progressive encephalopathy [108–110]. On the other hand, there are some cases of compound heterozygous, when somebody inherits a lipodystrophic-related mutation and R329X mutation (Figure 7(h)). In this case, the patient has both phenotypes of progressive encephalopathy and lipodystrophy. Recently, it was shown that metreleptin intervention, used for lipodystrophy treatment, could help a compound heterozygous patient with PELD [111].

Instead of losing its function, "Celia seipin" was described as a gain-of-toxic function related to extreme ER stress. Even

as a gain-of-function mutation [108–110], some heterozygous patients are completely asymptomatic (Figure 7(f)), and authors have discussed that wild-type seipin is capable of inhibiting toxic functions of "Celia seipin" [109]. Thus, this might be the rare case of gain-of-toxic-function mutation associated with recessive alleles.

8. Bioinformatics Predictions

Aiming to integrate the knowledge about seipin, we performed some bioinformatics analysis. During the reviewing process, we collected a list of different expressed genes in loss of seipin function (Supplementary Tables 1 and 2). Therefore, we built an interactive network with all of them in Figure 8 using the STRING database [41]. The parameters chosen were "Experiments," "Databases," "Coexpression," "Neighborhood," "Gene Fusion," "Cooccurrence," and "Minimum required interaction score = 0.1". It was possible to notice that seipin (referred to as *BSCL2*) has almost no relation to its own network, except for likely interacting with calnexin (referred to as *CANX*), an ER chaperone that was previously shown to bind to seipin during UPR [100]. This could mean that seipin affects gene transcription in an indirect way and UPR may have an important role in seipin loss-of-function expression genes. Cytoscape [42] was responsible for the organization of the network and as the circle comes to the center, the node tends to be more connected in the group (known as

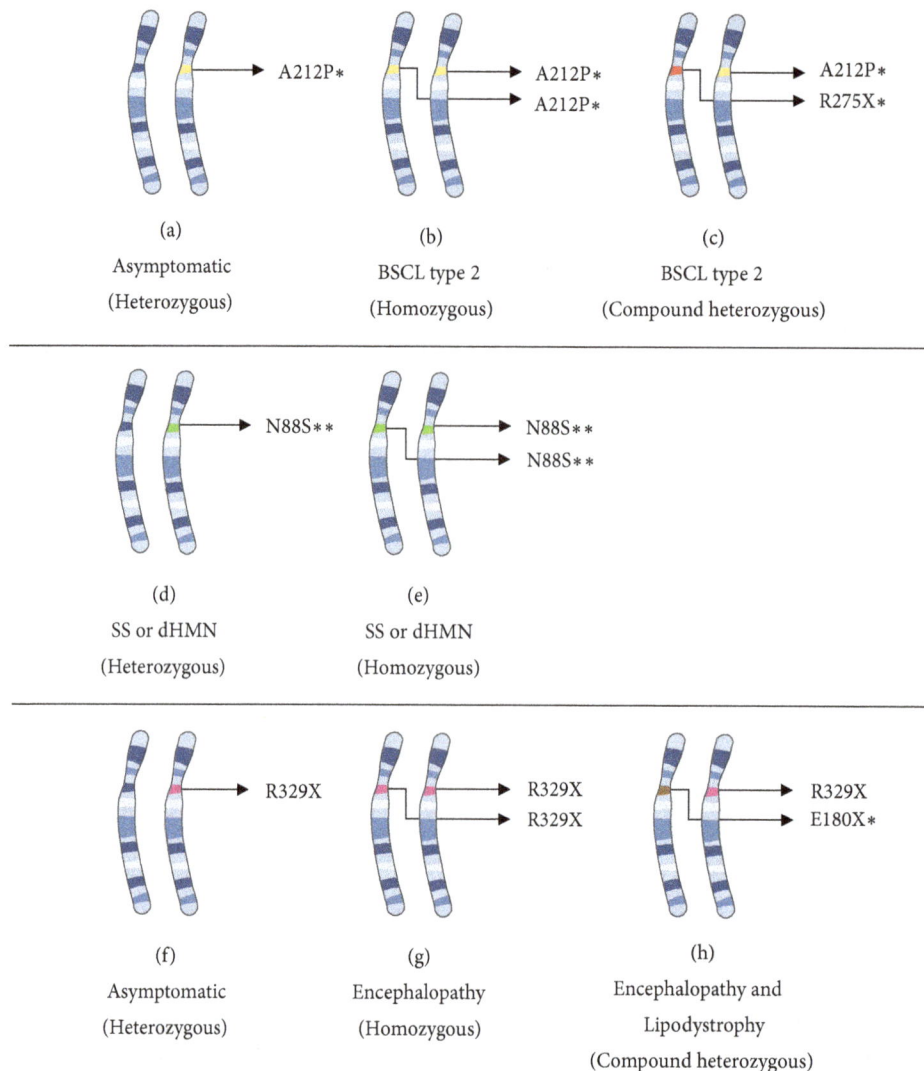

FIGURE 7: **Inheritance pattern of seipin-related diseases.** Here, we summarize the inheritance mechanism of seipin-related diseases. (a–c) BSCL type 2, a loss-of-function, and recessive disease. (d-e) Seipinopathies, gain-of-function, and dominant diseases. (f–h) Progressive encephalopathy with or without lipodystrophy (PELD) and a gain-of-function and recessive disease. Pieces of the illustrations are from the SMART website [40]. * Or other mutations related to BSCL type 2. ** Or other mutations related to seipinopathy. BSCL: Berardinelli-Seip congenital lipodystrophy; dHMN: distal hereditary motor neuropathies; SS: Silver syndrome.

the 'hub'). We can observe the *PPARG* gene in the center of visualization, allowing us to confirm its important role in seipin loss-of-function. As we also performed iRegulon [44], it was possible to notice that *PPARG*, *CEBPB*, and *PPARA* are transcription factors that are important to the expression of some genes. They have high enrichment scores (called "NES"): 7.278, 6.333, and 5.270, respectively.

Figure 9 summarizes graphically the results from the grouping of the reviewed genes in Supplementary Tables 1 and 2. A Venn diagram was performed through InteractVenn [43] to allow us to observe gene expression in cells with seipin loss-of-function. It is possible to notice that the differentially expressed genes are mostly in four different clusters (circled numbers). It means that we have different gene transcription patterns in seipin loss-of-function depending on the presence of adipogenesis stimuli (yellow boxes) or not (blue boxes). Therefore, this corroborates the discussion that the consequences for the lack of seipin might be different from an adipogenic compared to a nonadipogenic cell. Of all the 77 genes reviewed (100%), only 14 (19%) are differently expressed in more than one situation. The remaining 63 (81%) are expressed in only one situation. iRegulon [44] also allowed observing important transcription factors for the expression of some genes of each one of the four clusters. Interestingly, *CEBPB* is important for the transcription of many genes upregulated in nonadipogenic cells and this gene was also already seen higher expressed in these situations [45]. Therefore, seipin may influence expression of such genes through *CEBPB* upregulation. We bring attention to *HNF4A*, *ZNF740,* and *RORC* as names for future works that might be interesting to better understand how seipin affects gene transcription. Figure 10 also focuses on gene ontology for

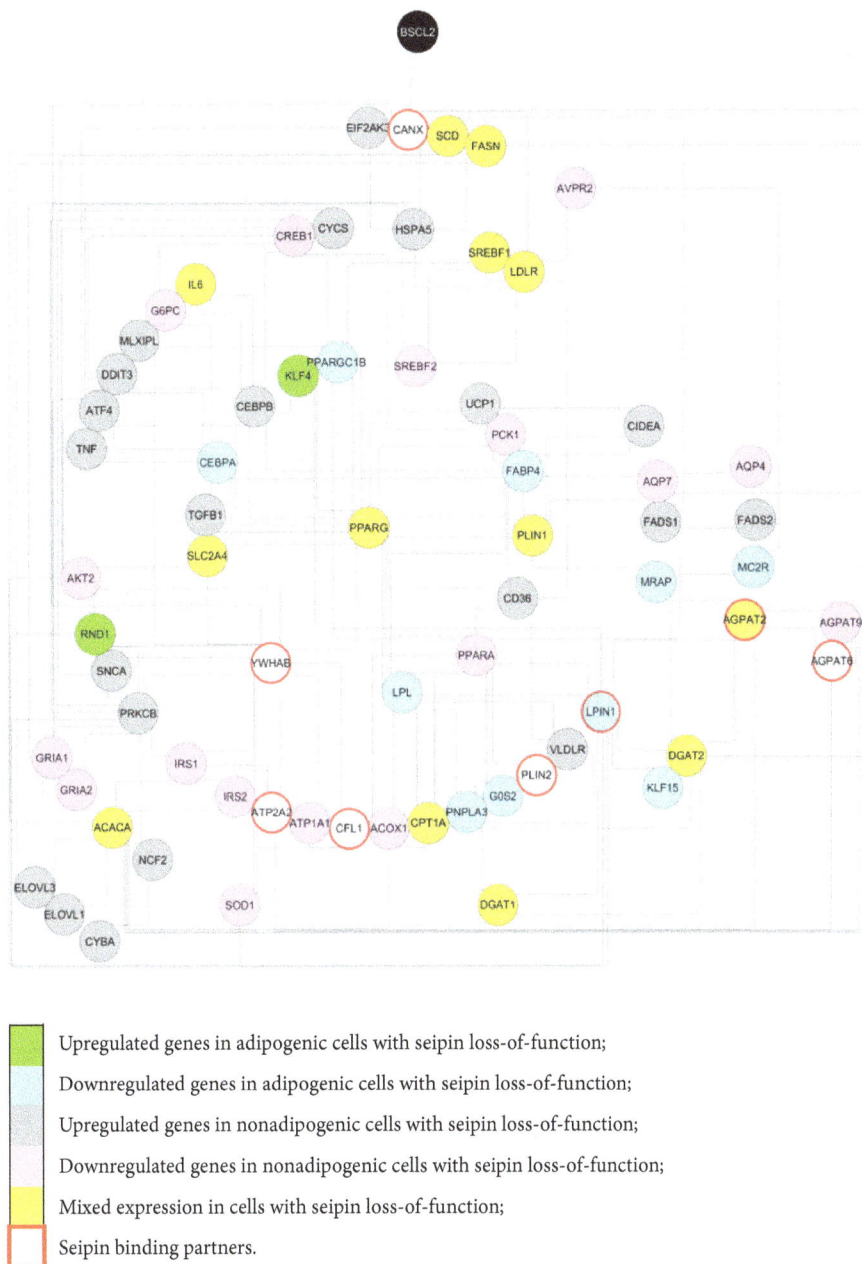

FIGURE 8: **Biological integration**. A network with genes differently regulated in seipin loss-of-function situations was created using STRING [41]. The gene list submitted is in accordance with Supplementary Tables 1 and 2. Genes painted with green are upregulated in adipogenic cells with seipin loss-of-function. Genes painted with blue are downregulated in adipogenic cells with seipin loss-of-function. Genes painted in gray are upregulated in nonadipogenic cells with seipin loss-of-function and genes painted in pink are downregulated in nonadipogenic cells with seipin loss-of-function. Genes that were observed either downregulated or upregulated in the same cell situations but in different papers are painted in yellow (mixed expression). Proteins already described as seipin physical binders are surrounded by a red circumference. The parameters chosen were "Experiments," "Databases," "Coexpression," "Neighborhood," "Gene Fusion," "Cooccurrence," and "Minimum required interaction score = 0.1." Cytoscape [42] classifies the genes based on their connectivity. As the circle comes to the center, the node tends to be more connected with the network.

biological processes affected by seipin loss-of-function. We used Panther [112] to input the gene list in Supplementary Table 1 and observed what would be the processes affected by the absence of seipin function. As we have already discussed, the results point to different biological processes: in adipogenic cells, the triglyceride biosynthetic process is inhibited because of the downregulation of genes important for that process and the white fat cell differentiation is also diminished. However, in nonadipogenic cells, the positive regulation of fatty acid β-oxidation is also inhibited, favoring the accumulation of lipids. These are some examples that corroborate our previous discussion.

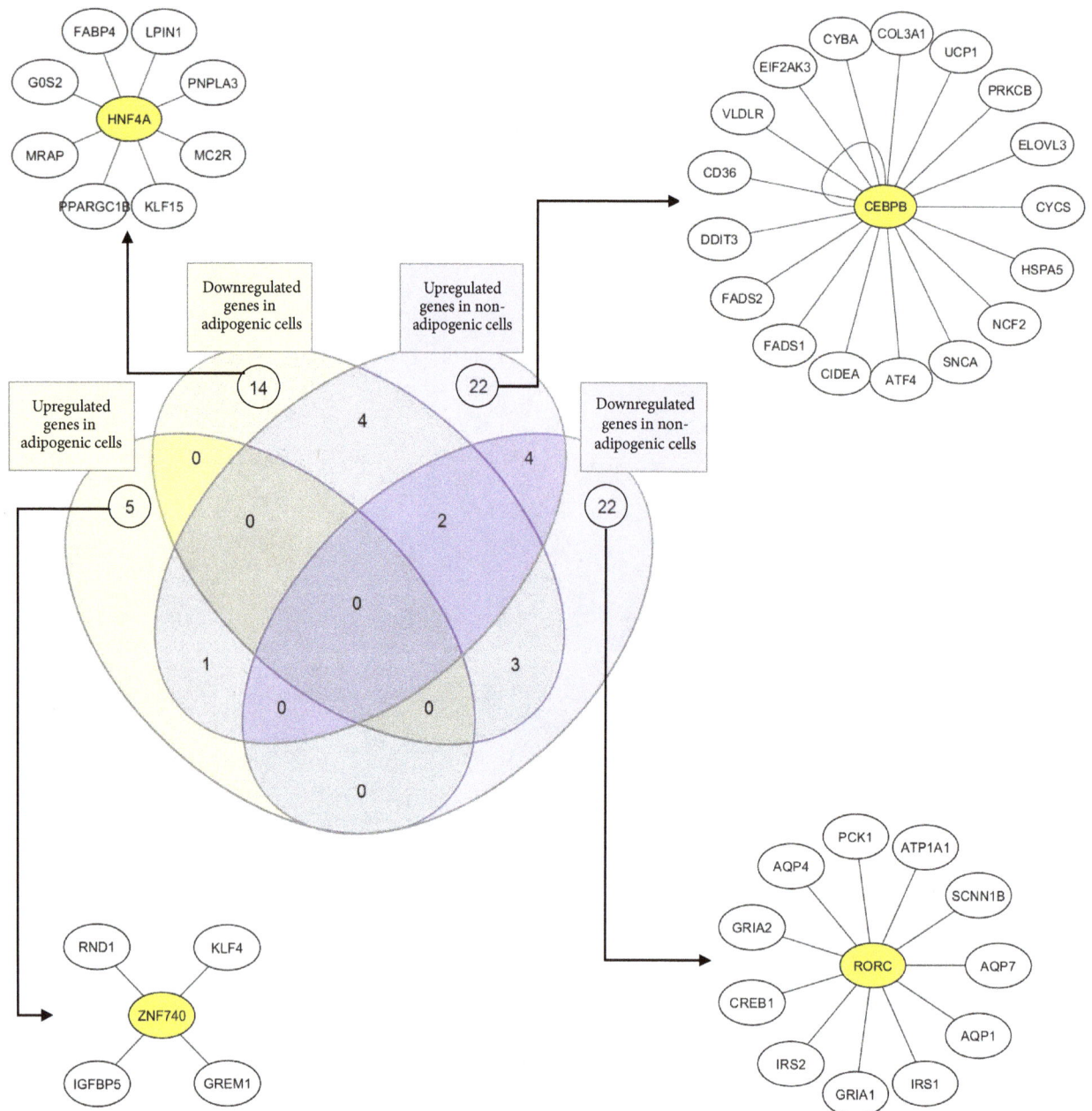

FIGURE 9: **Differently expressed genes during seipin loss-of-function**. InteractVenn [43] allowed us to see four different clusters formed when seipin loses its function in different cell types. iRegulon [44] shows *HNF4A*, *CEBPB*, *ZNF740*, and *RORC* as important transcription factors that may interfere with the expression of the observed genes for each cluster. Among them, only *CEBPB* was studied in a loss of seipin function [45].

9. Seipin Partners

Seipin interacts physically or functionally with many proteins of ER, revealing its functions and related pathways. Nonetheless, many seipin interactions seem to be preserved in adipogenic and nonadipogenic situations, and most of the studies were performed in heterologous overexpression systems, which requires caution during the interpretation of the physiological data and extrapolation to clinical relevance. Besides, even as important findings, they still do not clearly

explain the different phenotypes observed in nonadipogenic versus adipogenic cells. Thus, it is still difficult to understand the tissue-specific functions and biochemistry of seipin.

One crucial binding partner of seipin is 14-3-3β (UniProt gene name: *YWHAB*). This protein modulates many pathways and binds with seipin N and C terminus sequences. This phenomenon was found during 3T3-L1 mouse preadipocyte stem cell maturation, together with 14-3-3β and cofilin-1 interaction. The authors discussed that seipin-14-3-3β-cofilin-1 binding is important to actin

- Triglyceride biosynthetic process;
- Acylglycerol biosynthetic process;
- Neutral lipid biosynthetic process;
- White fat cell differentiation;
- Fatty acid homeostasis.

- Fatty acid elongation, unsaturated fatty acid;
- Fatty acid elongation, saturated fatty acid;
- Fatty acid elongation, polyunsaturated fatty acid;
- Fatty acid elongation, monounsaturated fatty acid;
- PERK-mediated unfolded protein response;
- ER overload response;
- ATF6-mediated unfolded protein response;
- Chronic inflammatory response.

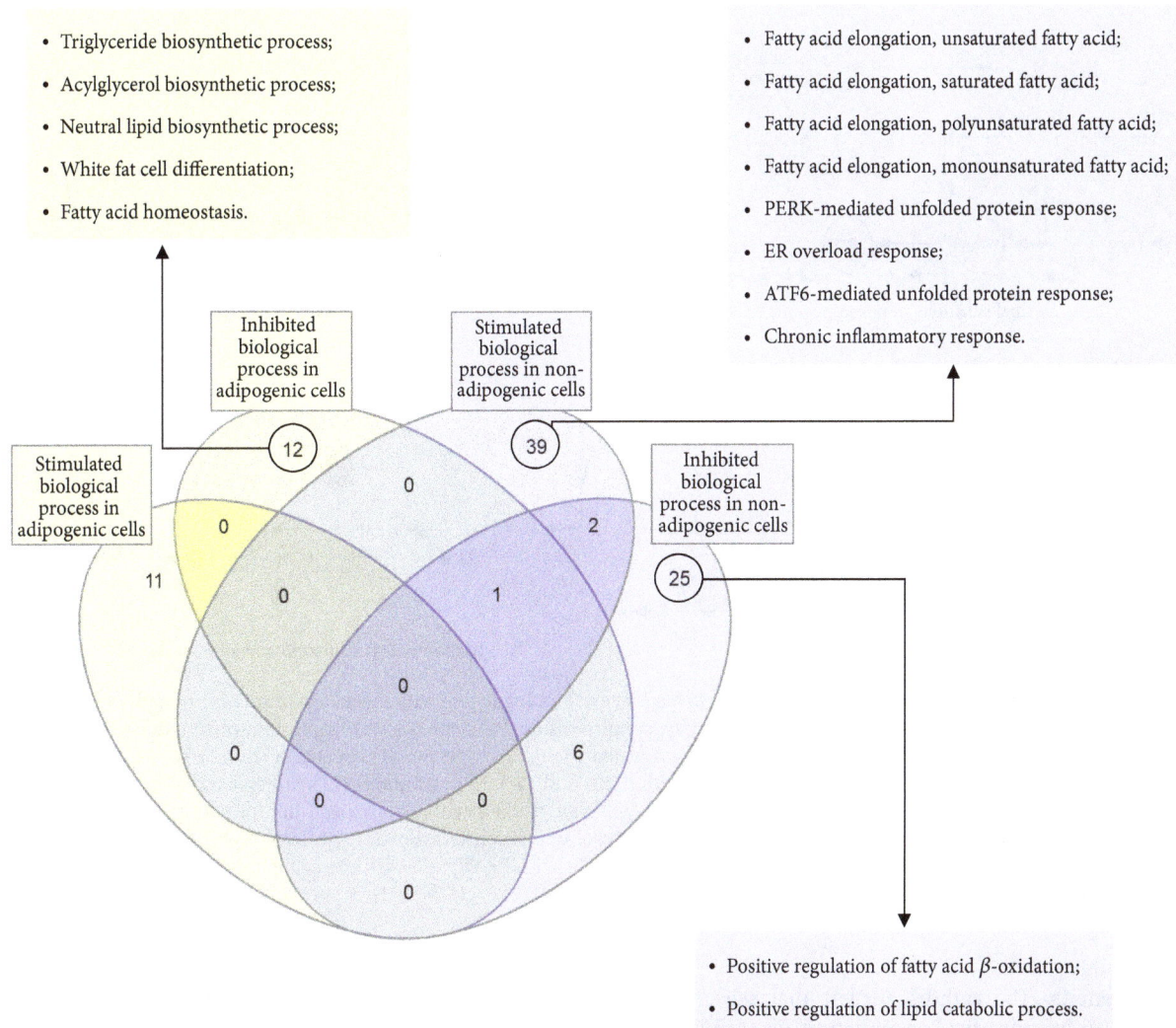

- Positive regulation of fatty acid β-oxidation;
- Positive regulation of lipid catabolic process.

FIGURE 10: **Gene ontology for seipin loss-of-function**. InteractVenn [43] allowed us to see four different clusters formed when seipin loses its function in different cell types. All of the biological processes had a fold enrichment >100.

cytoskeleton remodeling, which contributes to adipogenesis. Indeed, both 14-3-3β and cofilin-1 knockdown can also impair adipocyte development. However, seipin and 14-3-3β binding seem to be a constitutive event, also observed in nonadipogenic cells, together with 14-3-3β and cofilin-1 interaction [113].

Seipin was also demonstrated to bind sarco/endoplasmic reticulum Ca^{2+}-ATPase (SERCA, UniProt gene name: *ATP2A2*) protein both in the *Drosophila* fat body model and in human HEK293 cells. The loss of SERCA functions is able to generate ER stress because the enzyme transports cytosolic calcium (Ca^{2+}) into the ER lumen, a process that is important in adipogenesis. Seipin seems to regulate its activity and the authors discussed that the loss of *BSCL2* might prevent the increase of ER Ca^{2+} concentration and adipocyte maturation [114].

Seipin also directly interacts with adipocyte differentiation-related protein (ADRP, UniProt gene name: *PLIN2*) during adipogenesis stimulation. ADRP is important for

the development and maintenance of adipose tissue. The C terminus domain of seipin is important for that binding. Yet, that interaction is not specific for adipogenic cells, but happens in HEK293 cells too. Additionally, the authors discussed how seipin loss-of-function mutations change the intracellular distribution of ADRP and how this is important in adipogenesis [67].

Gao et al. found functional interaction with a yeast perilipin protein (Pet10), which stabilizes LDs and promotes their assembly [115]. In another paper, seipin also coimmunoprecipitated with the Reep1 protein in the NIH3T3 murine model. Reep1 is necessary for adipogenesis, ER stress resistance, and ER tubular network organization [116]. However, it is still not clear how Reep1 can contribute to findings observed during seipin disorders.

Another research group also proved that seipin physically interacts with AGPAT2 and lipin1 proteins during maturation of 3T3-L1 mouse preadipocyte stem cells. These enzymes are extremely important for TG synthesis (Figure 3)

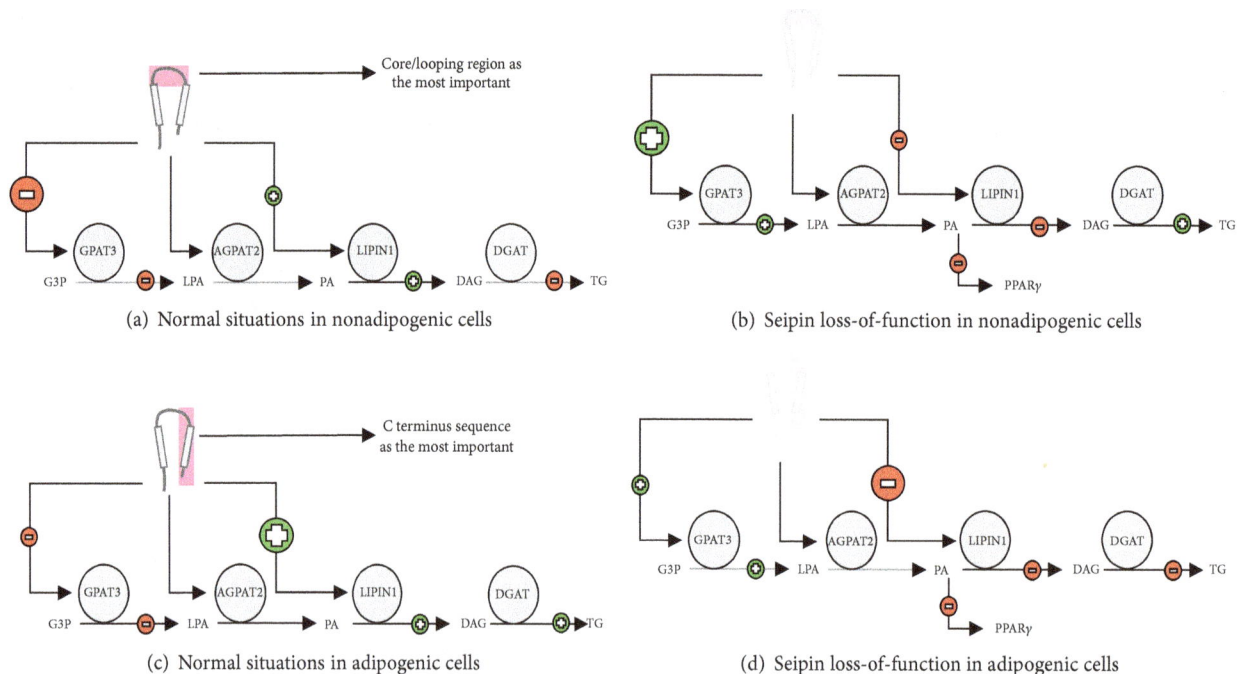

(a) Normal situations in nonadipogenic cells

(b) Seipin loss-of-function in nonadipogenic cells

(c) Normal situations in adipogenic cells

(d) Seipin loss-of-function in adipogenic cells

FIGURE 11: **Theoretical model for seipin**. (a) Seipin seems to have a core/looping region with more importance for nonadipogenic cells. This region can interact with GPAT3 and downregulate its activity, compromising TG synthesis. (b) Seipin loss-of-function in nonadipogenic cells seems to have the opposite behavior of the situation proposed in "(a)". (c) Seipin also seems to have a C terminus region with more importance for adipogenesis. This region can interact with lipin1 and 14-3-3β to promote TG synthesis and adipogenesis. (d) Seipin loss-of-function in adipogenic cells seems to have the opposite behavior of the situation proposed in "(c)". AGPAT: 1-acyl-sn-glycerol-3-phosphate acyltransferase; DG: diacylglycerol; DGAT: diacylglycerol o-acyltransferases; G3P: glycerol-3-phosphate; GPAT: glycerol-3-phosphate acyltransferase; LPA: lysophosphatidic acid; PA: phosphatidic acid; PPARγ: peroxisome proliferator-activated receptor gamma; TG: triacylglycerol.

and adipogenesis. The authors suggest that seipin might be required to increase the concentration of AGPAT2 and lipin1 in domains of the ER membrane. Secondly, seipin might interact with both proteins and act as a docking or scaffolding site for that complex, with a possible contribution to their activities. Seipin cytoplasmic N and C sequences showed importance to binding with lipin1, and its conserved core/looping or first transmembrane region to binding with AGPAT2 [117, 118]. Both AGPAT2 and lipin1 are important in adipocyte homeostasis, since disturbance of the first causes Berardinelli-Seip congenital lipodystrophy type 1 [119] and of the second results in *PPARG* gene downregulation [120]. However, it is important to highlight that the seipin-AGPAT2-lipin1 complex was also observed in nonadipogenic cells [117, 118]. Interestingly, Péterfy et al. observed that lipin1 is phosphorylated during adipocyte maturation and interacts with 14-3-3 proteins, which guarantee lipin1 cytoplasmic localization instead of nuclear [121]. Besides, diminution of lipin1 activity might accumulate PA, which can inhibit PPARγ activity [122].

GPAT3 is one of the enzymes that catalyzes the first step of TG synthesis (Figure 3), an important process for adipose tissue development. Even with low levels in preadipocytes [123], GPAT3 mRNA transcription is upregulated during adipogenesis of 3T3-L1 mouse cells. During the process, the authors observed that GPAT3 activity was important

for lipid accumulation [124], results confirmed by other groups [125]. They observed that *PPARG* is necessary to upregulate GPAT3 mRNA, increase TG accumulation, and contribute to adipogenesis [125]. Studies showed that *PPARG* mRNA silencing attenuated GPAT3 upregulation [125] and that PPARγ agonists were able to increase GPAT3 mRNA transcription [123, 126]. Besides, GPAT3 mRNA silencing also attenuated adipogenesis [125]. In WAT cells, upregulation of GPAT3 was associated with TG synthesis increase and enlarged LDs [127]. Taken together, GPAT3 seems to have an important function in TG synthesis and adipogenesis. However, there is a considerable lack of information about this protein and more studies are needed to elucidate its function. Not surprisingly, GPAT3 belongs to ER, where it interacts with seipin in yeast, mammalian cells, and adipogenic and nonadipogenic tissues. Nonetheless, the authors proposed that, during adipocyte differentiation, seipin negatively regulates GPAT3 activity. They observed that increased GPAT3 activity, through its overexpression, can impair adipogenesis in seipin-deficient cells. They also observed that knocking down GPAT3 enhanced adipocyte differentiation in seipin-deficient cells [128]. The ideas presented here are not mutually exclusive, as the GPAT3 downregulation might impair adipogenesis due to diminished TG synthesis (as previously reviewed), while overactivation can activate PA, a dangerous event in adipogenesis too [128].

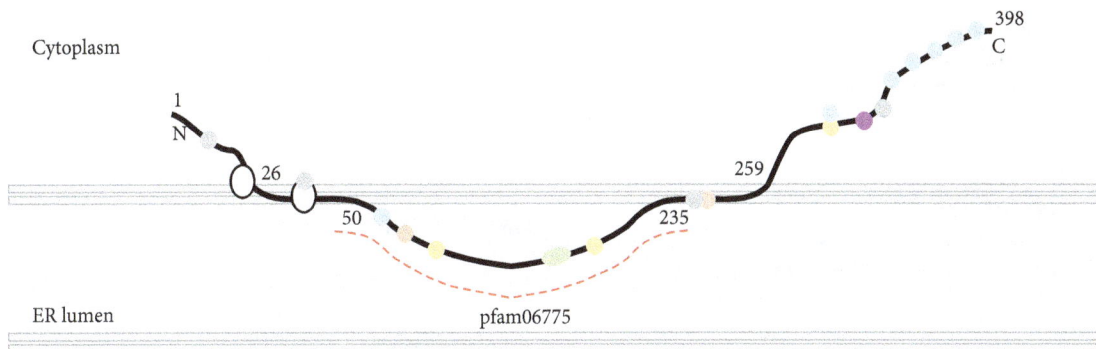

(a)

Modification	Amino acid position	Colour
N-myristoylation site	17-22; 32-37; 238-243; 297-302	
Leucine zipper pattern	20-41; 27-48	
Casein kinase II phosphorylation site	72-75; 280-283; 313-316; 318-321; 325-328; 336-339; 351-354	
N-glycosylation site	88-91; 242-245	
Protein kinase C phosphorylation site	143-145; 204-206; 280-282	
Tyrosine kinase phosphorylation site	176-184	
cAMP- and cGMP-dependent protein kinase phosphorylation site	286-289	

(b)

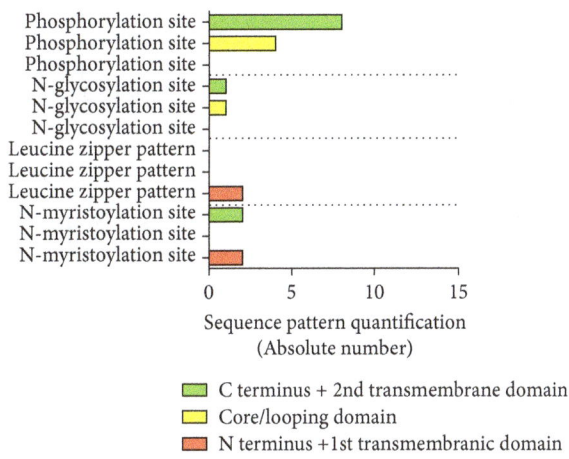

(c)

FIGURE 12: **Theoretical posttranslational modifications of seipin.** (a) Prosite [46] and Pfam [47] predictions for posttranslational modifications and conserved domains of seipin, respectively. (b) Residue patterns highlighted in (a). (c) Quantification of patterns. Even as theoretical predictions, the N-glycosylation site was already proven to occur and to be affected by the mutations N88S and S90L [48].

9.1. Are We All Touching Different Parts of the Same Elephant? We would like to propose a model to understand seipin's normal function based on what Yang et al. [61] and others proposed. They showed that seipin has a conserved core/looping that is important for nonadipogenic cells. However, they also observed that seipin gained a C terminus region during evolution, which is important for adipogenesis [61]. Interestingly, the C terminus is important for binding with lipin1 and 14-3-3β [113, 118], while the core/looping region is fundamental to GPAT3/4 interaction [128].

Here, we hypothesize that these papers are all "touching the same elephant." This popular analogy refers to a situation

in which blind men declared that they were touching a different animal when they were, in fact, all touching the same elephant. In every way, the following propositions need validation as a future perspective since our theoretical model is different from the others, because we are trying to integrate some previous publications about seipin in one basic explanation. As represented in Figure 11(a), we believe that the core/looping region of seipin is the most important in nonadipogenic cells. This sequence interacts with GPAT3/4 and negatively regulates their activities [128]. Yet, the remaining presence of the C terminus still allows the binding with lipin1 and 14-3-3β [113, 118]. In this way, if that C terminus domain is not the most important for these cells, lipin1 and 14-3-3β binding might not be enough for TG anabolism in the end. This could result in more negative stimulus to TG synthesis through more GPAT3/4 inhibition. These ideas are supported by the results of previous independent studies [61, 113, 118, 127, 128]. The main events that could happen during seipin absence caused by loss-of-function mutations are shown in Figure 11(b): TG synthesis could restart because of GPAT3/4 super stimulation [128] in nonadipogenic cells. This factor may be more important to TG synthesis than lipin1 defective-membrane association caused by the absence of seipin [117, 118]. In the same way, adipogenesis does not happen, because 14-3-3β does not interact with seipin in this situation to promote cytoskeleton remodeling, which contributes to adipogenesis [113]. Besides, GPAT3 super stimulation might increase PA concentration and impair PPARγ, as observed by more than one study [82, 128]. PPARγ seems to also protect nonadipose tissue against excessive lipid overload and maintain liver and skeletal muscle organ function, as reviewed by Kintscher and Janani [62, 129].

During adipogenesis of healthy cells (Figure 11(c)), we believe that the C terminus region of seipin might assume the control in a core/looping-almost-independent way. The C terminus sequence is not essential to GPAT3/4 and this might negatively impair the regulation of GPAT3/4 in a soft manner. However, the remaining presence of the core/looping domain still guarantees that binding. In this way, if now that the C terminus domain is the most important for these cells, lipin1 and 14-3-3β binding might be enough for adipogenesis and TG anabolism. This could result in more positive stimulus to TG synthesis, through intensified lipin1 and 14-3-4β activation, as shown by different and independent studies [61, 113, 118, 128]. During adipogenesis and seipin absence due to loss-of-function mutations (Figure 11(d)), TG synthesis can be impaired because of lipin1 activity commitment. This could increase PA production, associated with diminution of GPAT3/4 inhibition. Together, this could inhibit PPARγ actions, as supported by several studies [61, 62, 128–130].

When we put the seipin sequence on the Prosite [46] and Pfam [47] websites, it is possible to observe the predicted posttranslational modifications and the main conserved domain: pfam06775 (Figure 12). It is true that most of these modifications were never proved to happen with seipin. However, some regions of the protein present a pattern that was recognized by system algorithms and future research can work on biochemical validation or refutation. It is possible to observe that the C terminus sequence of seipin has more phosphorylation site patterns than the core region. This may indicate that these evolutionary acquired terminus residues are regulated by the cell. Perhaps, the posttranslational modification differences between the core and C terminus sequences are responsible for the seipin function in many tissues.

This model is only theoretical and we agree with Agarwal and Garg [34] that seipin is still a mysterious protein with many biochemistry functions to be discovered and/or better characterized. Here, we tried to understand seipin proposals and suggested a unified model. Nonetheless, we are aware of the biochemistry complexity and we proposed the model to be validated or complemented in future works.

Conflicts of Interest

There are no conflicts of interest associated with this paper.

Acknowledgments

We first want to honor seipin researchers cited in our work. Likewise, we would like to apologize to those not considered in this review. We also thank the website SMART, https://smart.servier.com/ [40], for supplying us with some pieces of our illustrations.

References

[1] J. G. Lima, L. H. Nobrega, N. N. de Lima, M. G. do Nascimento Santos, M. F. Baracho, and S. M. Jeronimo, "Clinical and laboratory data of a large series of patients with congenital generalized lipodystrophy," *Diabetology & Metabolic Syndrome*, vol. 8, article 23, 2016.

[2] W. Berardinelli, "An undiagnosed endocrinometabolic syndrome: report of 2 cases," *The Journal of Clinical Endocrinology & Metabolism*, vol. 14, no. 2, pp. 193–204, 1954.

[3] M. Seip and O. Trygstad, "Generalized lipodystrophy," *Archives of Disease in Childhood*, vol. 38, no. 201, pp. 447–453, 1963.

[4] J. Magré, M. Delépine, E. Khallouf et al., "Identification of the gene altered in Berardinelli-Seip congenital lipodystrophy on chromosome 11q13," *Nature Genetics*, vol. 28, no. 4, pp. 365–370, 2001.

[5] C. Windpassinger, M. Auer-Grumbach, J. Irobi et al., "Heterozygous missense mutations in BSCL2 are associated with distal hereditary motor neuropathy and Silver syndrome," *Nature Genetics*, vol. 36, no. 3, pp. 271–276, 2004.

[6] D. Ito, T. Fujisawa, H. Iida, and N. Suzuki, "Characterization of seipin/BSCL2, a protein associated with spastic paraplegia 17," *Neurobiology of Disease*, vol. 31, no. 2, pp. 266–277, 2008.

[7] A. S. Garfield, W. S. Chan, R. J. Dennis, D. Ito, L. K. Heisler, and J. J. Rochford, "Neuroanatomical characterisation of the expression of the lipodystrophy and motor-neuropathy gene Bscl2 in adult mouse brain," *PLoS ONE*, vol. 7, no. 9, 2012.

[8] X. Liu, B. Xie, Y. Qi et al., "The expression of SEIPIN in the mouse central nervous system," *Brain Structure & Function*, vol. 221, no. 8, pp. 4111–4127, 2016.

[9] M. Uhlen, P. Oksvold, L. Fagerberg et al., "Towards a knowledge-based Human Protein Atlas," *Nature Biotechnology*, vol. 28, no. 12, pp. 1248–1250, 2010.

[10] M. Uhlen, L. Fagerberg, B. M. Hallstrom et al., "Tissue-based map of the human proteome," *Science*, vol. 347, Article ID 1260419, 2015.

[11] W. Chen, V. K. Yechoor, B. H.-J. Chang, M. V. Li, K. L. March, and L. Chan, "The human lipodystrophy gene product

Berardinelli-Seip congenital lipodystrophy 2/seipin plays a key role in adipocyte differentiation," *Endocrinology*, vol. 150, no. 10, pp. 4552–4561, 2009.

[12] V. A. Payne, N. Grimsey, A. Tuthill et al., "The human lipodystrophy gene BSCL2/Seipin may be essential for normal adipocyte differentiation," *Diabetes*, vol. 57, no. 8, pp. 2055–2060, 2008.

[13] NCBI Resource Coordinators, "Database Resources of the National Center for Biotechnology Information," *Nucleic Acids Research*, vol. 45, no. D1, pp. D12–D17, 2017.

[14] R. Apweiler, A. Bairoch, C. H. Wu et al., "UniProt: the universal protein knowledgebase," *Nucleic Acids Research*, vol. 32, pp. D115–D119, 2004.

[15] M. Fu, R. Kazlauskaite, M. d. Paiva Baracho et al., "Mutations in Gng3lg and AGPAT2 in Berardinelli-Seip congenital lipodystrophy and Brunzell syndrome: Phenotype variability suggests important modifier effects," *The Journal of Clinical Endocrinology & Metabolism*, vol. 89, no. 6, pp. 2916–2922, 2004.

[16] J. Schuster, T. N. Khan, M. Tariq et al., "Exome sequencing circumvents missing clinical data and identifies a BSCL2 mutation in congenital lipodystrophy," *BMC Medical Genetics*, vol. 15, article 71, 2014.

[17] K. Mandal, S. Aneja, A. Seth, and A. Khan, "Berardinelli-Seip congenital lipodystrophy," *Indian Pediatrics*, vol. 43, no. 5, pp. 440–445, 2006.

[18] O. U. Rahman, N. Khawar, M. A. Khan et al., "Deletion mutation in BSCL2 gene underlies congenital generalized lipodystrophy in a Pakistani family," *Diagnostic Pathology*, vol. 8, article 78, pp. 1–7, 2013.

[19] H. U. Shirwalkar, Z. M. Patel, J. Magre et al., "Congenital generalized lipodystrophy in an Indian patient with a novel mutation in BSCL2 gene," *Journal of Inherited Metabolic Disease*, vol. 31, no. 2, pp. S317–S322, 2008.

[20] L. Van Maldergem, J. Magré, T. E. Khallouf et al., "Genotype-phenotype relationships in Berardinelli-Seip congenital lipodystrophy," *Journal of Medical Genetics*, vol. 39, no. 10, pp. 722–733, 2002.

[21] A. K. Agarwal, V. Simha, E. A. Oral et al., "Phenotypic and genetic heterogeneity in congenital generalized lipodystrophy," *The Journal of Clinical Endocrinology & Metabolism*, vol. 88, no. 10, pp. 4840–4847, 2003.

[22] K. Ebihara, T. Kusakabe, H. Masuzaki et al., "Gene and phenotype analysis of congenital generalized lipodystrophy in Japanese: a novel homozygous nonsense mutation in Seipin gene," *The Journal of Clinical Endocrinology & Metabolism*, vol. 89, no. 5, pp. 2360–2364, 2004.

[23] H. Huang, T. Chen, H. Hsiao et al., "A Taiwanese boy with congenital generalized lipodystrophy caused by homozygous Ile262fs mutation in the BSCL2 gene," *Kaohsiung Journal of Medical Sciences*, vol. 26, no. 11, pp. 615–620, 2010.

[24] A. Haghighi, Z. Kavehmanesh, A. Haghighi et al., "Congenital generalized lipodystrophy: Identification of novel variants and expansion of clinical spectrum," *Clinical Genetics*, vol. 89, no. 4, pp. 434–441, 2016.

[25] E. H. Jeninga, M. de Vroede, N. Hamers et al., "A patient with congenital generalized lipodystrophy due to a novel mutation in BSCL2: indications for secondary mitochondrial dysfunction," in *JIMD Reports - Case and Research Reports, 2012/1*, vol. 4 of *JIMD Reports*, pp. 47–54, Springer, Berlin, Germany, 2012.

[26] B. Akinci, H. Onay, T. Demir et al., "Natural history of congenital generalized lipodystrophy: a nationwide study from Turkey," *The Journal of Clinical Endocrinology & Metabolism*, vol. 101, no. 7, pp. 2759–2767, 2016.

[27] D. M. Miranda, B. L. Wajchenberg, M. R. Calsolari et al., "Novel mutations of the BSCL2 and AGPAT2 genes in 10 families with Berardinelli-Seip congenital generalized lipodystrophy syndrome," *Clinical Endocrinology*, vol. 71, no. 4, pp. 512–517, 2009.

[28] X. Su, R. Lin, Y. Huang et al., "Clinical and mutational features of three Chinese children with congenital generalized lipodystrophy," *Journal of Clinical Research in Pediatric Endocrinology*, vol. 9, no. 1, pp. 52–57, 2017.

[29] A. Nishiyama, M. Yagi, H. Awano et al., "Two Japanese infants with congenital generalized lipodystrophy due to BSCL2 mutations," *Pediatrics International*, vol. 51, no. 6, pp. 775–779, 2009.

[30] J. Jin, L. Cao, Z. Zhao et al., "Novel BSCL2 gene mutation E189X in Chinese congenital generalized lipodystrophy child with early onset diabetes mellitus," *European Journal of Endocrinology*, vol. 157, no. 6, pp. 783–787, 2007.

[31] B. Friguls, W. Coroleu, D. Alcazar, P. Hilbert, L. Van Maldergem, and G. Pintos-Morell, "Severe cardiac phenotype of Berardinelli-Seip congenital lipodystrophy in an infant with a E158X BSCL2 mutation," *European Journal of Medical Genetics*, vol. 52, no. 1, pp. 14–16, 2009.

[32] P. Di Tommaso, S. Moretti, I. Xenarios et al., "T-Coffee: a web server for the multiple sequence alignment of protein and RNA sequences using structural information and homology extension," *Nucleic Acids Research*, vol. 39, no. 2, pp. W13–W17, 2011.

[33] C. Lundin, R. Nordström, K. Wagner et al., "Membrane topology of the human seipin protein," *FEBS Letters*, vol. 580, no. 9, pp. 2281–2284, 2006.

[34] A. K. Agarwal and A. Garg, "Seipin: A mysterious protein," *Trends in Molecular Medicine*, vol. 10, no. 9, pp. 440–444, 2004.

[35] A. K. Agarwal and A. Garg, "Congenital generalized lipodystrophy: significance of triglyceride biosynthetic pathways," *Trends in Endocrinology & Metabolism*, vol. 14, no. 5, pp. 214–221, 2003.

[36] D. W. Leung, "The structure and functions of human lysophosphatidic acid acyltransferases," *Frontiers in Bioscience*, vol. 6, no. 3, pp. d944–953, 2001.

[37] W. Fei, G. Shui, B. Gaeta et al., "Fld1p, a functional homologue of human seipin, regulates the size of lipid droplets in yeast," *The Journal of Cell Biology*, vol. 180, no. 3, pp. 473–482, 2008.

[38] K. M. Szymanski, D. Binns, R. Bartz et al., "The lipodystrophy protein seipin is found at endoplasmic reticulum lipid droplet junctions and is important for droplet morphology," *Proceedings of the National Acadamy of Sciences of the United States of America*, vol. 104, no. 52, pp. 20890–20895, 2007.

[39] H. Wang, M. Becuwe, B. E. Housden et al., "Seipin is required for converting nascent to mature lipid droplets," *eLife*, vol. 5, 2016.

[40] SMART, "Servier Medical ART," https://smart.servier.com/.

[41] D. Szklarczyk, A. Franceschini, S. Wyder et al., "STRING v10: protein-protein interaction networks, integrated over the tree of life," *Nucleic Acids Research*, vol. 43, pp. D447–D452, 2015.

[42] P. Shannon, A. Markiel, O. Ozier et al., "Cytoscape: a software Environment for integrated models of biomolecular interaction networks," *Genome Research*, vol. 13, no. 11, pp. 2498–2504, 2003.

[43] H. Heberle, V. G. Meirelles, F. R. da Silva, G. P. Telles, and R. Minghim, "InteractiVenn: a web-based tool for the analysis of sets through Venn diagrams," *BMC Bioinformatics*, vol. 16, no. 1, article 169, 2015.

[44] R. Janky, A. Verfaillie, H. Imrichová et al., "iRegulon: from a gene list to a gene regulatory network using large motif and track collections," *PLoS Computational Biology*, vol. 10, no. 7, Article ID e1003731, 2014.

[45] W. Chen, B. Chang, P. Saha et al., "Berardinelli-seip congenital lipodystrophy 2/seipin is a cell-autonomous regulator of lipolysis essential for adipocyte differentiation," *Molecular and Cellular Biology*, vol. 32, no. 6, pp. 1099–1111, 2012.

[46] C. J. A. Sigrist, E. de Castro, L. Cerutti et al., "New and continuing developments at PROSITE," *Nucleic Acids Research*, vol. 41, no. 1, pp. D344–D347, 2013.

[47] R. D. Finn, P. Coggill, R. Y. Eberhardt et al., "The Pfam protein families database: towards a more sustainable future," *Nucleic Acids Research*, vol. 44, no. 1, pp. D279–D285, 2016.

[48] H.-D. Fan, S.-P. Chen, Y.-X. Sun, S.-H. Xu, and L.-J. Wu, "Seipin mutation at glycosylation sites activates autophagy in transfected cells via abnormal large lipid droplets generation," *Acta Pharmacologica Sinica*, vol. 36, no. 4, pp. 497–506, 2015.

[49] D. Binns, S. Lee, C. L. Hilton, Q.-X. Jiang, and J. M. Goodman, "Seipin is a discrete homooligomer," *Biochemistry*, vol. 49, no. 50, pp. 10747–10755, 2010.

[50] M. F. Michelle Sim, M. Mesbah Uddin Talukder, R. J. Dennis, S. O'Rahilly, J. Michael Edwardson, and J. J. Rochford, "Analysis of naturally occurring mutations in the human lipodystrophy protein seipin reveals multiple potential pathogenic mechanisms," *Diabetologia*, vol. 56, no. 11, pp. 2498–2506, 2013.

[51] M. F. M. Sim, M. M. U. Talukder, R. J. Dennis, J. M. Edwardson, and J. J. Rochford, "Analyzing the functions and structure of the human lipodystrophy protein seipin," *Methods in Enzymology*, vol. 537, pp. 161–175, 2014.

[52] M. Wang and R. J. Kaufman, "Protein misfolding in the endoplasmic reticulum as a conduit to human disease," *Nature*, vol. 529, no. 7586, pp. 326–335, 2016.

[53] G. Jing, J. J. Wang, and S. X. Zhang, "ER stress and apoptosis: a new mechanism for retinal cell death," *Journal of Diabetes Research*, vol. 2012, Article ID 589589, 11 pages, 2012.

[54] D. T. Rutkowski, J. Wu, S.-H. Back et al., "UPR pathways combine to prevent hepatic steatosis caused by ER stress-mediated suppression of transcriptional master regulators," *Developmental Cell*, vol. 15, no. 6, pp. 829–840, 2008.

[55] C. M. Oslowski and F. Urano, "Measuring ER stress and the unfolded protein response using mammalian tissue culture system," *Methods in Enzymology*, vol. 490, pp. 71–92, 2011.

[56] Y. Wang, J. Viscarra, S.-J. Kim, and H. S. Sul, "Transcriptional regulation of hepatic lipogenesis," *Nature Reviews Molecular Cell Biology*, vol. 16, no. 11, pp. 678–689, 2015.

[57] N. L. Gluchowski, M. Becuwe, T. C. Walther, and R. V. Farese, "Lipid droplets and liver disease: From basic biology to clinical implications," *Nature Reviews Gastroenterology & Hepatology*, vol. 14, no. 6, pp. 343–355, 2017.

[58] A. S. Joshi, H. Zhang, and W. A. Prinz, "Organelle biogenesis in the endoplasmic reticulum," *Nature Cell Biology*, vol. 19, no. 8, pp. 876–882, 2017.

[59] Y. Ohsaki, K. Sołtysik, and T. Fujimoto, "The lipid droplet and the endoplasmic reticulum," in *Organelle Contact Sites*, vol. 997 of *Advances in Experimental Medicine and Biology*, pp. 111–120, Springer, Singapore, Singapore, 2017.

[60] R. Zechner, F. Madeo, and D. Kratky, "Cytosolic lipolysis and lipophagy: two sides of the same coin," *Nature Reviews Molecular Cell Biology*, vol. 18, no. 11, pp. 671–684, 2017.

[61] W. Yang, S. Thein, X. Guo et al., "Seipin differentially regulates lipogenesis and adipogenesis through a conserved core sequence and an evolutionarily acquired C-terminus," *Biochemical Journal*, vol. 452, no. 1, pp. 37–44, 2013.

[62] C. Janani and B. D. Ranjitha Kumari, "PPAR gamma gene—a review," *Diabetes & Metabolic Syndrome: Clinical Research & Reviews*, vol. 9, no. 1, pp. 46–50, 2015.

[63] K. Sarjeant and J. M. Stephens, "Adipogenesis," *Cold Spring Harbor Perspectives in Biology*, vol. 4, no. 9, Article ID a008417, 2012.

[64] N. Patni and A. Garg, "Congenital generalized lipodystrophies - New insights into metabolic dysfunction," *Nature Reviews Endocrinology*, vol. 11, no. 9, pp. 522–534, 2015.

[65] X. Chen and J. M. Goodman, "The collaborative work of droplet assembly," *Molecular and Cell Biology of Lipids*, vol. 1862, no. 10, pp. 1205–1211, 2017.

[66] J. Han, R. Murthy, B. Wood et al., "ER stress signalling through eIF2α and CHOP, but not IRE1α, attenuates adipogenesis in mice," *Diabetologia*, vol. 56, no. 4, pp. 911–924, 2013.

[67] E. Mori, J. Fujikura, M. Noguchi et al., "Impaired adipogenic capacity in induced pluripotent stem cells from lipodystrophic patients with BSCL2 mutations," *Metabolism*, vol. 65, no. 4, pp. 543–556, 2016.

[68] W. Qiu, K. Wee, K. Takeda et al., "Suppression of adipogenesis by pathogenic seipin mutant is associated with inflammatory response," *PLoS ONE*, vol. 8, no. 3, Article ID e57874, 2013.

[69] S. Bhayana, V. M. Siu, G. I. Joubert, C. L. Clarson, H. Cao, and R. A. Hegele, "Cardiomyopathy in congenital complete lipodystrophy," *Clinical Genetics*, vol. 61, no. 4, pp. 283–287, 2002.

[70] V. Simha and A. Garg, "Phenotypic heterogeneity in body fat distribution in patients with congenital generalized lipodystrophy caused by mutations in the AGPAT2 or Seipin genes," *The Journal of Clinical Endocrinology & Metabolism*, vol. 88, no. 11, pp. 5433–5437, 2003.

[71] W. Fei, H. Li, G. Shui et al., "Molecular characterization of seipin and its mutants: Implications for seipin in triacylglycerol synthesis," *Journal of Lipid Research*, vol. 52, no. 12, pp. 2136–2147, 2011.

[72] Y. Tian, J. Bi, G. Shui et al., "Tissue-autonomous function of drosophila seipin in preventing ectopic lipid droplet formation," *PLoS Genetics*, vol. 7, no. 4, Article ID e1001364, 2011.

[73] M. A. Lounis, S. Lalonde, S. A. Rial et al., "Hepatic BSCL2 (Seipin) deficiency disrupts lipid droplet homeostasis and increases lipid metabolism via SCD1 activity," *Lipids*, vol. 52, no. 2, pp. 129–150, 2017.

[74] L. Liu, Q. Jiang, X. Wang et al., "Adipose-specific knockout of Seipin/Bscl2 results in progressive lipodystrophy," *Diabetes*, vol. 63, no. 7, pp. 2320–2331, 2014.

[75] W. Chen, H. Zhou, S. Liu et al., "Altered lipid metabolism in residual white adipose tissues of Bscl2 deficient mice," *PLoS ONE*, vol. 8, no. 12, Article ID e82526, 2013.

[76] L. Dollet, C. Levrel, T. Coskun et al., "FGF21 improves the adipocyte dysfunction related to seipin deficiency," *Diabetes*, vol. 65, no. 11, pp. 3410–3417, 2016.

[77] C. Ebihara, K. Ebihara, M. Aizawa-Abe et al., "Seipin is necessary for normal brain development and spermatogenesis in addition to adipogenesis," *Human Molecular Genetics*, vol. 24, no. 15, pp. 4238–4249, 2015.

[78] H. Zhou, S. M. Black, T. W. Benson, N. L. Weintraub, and W. Chen, "Berardinelli-seip congenital lipodystrophy 2/seipin is not required for brown adipogenesis but regulates brown adi-

pose tissue development and function," *Molecular and Cellular Biology*, vol. 36, no. 15, pp. 2027–2038, 2016.

[79] L. Dollet, J. Magré, M. Joubert et al., "Seipin deficiency alters brown adipose tissue thermogenesis and insulin sensitivity in a non-cell autonomous mode," *Scientific Reports*, vol. 6, no. 1, 2016.

[80] V. T. Salo, I. Belevich, S. Li et al., "Seipin regulates ER–lipid droplet contacts and cargo delivery," *EMBO Journal*, vol. 35, no. 24, pp. 2699–2716, 2016.

[81] M. Jiang, M. Gao, C. Wu et al., "Lack of testicular seipin causes teratozoospermia syndrome in men," *Proceedings of the National Acadamy of Sciences of the United States of America*, vol. 111, no. 19, pp. 7054–7059, 2014.

[82] L. Zhou, J. Yin, C. Wang, J. Liao, G. Liu, and L. Chen, "Lack of seipin in neurons results in anxiety- and depression-like behaviors via down regulation of PPARγ," *Human Molecular Genetics*, vol. 23, no. 15, pp. 4094–4102, 2014.

[83] G. Li, L. Zhou, Y. Zhu et al., "Seipin knockout in mice impairs stem cell proliferation and progenitor cell differentiation in the adult hippocampal dentate gyrus via reduced levels of PPARγ," *Disease Models & Mechanisms*, vol. 8, no. 12, pp. 1615–1624, 2015.

[84] L. Zhou, T. Chen, G. Li et al., "Activation of PPARγ ameliorates spatial cognitive deficits through restoring expression of AMPA receptors in seipin knock-out mice," *The Journal of Neuroscience*, vol. 36, no. 4, pp. 1242–1253, 2016.

[85] Y. Qian, J. Yin, J. Hong et al., "Neuronal seipin knockout facilitates Aβ-induced neuroinflammation and neurotoxicity via reduction of PPARγ in hippocampus of mouse," *Journal of Neuroinflammation*, vol. 13, article 145, 2016.

[86] L. Wang, J. Hong, Y. Wu, G. Liu, W. Yu, and L. Chen, "Seipin deficiency in mice causes loss of dopaminergic neurons via aggregation and phosphorylation of α-synuclein and neuroinflammation," *Cell Death & Disease*, vol. 9, article 440, 2018.

[87] A. E. El Zowalaty, R. Li, W. Chen, and X. Ye, "Seipin deficiency leads to increased endoplasmic reticulum stress and apoptosis in mammary gland alveolar epithelial cells during lactation," *Biology of Reproduction*, vol. 98, no. 4, pp. 570–578, 2018.

[88] Y. Chen, L. Wei, J. Tian, Y.-H. Wang, G. Liu, and C. Wang, "Seinpin knockout exacerbates cerebral ischemia/reperfusion damage in mice," *Biochemical and Biophysical Research Communications*, vol. 474, no. 2, pp. 377–383, 2016.

[89] W. Fei, H. Wang, C. Bielby, and H. Yang, "Conditions of endoplasmic reticulum stress stimulate lipid droplet formation in Saccharomyces cerevisiae," *Biochemical Journal*, vol. 424, no. 1, pp. 61–67, 2009.

[90] E. Boutet, H. El Mourabit, M. Prot et al., "Seipin deficiency alters fatty acid Δ9 desaturation and lipid droplet formation in Berardinelli-Seip congenital lipodystrophy," *Biochimie*, vol. 91, no. 6, pp. 796–803, 2009.

[91] X. Cui, Y. Wang, Y. Tang et al., "Seipin ablation in mice results in severe generalized lipodystrophy," *Human Molecular Genetics*, vol. 20, no. 15, Article ID ddr205, pp. 3022–3030, 2011.

[92] P. Xu, H. Wang, A. Kayoumu, M. Wang, W. Huang, and G. Liu, "Diet rich in Docosahexaenoic Acid/Eicosapentaenoic Acid robustly ameliorates hepatic steatosis and insulin resistance in seipin deficient lipodystrophy mice," *Nutrition & Metabolism*, vol. 12, article 58, 2015.

[93] W. Chen, H. Zhou, P. Saha, L. Li, and L. Chan, "Molecular mechanisms underlying fasting modulated liver insulin sensitivity and metabolism in male lipodystrophic Bscl2/seipin-deficient mice," *Endocrinology*, vol. 155, no. 11, pp. 4215–4225, 2014.

[94] M. Gao, M. Wang, X. Guo et al., "Expression of seipin in adipose tissue rescues lipodystrophy, hepatic steatosis and insulin resistance in seipin null mice," *Biochemical and Biophysical Research Communications*, vol. 460, no. 2, pp. 143–150, 2015.

[95] M. Wang, M. Gao, J. Liao, Y. Han, Y. Wang, and G. Liu, "Dysfunction of lipid metabolism in lipodystrophic Seipin-deficient mice," *Biochemical and Biophysical Research Communications*, vol. 461, no. 2, pp. 206–210, 2015.

[96] G. D. Mcilroy, K. Suchacki, A. J. Roelofs et al., "Adipose specific disruption of seipin causes early-onset generalised lipodystrophy and altered fuel utilisation without severe metabolic disease," *Molecular Metabolism*, vol. 10, pp. 55–65, 2018.

[97] X. Cui, Y. Wang, L. Meng et al., "Overexpression of a short human seipin/BSCL2 isoform in mouse adipose tissue results in mild lipodystrophy," *American Journal of Physiology-Endocrinology and Metabolism*, vol. 302, no. 6, pp. E705–E713, 2012.

[98] J. R. Silver, "Familial spastic paraplegia with amyotrophy of the hands," *Annals of Human Genetics*, vol. 30, no. 1, pp. 69–73, 1966.

[99] H.-J. Cho, D.-H. Sung, and C.-S. Ki, "Identification of de novo BSCL2 Ser90Leu mutation in a Korean family with Silver syndrome and distal hereditary motor neuropathy," *Muscle & Nerve*, vol. 36, no. 3, pp. 384–386, 2007.

[100] D. Ito and N. Suzuki, "Molecular pathogenesis of Seipin/BSCL2-related motor neuron diseases," *Annals of Neurology*, vol. 61, no. 3, pp. 237–250, 2007.

[101] T. Yagi, D. Ito, Y. Nihei, T. Ishihara, and N. Suzuki, "N88S seipin mutant transgenic mice develop features of seipinopathy/BSCL2-related motor neuron disease via endoplasmic reticulum stress," *Human Molecular Genetics*, vol. 20, no. 19, Article ID ddr304, pp. 3831–3840, 2011.

[102] J. Guo, W. Qiu, S. L. Soh et al., "Motor neuron degeneration in a mouse model of seipinopathy," *Cell Death & Disease*, vol. 4, no. 3, pp. e535–e535, 2013.

[103] M. Hölttä-Vuori, V. T. Salo, Y. Ohsaki, M. L. Suster, and E. Ikonen, "Alleviation of seipinopathy-related ER stress by triglyceride storage," *Human Molecular Genetics*, vol. 22, no. 6, pp. 1157–1166, 2013.

[104] J. G. Lima, L. H. Nobrega, N. N. Lima et al., "Causes of death in patients with Berardinelli-Seip congenital generalized lipodystrophy," *PLoS ONE*, vol. 13, no. 6, Article ID e0199052, 2018.

[105] M. Auer-Grumbach, B. Schlotter-Weigel, H. Lochmüller et al., "Phenotypes of the N88S Berardinelli-Seip congenital lipodystrophy 2 mutation," *Annals of Neurology*, vol. 57, no. 3, pp. 415–424, 2005.

[106] C. Windpassinger, K. Wagner, E. Petek, R. Fischer, and M. Auer-Grumbach, "Refinement of the 'Silver syndrome locus' on chromosome 11q12-q14 in four families and exclusion of eight candidate genes," *Human Genetics*, vol. 114, no. 1, pp. 99–109, 2003.

[107] J. Irobi, P. van den Bergh, L. Merlini et al., "The phenotype of motor neuropathies associated with BSCL2 mutations is broader than Silver syndrome and distal HMN type V," *Brain*, vol. 127, no. 9, pp. 2124–2130, 2004.

[108] M. R. Alaei, S. Talebi, M. Ghofrani, M. Taghizadeh, and M. Keramatipour, "Whole exome sequencing reveals a BSCL2 mutation causing progressive encephalopathy with lipodystrophy (PELD) in an Iranian pediatric patient," *Iranian Biomedical Journal*, vol. 20, no. 5, pp. 295–301, 2016.

[109] A. Ruiz-Riquelme, S. Sánchez-Iglesias, A. Rábano et al., "Larger aggregates of mutant seipin in Celia's Encephalopathy, a new

protein misfolding neurodegenerative disease," *Neurobiology of Disease*, vol. 83, pp. 44–53, 2015.

[110] E. Guillén-Navarro, S. Sánchez-Iglesias, R. Domingo-Jiménez et al., "A new seipin-associated neurodegenerative syndrome," *Journal of Medical Genetics*, vol. 50, no. 6, pp. 401–409, 2013.

[111] D. Araújo-Vilar, R. Domingo-Jiménez, Á. Ruibal et al., "Association of metreleptin treatment and dietary intervention with neurological outcomes in Celia's encephalopathy," *European Journal of Human Genetics*, vol. 26, no. 3, pp. 396–406, 2018.

[112] H. Mi, X. Huang, A. Muruganujan et al., "PANTHER version 11: Expanded annotation data from Gene Ontology and Reactome pathways, and data analysis tool enhancements," *Nucleic Acids Research*, vol. 45, no. 1, pp. D183–D189, 2017.

[113] W. Yang, S. Thein, X. Wang et al., "BSCL2/seipin regulates adipogenesis through actin cytoskeleton remodelling," *Human Molecular Genetics*, vol. 23, no. 2, pp. 502–513, 2014.

[114] J. Bi, W. Wang, Z. Liu et al., "Seipin promotes adipose tissue fat storage through the ER Ca 2+-ATPase SERCA," *Cell Metabolism*, vol. 19, no. 5, pp. 861–871, 2014.

[115] Q. Gao, D. D. Binns, L. N. Kinch et al., "Pet10p is a yeast perilipin that stabilizes lipid droplets and promotes their assembly," *The Journal of Cell Biology*, vol. 216, no. 10, pp. 3199–3217, 2017.

[116] B. Renvoisé, B. Malone, M. Falgairolle et al., "Reep1 null mice reveal a converging role for hereditary spastic paraplegia proteins in lipid droplet regulation," *Human Molecular Genetics*, vol. 25, no. 23, pp. 5111–5125, 2016.

[117] M. M. U. Talukder, M. F. M. Sim, S. O'Rahilly, J. M. Edwardson, and J. J. Rochford, "Seipin oligomers can interact directly with AGPAT2 and lipin 1, physically scaffolding critical regulators of adipogenesis," *Molecular Metabolism*, vol. 4, no. 3, pp. 199–209, 2015.

[118] M. F. M. Sim, R. J. Dennis, E. M. Aubry et al., "The human lipodystrophy protein seipin is an ER membrane adaptor for the adipogenic PA phosphatase lipin 1," *Molecular Metabolism*, vol. 2, no. 1, pp. 38–46, 2013.

[119] A. K. Agarwal, E. Arioglu, S. de Almeida et al., "AGPAT2 is mutated in congenital generalized lipodystrophy linked to chromosome 9q34," *Nature Genetics*, vol. 31, no. 1, pp. 21–23, 2002.

[120] P. Zhang, K. Takeuchi, L. S. Csaki, and K. Reue, "Lipin-1 phosphatidic phosphatase activity modulates phosphatidate levels to promote peroxisome proliferator-activated receptor γ (PPARγ) gene expression during adipogenesis," *The Journal of Biological Chemistry*, vol. 287, no. 5, pp. 3485–3494, 2012.

[121] M. Péterfy, T. E. Harris, N. Fujita, and K. Reue, "Insulin-stimulated interaction with 14-3-3 promotes cytoplasmic localization of lipin-1 in adipocytes," *The Journal of Biological Chemistry*, vol. 285, no. 6, pp. 3857–3864, 2010.

[122] T. Tsukahara, R. Tsukahara, Y. Fujiwara et al., "Phospholipase D2-Dependent Inhibition of the Nuclear Hormone Receptor PPARγ by Cyclic Phosphatidic Acid," *Molecular Cell*, vol. 39, no. 3, pp. 421–432, 2010.

[123] J. Cao, J. Li, D. Li, J. F. Tobin, and R. E. Gimeno, "Molecular identification of microsomal acyl-CoA:glycerol-3-phosphate acyltransferase, a key enzyme in de novo triacylglycerol synthesis," *Proceedings of the National Acadamy of Sciences of the United States of America*, vol. 103, no. 52, pp. 19695–19700, 2006.

[124] D. Shan, J.-L. Li, L. Wu et al., "GPAT3 and GPAT4 are regulated by insulin-stimulated phosphorylation and play distinct roles in adipogenesis," *Journal of Lipid Research*, vol. 51, no. 7, pp. 1971–1981, 2010.

[125] S. Ma, F. Jing, C. Xu et al., "Thyrotropin and obesity: increased adipose triglyceride content through glycerol-3-phosphate acyltransferase 3," *Scientific Reports*, vol. 5, article 7633, 2015.

[126] B. Lu, Y. J. Jiang, P. Kim et al., "Expression and regulation of GPAT isoforms in cultured human keratinocytes and rodent epidermis," *Journal of Lipid Research*, vol. 51, no. 11, pp. 3207–3216, 2010.

[127] M. Zhao and X. Chen, "Eicosapentaenoic acid promotes thermogenic and fatty acid storage capacity in mouse subcutaneous adipocytes," *Biochemical and Biophysical Research Communications*, vol. 450, no. 4, pp. 1446–1451, 2014.

[128] M. Pagac, D. E. Cooper, Y. Qi et al., "SEIPIN regulates lipid droplet expansion and adipocyte development by modulating the activity of glycerol-3-phosphate acyltransferase," *Cell Reports*, vol. 17, no. 6, pp. 1546–1559, 2016.

[129] U. Kintscher and R. E. Law, "PPARγ-mediated insulin sensitization: the importance of fat versus muscle," *American Journal of Physiology-Endocrinology and Metabolism*, vol. 288, pp. 287–291, 2005.

[130] A. Lüdtke, J. Buettner, H. H.-J. Schmidt, and H. J. Worman, "New PPARG mutation leads to lipodystrophy and loss of protein function that is partially restored by a synthetic ligand," *Journal of Medical Genetics*, vol. 44, no. 9, article e88, 2007.

8

Stem Cells Applications in Regenerative Medicine and Disease Therapeutics

Ranjeet Singh Mahla

Department of Biological Sciences, Indian Institute of Science Education and Research (IISER), Bhopal, Madhya Pradesh 462066, India

Correspondence should be addressed to Ranjeet Singh Mahla; ranjeet@iiserb.ac.in

Academic Editor: Paul J. Higgins

Regenerative medicine, the most recent and emerging branch of medical science, deals with functional restoration of tissues or organs for the patient suffering from severe injuries or chronic disease. The spectacular progress in the field of stem cell research has laid the foundation for cell based therapies of disease which cannot be cured by conventional medicines. The indefinite self-renewal and potential to differentiate into other types of cells represent stem cells as frontiers of regenerative medicine. The transdifferentiating potential of stem cells varies with source and according to that regenerative applications also change. Advancements in gene editing and tissue engineering technology have endorsed the ex vivo remodelling of stem cells grown into 3D organoids and tissue structures for personalized applications. This review outlines the most recent advancement in transplantation and tissue engineering technologies of ESCs, TSPSCs, MSCs, UCSCs, BMSCs, and iPSCs in regenerative medicine. Additionally, this review also discusses stem cells regenerative application in wildlife conservation.

1. Introduction

Regenerative medicine, the most recent and emerging branch of medical science, deals with functional restoration of specific tissue and/or organ of the patients suffering with severe injuries or chronic disease conditions, in the state where bodies own regenerative responses do not suffice [1]. In the present scenario donated tissues and organs cannot meet the transplantation demands of aged and diseased populations that have driven the thrust for search for the alternatives. Stem cells are endorsed with indefinite cell division potential, can transdifferentiate into other types of cells, and have emerged as frontline regenerative medicine source in recent time, for reparation of tissues and organs anomalies occurring due to congenital defects, disease, and age associated effects [1]. Stem cells pave foundation for all tissue and organ system of the body and mediates diverse role in disease progression, development, and tissue repair processes in host. On the basis of transdifferentiation potential, stem cells are of four types, that is, (1) unipotent, (2) multipotent, (3) pluripotent, and (4) totipotent [2]. Zygote, the only totipotent stem cell in human body, can give rise to whole organism through the process of transdifferentiation, while cells from inner cells mass (ICM) of embryo are pluripotent in their nature and can differentiate into cells representing three germ layers but do not differentiate into cells of extraembryonic tissue [2]. Stemness and transdifferentiation potential of the embryonic, extraembryonic, fetal, or adult stem cells depend on functional status of pluripotency factors like OCT4, cMYC, KLF44, NANOG, SOX2, and so forth [3–5]. Ectopic expression or functional restoration of endogenous pluripotency factors epigenetically transforms terminally differentiated cells into ESCs-like cells [3], known as induced pluripotent stem cells (iPSCs) [3, 4]. On the basis of regenerative applications, stem cells can be categorized as embryonic stem cells (ESCs), tissue specific progenitor stem cells (TSPSCs), mesenchymal stem cells (MSCs), umbilical cord stem cells (UCSCs), bone marrow stem cells (BMSCs), and iPSCs (Figure 1; Table 1). The transplantation of stem cells can be autologous, allogenic, and syngeneic for induction of tissue regeneration and immunolysis of pathogen or malignant cells. For avoiding the consequences of host-versus-graft rejections, tissue typing of human leucocyte antigens (HLA) for tissue and organ transplant as well as use of

Promises of stem cells in regenerative medicines

(1)
(i) Improvement of spinal cord injury
(ii) Regeneration of retinal sheet
(iii) Generation of retinal ganglion cells
(iv) Healing of heart defects
(v) Hepatic cell formation
(vi) Formation of insulin secreting β-cells
(vii) Cartilage lesion treatment
(viii) Regeneration of pacemaker
(ix) In vitro gametogenesis

Healthy donor

(4)
(i) T1DM and T2DM treatment
(ii) SLE (autoimmune disease) treatment
(iii) Application for HI treatment
(iv) Krabbe's disease treatment
(v) Hematopoiesis in neuroblastoma

(2)
(i) Treatment of diabetes and retinopathy
(ii) Neurodental therapeutic applications
(iii) Restoration of cognitive functions
(iv) Brain and cancer treatment
(v) Ear acoustic function restoration
(vi) Regeneration of intestinal mucosa
(vii) Treatment of vision defects
(viii) Muscle regeneration
(ix) Regeneration of fallopian tube

(1) ESCs, (2) TSPSCs,
(3) MSCs , (4) UCSCs,
(5) BMSCs , (6) IPSCs

(5)
(i) Treatment of anemia and blood cancer
(ii) Retroviral therapy
(iii) Correction of neuronal defects
(iv) Generation of functional platelets
(v) Alveolar bone regeneration
(vi) Regeneration of diaphragm tissue

(3)
(i) Regeneration of bladder tissue
(ii) Muscle regeneration
(iii) Regeneration of teeth tissue
(iv) Healing of orthopedic injuries
(v) Recovery from muscle injuries
(vi) Hear scar repair after attack

Patient

(6)
(i) Regeneration of kidney tissue
(ii) Vision restoration in AMD
(iii) Treatment of placental defects
(iv) Treatment of brain cortex defects
(v) ASD and autism treatment
(vi) Treatment of liver and lung disease
(vii) Generation of serotonin neurons
(viii) Regeneration of pacemaker

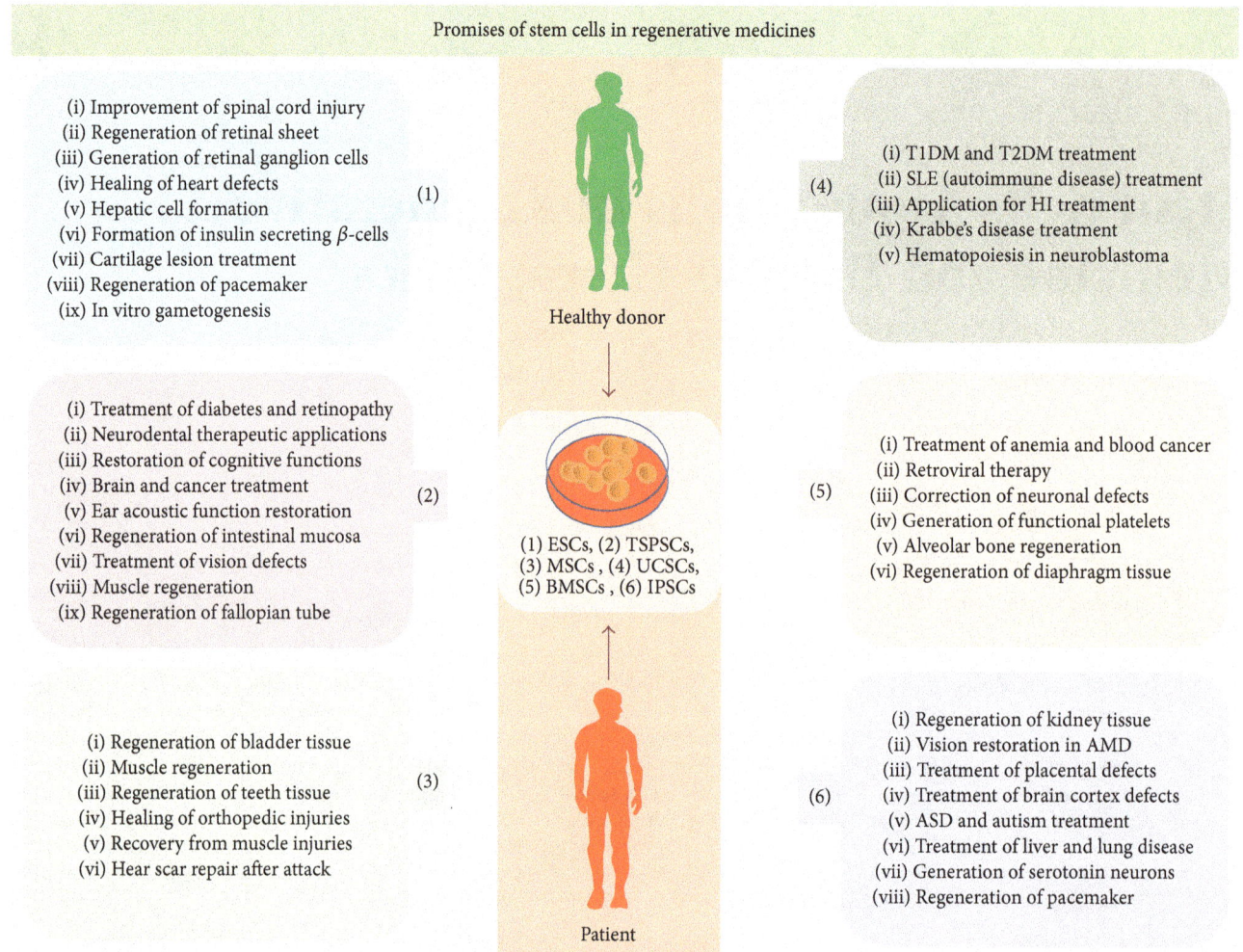

FIGURE 1: Promises of stem cells in regenerative medicine: the six classes of stem cells, that is, embryonic stem cells (ESCs), tissue specific progenitor stem cells (TSPSCs), mesenchymal stem cells (MSCs), umbilical cord stem cells (UCSCs), bone marrow stem cells (BMSCs), and induced pluripotent stem cells (iPSCs), have many promises in regenerative medicine and disease therapeutics.

immune suppressant is recommended [6]. Stem cells express major histocompatibility complex (MHC) receptor in low and secret chemokine that recruitment of endothelial and immune cells is enabling tissue tolerance at graft site [6]. The current stem cell regenerative medicine approaches are founded onto tissue engineering technologies that combine the principles of cell transplantation, material science, and microengineering for development of organoid; those can be used for physiological restoration of damaged tissue and organs. The tissue engineering technology generates nascent tissue on biodegradable 3D-scaffolds [7, 8]. The ideal scaffolds support cell adhesion and ingrowths, mimic mechanics of target tissue, support angiogenesis and neovascularisation for appropriate tissue perfusion, and, being nonimmunogenic to host, do not require systemic immune suppressant [9]. Stem cells number in tissue transplant impacts upon regenerative outcome [10]; in that case prior ex vivo expansion of transplantable stem cells is required [11]. For successful regenerative outcomes, transplanted stem cells must survive, proliferate, and differentiate in site specific

manner and integrate into host circulatory system [12]. This review provides framework of most recent (Table 1; Figures 1–8) advancement in transplantation and tissue engineering technologies of ESCs, TSPSCs, MSCs, UCSCs, BMSCs, and iPSCs in regenerative medicine. Additionally, this review also discusses stem cells as the tool of regenerative applications in wildlife conservation.

2. ESCs in Regenerative Medicine

For the first time in 1998, Thomson isolated human ESCs (hESCs) [13]. ESCs are pluripotent in their nature and can give rise to more than 200 types of cells and promises for the treatment of any kinds of disease [13]. The pluripotency fate of ESCs is governed by functional dynamics of transcription factors OCT4, SOX2, NANOG, and so forth, which are termed as pluripotency factors. The two alleles of the OCT4 are held apart in pluripotency state in ESCs; phase through homologues pairing during embryogenesis

TABLE 1: Application of stem cells in regenerative medicine: stem cells (ESCs, TSPSCs, MSCs, UCSCs, BMSCs, and iPSCs) have diverse applications in tissue regeneration and disease therapeutics.

SCs	Disease	Factors causing disease	Mode of stem cells application	Physiological and mechanistic aspects of stem cells therapeutics	Improvements in disease signatures & future use	References
ESCs	Spinal cord injuries	Infection, cancer, and accidents	ESCs transplantation to injury site	ESCs and secreted vasculogenic and neurogenic factor support tissue homing	Regeneration of spinal tissue and improved balance and sensation	[15]
	ARMD and glaucoma	Macular cones degeneration	ESCs-derived cones and RGCs transplantation to eye	COCO (activating TGF-β, BMP, and Wnt) & BRN3 (knock-in by CRISPER-Cas9) make ESCs become cones and RGCs form cells sheet & neuronal connection	Recovery from ARMD and macular defects & restoration of vision	[16, 17]
	Cardiovascular disease	Diabetes, drugs, genetic factor, and life style	ESCs-derived CMs & biomaterial coaxed ESCs	Cardiomyocytes express GCaMP3, secreting vasculogenic factors, and Tbox3 differentiates ESCs into SANPCs	Suppresses heart arrhythmias. CMs electrophysiologically integrate to heart as pacemaker	[18, 19, 28]
	Liver injuries	Toxins, drugs, genetic factors, and infection	Transplantation of ESCs-derived hepatocytes	ESCs-hepatocyte conversion is marked with expression of Cytp450, PXR, CYPA4&29, HNF4-α, and UGTA1; cells in transplant repopulate injured liver tissue	Regeneration of liver tissue can be used as model for screening of drugs	[20, 23, 24]
	Diabetes	Life style, heart defects, and genetics	Transplantation of ESCs-derived PPCs	Progenitors (CD24$^+$, CD49$^+$ & CD133$^+$) differentiate into β-cells, secrete insulin, and express PDXI, GCK, and GLUT2	Improvement in glucose level and obesity can be used for treatment of T1DM and T2DM	[25, 26]
	Osteoarthritis	When cartilage tissue wears away	Transplantation of chondrocytes organoids	Chondrocytes (SOX9$^+$ & collagen-II$^+$) form cells aggregate & remain active for 12 wks at transplantation site	Regeneration of cartilage tissue can be used for treatment of injuries faced by athletes	[27]
TSPSCs	Diabetes	Life style and genetic factors	Transplantation of SCs derived PPCs organoid	PPCs need niche supported active FGF & Notch signalling to become β-cell	PPCs occupancy as β-cell can treat T1DM & T2DM	[25, 29, 30]
	Neurodental problems	Accidents, age, and genetic factors	Transplantation of DSPSCs as neurons	Neurons express nestin, GFAP, βIII-tubulin, and L-type Ca^{2+} channels	Possible application in treatment of neurodental abnormalities	[31, 32]
	Acoustic problems	Age, noise, drugs, and infection	IESCs/IESCs-derived hair cells transplantation	γ-secretase shuts Notch by β-catenin & Atoh1 in Irg5$^+$TESCs to be hair cells	Cochlear regeneration leads to restoration of acoustic functions	[34, 35]
	Intestinal degeneration	Genetic factors and food borne infections	IPCs derived crypt-villi organoid transplantation	Mϕ, myofibroblasts, and bacteria signals IPCs to be crypt-villi organoid tissue	Regeneration of goblet mucosa can treat intestinal defects	[36–38]
	Corneal diseases	Burns, genetics, and inflammation	LPSCs transplantation to corneal tissue	LPSCs in transplant marked by ABCB5 differentiate into mature cornea	Regeneration of corneal tissue might treat multiple eye disease	[39, 40]
	Muscular deformities	Infection, drugs, and autoimmunity	Transplantation of PEG fibrinogen coaxed MABs	PDGF from MABs attract vasculogenic and neurogenic cells to transplant site	Muscle fibril regeneration; skeletal muscle defects treatment	[41, 42]
	Eye disease & retinopathy	Toxins, burns, and genetic factors	AdSCs intravitreal transplantation	AdSCs from healthy donor produce higher vasoprotective factors	Restoration of vascularisation, diabetic retinopathy treatment	[44, 45]
	Cardiac dysfunctions	Age, genetic factors, and toxins	Systemic infusion of CA-AdSCs myocardium	CA-AdSCs to epithelium differentiation are superior to AdSCs	Regeneration of ischemic myocardium	[47, 48]

TABLE 1: Continued.

SCs	Disease	Factors causing disease	Mode of stem cells application	Physiological and mechanistic aspects of stem cells therapeutics	Improvements in disease signatures & future use	References
	Bladder deformities	Cystitis, cancer, and infection,	Transplantation of BD-MSCs to bladder	BDMSCs (CD105$^+$, CD73$^+$, CD34$^-$, and CD45$^-$) with SIS heal bladder in 10 wks	Bladder regeneration from different origins MSCs	[50, 51]
	Dental problems	Infection, cancer age, and accidents	Transplants of EMSCs + DSCs biopolymer tissue	EMSC-DSCs and vasculogenic factors in biopolymer give rise to mature teeth units	Regeneration of oral tissue and application in periodontics	[31, 52]
MSCs	Bone degeneration	Injuries and tumor autoimmunity	Coaxed MSCs transplant & MSCs infusion	Actin modelling by cytochalasin-D transforms MSCs into osteoblasts	Regeneration of bones, reduction in injury pain	[53–55]
	Muscle degeneration	Genetic factors and work stress	Coaxed MSCs transplant and MSCs infusion	Alginate gel protects MSCs from immune attack and controls GFs release	Regeneration of heart scar and muscle tissue in controlled way	[56, 57]
	Alopecia	Age, disease, and medicine use	Transplantation of GAG-coated DPCs	GAG coating mimics ECM microenvironment, promoting DPCs regeneration	Regeneration of hair follicle for treatment of alopecia	[58]
	Congenital heart defects	Developmental errors	Transplantation of fibrin coaxed AFSCs	Addition of VEFG to PEG coaxed AFSCs promotes organogenesis	Regeneration of tissue repair for treatment of heart defects	[59, 60]
	Diabetes	Life style and genetic factors	WJ-SCs, transplantation, and intravenous injection	WJ-factors & Mφ differentiate WJ-SCs into β-cells, decreasing IL6 & IL1β	Improvement in function of β-cells leads to treatment of diabetes	[7, 9, 61–63]
	SLE	Autoimmunity	Intravenous infusion of WJ-SCs	WJ-SCs decrease SLEDAI & BILAG; reinfusion protects from disease relapse	Improvement in renal functions & stopping degeneration of tissues	[64]
UCSCs	LSD & neurodegenerative diseases	Genetics, tumor, age, and life style	Allogenic UCSCs cells and biomaterial coaxed UCSCs organoids	Organoids consisted of neuroblasts (GFAP$^+$, Nestin$^+$, and Ki67$^+$) & SCs (OCT4$^+$, SOC2$^+$); UCSCs recover from MSE deficiency and improve cognition	Treatment of Krabbe's disease, hurler syndrome, MLD, TSD, ALD, AD, ALS, SCI, SCI, TBI, Parkinson's, stroke, and so forth	[65–67]
	Cartilage and tendon injuries	Accident	Transplantation of UCB-SCs, UCB-SCs-HA gel	HA gel factors promote regeneration of hyaline cartilage & tendons in wks time	Recovery from tendons and cartilage injuries	[68, 69]
	Hodgkin's lymphoma	Genetic and environmental	Transplantation of UCSCs	Second dose infection of allogenic UCSCs improves patients life by 30%	Treatment of Hodgkin's lymphoma and other cancers	[10]
	Peritoneal fibrosis	Long term renal dialysis and fibrosis	WJ-SCs, transplantation by IP injection	WJ-SCs prevent programmed cells death and peritoneal wall thickness	Effective in treatment of encapsulating peritoneal fibrosis	[70]

Table 1: Continued.

SCs	Disease	Factors causing disease	Mode of stem cells application	Physiological and mechanistic aspects of stem cells therapeutics	Improvements in disease signatures & future use	References
BMSCs	Anaemia and blood cancer	Injury, genetics autoimmunity	Two-step infusion of lymphoid and myeloid	Haplo identical BMSCs can reconstruct immunity, which is major process for minority	Treatment of aplastic anaemia & haematological malignancies	[71]
	AIDS	HIV1 infection	Transplantation of HIV1 resistant CD4$^+$ cells	Anti-HIV1 CD4$^+$ cells express HIV1 anti-RNA, which restrict HIV infection	Treatment of AIDS as an alternative of antiretroviral	[72, 73]
	Blood clotting disorders	Lack of platelets	Transplantation of megakaryocyte organoids	GFs in silk sponge, microtubule 3D scaffolds mimic bone marrow	Therapeutics of burns and blood clotting diseases	[74, 75]
	Neurodegenerative diseases	Accidents, age, trauma, and stroke	Focal transplant of BMSCs with LA	LA$^+$ BMSCs induce neovascularisation that directs microglia for colonization	Treatment of neuronal damage disorders and cognitive restoration	[76]
	Orodental deformities	Trauma, disease, and birth defects	Bone marrow derived stem & progenitor (TRC)	CD14$^+$ & CD90$^+$ TRC accelerate alveolar jaw bone regeneration	Regeneration of defects in oral bone, skin, and gum	[77]
	Diaphragm abnormalities	Accidents & birth defects	Implantation of decellularized diaphragm	BMSCs niche perfused hemidiaphragm has similar myography & spirometry	Replacement therapy by donor derived niched diaphragm	[8]
iPSCs	Eye defects	Age, genetics, and birth defects	iPSCs derived NPCs transplantation	NPCs form 5-6 layers of photoreceptor nuclei, restoring visual acuity	Treatment of ARMD and other age-related eye defects	[78–80]
	Neurodegenerative disorders	Accidents, age, trauma, and stroke	iGABA-INs and cortical spheroid transplantation	(iGABA-INs) secrete GABA; FOXIG cause ASD, spheroid mimics to brain	ASD, Alzheimer's, seizer, and obstinate epilepsies treatment	[81–84]
	Liver & lung diseases	A1AD deficiency	Transplantation of A1AD mutation corrected iPSCs	A1AD is encoded by SERPINA1 in liver, and mutation leads to drugs sensitivity	Treatment of COPD causing lungs and liver degeneration	[85]
	Diabetes	Life style and genetic factors	iPSCs derived β-cells transplantation	Skin to β-cells reprogramming phase through cDE & cPF requires GPs	Treatment of T1DM and T2DM and insulin production	[86]
	Lung degeneration	Tuberculosis, cancer, and fibrosis	Biomaterial coaxed iPSCs transplantation	Miniature iPSCs lung resembles airways and alveoli, model drug testing	Regeneration of lung tissue	[87]
	SIDs and AIDS	Age, genetic factors, and infection	Transplantation of Oct4 and Nanog corrected iPSCs	CRISPER-Cas9 generate iPSCs in single step; iPSCs-Mϕ resists HIV1	Immunotherapy of SIDs, HIV1, and other immune diseases	[80, 88, 89]

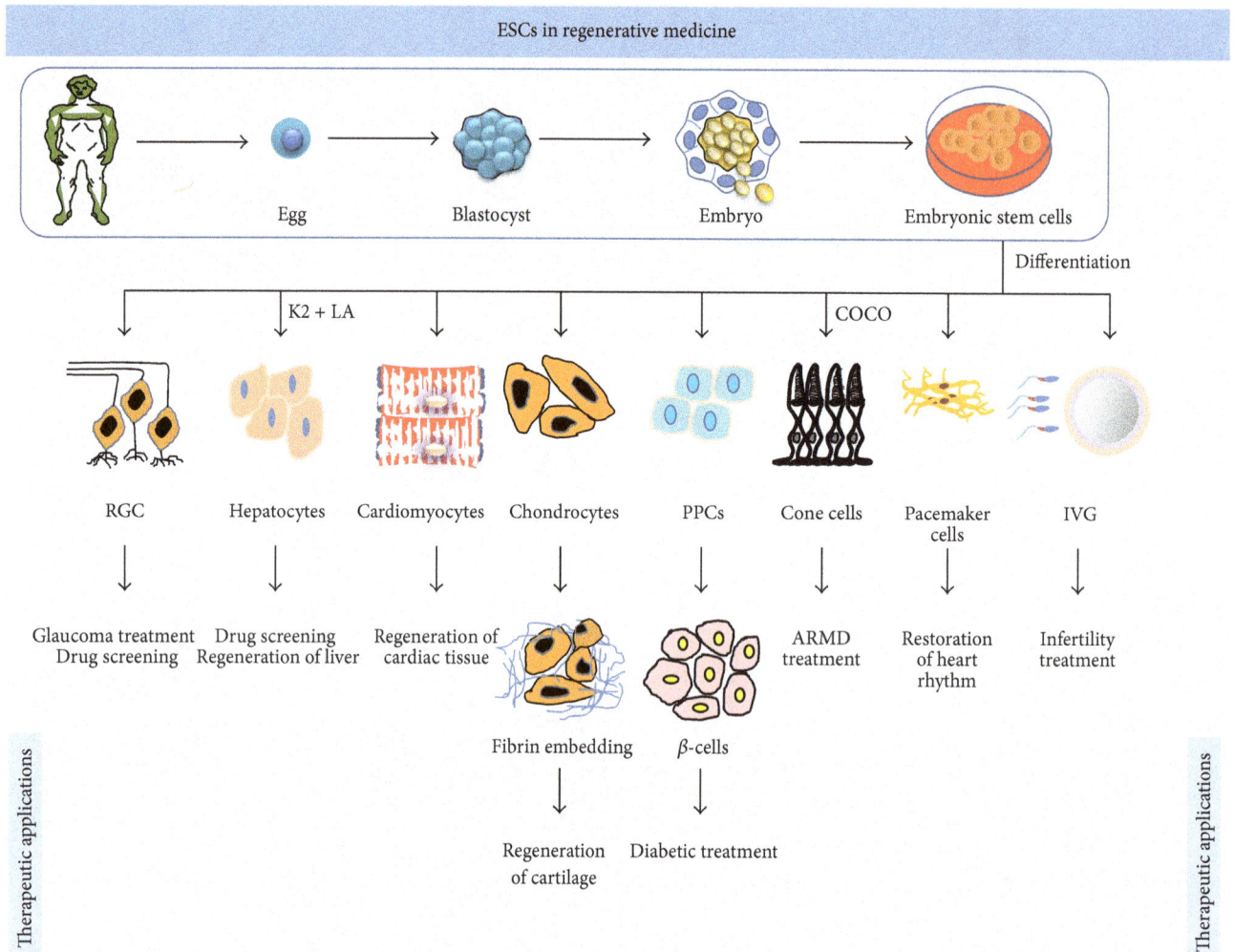

FIGURE 2: ESCs in regenerative medicine: ESCs, sourced from ICM of gastrula, have tremendous promises in regenerative medicine. These cells can differentiate into more than 200 types of cells representing three germ layers. With defined culture conditions, ESCs can be transformed into hepatocytes, retinal ganglion cells, chondrocytes, pancreatic progenitor cells, cone cells, cardiomyocytes, pacemaker cells, eggs, and sperms which can be used in regeneration of tissue and treatment of disease in tissue specific manner.

and transdifferentiation processes [14] has been considered as critical regulatory switch for lineage commitment of ESCs. The diverse lineage commitment potential represents ESCs as ideal model for regenerative therapeutics of disease and tissue anomalies. This section of review on ESCs discusses transplantation and transdifferentiation of ESCs into retinal ganglion, hepatocytes, cardiomyocytes, pancreatic progenitors, chondrocytes, cones, egg sperm, and pacemaker cells (Figure 2; Table 1). Infection, cancer treatment, and accidents can cause spinal cord injuries (SCIs). The transplantation of hESCs to paraplegic or quadriplegic SCI patients improves body control, balance, sensation, and limbal movements [15], where transplanted stem cells do homing to injury sites. By birth, humans have fixed numbers of cone cells; degeneration of retinal pigment epithelium (RPE) of macula in central retina causes age-related macular degeneration (ARMD). The genomic incorporation of COCO gene (expressed during embryogenesis) in the developing embryo leads lineage commitment of ESCs into cone cells, through suppression of

TGFβ, BMP, and Wnt signalling pathways. Transplantation of these cone cells to eye recovers individual from ARMD phenomenon, where transplanted cone cells migrate and form sheet-like structure in host retina [16]. However, establishment of missing neuronal connection of retinal ganglion cells (RGCs), cones, and PRE is the most challenging aspect of ARMD therapeutics. Recently, Donald Z Jacks group at John Hopkins University School of Medicine has generated RGCs from CRISPER-Cas9-m-Cherry reporter ESCs [17]. During ESCs transdifferentiation process, CRIPER-Cas9 directs the knock-in of m-Cherry reporter into 3′UTR of BRN3B gene, which is specifically expressed in RGCs and can be used for purification of generated RGCs from other cells [17]. Furthermore, incorporation of forskolin in transdifferentiation regime boosts generation of RGCs. Coaxing of these RGCs into biomaterial scaffolds directs axonal differentiation of RGCs. Further modification in RGCs generation regime and composition of biomaterial scaffolds might enable restoration of vision for ARMD and glaucoma patients [17].

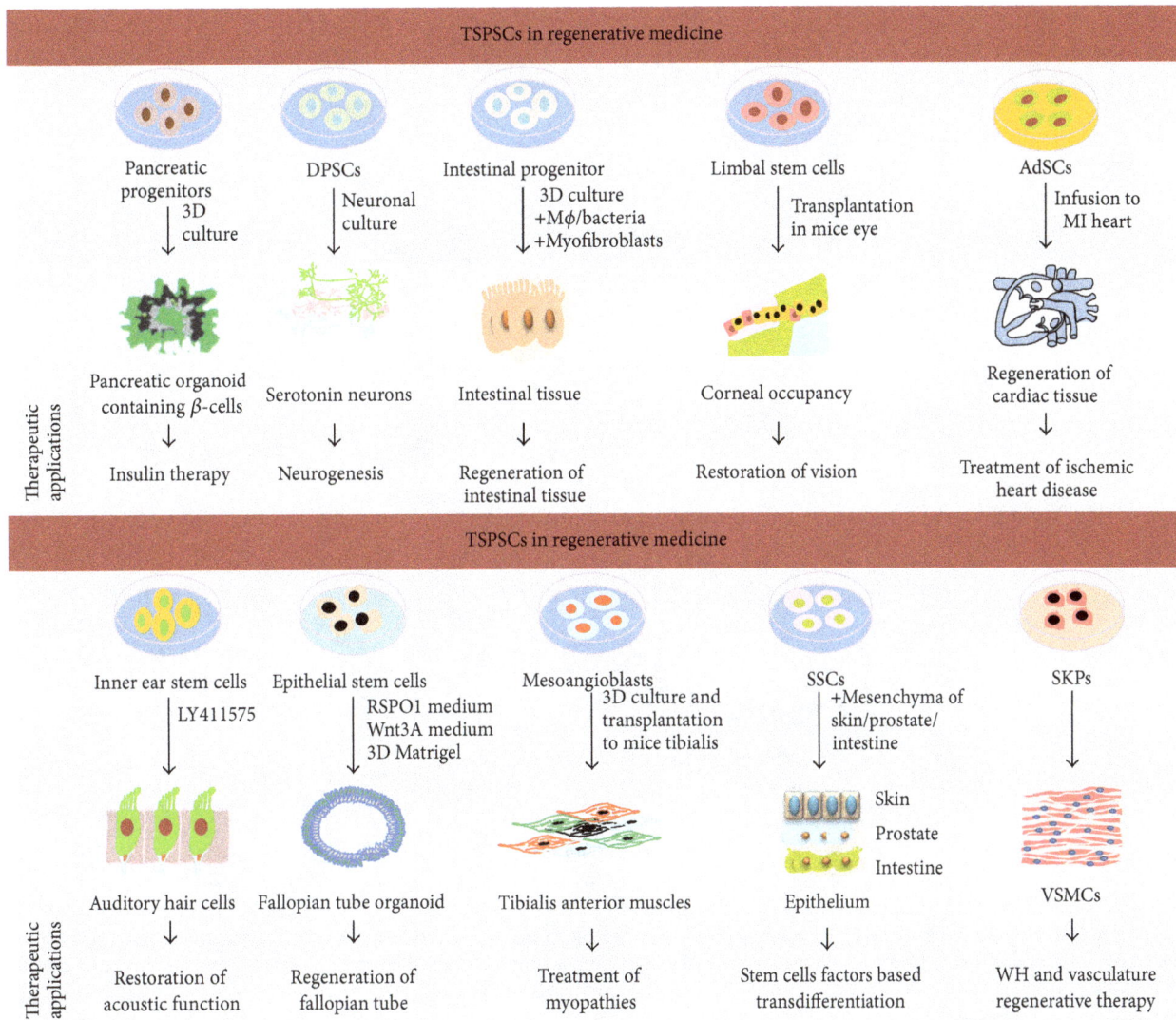

FIGURE 3: TSPSCs in regenerative medicine: tissue specific stem and progenitor cells have potential to differentiate into other cells of the tissue. Characteristically inner ear stem cells can be transformed into auditory hair cells, skin progenitors into vascular smooth muscle cells, mesoangioblasts into tibialis anterior muscles, and dental pulp stem cells into serotonin cells. The 3D-culture of TSPSCs in complex biomaterial gives rise to tissue organoids, such as pancreatic organoid from pancreatic progenitor, intestinal tissue organoids from intestinal progenitor cells, and fallopian tube organoids from fallopian tube epithelial cells. Transplantation of TSPSCs regenerates targets tissue such as regeneration of tibialis muscles from mesoangioblasts, cardiac tissue from AdSCs, and corneal tissue from limbal stem cells. Cell growth and transformation factors secreted by TSPSCs can change cells fate to become other types of cell, such that SSCs coculture with skin, prostate, and intestine mesenchyme transforms these cells from MSCs into epithelial cells fate.

Globally, especially in India, cardiovascular problems are a more common cause of human death, where biomedical therapeutics require immediate restoration of heart functions for the very survival of the patient. Regeneration of cardiac tissue can be achieved by transplantation of cardiomyocytes, ESCs-derived cardiovascular progenitors, and bone marrow derived mononuclear cells (BMDMNCs); however healing by cardiomyocytes and progenitor cells is superior to BMDMNCs but mature cardiomyocytes have higher tissue healing potential, suppress heart arrhythmias, couple electromagnetically into hearts functions, and provide mechanical and electrical repair without any associated tumorigenic effects

[18, 19]. Like CM differentiation, ESCs derived liver stem cells can be transformed into Cytp450-hepatocytes, mediating chemical modification and catabolism of toxic xenobiotic drugs [20]. Even today, availability and variability of functional hepatocytes are a major a challenge for testing drug toxicity [20]. Stimulation of ESCs and ex vivo VitK12 and lithocholic acid (a by-product of intestinal flora regulating drug metabolism during infancy) activates pregnane X receptor (PXR), CYP3A4, and CYP2C9, which leads to differentiation of ESCs into hepatocytes; those are functionally similar to primary hepatocytes, for their ability to produce albumin and apolipoprotein B100 [20]. These hepatocytes

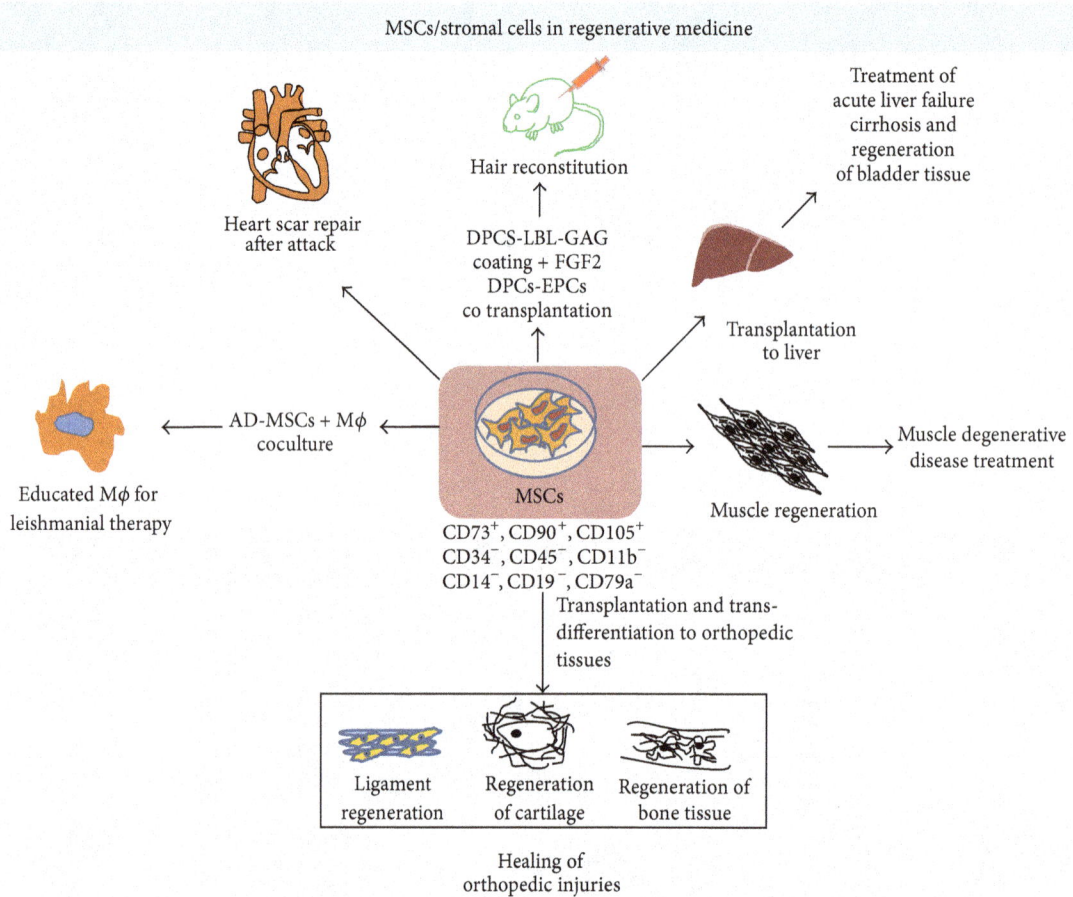

FIGURE 4: MSCs in regenerative medicine: mesenchymal stem cells are $CD73^+$, $CD90^+$, $CD105^+$, $CD34^-$, $CD45^-$, $CD11b^-$, $CD14^-$, $CD19^-$, and $CD79a^-$ cells, also known as stromal cells. These bodily MSCs represented here do not account for MSCs of bone marrow and umbilical cord. Upon transplantation and transdifferentiation these bodily MSCs regenerate into cartilage, bones, and muscles tissue. Heart scar formed after heart attack and liver cirrhosis can be treated from MSCs. ECM coating provides the niche environment for MSCs to regenerate into hair follicle, stimulating hair growth.

are excellent source for the endpoint screening of drugs for accurate prediction of clinical outcomes [20]. Generation of hepatic cells from ESCs can be achieved in multiple ways, as serum-free differentiation [21], chemical approaches [20, 22], and genetic transformation [23, 24]. These ESCs-derived hepatocytes are long lasting source for treatment of liver injuries and high throughput screening of drugs [20, 23, 24]. Transplantation of the inert biomaterial encapsulated hESCs-derived pancreatic progenitors ($CD24^+$, $CD49^+$, and $CD133^+$) differentiates into β-cells, minimizing high fat diet induced glycemic and obesity effects in mice [25] (Table 1). Addition of antidiabetic drugs into transdifferentiation regime can boost ESCs conservation into β-cells [25], which theoretically can cure T2DM permanently [25]. ESCs can be differentiated directly into insulin secreting β-cells (marked with GLUT2, INS1, GCK, and PDX1) which can be achieved through PDX1 mediated epigenetic reprogramming [26]. Globally, osteoarthritis affects millions of people and occurs when cartilage at joints wears away, causing stiffness of the joints. The available therapeutics for arthritis relieve symptoms but do not initiate reverse generation of cartilage. For young individuals and athletes replacement of joints is not feasible like old populations; in that case transplantation of stem cells represents an alternative for healing cartilage injuries [27]. Chondrocytes, the cartilage forming cells derived from hESC, embedded in fibrin gel effectively heal defective cartilage within 12 weeks, when transplanted to focal cartilage defects of knee joints in mice without any negative effect [27]. Transplanted chondrocytes form cell aggregates, positive for SOX9 and collagen II, and defined chondrocytes are active for more than 12 wks at transplantation site, advocating clinical suitability of chondrocytes for treatment of cartilage lesions [27]. The integrity of ESCs to integrate and differentiate into electrophysiologically active cells provides a means for natural regulation of heart rhythm as biological pacemaker. Coaxing of ESCs into inert biomaterial as well as propagation in defined culture conditions leads to transdifferentiation of ESCs to become sinoatrial node (SAN) pacemaker cells (PCs) [28]. Genomic incorporation TBox3 into ESCs ex vivo leads to generation of PCs-like cells; those express activated leukocyte cells adhesion molecules (ALCAM) and exhibit similarity to PCs for gene expression and immune

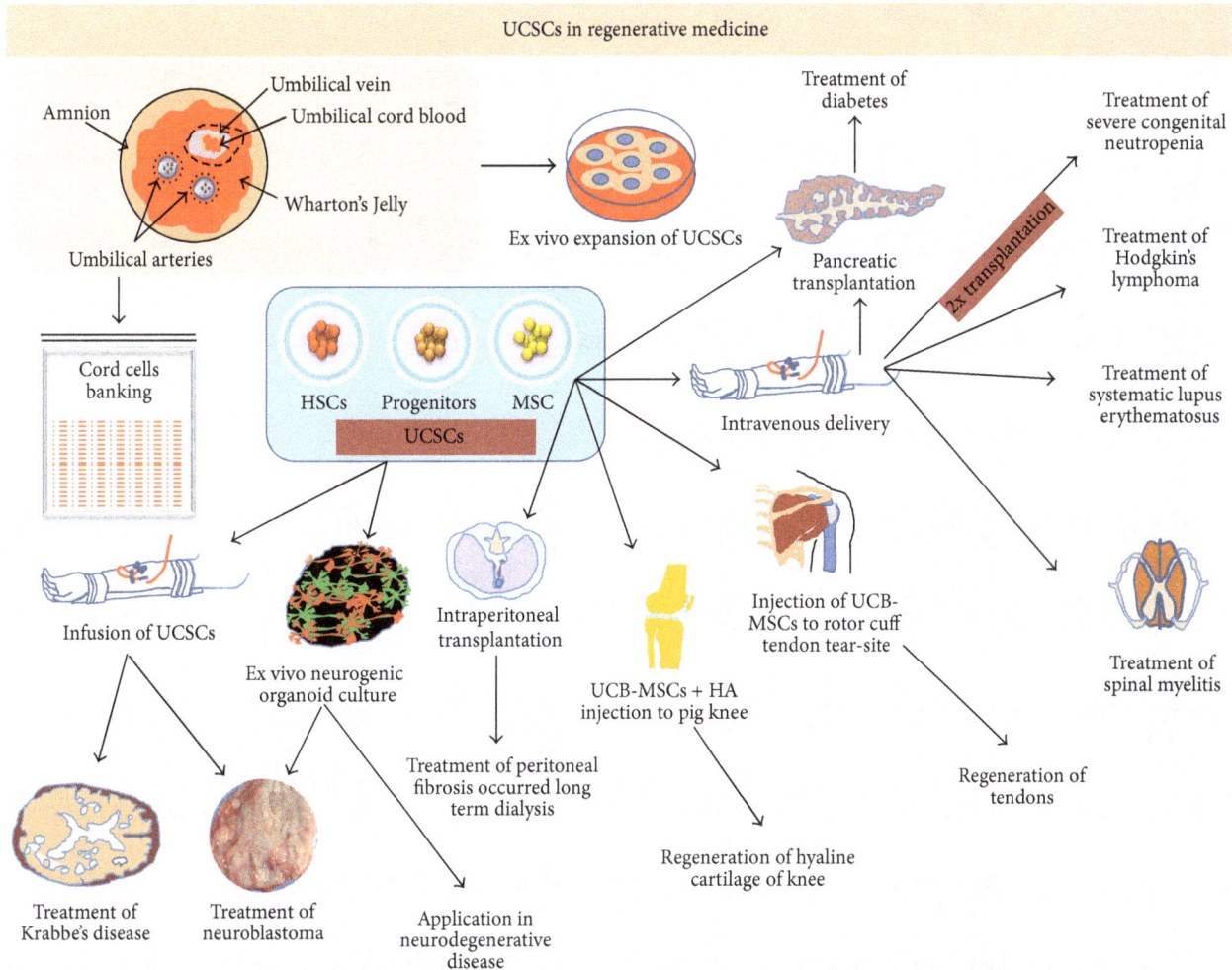

FIGURE 5: UCSCs in regenerative medicine: umbilical cord, the readily available source of stem cells, has emerged as futuristic source for personalized stem cell therapy. Transplantation of UCSCs to Krabbe's disease patients regenerates myelin tissue and recovers neuroblastoma patients through restoring tissue homeostasis. The UCSCs organoids are readily available tissue source for treatment of neurodegenerative disease. Peritoneal fibrosis caused by long term dialysis, tendon tissue degeneration, and defective hyaline cartilage can be regenerated by UCSCs. Intravenous injection of UCSCs enables treatment of diabetes, spinal myelitis, systemic lupus erythematosus, Hodgkin's lymphoma, and congenital neuropathies. Cord blood stem cells banking avails long lasting source of stem cells for personalized therapy and regenerative medicine.

functions [28]. Transplantation of PCs can restore pacemaker functions of the ailing heart [28]. In summary, ESCs can be transdifferentiated into any kinds of cells representing three germ layers of the body, being most promising source of regenerative medicine for tissue regeneration and disease therapy (Table 1). Ethical concerns limit the applications of ESCs, where set guidelines need to be followed; in that case TSPSCs, MSCs, UCSCs, BMSCs, and iPSCs can be explored as alternatives.

3. TSPSCs in Regenerative Medicine

TSPSCs maintain tissue homeostasis through continuous cell division, but, unlike ESCs, TSPSCs retain stem cells plasticity and differentiation in tissue specific manner, giving rise to few types of cells (Table 1). The number of TSPSCs population to total cells population is too low; in that case their harvesting as well as in vitro manipulation is really a tricky task [29], to explore them for therapeutic scale. Human body has foundation from various types of TSPSCs; discussing the therapeutic application for all types is not feasible. This section of review discusses therapeutic application of pancreatic progenitor cells (PPCs), dental pulp stem cells (DPSCs), inner ear stem cells (IESCs), intestinal progenitor cells (IPCs), limbal progenitor stem cells (LPSCs), epithelial progenitor stem cells (EPSCs), mesoangioblasts (MABs), spermatogonial stem cells (SSCs), the skin derived precursors (SKPs), and adipose derived stem cells (AdSCs) (Figure 3; Table 1). During embryogenesis PPCs give rise to insulin-producing β-cells. The differentiation of PPCs to become β-cells is negatively regulated by insulin [30]. PPCs require active FGF and Notch signalling; growing more rapidly in community than in single cell populations

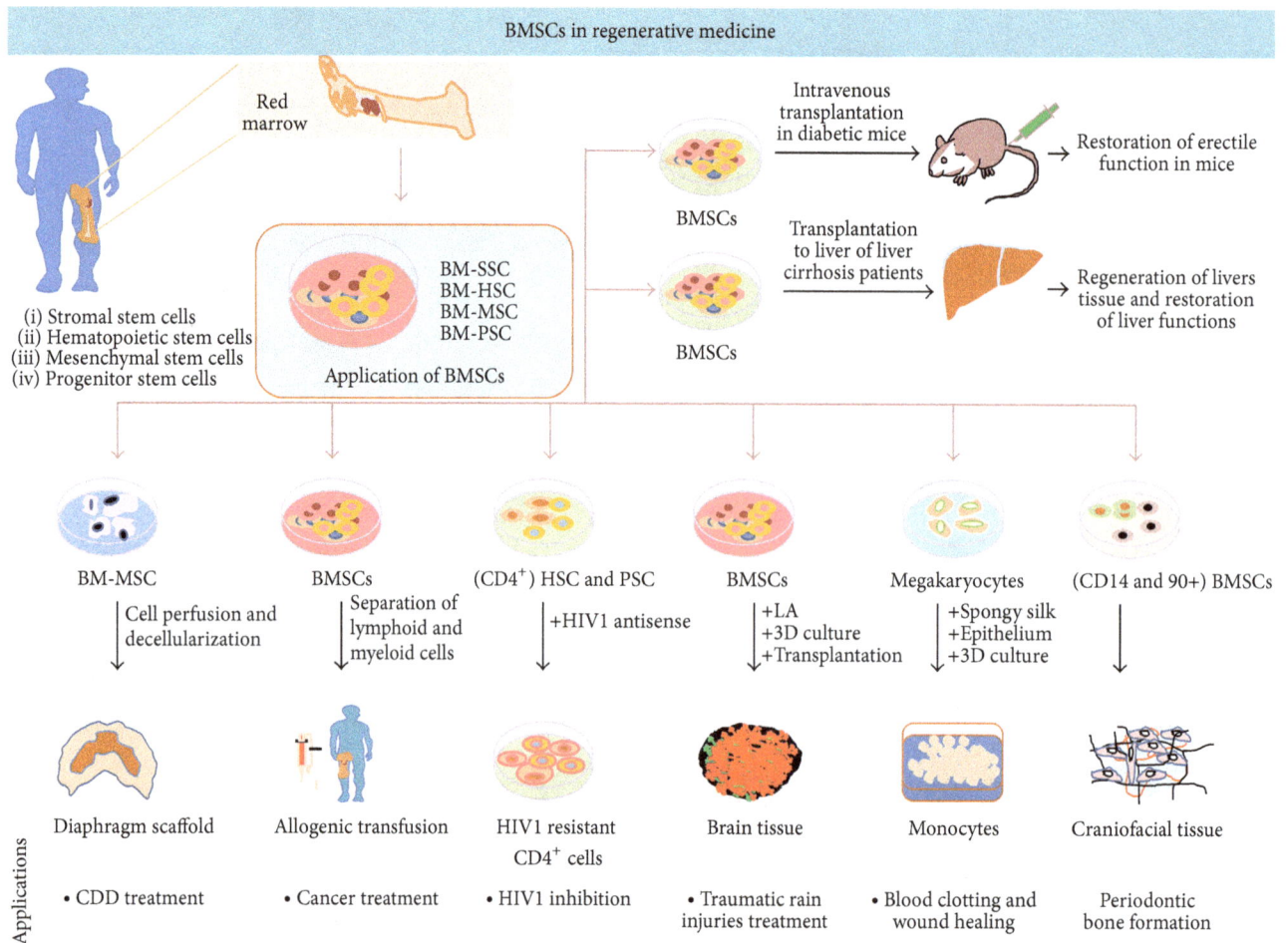

FIGURE 6: BMSCs in regenerative medicine: bone marrow, the soft sponge bone tissue that consisted of stromal, hematopoietic, and mesenchymal and progenitor stem cells, is responsible for blood formation. Even halo-HLA matched BMSCs can cure from disease and regenerate tissue. BMSCs can regenerate craniofacial tissue, brain tissue, diaphragm tissue, and liver tissue and restore erectile function and transdifferentiation monocytes. These multipotent stem cells can cure host from cancer and infection of HIV and HCV.

advocates the functional importance of niche effect in self-renewal and transdifferentiation processes. In 3D-scaffold culture system, mice embryo derived PPCs grow into hollow organoid spheres; those finally differentiate into insulin-producing β-cell clusters [29]. The DSPSCs, responsible for maintenance of teeth health status, can be sourced from apical papilla, deciduous teeth, dental follicle, and periodontal ligaments, have emerged as regenerative medicine candidate, and might be explored for treatment of various kinds of disease including restoration neurogenic functions in teeth [31, 32]. Expansion of DSPSCs in chemically defined neuronal culture medium transforms them into a mixed population of cholinergic, GABAergic, and glutaminergic neurons; those are known to respond towards acetylcholine, GABA, and glutamine stimulations in vivo. These transformed neuronal cells express nestin, glial fibrillary acidic protein (GFAP), βIII-tubulin, and voltage gated L-type Ca^{2+} channels [32]. However, absence of Na^+ and K^+ channels does not support spontaneous action potential generation, necessary for response generation against environmental stimulus. All

together, these primordial neuronal stem cells have possible therapeutic potential for treatment of neurodental problems [32]. Sometimes, brain tumor chemotherapy can cause neurodegeneration mediated cognitive impairment, a condition known as chemobrain [33]. The intrahippocampal transplantation of human derived neuronal stem cells to cyclophosphamide behavioural decremented mice restores cognitive functions in a month time. Here the transplanted stem cells differentiate into neuronal and astroglial lineage, reduce neuroinflammation, and restore microglial functions [33]. Furthermore, transplantation of stem cells, followed by chemotherapy, directs pyramidal and granule-cell neurons of the gyrus and CA1 subfields of hippocampus which leads to reduction in spine and dendritic cell density in the brain. These findings suggest that transplantation of stem cells to cranium restores cognitive functions of the chemo-brain [33]. The hair cells of the auditory system produced during development are not postmitotic; loss of hair cells cannot be replaced by inner ear stem cells, due to active state of the Notch signalling [34]. Stimulation of inner ear

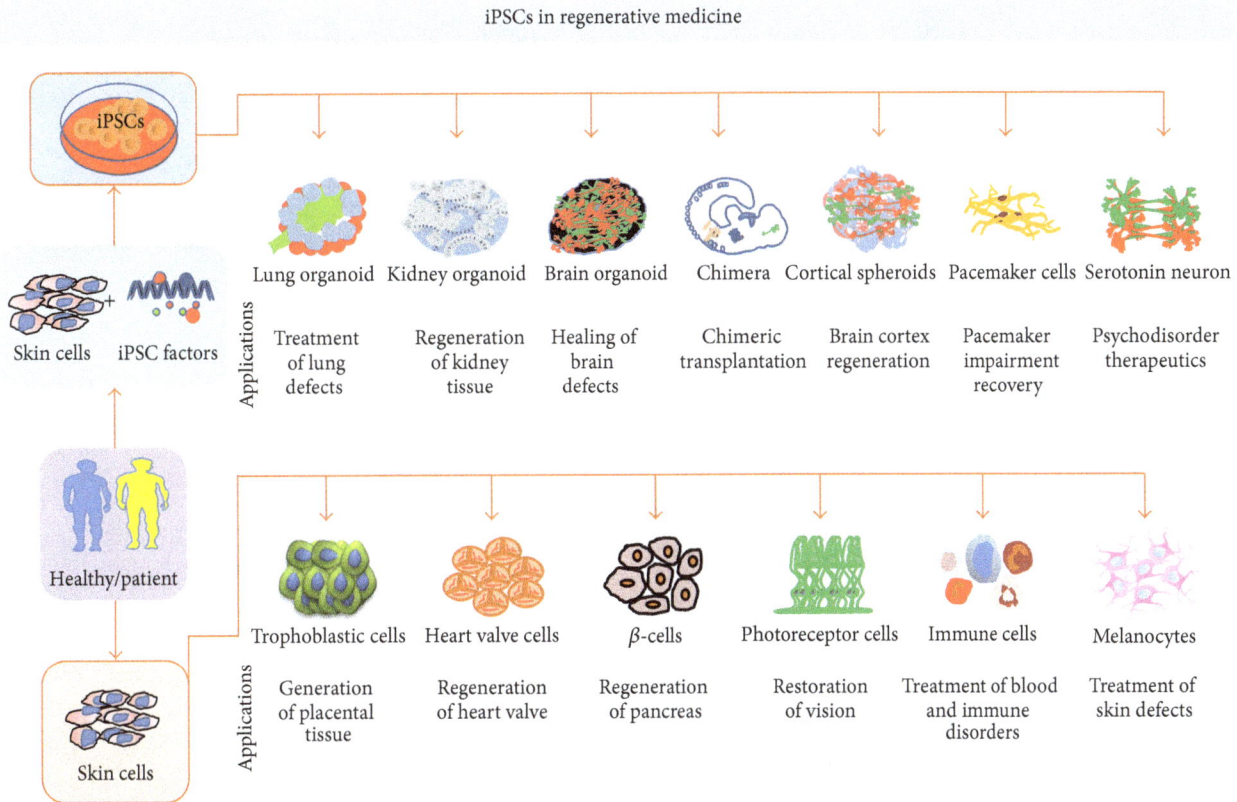

FIGURE 7: iPSCs in regenerative medicine: using the edge of iPSCs technology, skin fibroblasts and other adult tissues derived, terminally differentiated cells can be transformed into ESCs-like cells. It is possible that adult cells can be transformed into cells of distinct lineages bypassing the phase of pluripotency. The tissue specific defined culture can transform skin cells to become trophoblast, heart valve cells, photoreceptor cells, immune cells, melanocytes, and so forth. ECM complexation with iPSCs enables generation of tissue organoids for lung, kidney, brain, and other organs of the body. Similar to ESCs, iPSCs also can be transformed into cells representing three germ layers such as pacemaker cells and serotonin cells.

progenitors with Ɣ-secretase inhibitor (LY411575) abrogates Notch signalling through activation of transcription factor atonal homologue 1 (Atoh1) and directs transdifferentiation of progenitors into cochlear hair cells [34]. Transplantation of in vitro generated hair cells restores acoustic functions in mice, which can be the potential regenerative medicine candidates for the treatment of deafness [34]. Generation of the hair cells also can be achieved through overexpression of β-catenin and Atoh1 in Lrg5$^+$ cells in vivo [35]. Similar to ear progenitors, intestine of the digestive tract also has its own tissue specific progenitor stem cells, mediating regeneration of the intestinal tissue [34, 36]. Dysregulation of the common stem cells signalling pathways, Notch/BMP/TGF-β/Wnt, in the intestinal tissue leads to disease. Information on these signalling pathways [37] is critically important in designing therapeutics. Coaxing of the intestinal tissue specific progenitors with immune cells (macrophages), connective tissue cells (myofibroblasts), and probiotic bacteria into 3D-scaffolds of inert biomaterial, crafting biological environment, is suitable for differentiation of progenitors to occupy the crypt-villi structures into these scaffolds [36]. Omental implementation of these crypt-villi structures to dogs enhances intestinal mucosa through regeneration of goblet cells containing

intestinal tissue [36]. These intestinal scaffolds are close approach for generation of implantable intestinal tissue, divested by infection, trauma, cancer, necrotizing enterocolitis (NEC), and so forth [36]. In vitro culture conditions cause differentiation of intestinal stem cells to become other types of cells, whereas incorporation of valproic acid and CHIR-99021 in culture conditions avoids differentiation of intestinal stem cells, enabling generation of indefinite pool of stem cells to be used for regenerative applications [38]. The limbal stem cells of the basal limbal epithelium, marked with ABCB5, are essential for regeneration and maintenance of corneal tissue [39]. Functional status of ABCB5 is critical for survival and functional integrity of limbal stem cells, protecting them from apoptotic cell death [39]. Limbal stem cells deficiency leads to replacement of corneal epithelium with visually dead conjunctival tissue, which can be contributed by burns, inflammation, and genetic factors [40]. Transplanted human cornea stem cells to mice regrown into fully functional human cornea, possibly supported by blood eye barrier phenomena, can be used for treatment of eye diseases, where regeneration of corneal tissue is critically required for vision restoration [39]. Muscle degenerative disease like duchenne muscular dystrophy (DMD) can cause

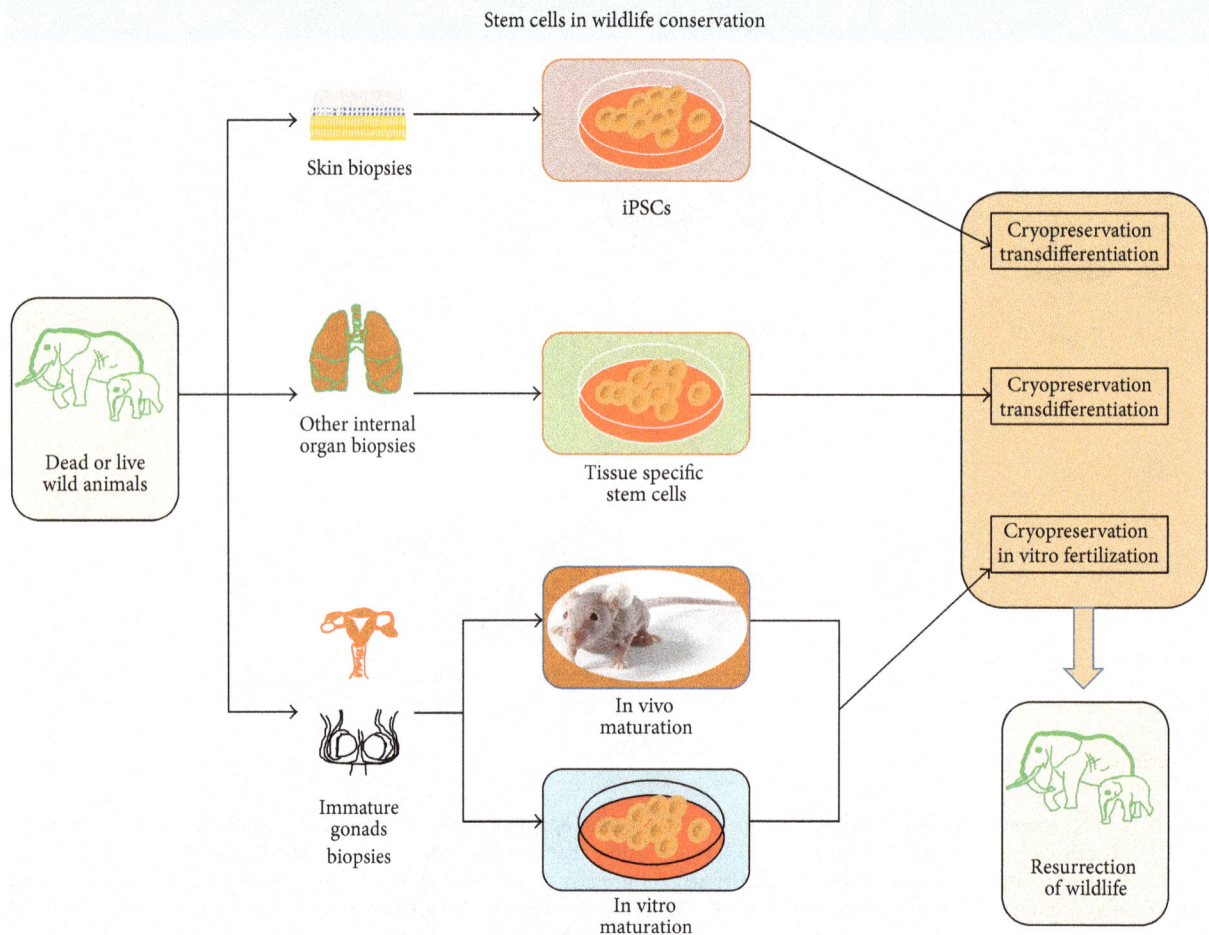

FIGURE 8: Stem cells in wildlife conservation: tissue biopsies obtained from dead and live wild animals can be either cryopreserved or transdifferentiated to other types of cells, through culture in defined culture medium or in vivo maturation. Stem cells and adult tissue derived iPSCs have great potential of regenerative medicine and disease therapeutics. Gonadal tissue procured from dead wild animals can be matured, ex vivo and in vivo for generation of sperm and egg, which can be used for assistive reproductive technology oriented captive breeding of wild animals or even for resurrection of wildlife.

extensive thrashing of muscle tissue, where tissue engineering technology can be deployed for functional restoration of tissue through regeneration [41]. Encapsulation of mouse or human derived MABs (engineered to express placental derived growth factor (PDGF)) into polyethylene glycol (PEG) fibrinogen hydrogel and their transplantation beneath the skin at ablated tibialis anterior form artificial muscles, which are functionally similar to those of normal tibialis anterior muscles [41]. The PDGF attracts various cell types of vasculogenic and neurogenic potential to the site of transplantation, supporting transdifferentiation of mesoangioblasts to become muscle fibrils [41]. The therapeutic application of MABs in skeletal muscle regeneration and other therapeutic outcomes has been reviewed by others [42]. One of the most important tissue specific stem cells, the male germline stem cells or spermatogonial stem cells (SSCs), produces spermatogenic lineage through mesenchymal and epithets cells [43] which itself creates niche effect on other cells. In vivo transplantation of SSCs with prostate, skin, and uterine mesenchyme leads to differentiation of these

cells to become epithelia of the tissue of origin [43]. These newly formed tissues exhibit all physical and physiological characteristics of prostate and skin and the physical characteristics of prostate, skin, and uterus, express tissue specific markers, and suggest that factors secreted from SSCs lead to lineage conservation which defines the importance of niche effect in regenerative medicine [43]. According to an estimate, more than 100 million people are suffering from the condition of diabetic retinopathy, a progressive dropout of vascularisation in retina that leads to loss of vision [44]. The intravitreal injection of adipose derived stem cells (AdSCs) to the eye restores microvascular capillary bed in mice. The AdSCs from healthy donor produce higher amounts of vasoprotective factors compared to glycemic mice, enabling superior vascularisation [44]. However use of AdSCs for disease therapeutics needs further standardization for cell counts in dose of transplant and monitoring of therapeutic outcomes at population scale [44]. Apart from AdSCs, other kinds of stem cells also have therapeutic potential in regenerative medicine for treatment of eye defects, which has been reviewed by others [45]. Fallopian tubes, connecting ovaries to uterus, are the sites where fertilization of the egg takes place. Infection in fallopian tubes can lead to inflammation, tissue scarring, and closure of the fallopian tube which often leads to infertility and ectopic pregnancies. Fallopian is also the site where onset of ovarian cancer takes place. The studies on origin and etiology of ovarian cancer are restricted due to lack of technical advancement for culture of epithelial cells. The in vitro 3D organoid culture of clinically obtained fallopian tube epithelial cells retains their tissue specificity, keeps cells alive, which differentiate into typical ciliated and secretory cells of fallopian tube, and advocates that ectopic examination of fallopian tube in organoid culture settings might be the ideal approach for screening of cancer [46]. The sustained growth and differentiation of fallopian TSPSCs into fallopian tube organoid depend both on the active state of the Wnt and on paracrine Notch signalling [46]. Similar to fallopian tube stem cells, subcutaneous visceral tissue specific cardiac adipose (CA) derived stem cells (AdSCs) have the potential of differentiation into cardiovascular tissue [47]. Systemic infusion of CA-AdSCs into ischemic myocardium of mice regenerates heart tissue and improves cardiac function through differentiation to endothelial cells, vascular smooth cells, and cardiomyocytes and vascular smooth cells. The differentiation and heart regeneration potential of CA-AdSCs are higher than AdSCs [48], representing CA-AdSCs as potent regenerative medicine candidates for myocardial ischemic therapy [47]. The skin derived precursors (SKPs), the progenitors of dermal papilla/hair/hair sheath, give rise to multiple tissues of mesodermal and/or ectodermal origin such as neurons, Schwann cells, adipocytes, chondrocytes, and vascular smooth muscle cells (VSMCs). VSMCs mediate wound healing and angiogenesis process can be derived from human foreskin progenitor SKPs, suggesting that SKPs derived VSMCs are potential regenerative medicine candidates for wound healing and vasculature injuries treatments [49]. In summary, TSPSCs are potentiated with tissue regeneration, where advancement in organoid culture (Figure 3; Table 1) technologies defines the importance of niche effect in tissue regeneration and therapeutic outcomes of ex vivo expanded stem cells.

4. MSCs/Stromal Cells in Regenerative Medicine

MSCs, the multilineage stem cells, differentiate only to tissue of mesodermal origin, which includes tendons, bone, cartilage, ligaments, muscles, and neurons [50]. MSCs are the cells which express combination of markers: $CD73^+$, $CD90^+$, $CD105^+$, $CD11b^-$, $CD14^-$, $CD19^-$, $CD34^-$, $CD45^-$, $CD79a^-$, and HLA-DR, reviewed elsewhere [50]. The application of MSCs in regenerative medicine can be generalized from ongoing clinical trials, phasing through different state of completions, reviewed elsewhere [90]. This section of review outlines the most recent representative applications of MSCs (Figure 4; Table 1). The anatomical and physiological characteristics of both donor and receiver have equal impact on therapeutic outcomes. The bone marrow derived MSCs

(BMDMSCs) from baboon are morphologically and phenotypically similar to those of bladder stem cells and can be used in regeneration of bladder tissue. The BMDMSCs ($CD105^+$, $CD73^+$, $CD34^-$, and $CD45^-$), expressing GFP reporter, coaxed with small intestinal submucosa (SIS) scaffolds, augment healing of degenerated bladder tissue within 10 wks of the transplantation [51]. The combinatorial CD characterized MACs are functionally active at transplantation site, which suggests that CD characterization of donor MSCs yields superior regenerative outcomes [51]. MSCs also have potential to regenerate liver tissue and treat liver cirrhosis, reviewed elsewhere [91]. The regenerative medicinal application of MSCs utilizes cells in two formats as direct transplantation or first transdifferentiation and then transplantation; ex vivo transdifferentiation of MSCs deploys retroviral delivery system that can cause oncogenic effect on cells. Nonviral, NanoScript technology, comprising utility of transcription factors (TFs) functionalized gold nanoparticles, can target specific regulatory site in the genome effectively and direct differentiation of MSCs into another cell fate, depending on regime of TFs. For example, myogenic regulatory factor containing NanoScript-MRF differentiates the adipose tissue derived MSCs into muscle cells [92]. The multipotency characteristics represent MSCs as promising candidate for obtaining stable tissue constructs through coaxed 3D organoid culture; however heterogeneous distribution of MSCs slows down cell proliferation, rendering therapeutic applications of MSCs. Adopting two-step culture system for MSCs can yield homogeneous distribution of MSCs in biomaterial scaffolds. For example, fetal-MSCs coaxed in biomaterial when cultured first in rotating bioreactor followed with static culture lead to homogeneous distribution of MSCs in ECM components [7]. Occurrence of dental carries, periodontal disease, and tooth injury can impact individual's health, where bioengineering of teeth can be the alternative option. Coaxing of epithelial-MSCs with dental stem cells into synthetic polymer gives rise to mature teeth unit, which consisted of mature teeth and oral tissue, offering multiple regenerative therapeutics, reviewed elsewhere [52]. Like the tooth decay, both human and animals are prone to orthopedic injuries, affecting bones, joint, tendon, muscles, cartilage, and so forth. Although natural healing potential of bone is sufficient to heal the common injuries, severe trauma and tumor-recession can abrogate germinal potential of bone-forming stem cells. In vitro chondrogenic, osteogenic, and adipogenic potential of MSCs advocates therapeutic applications of MSCs in orthopedic injuries [53]. Seeding of MSCs, coaxed into biomaterial scaffolds, at defective bone tissue, regenerates defective bone tissues, within four wks of transplantation; by the end of 32 wks newly formed tissues integrate into old bone [54]. Osteoblasts, the bone-forming cells, have lesser actin cytoskeleton compared to adipocytes and MSCs. Treatment of MSCs with cytochalasin-D causes rapid transportation of G-actin, leading to osteogenic transformation of MSCs. Furthermore, injection of cytochalasin-D to mice tibia also promotes bone formation within a wk time frame [55]. The bone formation processes in mice, dog, and human are fundamentally similar, so outcomes of research on mice and dogs can be directional for regenerative application to human. Injection of MSCs to femur head of Legg-Calve-Perthes suffering dog heals the bone very fast and reduces the injury associated pain [55]. Degeneration of skeletal muscle and muscle cramps are very common to sledge dogs, animals, and individuals involved in adventurous athletics activities. Direct injection of adipose tissue derived MSCs to tear-site of semitendinosus muscle in dogs heals injuries much faster than traditional therapies [56]. Damage effect treatment for heart muscle regeneration is much more complex than regeneration of skeletal muscles, which needs high grade fine-tuned coordination of neurons with muscles. Coaxing of MSCs into alginate gel increases cell retention time that leads to releasing of tissue repairing factors in controlled manner. Transplantation of alginate encapsulated cells to mice heart reduces scar size and increases vascularisation, which leads to restoration of heart functions. Furthermore, transplanted MSCs face host inhospitable inflammatory immune responses and other mechanical forces at transplantation site, where encapsulation of cells keeps them away from all sorts of mechanical forces and enables sensing of host tissue microenvironment, and respond accordingly [57]. Ageing, disease, and medicine consumption can cause hair loss, known as alopecia. Although alopecia has no life threatening effects, emotional catchments can lead to psychological disturbance. The available treatments for alopecia include hair transplantation and use of drugs, where drugs are expensive to afford and generation of new hair follicle is challenging. Dermal papillary cells (DPCs), the specialized MSCs localized in hair follicle, are responsible for morphogenesis of hair follicle and hair cycling. The layer-by-layer coating of DPCs, called GAG coating, consists of coating of geletin as outer layer, middle layer of fibroblast growth factor 2 (FGF2) loaded alginate, and innermost layer of geletin. GAG coating creates tissue microenvironment for DPCs that can sustain immunological and mechanical obstacles, supporting generation of hair follicle. Transplantation of GAG-coated DPCs leads to abundant hair growth and maturation of hair follicle, where GAG coating serves as ECM, enhancing intrinsic therapeutic potential of DPCs [58]. During infection, the inflammatory cytokines secreted from host immune cells attract MSCs to the site of inflammation, which modulates inflammatory responses, representing MSCs as key candidate of regenerative medicine for infectious disease therapeutics. Coculture of macrophages (Mϕ) and adipose derived MSCs from *Leishmania major* (LM) susceptible and resistant mice demonstrates that AD-MSCs educate Mϕ against LM infection, differentially inducing M1 and M2 phenotype that represents AD-MSC as therapeutic agent for leishmanial therapy [93]. In summary, the multilineage differentiation potential of MSCs, as well as adoption of next-generation organoid culture system, avails MSCs as ideal regenerative medicine candidate.

5. UCSCs in Regenerative Medicine

Umbilical cord, generally thrown at the time of child birth, is the best known source for stem cells, procured in noninvasive manner, having lesser ethical constraints than ESCs.

Umbilical cord is rich source of hematopoietic stem cells (HSCs) and MSCs, which possess enormous regeneration potential [94] (Figure 5; Table 1). The HSCs of cord blood are responsible for constant renewal of all types of blood cells and protective immune cells. The proliferation of HSCs is regulated by Musashi-2 protein mediated attenuation of Aryl hydrocarbon receptor (AHR) signalling in stem cells [95]. UCSCs can be cryopreserved at stem cells banks (Figure 5; Table 1), in operation by both private and public sector organization. Public stem cells banks operate on donation formats and perform rigorous screening for HLA typing and donated UCSCs remain available to anyone in need, whereas private stem cell banks operation is more personalized, availing cells according to donor consent. Stem cell banking is not so common, even in developed countries. Survey studies find that educated women are more eager to donate UCSCs, but willingness for donation decreases with subsequent deliveries, due to associated cost and safety concerns for preservation [96]. FDA has approved five HSCs for treatment of blood and other immunological complications [97]. The amniotic fluid, drawn during pregnancy for standard diagnostic purposes, is generally discarded without considering its vasculogenic potential. UCSCs are the best alternatives for those patients who lack donors with fully matched HLA typing for peripheral blood and PBMCs and bone marrow [98]. One major issue with UCSCs is number of cells in transplant, fewer cells in transplant require more time for engraftment to mature, and there are also risks of infection and mortality; in that case ex vivo propagation of UCSCs can meet the demand of desired outcomes. There are diverse protocols, available for ex vivo expansion of UCSCs, reviewed elsewhere [99]. Amniotic fluid stem cells (AFSCs), coaxed to fibrin (required for blood clotting, ECM interactions, wound healing, and angiogenesis) hydrogel and PEG supplemented with vascular endothelial growth factor (VEGF), give rise to vascularised tissue, when grafted to mice, suggesting that organoid cultures of UCSCs have promise for generation of biocompatible tissue patches, for treating infants born with congenital heart defects [59]. Retroviral integration of OCT4, KLF4, cMYC, and SOX2 transforms AFSCs into pluripotency stem cells known as AFiPSCs which can be directed to differentiate into extraembryonic trophoblast by BMP2 and BMP4 stimulation, which can be used for regeneration of placental tissues [60]. Wharton's jelly (WJ), the gelatinous substance inside umbilical cord, is rich in mucopolysaccharides, fibroblast, macrophages, and stem cells. The stem cells from UCB and WJ can be transdifferentiated into β-cells. Homogeneous nature of WJ-SCs enables better differentiation into β-cells; transplantation of these cells to streptozotocin induced diabetic mice efficiently brings glucose level to normal [7]. Easy access and expansion potential and plasticity to differentiate into multiple cell lineages represent WJ as an ideal candidate for regenerative medicine but cells viability changes with passages with maximum viable population at 5th-6th passages. So it is suggested to perform controlled expansion of WJ-MSCS for desired regenerative outcomes [9]. Study suggests that CD34$^+$ expression leads to the best regenerative outcomes, with less chance of host-versus-graft rejection. In vitro expansion of UCSCs, in presence of StemRegenin-1 (SR-1), conditionally expands CD34$^+$ cells [61]. In type I diabetic mellitus (T1DM), T-cell mediated autoimmune destruction of pancreatic β-cells occurs, which has been considered as tough to treat. Transplantation of WJ-SCs to recent onset-T1DM patients restores pancreatic function, suggesting that WJ-MSCs are effective in regeneration of pancreatic tissue anomalies [62]. WJ-MSCs also have therapeutic importance for treatment of T2DM. A non-placebo controlled phase I/II clinical trial demonstrates that intravenous and intrapancreatic endovascular injection of WJ-MSCs to T2DM patients controls fasting glucose and glycated haemoglobin through improvement of β-cells functions, evidenced by enhanced c-peptides and reduced inflammatory cytokines (IL-1β and IL-6) and T-cells counts [63]. Like diabetes, systematic lupus erythematosus (SLE) also can be treated with WJ-MSCs transplantation. During progression of SLE host immune system targets its own tissue leading to degeneration of renal, cardiovascular, neuronal, and musculoskeletal tissues. A non-placebo controlled follow-up study on 40 SLE patients demonstrates that intravenous infusion of WJ-MSC improves renal functions and decreases systematic lupus erythematosus disease activity index (SLEDAI) and British Isles Lupus Assessment Group (BILAG), and repeated infusion of WJ-MSCs protects the patient from relapse of the disease [64]. Sometimes, host inflammatory immune responses can be detrimental for HSCs transplantation and blood transfusion procedures. Infusion of WJ-MSC to patients, who had allogenic HSCs transplantation, reduces haemorrhage inflammation (HI) of bladder, suggesting that WJ-MSCs are potential stem cells adjuvant in HSCs transplantation and blood transfusion based therapies [100]. Apart from WJ, umbilical cord perivascular space and cord vein are also rich source for obtaining MSCs. The perivascular MSCs of umbilical cord are more primitive than WJ-MSCs and other MSCs from cord suggest that perivascular MSCs might be used as alternatives for WJ-MSCs for regenerative therapeutics outcome [101]. Based on origin, MSCs exhibit differential in vitro and in vivo properties and advocate functional characterization of MSCs, prior to regenerative applications. Emerging evidence suggests that UCSCs can heal brain injuries, caused by neurodegenerative diseases like Alzheimer's, Krabbe's disease, and so forth. Krabbe's disease, the infantile lysosomal storage disease, occurs due to deficiency of myelin synthesizing enzyme (MSE), affecting brain development and cognitive functions. Progression of neurodegeneration finally leads to death of babies aged two. Investigation shows that healing of peripheral nervous system (PNS) and central nervous system (CNS) tissues with Krabbe's disease can be achieved by allogenic UCSCs. UCSCs transplantation to asymptomatic infants with subsequent monitoring for 4–6 years reveals that UCSCs recover babies from MSE deficiency, improving myelination and cognitive functions, compared to those of symptomatic babies. The survival rate of transplanted UCSCs in asymptomatic and symptomatic infants was 100% and 43%, respectively, suggesting that early diagnosis and timely treatment are critical for UCSCs acceptance for desired therapeutic outcomes. UCSCs are more primitive than BMSCs, so perfect HLA

typing is not critically required, representing UCSCs as an excellent source for treatment of all the diseases involving lysosomal defects, like Krabbe's disease, hurler syndrome, adrenoleukodystrophy (ALD), metachromatic leukodystrophy (MLD), Tay-Sachs disease (TSD), and Sandhoff disease [65]. Brain injuries often lead to cavities formation, which can be treated from neuronal parenchyma, generated ex vivo from UCSCs. Coaxing of UCSCs into human originated biodegradable matrix scaffold and in vitro expansion of cells in defined culture conditions lead to formation of neuronal organoids, within three wks' time frame. These organoids structurally resemble brain tissue and consisted of neuroblasts (GFAP$^+$, Nestin$^+$, and Ki67$^+$) and immature stem cells (OCT4$^+$ and SOX2$^+$). The neuroblasts of these organoids further can be differentiated into mature neurons (MAP2$^+$ and TUJ1$^+$) [66]. Administration of high dose of drugs in divesting neuroblastoma therapeutics requires immediate restoration of hematopoiesis. Although BMSCs had been promising in restoration of hematopoiesis UCSCs are sparely used in clinical settings. A case study demonstrates that neuroblastoma patients who received autologous UCSCs survive without any associated side effects [12]. During radiation therapy of neoplasm, spinal cord myelitis can occur, although occurrence of myelitis is a rare event and usually such neurodegenerative complication of spinal cord occurs 6–24 years after exposure to radiations. Transplantation of allogenic UC-MSCs in laryngeal patients undergoing radiation therapy restores myelination [102]. For treatment of neurodegenerative disease like Alzheimer's disease (AD), amyotrophic lateral sclerosis (ALS), traumatic brain injuries (TBI), Parkinson's, SCI, stroke, and so forth, distribution of transplanted UCSCs is critical for therapeutic outcomes. In mice and rat, injection of UCSCs and subsequent MRI scanning show that transplanted UCSCs migrate to CNS and multiple peripheral organs [67]. For immunomodulation of tumor cells disease recovery, transplantation of allogenic DCs is required. The CD11c$^+$DCs, derived from UCB, are morphologically and phenotypically similar to those of peripheral blood derived CTLs-DCs, suggesting that UCB-DCs can be used for personalized medicine of cancer patient, in need for DCs transplantation [103]. Coculture of UCSCs with radiation exposed human lung fibroblast stops their transdifferentiation, which suggests that factors secreted from UCSCs may restore niche identity of fibroblast, if they are transplanted to lung after radiation therapy [104]. Tearing of shoulder cuff tendon can cause severe pain and functional disability, whereas ultrasound guided transplantation of UCB-MSCs in rabbit regenerates subscapularis tendon in four wks' time frame, suggesting that UCB-MSCs are effective enough to treat tendons injuries when injected to focal points of tear-site [68]. Furthermore, transplantation of UCB-MSCs to chondral cartilage injuries site in pig knee along with HA hydrogel composite regenerates hyaline cartilage [69], suggesting that UCB-MSCs are effective regenerative medicine candidate for treating cartilage and ligament injuries. Physiologically circulatory systems of brain, placenta, and lungs are similar. Infusion of UCB-MSCs to preeclampsia (PE) induced hypertension mice reduces the endotoxic effect, suggesting that UC-MSCs are potential source for treatment of endotoxin induced hypertension during pregnancy, drug abuse, and other kinds of inflammatory shocks [105]. Transplantation of UCSCs to severe congenital neutropenia (SCN) patients restores neutrophils count from donor cells without any side effect, representing UCSCs as potential alternative for SCN therapy, when HLA matched bone marrow donors are not accessible [106]. In clinical settings, the success of myocardial infarction (MI) treatment depends on ageing, systemic inflammation in host, and processing of cells for infusion. Infusion of human hyaluronan hydrogel coaxed UCSCs in pigs induces angiogenesis, decreases scar area, improves cardiac function at preclinical level, and suggests that the same strategy might be effective for human [107]. In stem cells therapeutics, UCSCs transplantation can be either autologous or allogenic. Sometimes, the autologous UCSCs transplants cannot combat over tumor relapse, observed in Hodgkin's lymphoma (HL), which might require second dose transplantation of allogenic stem cells, but efficacy and tolerance of stem cells transplant need to be addressed, where tumor replace occurs. A case study demonstrates that second dose allogenic transplants of UCSCs effective for HL patients, who had heavy dose in prior transplant, increase the long term survival chances by 30% [10]. Patients undergoing long term peritoneal renal dialysis are prone to peritoneal fibrosis and can change peritoneal structure and failure of ultrafiltration processes. The intraperitoneal (IP) injection of WJ-MSCs prevents methylglyoxal induced programmed cell death and peritoneal wall thickening and fibrosis, suggesting that WJ-MSCs are effective in therapeutics of encapsulating peritoneal fibrosis [70]. In summary, UCB-HSCs, WJ-MSCs, perivascular MSCs, and UCB-MSCs have tissue regeneration potential.

6. BMSCs in Regenerative Medicine

Bone marrow found in soft spongy bones is responsible for formation of all peripheral blood and comprises hematopoietic stem cells (producing blood cells) and stromal cells (producing fat, cartilage, and bones) [108] (Figure 6; Table 1). Visually bone marrow has two types, red marrow (myeloid tissue; producing RBC, platelets, and most of WBC) and yellow marrow (producing fat cells and some WBC) [108]. Imbalance in marrow composition can culminate to the diseased condition. Since 1980, bone marrow transplantation is widely accepted for cancer therapeutics [109]. In order to avoid graft rejection, HLA typing of donors is a must, but completely matched donors are limited to family members, which hampers allogenic transplantation applications. Since matching of all HLA antigens is not critically required, in that case defining the critical antigens for haploidentical allogenic donor for patients, who cannot find fully matched donor, might relieve from donor constraints. Two-step administration of lymphoid and myeloid BMSCs from haploidentical donor to the patients of aplastic anaemia and haematological malignancies reconstructs host immune system and the outcomes are almost similar to fully matched transplants, which recommends that profiling of critically important HLA

is sufficient for successful outcomes of BMSCs transplantation. Haploidentical HLA matching protocol is the major process for minorities and others who do not have access to matched donor [71]. Furthermore, antigen profiling is not the sole concern for BMSCs based therapeutics. For example, restriction of HIV1 (human immune deficiency virus) infection is not feasible through BMSCs transplantation because HIV1 infection is mediated through CD4$^+$ receptors, chemokine CXC motif receptor 4 (CXCR4), and chemokine receptor 5 (CCR5) for infecting and propagating into T helper (Th), monocytes, macrophages, and dendritic cells (DCs). Genetic variation in CCR2 and CCR5 receptors is also a contributory factor; mediating protection against infection has been reviewed elsewhere [110]. Engineering of hematopoietic stem and progenitor cells (HSPCs) derived CD4$^+$ cells to express HIV1 antagonistic RNA, specifically designed for targeting HIV1 genome, can restrict HIV1 infection, through immune elimination of latently infected CD4$^+$ cells. A single dose infusion of genetically modified (GM), HIV1 resistant HSPCs can be the alternative of HIV1 retroviral therapy. In the present scenario stem cells source, patient selection, transplantation-conditioning regimen, and postinfusion follow-up studies are the major factors, which can limit application of HIV1 resistant GM-HSPCs (CD4$^+$) cells application in AIDS therapy [72, 73]. Platelets, essential for blood clotting, are formed from megakaryocytes inside the bone marrow [74]. Due to infection, trauma, and cancer, there are chances of bone marrow failure. To an extent, spongy bone marrow microenvironment responsible for lineage commitment can be reconstructed ex vivo [75]. The ex vivo constructed 3D-scaffolds consisted of microtubule and silk sponge, flooded with chemically defined organ culture medium, which mimics bone marrow environment. The coculture of megakaryocytes and embryonic stem cells (ESCs) in this microenvironment leads to generation of functional platelets from megakaryocytes [75]. The ex vivo 3D-scaffolds of bone microenvironment can stride the path for generation of platelets in therapeutic quantities for regenerative medication of burns [75] and blood clotting associated defects. Accidents, traumatic injuries, and brain stroke can deplete neuronal stem cells (NSCs), responsible for generation of neurons, astrocytes, and oligodendrocytes. Brain does not repopulate NSCs and heal traumatic injuries itself and transplantation of BMSCs also can heal neurodegeneration alone. Lipoic acid (LA), a known pharmacological antioxidant compound used in treatment of diabetic and multiple sclerosis neuropathy when combined with BMSCs, induces neovascularisation at focal cerebral injuries, within 8 wks of transplantation. Vascularisation further attracts microglia and induces their colonization into scaffold, which leads to differentiation of BMSCs to become brain tissue, within 16 wks of transplantation. In this approach, healing of tissue directly depends on number of BMSCs in transplantation dose [76]. Dental caries and periodontal disease are common craniofacial disease, often requiring jaw bone reconstruction after removal of the teeth. Traditional therapy focuses on functional and structural restoration of oral tissue, bone, and teeth rather than biological restoration, but BMSCs based therapies promise for regeneration of craniofacial bone defects, enabling replacement of missing teeth in restored bones with dental implants. Bone marrow derived CD14[+] and CD90[+] stem and progenitor cells, termed as tissue repair cells (TRC), accelerate alveolar bone regeneration and reconstruction of jaw bone when transplanted in damaged craniofacial tissue, earlier to oral implants. Hence, TRC therapy reduces the need of secondary bone grafts, best suited for severe defects in oral bone, skin, and gum, resulting from trauma, disease, or birth defects [77]. Overall, HSCs have great value in regenerative medicine, where stem cells transplantation strategies explore importance of niche in tissue regeneration. Prior to transplantation of BMSCs, clearance of original niche from target tissue is necessary for generation of organoid and organs without host-versus-graft rejection events. Some genetic defects can lead to disorganization of niche, leading to developmental errors. Complementation with human blastocyst derived primary cells can restore niche function of pancreas in pigs and rats, which defines the concept for generation of clinical grade human pancreas in mice and pigs [111]. Similar to other organs, diaphragm also has its own niche. Congenital defects in diaphragm can affect diaphragm functions. In the present scenario functional restoration of congenital diaphragm defects by surgical repair has risk of reoccurrence of defects or incomplete restoration [8]. Decellularization of donor derived diaphragm offers a way for reconstruction of new and functionally compatible diaphragm through niche modulation. Tissue engineering technology based decellularization of diaphragm and simultaneous perfusion of bone marrow mesenchymal stem cells (BM-MSCs) facilitates regeneration of functional scaffolds of diaphragm tissues [8]. In vivo replacement of hemidiaphragm in rats with reseeded scaffolds possesses similar myography and spirometry as it has in vivo in donor rats. These scaffolds retaining natural architecture are devoid of immune cells, retaining intact extracellular matrix that supports adhesion, proliferation, and differentiation of seeded cells [8]. These findings suggest that cadaver obtained diaphragm, seeded with BM-MSCs, can be used for curing patients in need for restoration of diaphragm functions (Figure 6; Table 1). However, BMSCs are heterogeneous population, which might result in differential outcomes in clinical settings; however clonal expansion of BMSCs yields homogenous cells population for therapeutic application [8]. One study also finds that intracavernous delivery of single clone BMSCs can restore erectile function in diabetic mice [112] and the same strategy might be explored for adult human individuals. The infection of hepatitis C virus (HCV) can cause liver cirrhosis and degeneration of hepatic tissue. The intraparenchymal transplantation of bone marrow mononuclear cells (BMMNCs) into liver tissue decreases aspartate aminotransferase (AST), alanine transaminase (ALT), bilirubin, CD34, and α-SMA, suggesting that transplanted BMSCs restore hepatic functions through regeneration of hepatic tissues [113]. In order to meet the growing demand for stem cells transplantation therapy, donor encouragement is always required [8]. The stem cells donation procedure is very simple; with consent donor gets an injection of granulocyte-colony stimulating factor (G-CSF) that increases BMSCs population. Bone marrow

collection is done from hip bone using syringe in 4-5 hrs, requiring local anaesthesia and within a wk time frame donor gets recovered donation associated weakness.

7. iPSCs in Regenerative Medicine

The field of iPSCs technology and research is new to all other stem cells research, emerging in 2006 when, for the first time, Takahashi and Yamanaka generated ESCs-like cells through genetic incorporation of four factors, Sox2, Oct3/4, Klf4, and c-Myc, into skin fibroblast [3]. Due to extensive nuclear reprogramming, generated iPSCs are indistinguishable from ESCs, for their transcriptome profiling, epigenetic markings, and functional competence [3], but use of retrovirus in transdifferentiation approach has questioned iPSCs technology. Technological advancement has enabled generation of iPSCs from various kinds of adult cells phasing through ESCs or direct transdifferentiation. This section of review outlines most recent advancement in iPSC technology and regenerative applications (Figure 7; Table 1). Using the new edge of iPSCs technology, terminally differentiated skin cells directly can be transformed into kidney organoids [114], which are functionally and structurally similar to those of kidney tissue in vivo. Up to certain extent kidneys heal themselves; however natural regeneration potential cannot meet healing for severe injuries. During kidneys healing process, a progenitor stem cell needs to become 20 types of cells, required for waste excretion, pH regulation, and restoration of water and electrolytic ions. The procedure for generation of kidney organoids ex vivo, containing functional nephrons, has been identified for human. These ex vivo kidney organoids are similar to fetal first-trimester kidneys for their structure and physiology. Such kidney organoids can serve as model for nephrotoxicity screening of drugs, disease modelling, and organ transplantation. However generation of fully functional kidneys is a far seen event with today's scientific technologies [114]. Loss of neurons in age-related macular degeneration (ARMD) is the common cause of blindness. At preclinical level, transplantation of iPSCs derived neuronal progenitor cells (NPCs) in rat limits progression of disease through generation of 5-6 layers of photoreceptor nuclei, restoring visual acuity [78]. The various approaches of iPSCs mediated retinal regeneration including ARMD have been reviewed elsewhere [79]. Placenta, the cordial connection between mother and developing fetus, gets degenerated in certain pathophysiological conditions. Nuclear programming of OCT4 knock-out (KO) and wild type (WT) mice fibroblast through transient expression of GATA3, EOMES, TFAP2C, and +/− cMYC generates transgene independent trophoblast stem-like cells (iTSCs), which are highly similar to blastocyst derived TSCs for DNA methylation, H3K7ac, nucleosome deposition of H2A.X, and other epigenetic markings. Chimeric differentiation of iTSCs specifically gives rise to haemorrhagic lineages and placental tissue, bypassing pluripotency phase, opening an avenue for generation of fully functional placenta for human [115]. Neurodegenerative disease like Alzheimer's and obstinate epilepsies can degenerate cerebrum, controlling excitatory and inhibitory signals of the brain. The inhibitory tones in cerebral cortex and hippocampus are accounted by γ-amino butyric acid secreting (GABAergic) interneurons (INs). Loss of these neurons often leads to progressive neurodegeneration. Genomic integration of Ascl1, Dlx5, Foxg1, and Lhx6 to mice and human fibroblast transforms these adult cells into GABAergic-INs (iGABA-INs). These cells have molecular signature of telencephalic INs, release GABA, and show inhibition to host granule neuronal activity [81]. Transplantation of these INs in developing embryo cures from genetic and acquired seizures, where transplanted cells disperse and mature into functional neuronal circuits as local INs [82]. Dorsomorphin and SB-431542 mediated inhibition of TGF-β and BMP signalling direct transformation of human iPSCs into cortical spheroids. These cortical spheroids consisted of both peripheral and cortical neurons, surrounded by astrocytes, displaying transcription profiling and electrophysiology similarity with developing fetal brain and mature neurons, respectively [83]. The underlying complex biology and lack of clear etiology and genetic reprogramming and difficulty in recapitulation of brain development have barred understanding of pathophysiology of autism spectrum disorder (ASD) and schizophrenia. 3D organoid cultures of ASD patient derived iPSC generate miniature brain organoid, resembling fetal brain few months after gestation. The idiopathic conditions of these organoids are similar with brain of ASD patients; both possess higher inhibitory GABAergic neurons with imbalanced neuronal connection. Furthermore these organoids express forkhead Box G1 (FOXG1) much higher than normal brain tissue, which explains that FOXG1 might be the leading cause of ASD [84]. Degeneration of other organs and tissues also has been reported, like degeneration of lungs which might occur due to tuberculosis infection, fibrosis, and cancer. The underlying etiology for lung degeneration can be explained through organoid culture. Coaxing of iPSC into inert biomaterial and defined culture leads to formation of lung organoids that consisted of epithelial and mesenchymal cells, which can survive in culture for months. These organoids are miniature lung, resemble tissues of large airways and alveoli, and can be used for lung developmental studies and screening of antituberculosis and anticancer drugs [87]. The conventional multistep reprogramming for iPSCs consumes months of time, while CRISPER-Cas9 system based episomal reprogramming system that combines two steps together enables generation of ESCs-like cells in less than two wks, reducing the chances of culture associated genetic abrasions and unwanted epigenetic [80]. This approach can yield single step ESCs-like cells in more personalized way from adults with retinal degradation and infants with severe immunodeficiency, involving correction for genetic mutation of OCT4 and DNMT3B [80]. The iPSCs expressing anti-CCR5-RNA, which can be differentiated into HIV1 resistant macrophages, have applications in AIDS therapeutics [88]. The diversified immunotherapeutic application of iPSCs has been reviewed elsewhere [89]. The α-1 antitrypsin deficiency (A1AD) encoded by serpin peptidase inhibitor clade A member 1 (SERPINA1) protein synthesized in liver protects lungs from neutrophils elastase, the enzyme causing

disruption of lungs connective tissue. A1AD deficiency is common cause of both lung and liver disease like chronic obstructive pulmonary disease (COPD) and liver cirrhosis. Patient specific iPSCs from lung and liver cells might explain pathophysiology of A1AD deficiency. COPD patient derived iPSCs show sensitivity to toxic drugs which explains that actual patient might be sensitive in similar fashion. It is known that A1AD deficiency is caused by single base pair mutation and correction of this mutation fixes the A1AD deficiency in hepatic-iPSCs [85]. The high order brain functions, like emotions, anxiety, sleep, depression, appetite, breathing heartbeats, and so forth, are regulated by serotonin neurons. Generation of serotonin neurons occurs prior to birth, which are postmitotic in their nature. Any sort of developmental defect and degeneration of serotonin neurons might lead to neuronal disorders like bipolar disorder, depression, and schizophrenia-like psychiatric conditions. Manipulation of Wnt signalling in human iPSCs in defined culture conditions leads to an in vitro differentiation of iPSCs to serotonin-like neurons. These iPSCs-neurons primarily localize to rhombomere 2-3 segment of rostral raphe nucleus, exhibit electrophysiological properties similar to serotonin neurons, express hydroxylase 2, the developmental marker, and release serotonin in dose and time dependent manner. Transplantation of these neurons might cure from schizophrenia, bipolar disorder, and other neuropathological conditions [116]. The iPSCs technology mediated somatic cell reprogramming of ventricular monocytes results in generation of cells, similar in morphology and functionality with PCs. SA note transplantation of PCs to large animals improves rhythmic heart functions. Pacemaker needs very reliable and robust performance so understanding of transformation process and site of transplantation are the critical aspect for therapeutic validation of iPSCs derived PCs [28]. Diabetes is a major health concern in modern world, and generation of β-cells from adult tissue is challenging. Direct reprogramming of skin cells into pancreatic cells, bypassing pluripotency phase, can yield clinical grade β-cells. This reprogramming strategy involves transformation of skin cells into definitive endodermal progenitors (cDE) and foregut like progenitor cells (cPF) intermediates and subsequent in vitro expansion of these intermediates to become pancreatic β-cells (cPB). The first step is chemically complex and can be understood as nonepisomal reprogramming on day one with pluripotency factors (OCT4, SOX2, KLF4, and hair pin RNA against p53), then supplementation with GFs and chemical supplements on day seven (EGF, bFGF, CHIR, NECA, NaB, Par, and RG), and two weeks later (Activin-A, CHIR, NECA, NaB, and RG) yielding DE and cPF [86]. Transplantation of cPB yields into glucose stimulated secretion of insulin in diabetic mice defines that such cells can be explored for treatment of T1DM and T2DM in more personalized manner [86]. iPSCs represent underrated opportunities for drug industries and clinical research laboratories for development of therapeutics, but safety concerns might limit transplantation applications (Figure 7; Table 1) [117]. Transplantation of human iPSCs into mice gastrula leads to colonization and differentiation of cells into three germ layers, evidenced with clinical developmental fat measurements. The acceptance of human iPSCs by mice gastrula suggests that correct timing and appropriate reprogramming regime might delimit human mice species barrier. Using this fact of species barrier, generation of human organs in closely associated primates might be possible, which can be used for treatment of genetic factors governed disease at embryo level itself [118]. In summary, iPSCs are safe and effective for treatment of regenerative medicine.

8. Stem Cells in Wildlife Conservation

The unstable growth of human population threatens the existence of wildlife, through overexploitation of natural habitats and illegal killing of wild animals, leading many species to face the fate of being endangered and go for extinction. For wildlife conservation, the concept of creation of frozen zoo involves preservation of gene pool and germ plasm from threatened and endangered species (Figure 8). The frozen zoo tissue samples collection from dead or live animal can be DNA, sperms, eggs, embryos, gonads, skin, or any other tissue of the body [119]. Preserved tissue can be reprogrammed or transdifferentiated to become other types of tissues and cells, which opens an avenue for conservation of endangered species and resurrection of life (Figure 8). The gonadal tissue from young individuals harbouring immature tissue can be matured in vivo and ex vivo for generation of functional gametes. Transplantation of SSCs to testis of male from the same different species can give rise to spermatozoa of donor cells [120], which might be used for IVF based captive breeding of wild animals. The most dangerous fact in wildlife conservation is low genetic diversity, too few reproductively capable animals which cannot maintain adequate genetic diversity in wild or captivity. Using the edge of iPSC technology, pluripotent stem cells can be generated from skin cells. For endangered drill, *Mandrillus leucophaeus,* and nearly extinct white rhinoceros, *Ceratotherium simum cottoni,* iPSC has been generated in 2011 [121]. The endangered animal drill (*Mandrillus leucophaeus*) is genetically very close to human and often suffers from diabetes, while rhinos are genetically far removed from other primates. The progress in iPSCs, from the human point of view, might be transformed for animal research for recapturing reproductive potential and health in wild animals. However, stem cells based interventions in wild animals are much more complex than classical conservation planning and biomedical research has to face. Conversion of iPSC into egg or sperm can open the door for generation of IVF based embryo; those might be transplanted in womb of live counterparts for propagation of population. Recently, iPSCs have been generated for snow leopard (*Panthera uncia*), native to mountain ranges of central Asia, which belongs to cat family; this breakthrough has raised the possibilities for cryopreservation of genetic material for future cloning and other assisted reproductive technology (ART) applications, for the conservation of cat species and biodiversity. Generation of leopard iPSCs has been achieved through retroviral-system based genomic integration of OCT4, SOX2, KLF4, cMYC, and NANOG. These iPSCs from snow leopard also open an avenue for further transformation of iPSCs

into gametes [122]. The in vivo maturation of grafted tissue depends both on age and on hormonal status of donor tissue. These facts are equally applicable to accepting host. Ectopic xenografts of cryopreserved testis tissue from Indian spotted deer (*Moschiola indica*) to nude mice yielded generation of spermatocytes [123], suggesting that one-day procurement of functional sperm from premature tissue might become a general technique in wildlife conservation. In summary, tissue biopsies from dead or live animals can be used for generation of iPSCs and functional gametes; those can be used in assisted reproductive technology (ART) for wildlife conservation.

9. Future Perspectives

The spectacular progress in the field of stem cells research represents great scope of stem cells regenerative therapeutics. It can be estimated that by 2020 or so we will be able to produce wide array of tissue, organoid, and organs from adult stem cells. Inductions of pluripotency phenotypes in terminally differentiated adult cells have better therapeutic future than ESCs, due to least ethical constraints with adult cells. In the coming future, there might be new pharmaceutical compounds; those can activate tissue specific stem cells, promote stem cells to migrate to the side of tissue injury, and promote their differentiation to tissue specific cells. Except few countries, the ongoing financial and ethical hindrance on ESCs application in regenerative medicine have more chance for funding agencies to distribute funding for the least risky projects on UCSCs, BMSCs, and TSPSCs from biopsies. The existing stem cells therapeutics advancements are more experimental and high in cost; due to that application on broad scale is not feasible in current scenario. In the near future, the advancements of medical science presume using stem cells to treat cancer, muscles damage, autoimmune disease, and spinal cord injuries among a number of impairments and diseases. It is expected that stem cells therapies will bring considerable benefits to the patients suffering from wide range of injuries and disease. There is high optimism for use of BMSCs, TSPSCs, and iPSCs for treatment of various diseases to overcome the contradictions associated with ESCs. For advancement of translational application of stem cells, there is a need of clinical trials, which needs funding rejoinder from both public and private organizations. The critical evaluation of regulatory guidelines at each phase of clinical trial is a must to comprehend the success and efficacy in time frame.

Abbreviations

ESCs: Embryonic stem cells
TSPSCs: Tissue specific progenitor stem cells
UCSCs: Umbilical cord stem cells
BMSCs: Bone marrow stem cells
iPSCs: Induced pluripotent stem cells
MSCs: Mesenchymal stem cells
WJ-MSCs: Wharton's jelly mesenchymal stem cells
HSCs: Hematopoietic stem cells

RGCs: Retinal ganglion cells
T1DM: Type I diabetes mellitus
T2DM: Type 2 diabetes mellitus
A1AD: α-1 antitrypsin deficiency
COPD: Chronic obstructive pulmonary disease
HLA: Human leukocyte antigen
MHC: Major histocompatibility complex
3D: Three-dimensional
SCI: Spinal cord injury
ARMD: Age-related macular degeneration
RPE: Retinal pigment epithelium
PXR: Pregnane X receptor
DPSCs: Dental pulp stem cells
GFAP: Glial fibrillary acidic protein
Atoh1:   Activation of transcription factor atonal homologue 1
NEC: Necrotizing enterocolitis
DMD: Duchene muscular dystrophy
PDGF: Placental derived growth factor
PEG: Polyethylene glycol
SSCs: Spermatogonial stem cells
AdSCs: Adipose derived stem cells
HSCs: Hematopoietic stem cells
AFSCs: Amniotic fluid stem cells
VEGF: Vascular endothelial growth factor
UCB: Umbilical cord blood
SLEDAI: Systematic lupus erythematosus (SLE) disease activity index
HIV-1:   Human immunodeficiency virus-1
GM-HSPCs: Genetically modified hematopoietic stem and progenitor cells
Th: T helper
LA: Lipoic acid
TRC: Tissue repair cells
BM-MSCs: Bone marrow mesenchymal stem cells
PBSCs: Peripheral blood stem cells
G-CSF: Granulocyte-colony stimulating factor
SERPINA1:   Serpin peptidase inhibitor clade A member 1
ASD: Autism spectrum disorder
INs: Interneurons
GABAergic: Υ-amino butyric acid secreting
NPCs: Neuronal progenitor cells
iTSCs: Independent trophoblast stem-like cells
hCS: Human cortical spheroids
CMs: Cardiomyocytes
ALD: Adrenoleukodystrophy
MLD: Metachromatic leukodystrophy
TSD: Tay-Sachs disease
ALS: Amyotrophic lateral sclerosis
TBI: Traumatic brain injuries
AD: Alzheimer's disease
NSCs: Neuronal stem cells
SID: Severe immune deficiency.

Competing Interests

There are no competing interests associated with this paper.

Acknowledgments

Dr. Anuradha Reddy from Centre for Cellular and Molecular Biology Hyderabad and Mrs. Sarita Kumari from Department of Yoga Science, BU, Bhopal, India, are acknowledged for their critical suggestions and comments on paper.

References

[1] C. Mason and P. Dunnill, "A brief definition of regenerative medicine," *Regenerative Medicine*, vol. 3, no. 1, pp. 1–5, 2008.

[2] L. A. Fortier, "Stem cells: classifications, controversies, and clinical applications," *Veterinary Surgery*, vol. 34, no. 5, pp. 415–423, 2005.

[3] K. Takahashi and S. Yamanaka, "Induction of pluripotent stem cells from mouse embryonic and adult fibroblast cultures by defined factors," *Cell*, vol. 126, no. 4, pp. 663–676, 2006.

[4] J. Yu, M. A. Vodyanik, K. Smuga-Otto et al., "Induced pluripotent stem cell lines derived from human somatic cells," *Science*, vol. 318, no. 5858, pp. 1917–1920, 2007.

[5] M. Thomson, S. J. Liu, L.-N. Zou, Z. Smith, A. Meissner, and S. Ramanathan, "Pluripotency factors in embryonic stem cells regulate differentiation into germ layers," *Cell*, vol. 145, no. 6, pp. 875–889, 2011.

[6] E. W. Petersdorf, M. Malkki, T. A. Gooley, P. J. Martin, and Z. Guo, "MHC haplotype matching for unrelated hematopoietic cell transplantation," *PLoS Medicine*, vol. 4, no. 1, article e8, 2007.

[7] A. M. Leferink, Y. C. Chng, C. A. van Blitterswijk, and L. Moroni, "Distribution and viability of fetal and adult human bone marrow stromal cells in a biaxial rotating vessel bioreactor after seeding on polymeric 3D additive manufactured scaffolds," *Frontiers in Bioengineering and Biotechnology*, vol. 3, article 169, 2015.

[8] E. A. Gubareva, S. Sjöqvist, I. V. Gilevich et al., "Orthotopic transplantation of a tissue engineered diaphragm in rats," *Biomaterials*, vol. 77, pp. 320–335, 2016.

[9] I. Garzón, B. Pérez-Köhler, J. Garrido-Gómez et al., "Evaluation of the cell viability of human Wharton's Jelly stem cells for use in cell therapy," *Tissue Engineering Part C: Methods*, vol. 18, no. 6, pp. 408–419, 2012.

[10] P. A. Thompson, T. Perera, D. Marin et al., "Double umbilical cord blood transplant is effective therapy for relapsed or refractory Hodgkin lymphoma," *Leukemia & Lymphoma*, vol. 57, no. 7, pp. 1607–1615, 2016.

[11] S. S. Nathamgari, B. Dong, F. Zhou et al., "Isolating single cells in a neurosphere assay using inertial microfluidics," *Lab on a Chip—Miniaturisation for Chemistry and Biology*, vol. 15, no. 24, pp. 4591–4597, 2015.

[12] B. Ning, D. K. Cheuk, A. K. Chiang, P. P. Lee, S. Y. Ha, and G. C. Chan, "Autologous cord blood transplantation for metastatic neuroblastoma," *Pediatric Transplantation*, vol. 20, no. 2, pp. 290–296, 2015.

[13] J. A. Thomson, "Embryonic stem cell lines derived from human blastocysts," *Science*, vol. 282, no. 5391, pp. 1145–1147, 1998.

[14] M. S. Hogan, D.-E. Parfitt, C. J. Zepeda-Mendoza, M. M. Shen, and D. L. Spector, "Transient pairing of homologous Oct4 alleles accompanies the onset of embryonic stem cell differentiation," *Cell Stem Cell*, vol. 16, no. 3, pp. 275–288, 2015.

[15] G. Shroff and R. Gupta, "Human embryonic stem cells in the treatment of patients with spinal cord injury," *Annals of Neurosciences*, vol. 22, no. 4, pp. 208–216, 2015.

[16] S. Zhou, A. Flamier, M. Abdouh et al., "Differentiation of human embryonic stem cells into cone photoreceptors through simultaneous inhibition of BMP, TGFβ and Wnt signaling," *Development*, vol. 142, no. 19, pp. 3294–3306, 2015.

[17] V. M. Sluch, C.-H. O. Davis, V. Ranganathan et al., "Differentiation of human ESCs to retinal ganglion cells using a CRISPR engineered reporter cell line," *Scientific Reports*, vol. 5, Article ID 16595, 2015.

[18] Y. Shiba, S. Fernandes, W.-Z. Zhu et al., "Human ES-cell-derived cardiomyocytes electrically couple and suppress arrhythmias in injured hearts," *Nature*, vol. 489, no. 7415, pp. 322–325, 2012.

[19] S. Fernandes, J. J. H. Chong, S. L. Paige et al., "Comparison of human embryonic stem cell-derived cardiomyocytes, cardiovascular progenitors, and bone marrow mononuclear cells for cardiac repair," *Stem Cell Reports*, vol. 5, no. 5, pp. 753–762, 2015.

[20] Y. Avior, G. Levy, M. Zimerman et al., "Microbial-derived lithocholic acid and vitamin K_2 drive the metabolic maturation of pluripotent stem cells-derived and fetal hepatocytes," *Hepatology*, vol. 62, no. 1, pp. 265–278, 2015.

[21] K. Cameron, B. Lucendo-Villarin, D. Szkolnicka, and D. C. Hay, "Serum-free directed differentiation of human embryonic stem cells to hepatocytes," *Methods in Molecular Biology*, vol. 1250, pp. 105–111, 2015.

[22] M. Zhang, P. Sun, Y. Wang et al., "Generation of self-renewing hepatoblasts from human embryonic stem cells by chemical approaches," *Stem Cells Translational Medicine*, vol. 4, no. 11, pp. 1275–1282, 2015.

[23] L. Tolosa, J. Caron, Z. Hannoun et al., "Transplantation of hESC-derived hepatocytes protects mice from liver injury," *Stem Cell Research & Therapy*, vol. 6, article 246, 2015.

[24] A. Carpentier, I. Nimgaonkar, V. Chu, Y. Xia, Z. Hu, and T. J. Liang, "Hepatic differentiation of human pluripotent stem cells in miniaturized format suitable for high-throughput screen," *Stem Cell Research*, vol. 16, no. 3, pp. 640–650, 2016.

[25] J. E. Bruin, N. Saber, N. Braun et al., "Treating diet-induced diabetes and obesity with human embryonic stem cell-derived pancreatic progenitor cells and antidiabetic drugs," *Stem Cell Reports*, vol. 4, no. 4, pp. 605–620, 2015.

[26] C. Salguero-Aranda, R. Tapia-Limonchi, and G. M. Cahuana, "Differentiation of mouse embryonic stem cells towards functional pancreatic beta-cell surrogates through epigenetic regulation of Pdx1 by nitric oxide," *Cell Transplantation*, 2016.

[27] A. Cheng, Z. Kapacee, J. Peng et al., "Cartilage repair using human embryonic stem cell-derived chondroprogenitors," *Stem Cells Translational Medicine*, vol. 3, no. 11, pp. 1287–1295, 2014.

[28] V. Vedantham, "New approaches to biological pacemakers: links to sinoatrial node development," *Trends in Molecular Medicine*, vol. 21, no. 12, pp. 749–761, 2015.

[29] C. Greggio, F. De Franceschi, M. Figueiredo-Larsen et al., "Artificial three-dimensional niches deconstruct pancreas development in vitro," *Development*, vol. 140, no. 21, pp. 4452–4462, 2013.

[30] L. Ye, M. A. Robertson, T. L. Mastracci, and R. M. Anderson, "An insulin signaling feedback loop regulates pancreas progenitor cell differentiation during islet development and regeneration," *Developmental Biology*, vol. 409, no. 2, pp. 354–369, 2016.

[31] P. D. Potdar and Y. D. Jethmalani, "Human dental pulp stem cells: applications in future regenerative medicine," *World Journal of Stem Cells*, vol. 7, no. 5, pp. 839–851, 2015.

[32] K. M. Ellis, D. C. O'Carroll, M. D. Lewis, G. Y. Rychkov, and S. A. Koblar, "Neurogenic potential of dental pulp stem cells isolated from murine incisors," *Stem Cell Research & Therapy*, vol. 5, article 30, 2014.

[33] M. M. Acharya, V. Martirosian, N. N. Chmielewski et al., "Stem cell transplantation reverses chemotherapy-induced cognitive dysfunction," *Cancer Research*, vol. 75, no. 4, pp. 676–686, 2015.

[34] K. Mizutari, M. Fujioka, M. Hosoya et al., "Notch inhibition induces cochlear hair cell regeneration and recovery of hearing after acoustic trauma," *Neuron*, vol. 77, no. 1, pp. 58–69, 2013.

[35] B. R. Kuo, E. M. Baldwin, W. S. Layman, M. M. Taketo, and J. Zuo, "In vivo cochlear hair cell generation and survival by coactivation of β-catenin and Atoh1," *The Journal of Neuroscience*, vol. 35, no. 30, pp. 10786–10798, 2015.

[36] S. A. Shaffiey, H. Jia, T. Keane et al., "Intestinal stem cell growth and differentiation on a tubular scaffold with evaluation in small and large animals," *Regenerative Medicine*, vol. 11, no. 1, pp. 45–61, 2016.

[37] D. H. Scoville, T. Sato, X. C. He, and L. Li, "Current view: intestinal stem cells and signaling," *Gastroenterology*, vol. 134, no. 3, pp. 849–864, 2008.

[38] X. Yin, H. F. Farin, J. H. Van Es, H. Clevers, R. Langer, and J. M. Karp, "Niche-independent high-purity cultures of Lgr5$^+$ intestinal stem cells and their progeny," *Nature Methods*, vol. 11, no. 1, pp. 106–112, 2014.

[39] B. R. Ksander, P. E. Kolovou, B. J. Wilson et al., "ABCB5 is a limbal stem cell gene required for corneal development and repair," *Nature*, vol. 511, no. 7509, pp. 353–357, 2014.

[40] S. Ahmad, "Concise review: limbal stem cell deficiency, dysfunction, and distress," *Stem Cells Translational Medicine*, vol. 1, no. 2, pp. 110–115, 2012.

[41] C. Fuoco, R. Rizzi, A. Biondo et al., "In vivo generation of a mature and functional artificial skeletal muscle," *EMBO Molecular Medicine*, vol. 7, no. 4, pp. 411–422, 2015.

[42] S. E. Berry, "Concise review: mesoangioblast and mesenchymal stem cell therapy for muscular dystrophy: progress, challenges, and future directions," *Stem Cells Translational Medicine*, vol. 4, no. 1, pp. 91–98, 2015.

[43] L. Simon, G. C. Ekman, N. Kostereva et al., "Direct transdifferentiation of stem/progenitor spermatogonia into reproductive and nonreproductive tissues of all germ layers," *Stem Cells*, vol. 27, no. 7, pp. 1666–1675, 2009.

[44] S. M. Cronk, M. R. Kelly-Goss, H. C. Ray et al., "Adipose-derived stem cells from diabetic mice show impaired vascular stabilization in a murine model of diabetic retinopathy," *Stem Cells Translational Medicine*, vol. 4, no. 5, pp. 459–467, 2015.

[45] B. Mead, M. Berry, A. Logan, R. A. H. Scott, W. Leadbeater, and B. A. Scheven, "Stem cell treatment of degenerative eye disease," *Stem Cell Research*, vol. 14, no. 3, pp. 243–257, 2015.

[46] M. Kessler, K. Hoffmann, V. Brinkmann et al., "The Notch and Wnt pathways regulate stemness and differentiation in human fallopian tube organoids," *Nature Communications*, vol. 6, article 8989, 2015.

[47] H. Nagata, M. Ii, E. Kohbayashi, M. Hoshiga, T. Hanafusa, and M. Asahi, "Cardiac adipose-derived stem cells exhibit high differentiation potential to cardiovascular cells in C57BL/6 mice," *Stem Cells Translational Medicine*, vol. 5, no. 2, pp. 141–151, 2016.

[48] L. Chen, F. Qin, M. Ge, Q. Shu, and J. Xu, "Application of adipose-derived stem cells in heart disease," *Journal of Cardiovascular Translational Research*, vol. 7, no. 7, pp. 651–663, 2014.

[49] S. K. Steinbach and M. Husain, "Vascular smooth muscle cell differentiation from human stem/progenitor cells," *Methods*, vol. 101, pp. 85–92, 2016.

[50] M. Dominici, K. Le Blanc, I. Mueller et al., "Minimal criteria for defining multipotent mesenchymal stromal cells. The International Society for Cellular Therapy position statement," *Cytotherapy*, vol. 8, no. 4, pp. 315–317, 2006.

[51] A. K. Sharma, M. I. Bury, A. J. Marks et al., "A nonhuman primate model for urinary bladder regeneration using autologous sources of bone marrow-derived mesenchymal cells," *STEM CELLS*, vol. 29, no. 2, pp. 241–250, 2011.

[52] M. Oshima and T. Tsuji, "Whole tooth regeneration as a future dental treatment," *Advances in Experimental Medicine and Biology*, vol. 881, pp. 255–269, 2015.

[53] C. Csaki, U. Matis, A. Mobasheri, H. Ye, and M. Shakibaei, "Chondrogenesis, osteogenesis and adipogenesis of canine mesenchymal stem cells: a biochemical, morphological and ultrastructural study," *Histochemistry and Cell Biology*, vol. 128, no. 6, pp. 507–520, 2007.

[54] I. Ribitsch, J. Burk, U. Delling et al., "Basic science and clinical application of stem cells in veterinary medicine," *Advances in Biochemical Engineering/Biotechnology*, vol. 123, pp. 219–263, 2010.

[55] B. Sen, Z. Xie, G. Uzer et al., "Intranuclear actin regulates osteogenesis," *STEM CELLS*, vol. 33, no. 10, pp. 3065–3076, 2015.

[56] S. G. Brown, R. J. Harman, and L. L. Black, "Adipose-derived stem cell therapy for severe muscle tears in working German shepherds: two case reports," *Stem Cell Discovery*, vol. 2, no. 2, pp. 41–44, 2012.

[57] R. D. Levit, N. Landázuri, E. A. Phelps et al., "Cellular encapsulation enhances cardiac repair," *Journal of the American Heart Association*, vol. 2, no. 5, Article ID e000367, 2013.

[58] B.-J. Lin, J. Wang, Y. Miao et al., "Cytokine loaded layer-by-layer ultrathin matrices to deliver single dermal papilla cells for spot-by-spot hair follicle regeneration," *Journal of Materials Chemistry B*, vol. 4, no. 3, pp. 489–504, 2016.

[59] O. M. Benavides, A. R. Brooks, S. K. Cho, J. Petsche Connell, R. Ruano, and J. G. Jacot, "In situ vascularization of injectable fibrin/poly(ethylene glycol) hydrogels by human amniotic fluid-derived stem cells," *Journal of Biomedical Materials Research Part A*, vol. 103, no. 8, pp. 2645–2653, 2015.

[60] K. Wolfrum, Y. Wang, A. Prigione, K. Sperling, H. Lehrach, and J. Adjaye, "The LARGE principle of cellular reprogramming: lost, acquired and retained gene expression in foreskin and amniotic fluid-derived human iPS cells," *PLoS ONE*, vol. 5, no. 10, Article ID e13703, 2010.

[61] J. E. Wagner Jr., C. G. Brunstein, A. E. Boitano et al., "Phase I/II trial of stemregenin-1 expanded umbilical cord blood hematopoietic stem cells supports testing as a stand-alone graft," *Cell Stem Cell*, vol. 18, no. 1, pp. 144–155, 2016.

[62] J. Hu, X. Yu, Z. Wang et al., "Long term effects of the implantation of Wharton's jelly-derived mesenchymal stem cells from the umbilical cord for newly-onset type 1 diabetes mellitus," *Endocrine Journal*, vol. 60, no. 3, pp. 347–357, 2013.

[63] X. Liu, P. Zheng, X. Wang et al., "A preliminary evaluation of efficacy and safety of Wharton's jelly mesenchymal stem cell transplantation in patients with type 2 diabetes mellitus," *Stem Cell Research & Therapy*, vol. 5, no. 2, article 57, 2014.

[64] D. Wang, J. Li, Y. Zhang et al., "Umbilical cord mesenchymal stem cell transplantation in active and refractory systemic lupus erythematosus: A Multicenter Clinical Study," *Arthritis Research and Therapy*, vol. 16, no. 2, article R79, 2014.

[65] M. L. Escolar, M. D. Poe, J. M. Provenzale et al., "Transplantation of umbilical-cord blood in babies with infantile Krabbe's disease," *The New England Journal of Medicine*, vol. 352, no. 20, pp. 2069–2081, 2005.

[66] M. Jurga, A. W. Lipkowski, B. Lukomska et al., "Generation of functional neural artificial tissue from human umbilical cord blood stem cells," *Tissue Engineering Part C: Methods*, vol. 15, no. 3, pp. 365–372, 2009.

[67] J. Ehrhart, D. Darlington, N. Kuzmin-Nichols et al., "Biodistribution of infused human umbilical cord blood cells in Alzheimer's disease-like murine model," *Cell Transplantation*, vol. 25, no. 1, pp. 195–199, 2016.

[68] G.-Y. Park, D. R. Kwon, and S. C. Lee, "Regeneration of full-thickness rotator cuff tendon tear after ultrasound-guided injection with umbilical cord blood-derived mesenchymal stem cells in a Rabbit model," *Stem Cells Translational Medicine*, vol. 4, no. 11, pp. 1344–1351, 2015.

[69] C.-W. Ha, Y.-B. Park, J.-Y. Chung, and Y.-G. Park, "Cartilage repair using composites of human umbilical cord blood-derived mesenchymal stem cells and hyaluronic acid hydrogel in a minipig model," *Stem Cells Translational Medicine*, vol. 4, no. 9, pp. 1044–1051, 2015.

[70] Y. P. Fan, C. Hsia, K. Tseng et al., "The therapeutic potential of human umbilical mesenchymal stem cells from Wharton's jelly in the treatment of rat peritoneal dialysis-induced fibrosis," *Stem Cells Translational Medicine*, vol. 5, no. 2, pp. 235–247, 2016.

[71] S. Gaballa, N. Palmisiano, O. Alpdogan et al., "A two-step haploidentical versus a two-step matched related allogeneic myeloablative peripheral blood stem cell transplantation," *Biology of Blood and Marrow Transplantation*, vol. 22, no. 1, pp. 141–148, 2016.

[72] D. L. DiGiusto, R. Stan, A. Krishnan, H. Li, J. J. Rossi, and J. A. Zaia, "Development of hematopoietic stem cell based gene therapy for HIV-1 infection: considerations for proof of concept studies and translation to standard medical practice," *Viruses*, vol. 5, no. 11, pp. 2898–2919, 2013.

[73] E. Herrera-Carrillo and B. Berkhout, "Bone marrow gene therapy for HIV/AIDS," *Viruses*, vol. 7, no. 7, pp. 3910–3936, 2015.

[74] K. R. Machlus and J. E. Italiano Jr., "The incredible journey: from megakaryocyte development to platelet formation," *The Journal of Cell Biology*, vol. 201, no. 6, pp. 785–796, 2013.

[75] C. A. Di Buduo, L. S. Wray, L. Tozzi et al., "Programmable 3D silk bone marrow niche for platelet generation ex vivo and modeling of megakaryopoiesis pathologies," *Blood*, vol. 125, no. 14, pp. 2254–2264, 2015.

[76] S. Paradells, I. Zipancic, M. M. Martínez-Losa et al., "Lipoic acid and bone marrow derived cells therapy induce angiogenesis and cell proliferation after focal brain injury," *Brain Injury*, vol. 29, no. 3, pp. 380–395, 2015.

[77] D. Kaigler, G. Pagni, C. H. Park et al., "Stem cell therapy for craniofacial bone regeneration: a randomized, controlled feasibility trial," *Cell Transplantation*, vol. 22, no. 5, pp. 767–777, 2013.

[78] Y. Tsai, B. Lu, B. Bakondi et al., "Human iPSC-derived neural progenitors preserve vision in an AMD-like model," *STEM CELLS*, vol. 33, no. 8, pp. 2537–2549, 2015.

[79] J. Yang, B. Cai, P. Glencer, Z. Li, X. Zhang, and X. Li, "Induced pluripotent stem cells and outer retinal disease," *Stem Cells International*, vol. 2016, Article ID 2850873, 6 pages, 2016.

[80] S. E. Howden, J. P. Maufort, B. M. Duffin, A. G. Elefanty, E. G. Stanley, and J. A. Thomson, "Simultaneous reprogramming and

gene correction of patient fibroblasts," *Stem Cell Reports*, vol. 5, no. 6, pp. 1109–1118, 2015.

[81] G. Colasante, G. Lignani, A. Rubio et al., "Rapid conversion of fibroblasts into functional forebrain GABAergic interneurons by direct genetic reprogramming," *Cell Stem Cell*, vol. 17, no. 6, pp. 719–734, 2015.

[82] R. F. Hunt and S. C. Baraban, "Interneuron transplantation as a treatment for epilepsy," *Cold Spring Harbor Perspectives in Medicine*, vol. 5, no. 12, Article ID a022376, 2015.

[83] A. M. Pasca, S. A. Sloan, L. E. Clarke et al., "Functional cortical neurons and astrocytes from human pluripotent stem cells in 3D culture," *Nature Methods*, vol. 12, no. 7, pp. 671–678, 2015.

[84] J. Mariani, G. Coppola, P. Zhang et al., "FOXG1-dependent dysregulation of GABA/glutamate neuron differentiation in autism spectrum disorders," *Cell*, vol. 162, no. 2, pp. 375–390, 2015.

[85] A. A. Wilson, L. Ying, M. Liesa et al., "Emergence of a stage-dependent human liver disease signature with directed differentiation of alpha-1 antitrypsin-deficient iPS cells," *Stem Cell Reports*, vol. 4, no. 5, pp. 873–885, 2015.

[86] S. Zhu, H. A. Russ, X. Wang et al., "Human pancreatic beta-like cells converted from fibroblasts," *Nature Communications*, vol. 7, Article ID 10080, 2016.

[87] B. R. Dye, D. R. Hill, M. A. Ferguson et al., "In vitro generation of human pluripotent stem cell derived lung organoids," *eLife*, vol. 4, Article ID e05098, pp. 1–25, 2015.

[88] A. Kambal, G. Mitchell, W. Cary et al., "Generation of HIV-1 resistant and functional macrophages from hematopoietic stem cell-derived induced pluripotent stem cells," *Molecular Therapy*, vol. 19, no. 3, pp. 584–593, 2011.

[89] Z. Jiang, Y. Han, and X. Cao, "Induced pluripotent stem cell (iPSCs) and their application in immunotherapy," *Cellular and Molecular Immunology*, vol. 11, no. 1, pp. 17–24, 2014.

[90] T. Squillaro, G. Peluso, and U. Galderisi, "Clinical trials with mesenchymal stem cells: an update," *Cell Transplantation*, vol. 25, no. 5, pp. 829–848, 2016.

[91] V. Volarevic, J. Nurkovic, N. Arsenijevic, and M. Stojkovic, "Concise review: therapeutic potential of mesenchymal stem cells for the treatment of acute liver failure and cirrhosis," *STEM CELLS*, vol. 32, no. 11, pp. 2818–2823, 2014.

[92] S. Patel, P. T. Yin, H. Sugiyama, and K.-B. Lee, "Inducing stem cell myogenesis using nanoscript," *ACS Nano*, vol. 9, no. 7, pp. 6909–6917, 2015.

[93] S. Dameshghi, A. Zavaran-Hosseini, S. Soudi, F. J. Shirazi, S. Nojehdehi, and S. M. Hashemi, "Mesenchymal stem cells alter macrophage immune responses to Leishmania major infection in both susceptible and resistance mice," *Immunology Letters*, vol. 170, pp. 15–26, 2016.

[94] S. Shahrokhi, F. Menaa, K. Alimoghaddam, C. McGuckin, and M. Ebtekar, "Insights and hopes in umbilical cord blood stem cell transplantations," *Journal of Biomedicine and Biotechnology*, vol. 2012, Article ID 572821, 11 pages, 2012.

[95] S. Rentas, N. T. Holzapfel, M. S. Belew et al., "Musashi-2 attenuates AHR signalling to expand human haematopoietic stem cells," *Nature*, vol. 532, no. 7600, pp. 508–511, 2016.

[96] H. Lu, Y. Chen, Q. Lan et al., "Factors that influence a mother's willingness to preserve umbilical cord blood: a survey of 5120 Chinese mothers," *PLoS ONE*, vol. 10, no. 12, Article ID e0144001, 2015.

[97] A. Rosemann, "Why regenerative stem cell medicine progresses slower than expected," *Journal of Cellular Biochemistry*, vol. 115, no. 12, pp. 2073–2076, 2014.

[98] E. Gluckman, G. Koegler, and V. Rocha, "Human leukocyte antigen matching in cord blood transplantation," *Seminars in Hematology*, vol. 42, no. 2, pp. 85–90, 2005.

[99] R. S. Mehta, K. Rezvani, A. Olson et al., "Novel techniques for ex vivo expansion of cord blood: clinical trials," *Frontiers in Medicine*, vol. 2, article 89, 2015.

[100] Y. Wang, F. Chen, B. Gu, G. Chen, H. Chang, and D. Wu, "Mesenchymal stromal cells as an adjuvant treatment for severe late-onset hemorrhagic cystitis after allogeneic hematopoietic stem cell transplantation," *Acta Haematologica*, vol. 133, no. 1, pp. 72–77, 2015.

[101] B. Gokcinar-Yagci, O. Ozyuncu, and B. Celebi-Saltik, "Isolation, characterisation and comparative analysis of human umbilical cord vein perivascular cells and cord blood mesenchymal stem cells," *Cell and Tissue Banking*, vol. 17, no. 2, pp. 345–352, 2015.

[102] J. Liang, F. Wang, D. Wang et al., "Transplantation of mesenchymal stem cells in a laryngeal carcinoma patient with radiation myelitis," *Stem Cell Research and Therapy*, vol. 6, article 213, 2015.

[103] J. Kumar, V. Kale, and L. Limaye, "Umbilical cord blood-derived CD11c+ dendritic cells could serve as an alternative allogeneic source of dendritic cells for cancer immunotherapy," *Stem Cell Research & Therapy*, vol. 6, article 184, 2015.

[104] C. Zhang, Y. Zhu, Y. Zhang, L. Gao, N. Zhang, and H. Feng, "Therapeutic potential of umbilical cord mesenchymal stem cells for inhibiting myofibroblastic differentiation of irradiated human lung fibroblasts," *The Tohoku Journal of Experimental Medicine*, vol. 236, no. 3, pp. 209–217, 2015.

[105] L. Fu, Y. Liu, D. Zhang, J. Xie, H. Guan, and T. Shang, "Beneficial effect of human umbilical cord-derived mesenchymal stem cells on an endotoxin-induced rat model of preeclampsia," *Experimental and Therapeutic Medicine*, vol. 10, no. 5, pp. 1851–1856, 2015.

[106] S. Osone, T. Imamura, Y. Fukushima-Nakase et al., "Case reports of severe congenital neutropenia treated with unrelated cord blood transplantation with reduced-intensity conditioning," *Journal of Pediatric Hematology/Oncology*, vol. 38, no. 1, pp. 49–52, 2016.

[107] M.-Y. Chang, T.-T. Huang, C.-H. Chen, B. Cheng, S.-M. Hwang, and P. C. H. Hsieh, "Injection of human cord blood cells with hyaluronan improves postinfarction cardiac repair in pigs," *Stem Cells Translational Medicine*, vol. 5, no. 1, pp. 56–66, 2016.

[108] G. S. Travlos, "Normal structure, function, and histology of the bone marrow," *Toxicologic Pathology*, vol. 34, no. 5, pp. 548–565, 2006.

[109] E. Gschweng, S. De Oliveira, and D. B. Kohn, "Hematopoietic stem cells for cancer immunotherapy," *Immunological Reviews*, vol. 257, no. 1, pp. 237–249, 2014.

[110] R. Singh Mahla, "HIV-1 Infection: the functional importance of SDF1, CCR2 and CCR5 in protection and therapeutics," *Health Care: Current Reviews*, vol. 3, article 150, 2015.

[111] T. Miyamoto and H. Nakauchi, "Generation of functional organs from pluripotent stem cells," *The Japanese Journal of Clinical Hematology*, vol. 56, pp. 2213–2219, 2015.

[112] R. Morizane, A. Q. Lam, B. S. Freedman, S. Kishi, M. T. Valerius, and J. V. Bonventre, "Nephron organoids derived from human pluripotent stem cells model kidney development and injury," *Nature Biotechnology*, vol. 33, no. 11, pp. 1193–1200, 2015.

[113] S. P. Lukashyk, V. M. Tsyrkunov, Y. I. Isaykina et al., "Mesenchymal bone marrow-derived stem cells transplantation in patients with HCV related liver cirrhosis," *Journal of Clinical and Translational Hepatology*, vol. 2, no. 4, pp. 217–221, 2014.

[114] M. Takasato, P. X. Er, H. S. Chiu et al., "Kidney organoids from human iPS cells contain multiple lineages and model human nephrogenesis," *Nature*, vol. 526, no. 7574, pp. 564–568, 2015.

[115] H. Benchetrit, S. Herman, N. Van Wietmarschen et al., "Extensive nuclear reprogramming underlies lineage conversion into functional trophoblast stem-like cells," *Cell Stem Cell*, vol. 17, no. 5, pp. 543–556, 2015.

[116] J. Lu, X. Zhong, H. Liu et al., "Generation of serotonin neurons from human pluripotent stem cells," *Nature Biotechnology*, vol. 34, no. 1, pp. 89–94, 2016.

[117] K. Takahashi and S. Yamanaka, "Induced pluripotent stem cells in medicine and biology," *Development*, vol. 140, no. 12, pp. 2457–2461, 2013.

[118] V. L. Mascetti and R. A. Pedersen, "Human-mouse chimerism validates human stem cell pluripotency," *Cell Stem Cell*, vol. 18, no. 1, pp. 67–72, 2016.

[119] P. Comizzoli and W. V. Holt, "Recent advances and prospects in germplasm preservation of rare and endangered species," *Advances in Experimental Medicine and Biology*, vol. 753, pp. 331–356, 2014.

[120] I. Dobrinski and A. J. Travis, "Germ cell transplantation for the propagation of companion animals, non-domestic and endangered species," *Reproduction, Fertility and Development*, vol. 19, no. 6, pp. 732–739, 2007.

[121] I. F. Ben-Nun, S. C. Montague, M. L. Houck et al., "Induced pluripotent stem cells from highly endangered species," *Nature Methods*, vol. 8, no. 10, pp. 829–831, 2011.

[122] R. Verma, M. K. Holland, P. Temple-Smith, and P. J. Verma, "Inducing pluripotency in somatic cells from the snow leopard (*Panthera uncia*), an endangered felid," *Theriogenology*, vol. 77, no. 1, pp. 220–228.e2, 2012.

[123] L. Pothana, H. Makala, L. Devi, V. P. Varma, and S. Goel, "Germ cell differentiation in cryopreserved, immature, Indian spotted mouse deer (Moschiola indica) testes xenografted onto mice," *Theriogenology*, vol. 83, no. 4, pp. 625–633, 2015.

9

Pulling a Ligase out of a "HAT": pCAF Mediates Ubiquitination of the Class II Transactivator

Julie E. Morgan and Susanna F. Greer

Division of Cellular Biology and Immunology, Department of Biology, Georgia State University, Atlanta, GA 30302, USA

Correspondence should be addressed to Susanna F. Greer; sgreer@gsu.edu

Academic Editor: Paul J. Higgins

The Class II Transactivator (CIITA) is essential to the regulation of Major Histocompatibility Class II (MHC II) genes transcription. As the "master regulator" of MHC II transcription, CIITA regulation is imperative and requires various posttranslational modifications (PTMs) in order to facilitate its role. Previously we identified various ubiquitination events on CIITA. Monoubiquitination is important for CIITA transactivity, while K63 linked ubiquitination is involved in crosstalk with ERK1/2 phosphorylation, where together they mediate cellular movement from the cytoplasm to nuclear region. Further, CIITA is also modified by degradative K48 polyubiquitination. However, the E3 ligase responsible for these modifications was unknown. We show CIITA ubiquitination and transactivity are enhanced with the histone acetyltransferase (HAT), p300/CBP associated factor (pCAF), and the E3 ligase region within pCAF is necessary for both. Additionally, pCAF mediated ubiquitination is independent of pCAF's HAT domain, and acetylation deficient CIITA is K48 polyubiquitinated and degraded in the presence of pCAF. Lastly, we identify the histone acetyltransferase, pCAF, as the E3 ligase responsible for CIITA's ubiquitination.

1. Introduction

Major Histocompatibility Class II (MHC II) genes are essential for the initiation of adaptive immune responses to extracellular pathogens; thus their expression and activation are of critical importance and are tightly regulated [1–3]. Coordinated orchestration of multiple proteins accomplishes transcription of MHC II; however, one protein in particular, known as the "master regulator" of MHC II genes, the Class II Transactivator, is particularly important [4–7]. In addition to CIITA, various other chromatin-remodeling enzymes are required for the "opening" of the MHC II promoter, thus allowing the transcriptional machinery to bind. In particular, two histone acetyltransferases (HATs), the CREB binding protein (CBP/p300) and p300/CBP associated factor (pCAF), are recruited to the MHC II promoter where they assist in the remodeling of chromatin which occurs before and in the presence of CIITA [8, 9].

CIITA is 1130 amino acid protein and is dynamically regulated through an intricate series of posttranslational modifications (PTMs) [10]. PTMs on CIITA include phosphorylation, ubiquitination, and acetylation [11–18]. These modifications precisely regulate CIITA's location, function, and stability within the cell and increase CIITA activity at the MHC II promoter [8, 10, 13–15, 19–22]. HATs including pCAF and CBP are responsible for acetylation of CIITA at lysine(s) (K) 141 and 144 [14]. It has further been shown that acetylation plays important roles in the ubiquitination of CIITA [13, 14]. Located at the N-terminal region of pCAF, lies a domain containing ubiquitin E3 ligase activity [23]. Ubiquitination requires three enzymes: an E1 activating enzyme, an E2 conjugating enzyme, and E3 ligase, which is responsible for the ligation of ubiquitin onto a substrate in conjunction with the E2 [24]. Previously pCAF's intrinsic ubiquitination domain was identified and shown to play a role in the ubiquitination and stability of the critical cell cycle protein, human double minute 2 (the human ortholog of Mdm2) [23, 25], and Gli1, a transcription factor that mediates hedgehog signaling [26]. Thus, pCAF is not only HAT, but also ubiquitin E3 ligase. Presently, pCAF is shown to ubiquitinate only a few substrates: Hdm2, Gli1, and itself [23, 25, 26]. As pCAF is known to affect the activity of many transcription factors and cofactors through its HAT activities, it is likely that pCAF also has additional targets for its ubiquitin E3 ligase activities. As CIITA has previously been shown to be a substrate for pCAF's HAT activity and observations have been made of CIITA's increased ubiquitination in the presence of pCAF [13], we sought to determine if pCAF was potential E3 ligase for CIITA.

We hypothesized pCAF is playing a novel role as ubiquitin E3 ligase for CIITA in addition to its traditional role as HAT. We show here that both CIITA transactivity levels and global ubiquitination (all ubiquitin types) significantly decline in the absence of the pCAF E3 ligase domain. Further, we demonstrate CIITA ubiquitination does not rely on the HAT domain of pCAF. Acetylation null CIITA mutants lacking the signal to become nuclear bound are ubiquitinated in a K48 linked fashion leading to degradation. In vitro ubiquitination assays confirm pCAF's ability to facilitate CIITA ubiquitination. Lastly, we identify that CIITA mono, K63, and, K48 linked ubiquitination are mediated by pCAF in vivo. These results demonstrate pCAF's capacity to facilitate various topologies of CIITA ubiquitination. These results indicate that pCAF, via its E3 ligase activity, plays additional important roles in the regulation of CIITA activity and thus in regulating the expression of MHC II genes. Further, identification of the E3 ligase responsible for ubiquitination of CIITA is critical for gaining added understanding of CIITA regulation by PTMs. Identifying enzymes responsible for these PTMs allows for valuable insight into the regulation of the adaptive immune response and for the identification of potential therapeutic targets.

2. Materials and Methods

2.1. Cell Culture. COS cells (Monkey fibroblast) from ATCC (Manassas, VA) were maintained using high-glucose Dulbecco's modified Eagle medium (DMEM) (Mediatech, Inc., Herndon, VA) supplemented with 10% fetal bovine serum, 50 units/mL of penicillin, 50 μg/mL of streptomycin, and 2 mM of L-glutamine. Cells were maintained at 37° with 5% CO_2.

2.2. Plasmids and Purified Proteins. Flag-K141R, K144R, K141/144R, and Myc-CIITA were kindly provided by Dr. Jenny Ting. Flag-pCAF was a generous gift from Drs. Linares et al. [23]. Myc-pCAF was subcloned into Myc tagged pCMV-3 using the EcoR1 restriction site. HA-K48 Ub and K63 Ub were a gift from Dr. Ted Dawson. Both, HA-K48 and K63 ubiquitin have all internal lysine residues of ubiquitin mutated to arginine except K48 or K63, allowing polyubiquitination to only occur on those lysine residues. The HLA-DRA luciferase reporter construct was described previously [27]. The E1 activating enzyme UBE1 (Boston Biochem, Boston, MA), E2 conjugating enzyme UbcH5b (Boston Biochem), Flag ubiquitin (Boston Biochem), Hdm2 (Boston Biochem), and His-pCAF (Proteinone, Rockville, MD) were all obtained commercially.

2.3. GST-Protein Production and Purification. BL21 star (DE3) competent cells (Invitrogen, Carlsbad, CA) were transformed with pGEX constructs. Transformed colonies were selected and inoculated in 5 mL LB supplemented with AMP and bacteria were allowed to grow overnight at 37°. One mL preps was added to 100 mL fresh LB supplemented AMP and bacterial were allowed to grow for three and a half hours at 37° to OD_{600}, 0.8. IPTG was added to induce expression. Cells were centrifuged and the pellet was washed with chilled PBS and centrifuged again. The cell pellet was frozen for one hour at −80°; pellet was allowed to thaw on ice and was resuspended in buffer A (PBS + 1% Triton-X100 + 0.1 M NaCl + Protease Inhibitor (Roche)). Cells were sonicated on ice and were centrifuged to obtain the soluble fraction. The insoluble fraction was then resuspended in buffer B (buffer A + 25% (w/v) sucrose) and the mixture was centrifuged at 20,000 rpm for 20 minutes. The supernatant was then collected as the insoluble fraction. Solubilization and refolding of inclusion bodies were performed in 8 M urea + 5 mM DTT to dissolve the pellet. The protein-urea mixture was then dialyzed in PBS at 4° for two days. GST-CIITA protein was added to a Glutathione Resin column and the protein was eluted in 10 mM glutathione elution buffer (0.154 g reduced glutathione dissolved in 50 mL of 50 mM Tris-HCL, pH 8.0). GST-CIITA, flow through, wash, and elutes were analyzed by SDS-PAGE and then stained with Coomassie. Elutes were dialyzed to remove free glutathione.

2.4. Coimmunoprecipitations. COS cells were plated at cell density of 8×10^5/10 cm on tissue culture plates. Cells were transfected using GeneJuice (Merck Millipore, Darmstadt, Germany) as indicated with 5 μg of Myc-CIITA, Flag-pCAF, Flag-K141R, K144R, K141/144R CIITA, HA-K48 Ub, K63 Ub, HA-mono Ub, or pCDNA control. Twenty-four hours after transfections, cells were lysed in 1% NP40 buffer supplemented with EDTA-free protease inhibitors (Roche) on ice. Lysates were centrifuged, normalized for protein concentration, and precleared with Mouse IgG (Sigma-Aldrich) and Protein G (Thermo Fisher) followed by immunoprecipitation with either EZ view anti-c Myc affinity gel beads (Sigma-Aldrich) or with anti-Flag M2 affinity gel (Sigma-Aldrich). Immune complexes were denatured with Laemmli buffer, boiled, and separated by SDS-PAGE gel electrophoresis. Gels were transferred to nitrocellulose and were individually immunoblotted with anti-Myc (Abcam, Cambridge, MA), anti-Flag (Sigma-Aldrich), antiubiquitin (Life Sensors, Malvern, PA), anti-K48 ubiquitin (Cell Signaling, Danvers, MA), anti-K63 ubiquitin (Millipore), or with anti-GST (Abcam, Cambridge, MA). HRP conjugates were detected using HyGlo Chemiluminescent substrate (Denville). Protein normalization and equal loading were determined in lysates.

2.5. Luciferase Reporter Assays. COS cells were plated at 5 $\times 10^4$ cells/well density (70% confluency). Following adhesion, cells were cotransfected as indicated with HLA-DRA, Renilla, pcDNA, Myc-CIITA, and Flag-pCAF using GeneJuice (Merck Millipore, Darmstadt, Germany) according to the manufacturer's protocol. Twenty-four hours following transfections, cells were lysed with 1x Passive lysis buffer (Promega, Madison, WI) supplemented with EDTA-free protease inhibitor (Roche). Dual luciferase assays were performed using the Lmax II$_{384}$ (Molecular Devices, Sunnyvale,

CA) according to the manufacturer's instructions. Luciferase readings were normalized to Renilla readings for protein normalization.

2.6. In Vitro Ubiquitination Assay. The CIITA in vitro ubiquitination assay was carried out in 150 μL of reaction mixture containing 40 mM Tris-HCL (pH 7.5), 5 mM MgCl$_2$, 2 mM dithiothreitol, 1 mM Creatine Phosphate, 2 mM ATP, 400 ng of Recombinant GST-CIITA (substrate), 400 ng of GST-Hdm2 (Boston Biochem), 400 ng Recombinant His-pCAF (E3 ligase candidate) (Proteinone Rockville, MD), 500 ng Flag ubiquitin (Boston Biochem, Boston, MA), 200 ng E1 activating enzyme, UBE1 (Boston Biochem), 200 ng E2 conjugating enzyme, and UbcH5b (Boston Biochem). All components were added and incubated at 37°C for 60 minutes and were analyzed via SDS-PAGE. Ubiquitination was detected by immunoblot using Flag antibody (Sigma), and CIITA ubiquitination was verified with GST (Abcam), and pCAF was verified with monoclonal α pCAF antibody (Santa Cruz). Verification of Hdm2 ubiquitination was detected using GST (Abcam).

3. Results

3.1. CIITA and pCAF Coimmunoprecipitate. CIITA and pCAF previously have been shown to associate [14], and pCAF is known to acetylate CIITA on lysines (K) 141 and 144. These residues lie within a nuclear localization signal (NLS) region and acetylation is necessary to shuttle CIITA to the nucleus. Once pCAF acetylates CIITA, CIITA accumulates in the nucleus, where it binds to the enhanceosome complex at the MHC II promoter [14] and drives MHC II transcription. To confirm previous findings, we conducted coimmunoprecipitation assays to verify interactions between WT CIITA and WT pCAF. Lane one indicates association of WT Myc-CIITA and WT Flag-pCAF through coimmunoprecipitation analysis (Figure 1, top panel, lane one).

3.2. pCAF's E3 Ligase Domain Is Necessary for CIITA Transactivity. While pCAF is known primarily for its HAT role, pCAF is also considered to be an ubiquitination factor with intrinsic E3 ligase capabilities [23, 26]. Interestingly, pCAF does not have any homology to other known E3 ligases. Linares et al., performed a series of deletion mutations and identified a region (amino acids 121–242) that possesses E3 ligase capability [23]. Previous reports suggest pCAF is able to mediate both acetylation and ubiquitination of the same target [13, 23, 26, 28]. Thus, we next wanted to determine if pCAF's E3 ligase domain is necessary for CIITA's increased transactivity. Levels of transactivity were determined using a dual luciferase reporter assay. CIITA cotransfected with WT pCAF leads to a 2-fold increase in CIITA transactivity and ability to drive MHC II transcription; however, deletion of the E3 ligase domain of pCAF leads to a significant decrease in CIITA transactivity levels (Figure 2).

3.3. E3 Ligase Domain Deficient pCAF Is Unable to Ubiquitinate CIITA. We next investigated if CIITA ubiquitination would be impaired or altered in the absence of the E3

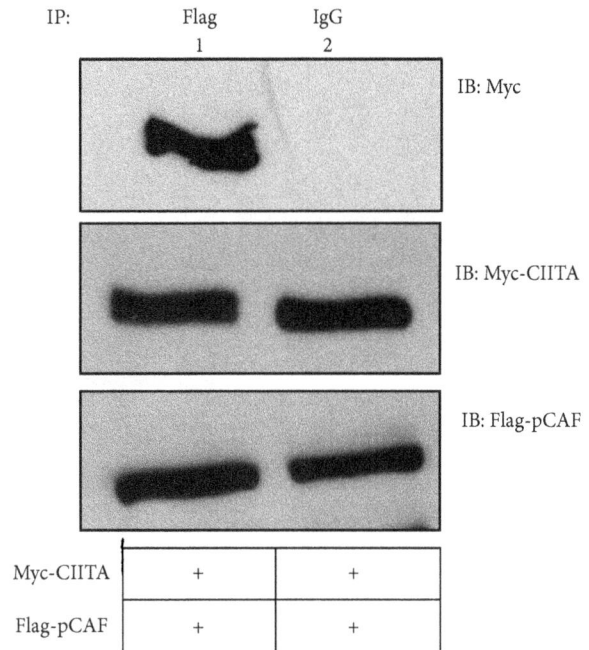

FIGURE 1: CIITA associates with the E3 ligase pCAF. *Coimmunoprecipitation of CIITA and pCAF.* COS cells were cotransfected with Myc-CIITA and Flag-pCAF. Cells were harvested, lysed, precleared, and immunoprecipitated with anti-Flag and anti-Mouse IgG. Western blots were performed and immunoprecipitated samples were immunoblotted using anti-Myc antibodies. Lysate controls demonstrate expressions of Myc-CIITA and Flag-pCAF. Data shown are cropped images from one immunoprecipitation gel and one lysate gel and are representative of three individual experiments.

ligase domain of pCAF. pCAF contains an autoubiquitination domain that is able to mediate self-ubiquitination but has not been shown to be involved in ubiquitination of other substrates [23]. To determine if the E3 ligase domain is necessary for facilitating CIITA ubiquitination, we performed an in vivo ubiquitination assay. WT-Myc-CIITA, WT-Flag-pCAF, or the ΔE3 pCAF mutant were cotransfected into COS cells. Ubiquitination levels of WT CIITA cotransfected with WT pCAF show a significant increase over those of WT CIITA transfected alone (Figure 3, compare lanes 1 and 2). However, the ubiquitination levels of CIITA cotransfected with the ΔE3 pCAF mutant show levels of ubiquitination that are significantly decreased when compared to those of CIITA cotransfected with WT pCAF (Figure 3, compare lanes 2 and 3). These data support that the E3 ligase domain of pCAF is important for CIITA ubiquitination and is involved in mediating CIITA ubiquitination.

3.4. pCAF Facilitates CIITA Ubiquitination Independent of Its HAT Domain. pCAF is a well-known HAT and is involved in many aspects of CIITA and MHC II regulation [14, 29, 30]. pCAF assists in remodeling the chromatin structure of the MHC II promoter where it acetylates histones H3 and H4 [30] and also regulates CIITA's nuclear relocation by acetylating CIITA K141 and K144, leading to increased activation of CIITA and increased expression of MHC II [14].

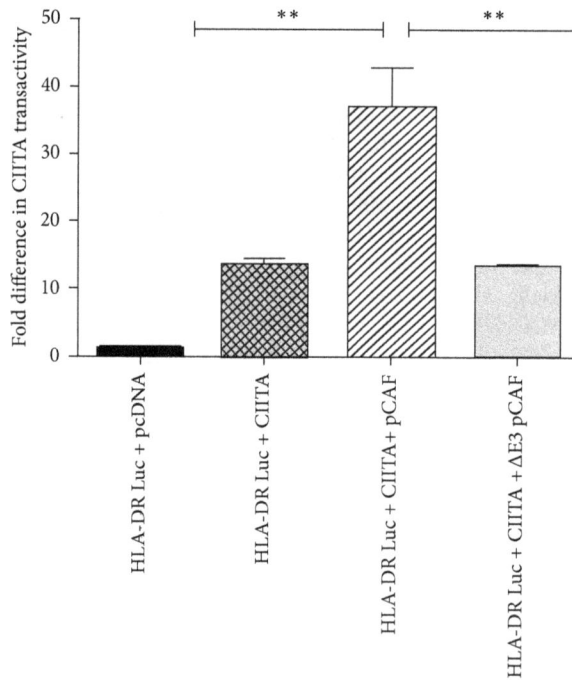

FIGURE 2: The pCAF E3 ligase domain is necessary for enhanced CIITA transactivity. CIITA transactivation increases in the presence of expressed WT pCAF. *Reporter Assay*. COS cells were transfected as indicated with the MHC II HLA-DR Luc reporter construct, Renilla, CIITA, pCAF, and Δ E3 pCAF. Luciferase assays were performed in triplicate in three independent experiments; data are presented as fold increase in luciferase activity. Results are standardized to Renilla values and represent the mean ± SD, $** < 0.01$.

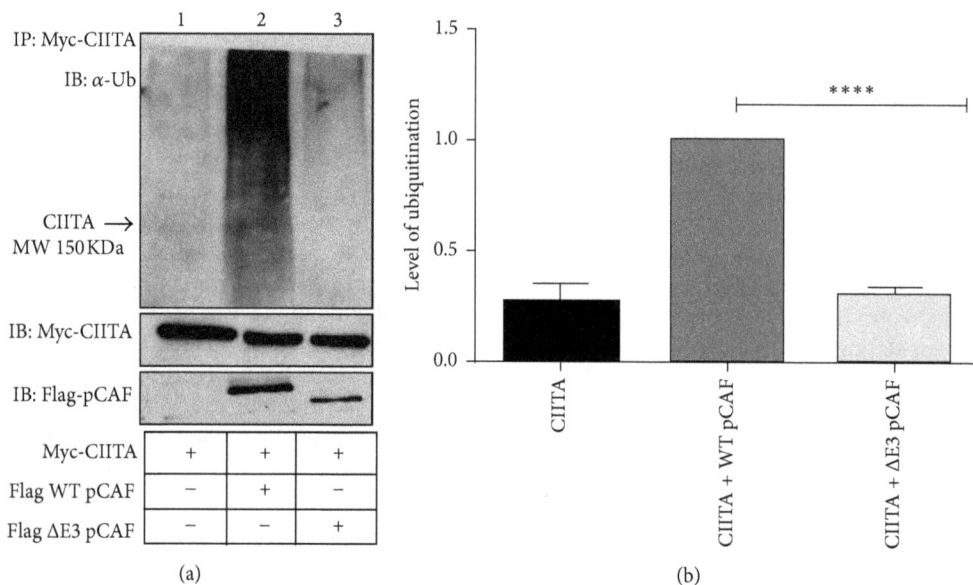

(a)

(b)

FIGURE 3: CIITA ubiquitination depends on the E3 ligase domain of pCAF. (a) The E3 ligase null mutant pCAF negates CIITA ubiquitination. In vivo *ubiquitination assay*. COS cells were cotransfected with Myc-CIITA, Flag-pCAF, and Flag-pCAF Δ E3 ligase mutants as indicated. Eighteen hours following transfections, cells were harvested, lysed, precleared, and immunoprecipitated with the anti-Myc antibody. Western blots were performed and immunoprecipitated samples were immunoblotted using antiubiquitin antibodies. Lysate controls (bottom two panels) demonstrate expression of Myc-CIITA, Flag-pCAF, and Flag-pCAF Δ E3. Data shown are cropped images from one immunoprecipitation gel and one lysate gel and are representative of three independent experiments. (b) Densitometry and quantification of data in Figure 3(a). Densitometry was performed on three independent experiments, mean ± SD, $**** < 0.0001$. Area for densitometry analysis was selected from CIITA's molecular weight of 150 KDa as labeled and above.

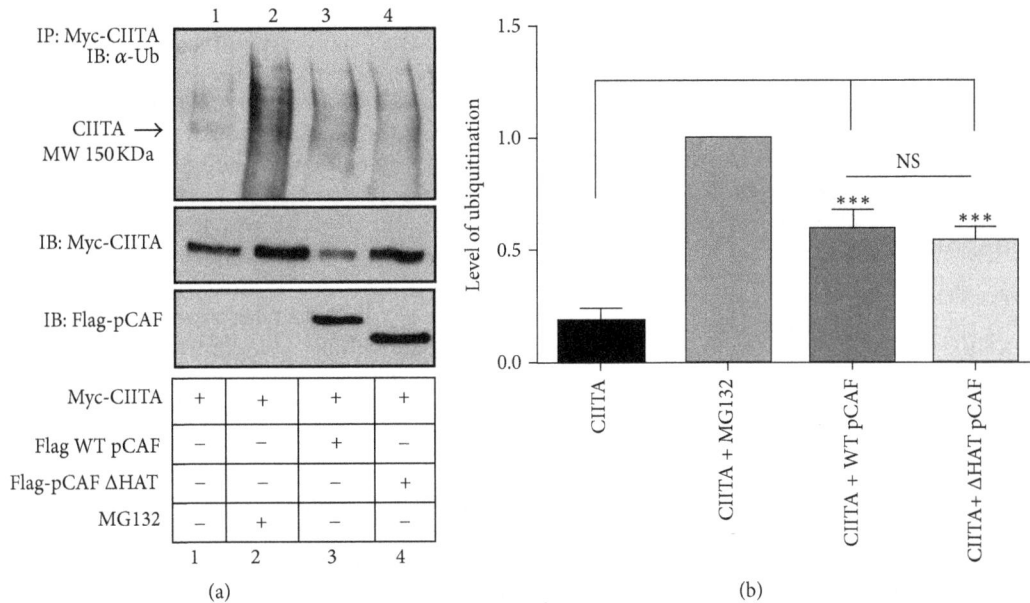

FIGURE 4: pCAF facilitates CIITA ubiquitination independent of its HAT domain. (a) CIITA is ubiquitinated by pCAF. In vivo *ubiquitination assay*. COS cells were cotransfected as indicated with Myc-CIITA, Flag-pCAF, and Flag-pCAF Δ HAT mutant. Eighteen hours following transfections, cells were harvested, lysed, precleared, and immunoprecipitated with the anti-Myc antibody. Western blots were performed and immunoprecipitated samples were immunoblotted using antiubiquitin antibodies. Lysate controls (bottom two panels) demonstrate expression of Myc-CIITA, Flag-pCAF, and Flag-pCAF Δ HAT. Data shown are cropped images from one immunoprecipitation gel and one lysate gel and are representative of three independent experiments. (b) Densitometry and quantification of data in (a). Densitometry was performed on three independent experiments, mean ± SD, *** < 0.001. Area for densitometry analysis was selected from CIITA's molecular weight of 150 KDa as labeled and above.

To determine if pCAF's role in ubiquitination of CIITA was independent of pCAF's HAT activities, we performed in vivo ubiquitination assays using WT CIITA and HAT deletion mutant of pCAF. In this assay, expression of WT CIITA alone indicates low level ubiquitination (Figure 4(a), lane 1), and ubiquitination levels significantly increase when WT pCAF is overexpressed (Figure 4(a), lane 3). However, in the absence of the pCAF HAT domain, there is no measureable difference in CIITA ubiquitination levels when CIITA ubiquitination is compared to that generated in the presence of WT pCAF (Figure 4(a), compare lanes 3 and 4). Thus, we conclude that pCAF's ability to ubiquitinate CIITA is independent of the pCAF HAT domain.

3.5. pCAF Associates with CIITA Acetylation Null Mutants. As pCAF acts in a dual way as both HAT and ubiquitin ligase, and as acetylation drives CIITA nuclear import, we next determined if acetylation of CIITA is necessary for CIITA ubiquitination mediated by pCAF. To begin to investigate the relationship between acetylation and ubiquitination on CIITA, we first wanted to identify if acetylation null CIITA mutants and pCAF are able to associate. Coimmunoprecipitation analysis indicates CIITA acetylation mutants deficient at K141R, K144R individually, or K141/144R double mutant sites all remain capable of interaction with pCAF (Figure 5).

3.6. pCAF Mediates K48 Linked Ubiquitination of Acetylation Null CIITA Mutants. Acetylation at K141 and K144 of CIITA

is critical in signaling the movement of CIITA from the cytoplasm to the nucleus [14]. To investigate pCAF's role in mediating acetylation independent ubiquitination, we utilized acetylation null mutants of CIITA. These CIITA mutants are incapable of being acetylated at K141 and K144 and thus likely targets for K48 linked polyubiquitination and degradation. Our results indicate K141R and K144R and K141/144R CIITA mutants display low levels of ubiquitination and the addition of pCAF yields significantly lower levels of detectable ubiquitination (Figure 6(a), top blot) as compared to WT CIITA coexpressed with pCAF (compare lanes 3, 6, 9, and 12). Further, lysate blots indicate CIITA transfections and show a decrease in CIITA protein expression (Figure 6(a), middle blot). These data suggest the lysine deficient CIITA is being degraded at greater rate with the overexpression of pCAF (Figure 6(a), compare lanes 5 and 6). Proteasome inhibition by MG132 indicates an accumulation of ubiquitinated CIITA when pCAF is present (Figure 6(a) lane 7), indicating the lack of ubiquitination smear seen in lane 6 is likely due to CIITA degradation. Further, the K144R mutant and the double K141/K144R mutants indicate similar trends, respectively (Figure 6(a) lanes 8–10 and 11–13). To further determine if pCAF mediates CIITA K48 linked ubiquitination, we took 15 uL of the same immunoprecipitation sample and immunoblotted it for K48 ubiquitination using specific antibodies recognizing K48 ubiquitination (Figure 6(b)). Similar trends were observed as seen in Figure 6(a), where CIITA acetylation mutants were

IP: Myc

IB: Flag

IB: Flag CIITA

IB: Myc-pCAF

Flag K141R CIITA	+	−	−
Flag K144R CIITA	−	+	−
Flag K141/144R CIITA	−	−	+
Myc-pCAF	+	+	+
	1	2	3

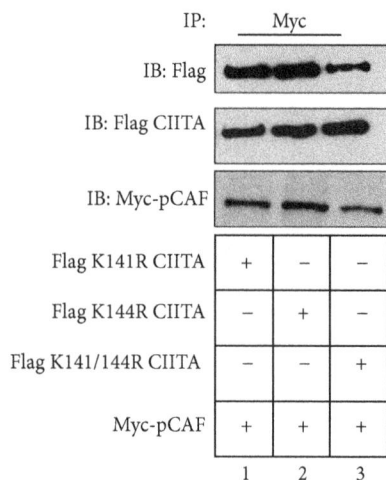

FIGURE 5: Acetylation null CIITA mutants coimmunoprecipitate with pCAF. *Coimmunoprecipitation of acetylation null CIITA mutants with pCAF.* COS cells were cotransfected with Myc-CIITA, K141R CIITA, K144R CIITA or K141/144R CIITA, and Flag-pCAF. Cells were harvested, lysed, precleared, and immunoprecipitated with anti-Myc or Mouse IgG antibodies as indicated. Western blots were performed and immunoprecipitated samples were immunoblotted using anti-Myc antibodies. Lysate controls demonstrate expression of Myc-CIITA and acetylation null mutants and Flag-pCAF (bottom two panels). Data shown are cropped images from one immunoprecipitation gel and one lysate gel and are representative of 3 individual experiments.

ubiquitinated; however, when pCAF is introduced we again observe that ubiquitination diminishes and samples treated with proteasome inhibitor show accumulation of ubiquitinated CIITA. Densitometry assays indicate significant differences in ubiquitination levels in both global ubiquitination and K48 specific ubiquitination (Figures 6(c) and 6(d)).

3.7. In Vitro Ubiquitination Assays Indicate pCAF's Ability to Mediate Ubiquitination of CIITA. To elucidate if CIITA was a substrate for pCAF as E3 ligase, we conducted an in vitro ubiquitination assay using purified human recombinant proteins. As a positive control, we used GST-Hdm2, which has previously been seen to be ubiquitinated by pCAF [23] (Figure 7(c)). Purified human recombinant proteins E1 (UbA-1), E2 (UbCH5B), Flag ubiquitin, GST-WT CIITA, and His-WT pCAF (E3) were all used in the presence of a reaction mixture (see Materials and Methods). As shown in Figures 7(a) and 7(b), when all ubiquitination components (E1, E2, pCAF (E3), and ubiquitin) were present along with the substrate in question, CIITA ubiquitination occurs (Figures 7(a) and 7(b), lane 5); however in the absence of any of these components, ubiquitination did not occur. We immunoblotted both anti-Flag ubiquitin and anti GST-CIITA. These data reveal that CIITA is a substrate for the ubiquitin E3 ligase pCAF.

3.8. CIITA Mono, K63, and K48 Linked Ubiquitination Is Increased In Vivo with pCAF Expression. We previously determined CIITA's ubiquitination status to include mono,

K63, and K48 ubiquitination [10, 11]. Bhat et al. showed the three sites of monoubiquitination on CIITA are necessary for CIITA stability and for increased CIITA transactivity. Additionally, we previously demonstrated CIITA modified by K63 linked ubiquitination plays a role in CIITA movement from the cytoplasm to the nucleus. Knowing that several types of ubiquitination linkages modify CIITA, we wanted to determine if pCAF was able to facilitate all three types of CIITA ubiquitination. Previously, we demonstrated pCAF is able to mediate K48 linked CIITA ubiquitination in the absence of CIITA acetylation. To investigate if pCAF is able to mediate various ubiquitin linkages on CIITA, we conducted in vivo ubiquitination assays. We demonstrate CIITA monoubiquitination, K63, and K48 linked ubiquitination all increase in the presence of WT pCAF as compared to assays performed when WT CIITA is expressed alone (Figures 8(a), 8(b), and 8(c), compare lanes 1 and 2). When the proteasome inhibitor MG132 inhibits the 26S proteasome, all three forms of ubiquitinated CIITA accumulated indicating maximum ubiquitination levels (Figures 8(a), 8(b), and 8(c), lane 3). These data indicate pCAF's ability to mediate multiple forms of ubiquitination of CIITA.

4. Discussion

We sought here to identify the ubiquitin E3 ligase mediating CIITA ubiquitination. CIITA is known as the "master regulator" of MHC II genes. MHC II is critically important for proper presentation of extracellular pathogen to $CD4^+$ T cells in adaptive immune responses. While MHC II is regulated at the level of transcription, CIITA is tightly regulated at the level of posttranslational modification. CIITA is heavily modified through a complex series of PTMs that dynamically regulate its location, function, stability, and activity.

Previous reports identify CIITA as being modified by mono, K63, and K48 linked ubiquitination [10, 11]. Ubiquitination of CIITA has been demonstrated as an essential posttranslational modification. Monoubiquitinated CIITA is more stable and active at the MHC II promoter [11, 13, 31]. K63 linked ubiquitination is also important as this particular ubiquitin linkage demonstrates enhanced crosstalk with phosphorylation and together these modifications are important for movement of CIITA from the cytoplasm to the nucleus [10]. Additionally, CIITA is modified by K48 linked ubiquitination, leading to recognition and degradation by the proteasome [11].

pCAF is a well-known histone acetyltransferase or HAT [32] and has previously been demonstrated to be recruited to and to activate transcription of the MHC II promoter via pCAF's HAT activities [2, 22]. pCAF must be localized to the MHC II promoter where pCAF cooperates with CIITA to drive MHC II transcription. The interaction of pCAF and CIITA is independent of pCAF's HAT domain [2, 8, 32]. We confirm here the addition of pCAF drives increased CIITA transactivation levels at the MHC II promoter (Figure 2).

Reports by several groups of pCAF's ability to act as E3 ubiquitin ligase, facilitating ubiquitination on targets such as Hdm2/Mdm2 and Gli1 [23, 26] where pCAF has

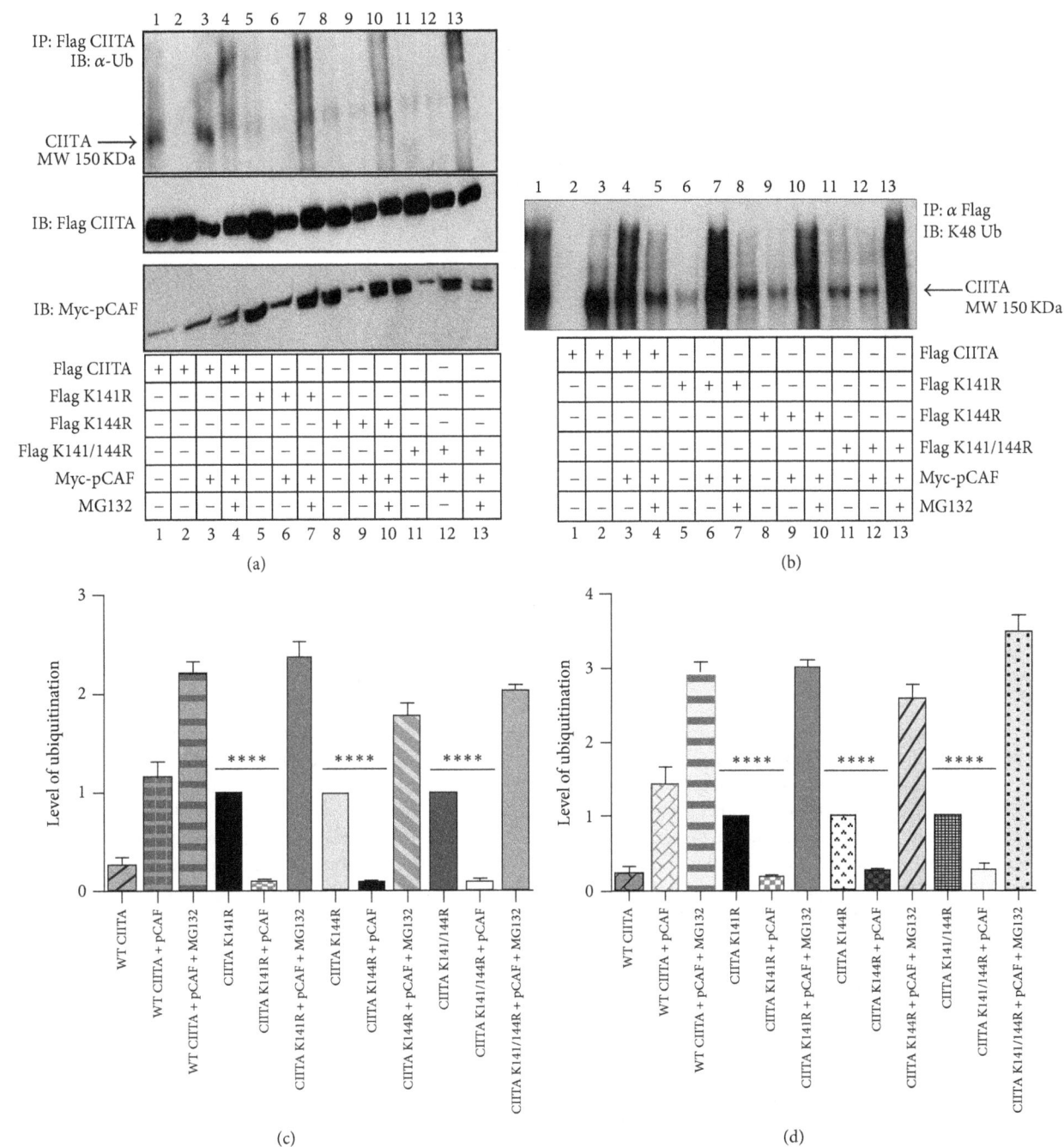

FIGURE 6: pCAF enhances ubiquitination of CIITA acetylation null mutants. (a) Acetylation null CIITA mutants have increased levels of degradative ubiquitination in the presence of pCAF. In vivo *ubiquitination assay*. COS cells were cotransfected with Flag-K141R, K144R, K141/144R CIITA, and Myc-pCAF. Eighteen hours following transfections, MG132 was added to indicated samples in order to inhibit the 26S proteasome. Cells were harvested, lysed, precleared, and immunoprecipitated with anti-Flag antibody (lane 2 negative control, immunoprecipitated with IgG). Western blots were performed and immunoprecipitated samples were immunoblotted using antiubiquitin antibodies. Lysate controls (bottom two panels) demonstrate expression of WT Flag CIITA, Flag-K141R, K144R, K141/144R, and Myc-pCAF. Data shown are cropped images from one immunoprecipitation gel and one lysate gel and are representative of three individual experiments. (b) CIITA acetylation null mutants have increased levels of K48 specific linkage ubiquitination in the presence of pCAF. 15 μL from the same sample was immunoprecipitated with anti-Flag antibody and immunoblotted for anti-K48 specific ubiquitin antibody. (c) Densitometry and quantification of data in (a). Densitometry was performed on three independent experiments, mean ± SD, ∗∗∗∗ < 0.0001. (d) Densitometry and quantification of data in (b). Densitometry was performed on three independent experiments, mean ± SD, ∗∗∗∗ < 0.0001. Area for densitometry analysis was selected from CIITA's molecular weight of 150 KDa as labeled and above.

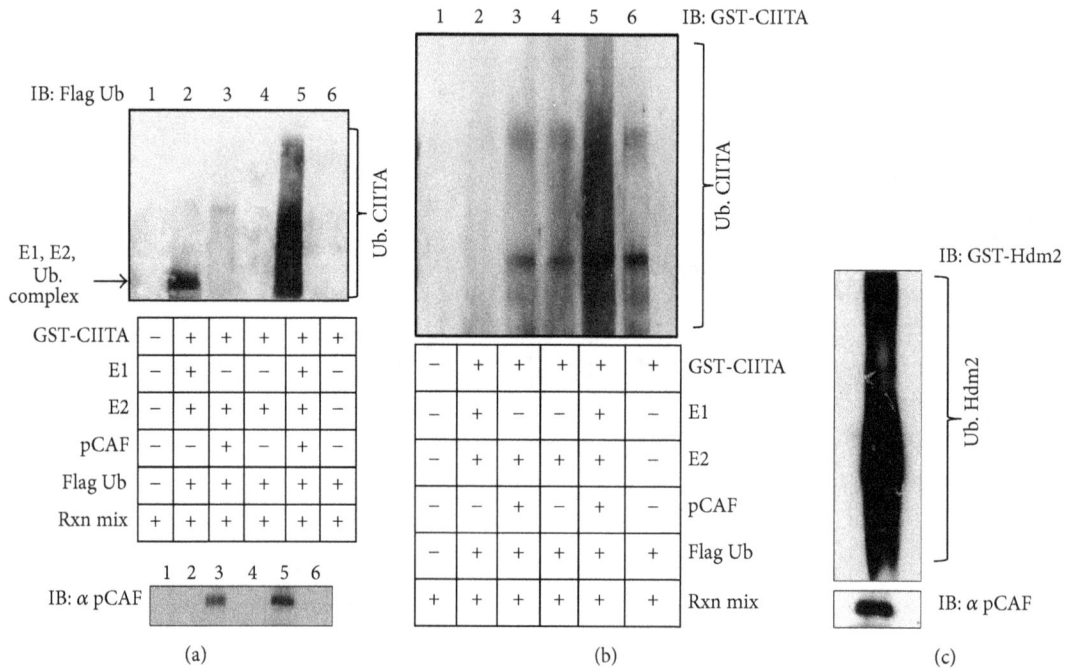

FIGURE 7: pCAF ubiquitinates CIITA in vitro. (a) Ubiquitination of CIITA by pCAF in vitro. Reaction components are as indicated and were performed as described in Material and Methods. Ubiquitination was detected with anti-Flag antibody (top panel); pCAF was detected using anti-pCAF antibody (bottom panel). (b) Ubiquitinated CIITA was further detected with anti-GST antibody. (c) Positive control demonstrating pCAF's ability to facilitate ubiquitination in vitro on Hdm2.

also been shown to act as HAT, raised the possibility that pCAF could have dual enzyme activity on CIITA. We next sought to determine if pCAF was participating as E3 ligase and able to mediate CIITA ubiquitination. Levels of CIITA ubiquitination significantly increase in the presence of WT pCAF, while in the absence of the pCAF E3 ligase region, ubiquitination is abolished (Figure 3). The region of pCAF containing the E3 ligase domain does not conform to any known E3 ligase structures; many questions remain as to how pCAF may function as E3 ligase. Possible mechanism is many E3 ligases are regulated through autoubiquitination [33]. pCAF's autoubiquitination domain has previously been shown to not be involved in substrate ubiquitination but does promote self-ubiquitination [23]. E3 ligases also can be regulated by phosphorylation, leading to either activation or deactivation [34, 35]. pCAF is phosphorylated at tyrosine (Y) 729 and threonine (T) 731 and the role for phosphorylation has yet to be determined [36]; however, phosphorylation at these residues could be involved in pCAF's E3 ligase function.

We further show CIITA ubiquitination by pCAF is independent of pCAF's HAT domain (Figure 4(a)). When CIITA acetylation is blocked, CIITA is ubiquitinated by K48 linked polyubiquitination and subsequently degraded. This data is in line with previous reports indicating treatment with Trichostatin A (TSA) and HDAC1 which show reduced levels of CIITA ubiquitination [13]. By blocking the proteasome, we were able to visualize accumulated levels of ubiquitination (Figures 6(a) and 6(b)). These data indicate pCAF's ability to ubiquitinate independently of its HAT function.

In vitro analysis confirmed pCAF's role in ubiquitination of CIITA when all necessary ubiquitin components were available and pCAF as the E3; ubiquitination of CIITA occurred (Figures 7(a) and 7(b)). Further, In vivo ubiquitination assays revealed pCAF's ability to mediate multiple types of CIITA ubiquitination, including mono, K63, and K48 linked polyubiquitination (Figures 8(a), 8(b), and 8(c)). pCAF utilizes the E2 conjugating enzyme UbcH5b, which is capable of synthesizing ubiquitin chains containing all of the possible seven linkages (K6, 11, 27, 29, 33, 48, and 63) of ubiquitin [37].

Our study increases understanding of the regulation of the Class II Transactivator, thus leading directly to the molecular events, which contribute to the regulation of MHC II genes. Ubiquitination has been shown to be one of the many important and necessary PTMs, which regulate CIITA. Our previous identification of both mono and K63 linked ubiquitination provided valuable insight that ubiquitination regulates the stability, location, and activity of CIITA [10, 11]. Here we identify a novel substrate for the E3 ligase function of pCAF and identify pCAF's ability to mediate various ubiquitination moieties on CIITA. Future work will focus on understanding the regulation of pCAF as E3 ligase. pCAF does not contain homologous domains of other known E3 ligases and it remains unknown how enzymes with "dual" HAT and E3 ligase activity are regulated. Understanding the mechanism controlling each of these functions and how they are simultaneously regulated will be important to further understand the regulation of CIITA and the adaptive immune response [38].

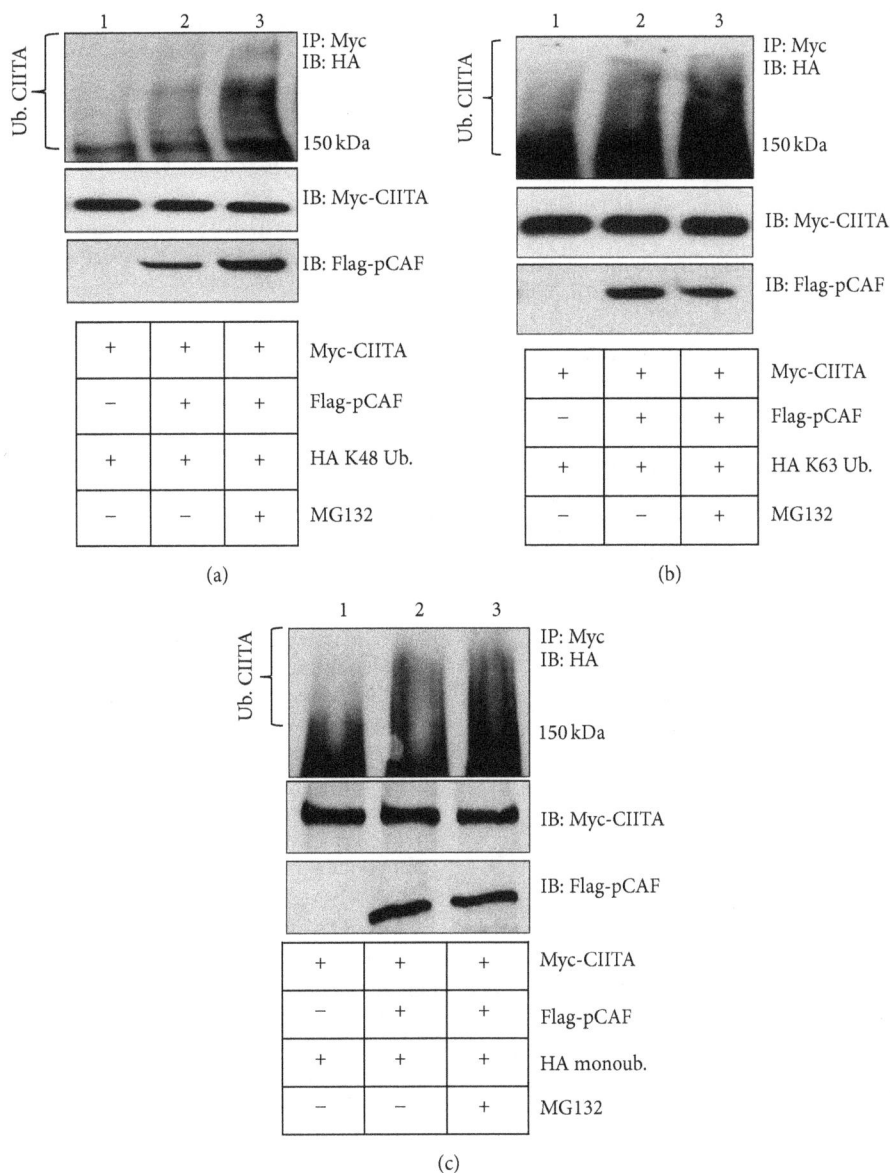

FIGURE 8: pCAF enhances K48, K63, and monoubiquitination of CIITA. (a) *In vivo ubiquitination assay*. COS cells were cotransfected with Myc-CIITA, Flag-pCAF, HA-K48 ubiquitin, HA-K63 ubiquitin, or HA-mono Ub. Eighteen hours following transfections, MG132 was added to indicated samples to inhibit the 26S proteasome in treated cells. Cells were harvested, lysed, precleared, and immunoprecipitated with anti-Myc antibody. Western blots were performed and immunoprecipitated samples were immunoblotted using anti-HA antibody. Lysate controls (bottom two panels) demonstrate expression of Myc-CIITA and Flag-pCAF. Data shown are cropped images from one immunoprecipitation gel and one lysate gel and are representative of three individual experiments.

Abbreviations

CIITA: Class II Transactivator
MHC II: Major Histocompatibility Class II
PTMs: Posttranslational modifications
pCAF: p300/CBP associated factor
HAT: Histone acetyltransferase
Ub: Ubiquitin.

Competing Interests

The authors declare that they have no competing interests.

Authors' Contributions

Julie E. Morgan performed all experiments within the paper. Manuscript was prepared and written by Julie E. Morgan under the guidance of Susanna F. Greer.

Acknowledgments

The authors would like to thank Dr. Ya-Shu Huang for assistance in protein purification of GST-CIITA and Ronald L. Shanderson for assistance in acetylation null CIITA experiments. Funding was provided by Georgia State University,

Molecular Basis of Disease Area of Focus Fellowship to Julie E. Morgan.

References

[1] A. K. Abbas and C. A. Janeway Jr., "Immunology: improving on nature in the twenty-first century," *Cell*, vol. 100, no. 1, pp. 129–138, 2000.

[2] G. Drozina, J. Kohoutek, N. Jabrane-Ferrat, and B. M. Peterlin, "Expression of MHC II genes," *Current Topics in Microbiology and Immunology*, vol. 290, pp. 147–170, 2005.

[3] F. Schnappauf, S. B. Hake, M. M. Camacho Carvajal, S. Bontron, B. Lisowska-Grospierre, and V. Steimle, "N-terminal destruction signals lead to rapid degradation of the major histocompatibility complex class II transactivator CIITA," *European Journal of Immunology*, vol. 33, no. 8, pp. 2337–2347, 2003.

[4] C. Benoist and D. Mathis, "Regulation of major histocompatibility complex class-II genes: X, Y and other letters of the alphabet," *Annual Review of Immunology*, vol. 8, pp. 681–715, 1990.

[5] J. M. Boss and P. E. Jensen, "Transcriptional regulation of the MHC class II antigen presentation pathway," *Current Opinion in Immunology*, vol. 15, no. 1, pp. 105–111, 2003.

[6] C.-H. Chang, J. D. Fontes, M. Peterlin, and R. A. Flavell, "Class II transactivator (CIITA) is sufficient for the inducible expression of major histocompatibility complex class II genes," *Journal of Experimental Medicine*, vol. 180, no. 4, pp. 1367–1374, 1994.

[7] J. D. Fontes, S. Kanazawa, N. Nekrep, and B. M. Peterlin, "The class II transactivator CIITA is a transcriptional integrator," *Microbes and Infection*, vol. 1, no. 11, pp. 863–869, 1999.

[8] J. A. Harton, E. Zika, and J. P.-Y. Ting, "The histone acetyltransferase domains of CREB-binding protein (CBP) and p300/CBP-associated factor are not necessary for cooperativity with the class II transactivator," *The Journal of Biological Chemistry*, vol. 276, no. 42, pp. 38715–38720, 2001.

[9] E. Zika, L. Fauquier, L. Vandel, and J. P.-Y. Ting, "Interplay among coactivator-associated arginine methyltransferase 1, CBP, and CIITA in IFN-γ-inducible MHC-II gene expression," *Proceedings of the National Academy of Sciences of the United States of America*, vol. 102, no. 45, pp. 16321–16326, 2005.

[10] J. E. Morgan, R. L. Shanderson, N. H. Boyd, E. Cacan, and S. F. Greer, "The class II transactivator (CIITA) is regulated by post-translational modification cross-talk between ERK1/2 phosphorylation, mono-ubiquitination and Lys^{63} ubiquitination," *Bioscience Reports*, vol. 35, no. 4, Article ID e00233, 2015.

[11] K. P. Bhat, A. D. Truax, and S. F. Greer, "Phosphorylation and ubiquitination of degron proximal residues are essential for class II transactivator (CIITA) transactivation and major histocompatibility class II expression," *The Journal of Biological Chemistry*, vol. 285, no. 34, pp. 25893–25903, 2010.

[12] S. F. Greer, J. A. Harton, M. W. Linhoff, C. A. Janczak, J. P.-Y. Ting, and D. E. Cressman, "Serine residues 286, 288, and 293 within the CIITA: a mechanism for down-regulating CIITA activity through phosphorylation," *Journal of Immunology*, vol. 173, no. 1, pp. 376–383, 2004.

[13] S. F. Greer, E. Zika, B. Conti, X.-S. Zhu, and J. P.-Y. Ting, "Enhancement of CIITA transcriptional function by ubiquitin," *Nature Immunology*, vol. 4, no. 11, pp. 1074–1082, 2003.

[14] C. Spilianakis, J. Papamatheakis, and A. Kretsovali, "Acetylation by PCAF enhances CIITA nuclear accumulation and transactivation of major histocompatibility complex class II genes,"

[15] G. Tosi, N. Jabrane-Ferrat, and B. M. Peterlin, "Phosphorylation of CIITA directs its oligomerization, accumulation and increased activity on MHCII promoters," *EMBO Journal*, vol. 21, no. 20, pp. 5467–5476, 2002.

[16] L. N. Voong, A. R. Slater, S. Kratovac, and D. E. Cressman, "Mitogen-activated protein kinase ERK1/2 regulates the class II transactivator," *Journal of Biological Chemistry*, vol. 283, no. 14, pp. 9031–9039, 2008.

[17] X. Wu, X. Kong, D. Chen et al., "SIRT1 links CIITA deacetylation to MHC II activation," *Nucleic Acids Research*, vol. 39, no. 22, pp. 9549–9558, 2011.

[18] X. Wu, X. Kong, L. Luchsinger, B. D. Smith, and Y. Xu, "Regulating the activity of class II transactivator by posttranslational modifications: exploring the possibilities," *Molecular and Cellular Biology*, vol. 29, no. 21, pp. 5639–5644, 2009.

[19] N. Jabrane-Ferrat, N. Nekrep, G. Tosi, L. Esserman, and B. M. Peterlin, "MHC class II enhanceosome: how is the class II transactivator recruited to DNA-bound activators?" *International Immunology*, vol. 15, no. 4, pp. 467–475, 2003.

[20] K. Masternak, A. Muhlethaler-Mottet, J. Villard, M. Zufferey, V. Steimle, and W. Reith, "CIITA is a transcriptional coactivator that is recruited to MHC class II promoters by multiple synergistic interactions with an enhanceosome complex," *Genes & Development*, vol. 14, no. 9, pp. 1156–1166, 2000.

[21] T. J. Sisk, K. Nickerson, R. P. S. Kwok, and C.-H. Chang, "Phosphorylation of class II transactivator regulates its interaction ability and transactivation function," *International Immunology*, vol. 15, no. 10, pp. 1195–1205, 2003.

[22] K. L. Wright and J. P.-Y. Ting, "Epigenetic regulation of MHC-II and CIITA genes," *Trends in Immunology*, vol. 27, no. 9, pp. 405–412, 2006.

[23] L. K. Linares, R. Kiernan, R. Triboulet et al., "Intrinsic ubiquitination activity of PCAF controls the stability of the oncoprotein Hdm2," *Nature Cell Biology*, vol. 9, no. 3, pp. 331–338, 2007.

[24] A. Ciechanover, "The ubiquitin-mediated proteolytic pathway: mechanisms of action and cellular physiology," *Biological Chemistry Hoppe-Seyler*, vol. 375, no. 9, pp. 565–582, 1994.

[25] X. Wang, J. Taplick, N. Geva, and M. Oren, "Inhibition of p53 degradation by Mdm2 acetylation," *FEBS Letters*, vol. 561, no. 1–3, pp. 195–201, 2004.

[26] D. Mazzà, P. Infante, V. Colicchia et al., "PCAF ubiquitin ligase activity inhibits Hedgehog/Gli1 signaling in p53-dependent response to genotoxic stress," *Cell Death and Differentiation*, vol. 20, no. 12, pp. 1688–1697, 2013.

[27] K. P. Bhat, J. D. Turner, S. E. Myers, A. D. Cape, J. P.-Y. Ting, and S. F. Greer, "The 19S proteasome ATPase Sug1 plays a critical role in regulating MHC class II transcription," *Molecular Immunology*, vol. 45, no. 8, pp. 2214–2224, 2008.

[28] E. M. Kass, M. V. Poyurovsky, Y. Zhu, and C. Prives, "Mdm2 and PCAF increase Chk2 ubiquitination and degradation independently of their intrinsic E3 ligase activities," *Cell Cycle*, vol. 8, no. 3, pp. 430–437, 2009.

[29] S.-D. Chou, A. N. H. Khan, W. J. Magner, and T. B. Tomasi, "Histone acetylation regulates the cell type specific CIITA promoters, MHC class II expression and antigen presentation in tumor cells," *International Immunology*, vol. 17, no. 11, pp. 1483–1494, 2005.

[30] G. Li, J. A. Harton, X. Zhu, and J. P.-Y. Ting, "Downregulation of CIITA function by protein kinase A (PKA)-mediated

Molecular and Cellular Biology, vol. 20, no. 22, pp. 8489–8498, 2000.

phosphorylation: mechanism of prostaglandin E, cyclic AMP, and PKA inhibition of class II major histocompatibility complex expression in monocytic lines," *Molecular and Cellular Biology*, vol. 21, no. 14, pp. 4626–4635, 2001.

[31] G. Drozina, J. Kohoutek, T. Nishiya, and B. M. Peterlin, "Sequential modifications in class II transactivator isoform 1 induced by lipopolysaccharide stimulate major histocompatibility complex class II transcription in macrophages," *Journal of Biological Chemistry*, vol. 281, no. 52, pp. 39963–39970, 2006.

[32] E. Zika and J. P.-Y. Ting, "Epigenetic control of MHC-II: interplay between CIITA and histone-modifying enzymes," *Current Opinion in Immunology*, vol. 17, no. 1, pp. 58–64, 2005.

[33] P. De Bie and A. Ciechanover, "Ubiquitination of E3 ligases: self-regulation of the ubiquitin system via proteolytic and nonproteolytic mechanisms," *Cell Death and Differentiation*, vol. 18, no. 9, pp. 1393–1402, 2011.

[34] T. Hunter, "The age of crosstalk: phosphorylation, ubiquitination, and beyond," *Molecular Cell*, vol. 28, no. 5, pp. 730–738, 2007.

[35] T. Woelk, S. Sigismund, L. Penengo, and S. Polo, "The ubiquitination code: a signalling problem," *Cell Division*, vol. 2, article 11, 2007.

[36] P. V. Hornbeck, B. Zhang, B. Murray, J. M. Kornhauser, V. Latham, and E. Skrzypek, "PhosphoSitePlus, 2014: mutations, PTMs and recalibrations," *Nucleic Acids Research*, vol. 43, no. 1, pp. D512–D520, 2015.

[37] W. Li and Y. Ye, "Polyubiquitin chains: functions, structures, and mechanisms," *Cellular and Molecular Life Sciences*, vol. 65, no. 15, pp. 2397–2406, 2008.

[38] J. E. Morgan, *Dynamic regulation of the class II transactivator by posttranslational modifications [dissertation]*, Georgia State University, 2015, http://scholarworks.gsu.edu/biology_diss/157.

10

Odontoblast-Like Cells Differentiated from Dental Pulp Stem Cells Retain their Phenotype after Subcultivation

Paula A. Baldión,[1] **Myriam L. Velandia-Romero,**[2] **and Jaime E. Castellanos** ⓘ [1,2]

[1]*Grupo de Investigaciones Básicas y Aplicadas en Odontología, Universidad Nacional de Colombia, Bogotá, Colombia*
[2]*Grupo de Virología, Universidad El Bosque, Bogotá, Colombia*

Correspondence should be addressed to Jaime E. Castellanos; jecastellanosp@unal.edu.co

Academic Editor: Maria Roubelakis

Odontoblasts, the main cell type in teeth pulp tissue, are not cultivable and they are responsible for the first line of response after dental restauration. Studies on dental materials cytotoxicity and odontoblast cells physiology require large quantity of homogenous cells retaining most of the phenotype characteristics. Odontoblast-like cells (OLC) were differentiated from human dental pulp stem cells using differentiation medium (containing TGF-β1), and OLC expanded after trypsinization (EXP-21) were evaluated and compared. Despite a slower cell growth curve, EXP-21 cells express similarly the odontoblast markers dentinal sialophosphoprotein and dentin matrix protein-1 concomitantly with RUNX2 transcripts and low alkaline phosphatase activity as expected. Both OLC and EXP-21 cells showed similar mineral deposition activity evidenced by alizarin red and von Kossa staining. These results pointed out minor changes in phenotype of subcultured EXP-21 regarding the primarily differentiated OLC, making the subcultivation of these cells a useful strategy to obtain odontoblasts for biocompatibility or cell physiology studies in dentistry.

1. Introduction

Odontoblasts are highly specialized cells that produce both collagen and noncollagen proteins to build the dentinal extracellular matrix. They are also the pulp first responders against exogenous stimulus or dental materials [1, 2]. Due to their postmitotic phenotype, odontoblasts are difficult to culture [3, 4]; thus, biochemical or toxicologic studies assessing their use in dental materials have been limited. To overcome such difficulties, studies on materials toxicity have been developed on primary gingival fibroblasts [5], primary human or mouse undifferentiated mesenchymal stem cells, and immortalized cell lines [6, 7].

However, the genotype, phenotype, or cell responses in these *in vitro* models could differ from actual odontoblast cell responses due to genetic changes [8] or environmental adaptations of the cell lines, which complicates the interpretation of cytotoxicity results.

Human dental pulp stem cells (hDPSC) have been isolated as adherent mononucleated cells with *in vitro* differentiation capacity toward several lineages, including odontoblasts, osteoblasts, adipocytes, and neural cells. Differentiated odontoblasts establish a structure known as the dentin-pulp complex; therefore, these cells have been proposed as tools in dental regeneration and repair [9].

Obtaining homogeneous odontoblast-like cells (OLC) differentiated from hDPSC that retain their phenotype after monolayer detachment, reseeding and expansion could benefit (I) studies on gene expression and the specific protein synthesis involved in the secretion and mineralization of the extracellular matrix (ECM) and (II) studies on the cellular response after challenge with dental biomaterials. Expanded cells enable a sufficient number of cells to make replicates for toxicity and cell metabolism studies, overcoming the heterogeneity and lack of reproducibility that frequently appear during the use of primarily differentiated odontoblasts.

Differentiated cultured cells are well known to revert to an undifferentiated phenotype after subcultivation [10], making it difficult to perform studies that require large numbers of cells. Therefore, data on culture conditions, proliferation rates, and the differentiation status of subcultured cells are important for comparing and validating experiments using primary differentiated cultures. This study aimed to standardize an *in vitro* cell model for differentiating hDPSC into OLC, as well as describe and compare the phenotype after the detachment and reexpansion of differentiated cells.

The results showed that hDPSC differentiate to OLC in the presence of mineralizing medium enriched with TGF-β1. The detachment and subcultivation of differentiated cells cause minor changes in protein and gene expression in relation to primary differentiated OLC; therefore, subculture in the differentiation medium allows the cells to retain the OLC phenotype. The reexpanded OLC may be a useful tool for studying the cell physiology of dentin-pulp complex formation and for use in biocompatibility studies of dental biomaterials.

2. Materials and Methods

2.1. Primary Culture of Human Dental Pulp. The study protocol was revised and approved by the Ethics Committee of Facultad de Odontología of Universidad Nacional de Colombia (CIE-233-14). Teeth collection, handling, and disposal were carried out in accordance with ethical standards of national and international legislation, and all volunteers signed the informed consent form before surgery. The third molars of young individuals (14–18 years old) were surgically extracted due to orthodontic considerations and processed following the protocol suggested by Gronthos et al. [11]. Teeth were decontaminated with 0.5% sodium hypochlorite and sectioned to obtain the pulp, which was cut into small fragments and digested overnight with collagenase (3 mg/mL) (Sigma-Aldrich; St Louis, MO, USA) and dispase (4 mg/mL) (Gibco; Thermo Fisher Scientific; Bremen, Germany) prepared in low-glucose DMEM supplemented with 10% fetal bovine serum (FBS) (Hyclone; Thermo Fisher Scientific; Bremen, Germany), penicillin (100 U/mL), and streptomycin (100 μg/mL) at 37°C in a 5% CO_2 incubator. The cell suspension was centrifuged, and the pellet was resuspended in low-glucose DMEM supplemented with 10% FBS and seeded in 25 cm^2 culture flasks until they reached 80% confluence. Cells were detached with trypsin plus EDTA and reseeded in 75 cm^2 culture flasks. At fourth passage (split), cells from independent donors were evaluated by flow cytometry to detect mesenchymal cell markers. Independent cultures from three donors and three replicas were analyzed ($n = 9$).

2.2. Flow Cytometry. The criteria of the International Society for Cell Therapy for mesenchymal stem cells [12] were used to determine the phenotype of hDPSC. A staining cocktail for phenotyping was used on 1×10^6 cells (Miltenyi Biotec; Bergisch Gladbach, Germany). Markers included antibodies targeting CD73, CD90, CD105, CD14, CD20, CD34, and CD45. Cells lacking primary antibodies were incubated with mouse isotype controls (FITC-IgG$_1$, PE-IgG$_1$, APC-IgG$_1$, PerCP-IgG$_1$, and PerCP-IgG$_{2a}$). Fluorescence of stained cells was captured in a FACSCalibur cytometer (BD Biosciences; San Jose, CA, USA), and after the acquisition of 100.000 events, data were analyzed using FCS Express software. Data are expressed as the percentage of cells positive for each marker.

2.3. Odontoblast-Like Cell Differentiation. Cell differentiation was performed using the protocol reported by Teti et al. (2013) with minor modifications [13]. Briefly, pulp mesenchymal stem cells were treated with odontogenic induction medium, DMEM supplemented with 10% FBS, antibiotics, 0.1 μM dexamethasone (Sigma-Aldrich), 5 mM β-glycerophosphate (Santa Cruz, CA, USA), 50 μg/mL ascorbic acid (Sigma-Aldrich), and 10 ng/mL TGF-β1 (Sigma-Aldrich) for 7, 14, or 21 days at 37°C in a 5% CO_2 incubator. After 21 days of treatment, differentiated odontoblast-like cells (OLC) were detached using trypsin/EDTA (0.25%/0.5 mM) and reseeded in new flasks using the same odontogenic induction media; these cells were named "expanded at 21 days" (EXP-21). To assess whether cryopreserved EXP-21 cells retain their differentiated phenotype, the cells were resuspended in 10% FBS supplemented DMEM containing 10% dimethyl sulfoxide at a concentration of 1-2 $\times 10^6$ cells/mL and then frozen at -80°C overnight and transferred to liquid nitrogen for undefined storage. The cells were thawed in a 37°C water bath in differentiation medium and cultured again for 24 h with differentiation medium. Both the OLC and EXP-21 cells (fresh and thawed) were characterized based on their phenotype as described below. Odontoblast markers were evaluated by immunohistochemistry at 7, 14, and 21 days after reseeding in fresh and thawed EXP-21 cells. For extracellular matrix formation and population doubling time, cells were maintained for 14 and 21 additional days, respectively, in the same manner as for OLC. Human dental pulp mesenchymal cells maintained with basal medium were considered negative controls.

2.4. Evaluation of Odontoblast Gene Expression by Quantitative RT-PCR. TRIzol (Ambion; Life Technologies, Carlsbad, California) was used to isolate total RNA from OLC at 7, 14, and 21 days after differentiation and from EXP-21 cells. With 200 ng of RNA per sample, RT-PCR was performed using the SYBR Green One-Step Real-Time RT-PCR Master Mix system (Invitrogen; Grand Island, NY, USA) in the CFX96 Real-Time Thermal Cycler detection system (Bio-Rad; Hercules, CA, USA). Amplification conditions were as follows: 15 min retrotranscription at 50°C, 4 min at 94°C and 40 amplification cycles of 20 s at 94°C, 20 s at 60–62°C, and 20 s at 72°C. Dentinogenic mRNA markers were quantitated using a specific primer set (Table 1); the genes evaluated were osteopontin (OPN), osterix (OSX), alkaline phosphatase (ALP), type I-collagen (COL-I), runt-related transcription factor 2 (RUNX2), dentin sialophosphoprotein (DSPP), dentin matrix protein-1 (DMP-1), and β-Actin (β-ACT). DSPP and DMP-1 as specific odontogenic markers were further evaluated in thawed EXP-21 cells to demonstrate phenotype maintenance after cryopreservation. PCR efficiencies were calculated using LinRegPCR (Academic Medical Center, AMC, Amsterdam, Netherlands), and relative gene quantitation was performed following Schefe's method [14].

2.5. Calculation of Cell Growth Curves and Population Doubling Time (PDT). The three cell types (hDPSC, OLC, and EXP-21 cells) were seeded in triplicate in 48-well microplates (8.000 cells/well) in the appropriate culture medium and were

TABLE 1: Primers used in this study.

Gene bank	Name	Forward Reverse
NM_001015051.3	RUNX2	5′-CATCTAATGACACCACCAGGC-3′ 5′-GCCTACAAAGGTGGGTTTGA-3′
NM_001040058.1	OPN	5′-TGAAACGAGTCAGCTGGATGACCA-3′ 5′-TGGCTGTGAAATTCATGGCTGTGG-3′
NM_001173467.1	OSX	5′-TGGGAAAAGGGAGGGTAATC-3′ 5′-CGGGACTCAACAACTCTGG-3′
NM_000478.4	ALP	5′-TCAGAAGCTCAACACCAACG-3′ 5′-GTCAGGGACCTGGGCATT-3′
NM_000088.3	COLI	5′-TGACCTCAAGATGTGCCACT-3′ 5′-ACCAGACATGCCTCTTGTCC-3′
NM_014208.3	DSPP	5′-GGCAGTGCATCAAAAGGAGC-3′ 5′-TGCTGTCACTGTCACTGCTG-3′
NM_004407.3	DMP-1	5-TGGAGTTGCTGTTTTCTGTAGAG-3 5′-ATTGCCGACAGGATGCAGA-3′
NM_001101.3	β-ACT	5′-ATTGCCGACAGGATGCAGA-3′ 5′-GAGTACTTGCGCTCAGGAGGA-3′

RUNX2: runt-related transcription factor 2; OPN: osteopontin; OSX: osterix; ALP: alkaline phosphatase; COL-I: collagen type I; DSPP: dentin sialophospho-protein; DMP-1: dentin matrix protein-1; β-ACT: β-actin.

trypsinized and counted in a hemocytometer with trypan blue daily for 21 days. Data regarding the cell number and the number of days in culture were plotted, and PDT was estimated using the formula $(t2 - t1)/3.32 \times (\log n2 - \log n1)$, where t is the number of days in culture and n is the cell number; data output was confirmed using Doubling Time software [15].

2.6. Immunocytochemical Detection of Odontogenic Markers. The immunocytochemistry protocol has been reported previously [16]. Cells were seeded on poly-L-lysine-treated glass coverslips (8.000 cells/well). After reaching 30% confluence, the cells were differentiated for 21 days as described above. Monolayers were fixed with paraformaldehyde (4%), permeabilized, and incubated with one of two primary antibodies: first, polyclonal anti-DSPP (Abcam, Cambridge, MA, USA), which is specific to the DSPP N-terminus corresponding to the natural cleavage fragment dentin sialoprotein (DSP); second, anti-DMP-1 (Sigma-Aldrich). Both were prepared in blocking buffer at 1:50 dilution. Goat anti-rabbit IgG biotinylated antibody was added at room temperature, the samples were washed, and peroxidase-coupled streptavidin (Thermo Fisher Scientific) was added. Specific binding was visualized using H_2O_2 and 3,3′-diaminobenzidine tetrahydrochloride chromogen. Nuclei were counterstained with hematoxylin, and the cells were photographed using a Zeiss Axio Imager A2 microscope (Gottingen, Germany). Other culture sets were processed for immunofluorescence using Alexa Fluor 594 coupled streptavidin (Thermo Fisher Scientific). The nuclei were counterstained with Hoechst, observed in an Axio Imager A2 microscope (Zeiss, Germany) and analyzed with the AxioVision software.

2.7. ECM Protein Adhesion Assay. The CytoSelect system (Cell Biolabs, Inc.; San Diego, CA, USA), which uses a 48-well microplate coated with five different ECM proteins, was used to test the cell substrate adhesion of the EXP-21 cultures. Trypsinized cells (7.000 cells/well) were seeded in serum-free medium in each of the coated wells at 37°C for 1 h, followed by washing with PBS and staining with Coomassie blue for 10 min. Cells were destained with acetic acid, and absorbance was measured at 560 nm in a microplate reader (Tecan, Infinite M200; Männedorf, Switzerland).

2.8. Measurement of Alkaline Phosphatase Activity. The hDPSC were cultivated in maintenance medium (low-glucose DMEM supplemented with 10% FBS and antibiotics) at 10.000 cells/well in a 24-well microplate until they reached 60% confluence. The EXP-21 cells were maintained for 24 h until adherence in differentiation media, and OLC were differentiated over 3, 7, 14, and 21 days. During each assay period, the cells were detached, lysed in assay buffer, and centrifuged. Supernatants were transferred to a 96-well microplate to quantify the ALP activity using an Abcam kit based on the fluorogenic substrate 4-methylumbelliferyl phosphate, resulting in an increased fluorescent signal ($Ex/Em = 360$ nm/440 nm) when the substrate is dephosphorylated.

2.9. Staining of Cultures. ECM components were stained using Masson's trichrome. Cells (hDPSC, OLC, and EXP-21 cells) were seeded on poly-L-lysine-treated glass coverslips. The OLC were differentiated for 7, 14, or 21 days, and the EXP-21 cells were maintained for 14 days before ethanol/acetone fixation. The collagen fibers were then stained with 1% phosphomolybdic/phosphotungstic acid for 5 min, followed by immersion in aniline blue stain, washing for 2 min with 1% acetic acid, mounting, and photography. To evaluate calcified nodes, the cultured cells were fixed with formaldehyde and stained with 2% alizarin red pH 4.1 for 15 min (Sigma-Aldrich) and photographed. The stain was extracted for 16 h with 5% (v/v) 2-isopropanol and 10% (v/v) acetic acid solution, and absorbance was measured using a microplate reader at 550 nm. Von Kossa staining was used to detect the calcium deposits in the formaldehyde-fixed cultures. A 5% silver nitrate solution was added to the cultures for 15 min; after washing, the cultures were exposed to UV light for 1 h, and calcium phosphate nodes were counted using a Zeiss Axiovert 40 CFL microscope.

2.10. Statistical Analysis. Data were collected in an Excel worksheet and exported to SPSS software, version 21.0 (SPSS, Chicago, IL, USA) for analysis. Data are presented as the mean and standard deviation. Differences with p values less than 0.05 were considered significant. Student's t-test was used in series with normal distributions, and the Mann–Whitney U test was used to compare data with a nonparametric distribution. All of the conditions were evaluated in three independent experiments with two, three, or six replicates.

3. Results

3.1. hDPSC Were Differentiated to Odontoblast-Like Cells and Retained Their Phenotype after Detachment. As expected, hDPSC showed the marker profile of mesenchymal cells (CD73+, CD90+, CD105+, CD14−, CD20−, CD34−, CD45−) (Figure 1(a)) and formed adherent clonogenic clusters after culture in flasks with maintenance medium. These fibroblast-like cells had a high proliferation rate between 7 and 21 days in culture, and differentiated cells (odontoblast-like cells, OLC) appeared with the same morphology throughout the entire culture period. The EXP-21 cells acquired a fusiform morphology and were capable of colony formation, significantly increasing the proliferation rate after reseeding for 7, 14, and 21 days of cultivation. After reaching confluence, the EXP-21 cells grew as an adherent multiple layer pattern arranged in parallel lines (Figure 1(b)). While hDPSC did not express the odontoblast phenotype markers DSPP and DMP-1, differentiating OLC highly expressed those markers beginning on day 14, and this differentiated pattern was maintained in detached and reseeded EXP-21 cells even after freeze-thaw processing (Figure 2). A low cell density per well seeding at the beginning of the differentiation stimuli enabled us to obtain a detailed protein marker location. We thus easily found in both OLC and EXP-21 cells (fresh and thawed) a cytoplasmic fluorescence and nuclear pattern for DSP, while DMP-1 was primarily located in the nuclei with less intensity in the cytoplasm. At the end of the differentiation treatment (21 days), we found both secreted DSP and DMP-1 outside the cell as part of the ECM (Figure 3).

3.2. Cell Growth and Proliferation Changed during Differentiation. To evaluate the role of the differentiation medium and the detachment process, growth curves and PDT values were obtained during 21 days in the three culture types. OLC proliferation during differentiation was slow (PDT 51.6 h),

causing a delay in reaching confluence and generating tightly multilayered colonies with a concomitantly higher number of cells per unit of culture area. However, the hDPSC and EXP-21 cells had similar PDT values (38.2 h and 38.8 h, resp.) when evaluated after the first 96 h of culture (Figure 4).

3.3. Differentiated Cells Expressed Odontogenic Genes and Markers. RUNX2 transcription factor mRNA increased by 12–18-fold in differentiated OLC and EXP-21 cells with respect to the undifferentiated hDPSC and remained high at all of the evaluated points, whereas the OSX transcripts reached a significant peak after 7 days of OLC differentiation (200-fold change). Similarly, COL-I and ALP transcript levels were significantly higher after 7 days of differentiation (125- and 160.000-fold differences, resp.) compared with OLC at later differentiation times (14 or 21 days) or even in EXP-21 cells, which had values equivalent to those of OLC at 21 days of differentiation (OLC-21) (Figure 5).

Interestingly, the transcript level of OSX was more than threefold higher in the EXP-21 cells compared to the OLC-21 cells. Regarding the odontoblast markers, the DSPP transcripts increased by 5-fold in the OLC-7 cells and 25-fold in the OLC-21 cells compared with hDPSC. The expression peak of DMP-1 was at 14 days of differentiation, showing 90-fold more expression than the hDPSC. The EXP-21 cells maintained high DSPP (32-fold increase over hDPSC) and DMP-1 (63-fold increase over hDPSC) transcription, giving expression values that were like those in the cells after being thawed. In turn, OPN transcription was 20-fold higher in the EXP-21 cells than the OLC-21 cells (Figure 5). These data were obtained from cells of three different donors in experiments performed independently in duplicate.

3.4. EXP-21 Cells Use Primarily Fibronectin, Collagen, and Fibrinogen for Substrate Adherence. The CytoSelect system showed increased adherence of EXP-21 to fibronectin and to type I- and IV-collagen (25-fold higher) compared with the control substrate (BSA). A similar finding was observed in fibrinogen-coated wells but not in laminin-coated substrates. Cells obtained from different donors had a similar adherence behavior (Figure 6).

3.5. Differentiated OLC Mineralized the ECM. ECM proteins, primarily COL-I, were synthesized beginning at seven days of differentiation as revealed by Masson's trichrome staining (Figure 7). As expected, the evaluation of ALP showed strong activity at early differentiation times (OLC-3 and -7) as a marker of early odontoblast differentiation but low expression (Figure 5) and activity in hDPSC, differentiated cells (OLC-21), and EXP-21 cells (Figure 8(b)). Calcium deposits and mineralization nodes progressively appeared during the differentiation process, as shown by alizarin red and von Kossa stainings, which revealed extensive calcified matrix in both the OLC and 14-day cultured EXP-21 cells (Figures 8(a) and 8(c)).

4. Discussion

This work showed that it is possible to differentiate dental pulp stem cells (or mesenchymal progenitor cells) into

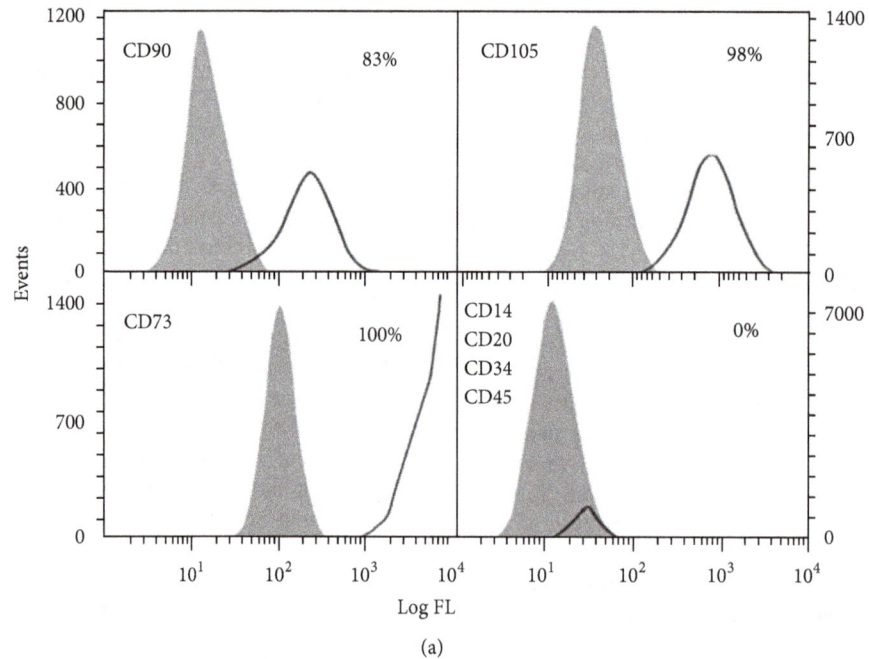

(a)

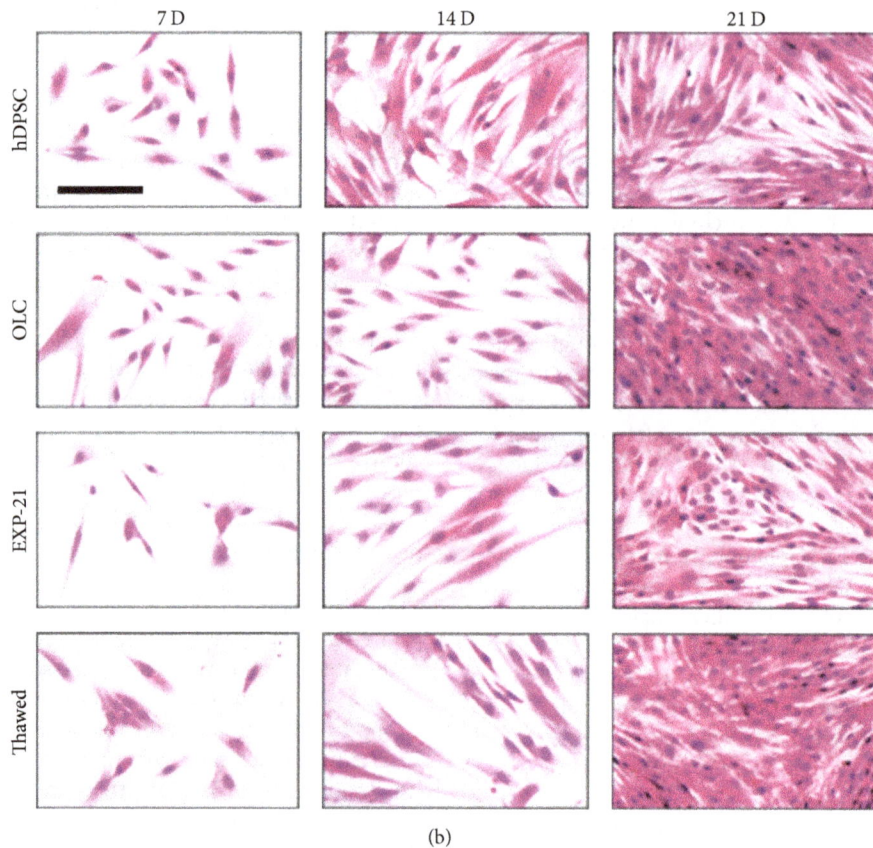

(b)

FIGURE 1: *hDPSC surface markers and morphology of the cultures*. The hDPSC were isolated from dental pulp of permanent teeth from three healthy donors. (a) The analysis by flow cytometry showed a homogeneous cell population positive for mesenchymal markers CD90, CD105, and CD73 and negative for early hematopoietic markers CD34, CD45, CD14, and CD20. Cells from fourth passage were recorded of each sample in triplicate. Data correspond to a representative experiment. (b) The hDPSC showed spindle-shape or fibroblast-like morphology with adherence to the surface and colony formation ability growing in a swirling-like pattern. During the differentiation process (OLC) and in the EXP-21, the cells maintained a typical fibroblastic-like shape with long cytoplasmic processes and were orderly arranged with a tendency to align themselves in parallel lines. The EXP-21 cells after cryopreservation, storage, and thawing showed similar morphology than their regular noncryopreserved cells. Scale bar: 200 μm.

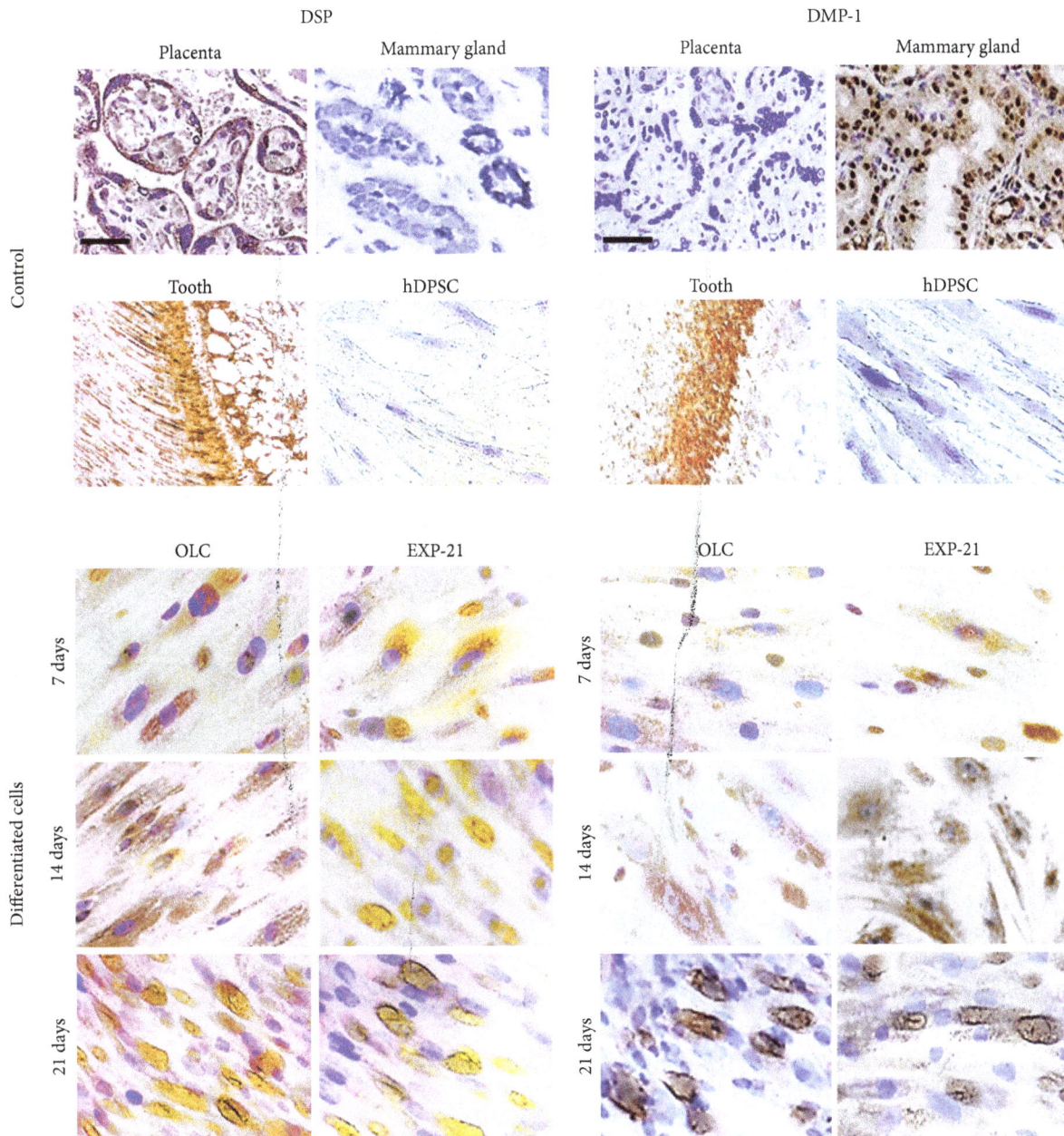

FIGURE 2: *Immunocytochemical detection of DSP and DMP-1 odontogenic markers.* Upper panel: the staining pattern in positive controls for DSP (placenta), DMP-1 (mammary gland), and dental pulp tissue positive for both markers. Mesenchymal cells are negative for these markers. Lower panel: DSP and DMP-1 staining pattern in OLC differentiating cultures for 7, 14, and 21 days and reseeded expanded cells (EXP-21) at 7, 14, and 21 days after differentiation. Note the cytoplasmic, nuclear, and late extracellular DSP staining pattern (OLC-21 and EXP-21, left panel). DMP-1 is expressed in both nuclei and cytoplasm and also in extracellular space as matrix vesicles (OLC-21 and EXP-21, right panel). Bar correspond to 100 μm.

odontoblast-like cells. We also showed that the detachment, reseeding, and expansion of these differentiated cells did not affect the proliferative and differentiation characteristics of cultured OLC, which represents a valuable characteristic for their use in studies of regeneration or dental material biocompatibility. The most interesting advantage of the reported protocol is the possibility of obtaining a primary culture of relevant cells for research in dentistry, with a homogeneous and uniform phenotype that remains after

detaching and reseeding, enabling researchers to perform reproducible experiments with many replicas and accurate cell number counting. The ability to freeze and thaw these cells for reuse on demand provides an additional benefit.

Many studies have reported that stem or progenitor cells of dental tissues are easy to obtain, can be expanded *in vitro*, and show remarkable plasticity; in addition, these cells differentiate into odontogenic cells [17]. The hDPSC isolated in this work showed classical mesenchymal stem cell

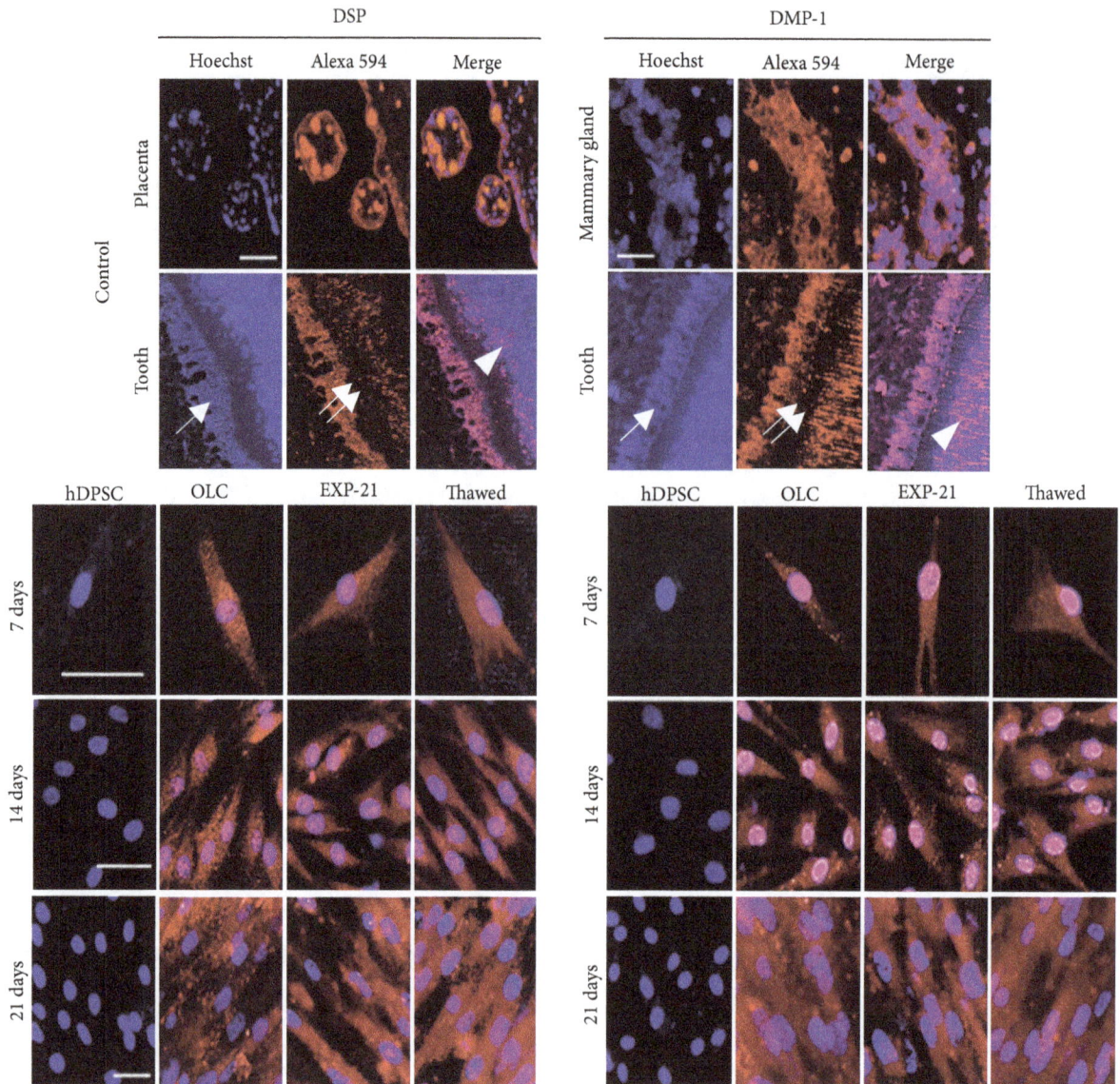

FIGURE 3: *Immunofluorescence analysis of DSP and DMP-1 localization.* Staining of respective positive controls for DSP (placenta) and DMP-1 (mammary gland). In the tooth, a layer of columnar odontoblasts (arrow) surrounds the inner surface of dentin and shows positive staining for both DSP and DMP-1; negative staining in the predentin area was observed (double arrow) and positive staining on the dentin mineralization front (arrowhead). To obtain a detailed protein marker location in cell cultures, they were processed for Alexa 594 immunofluorescence with antibodies to DSP or DMP-1 and Hoechst to stain nuclei after 7, 14, and 21 days of differentiation period (OLC) and 7, 14, and 21 days after reseeding process in fresh cells (EXP-21) and thawed EXP-21 cells (Thawed). Note the intense nuclear staining of DMP-1 and the absence of staining in mesenchymal cells. Bar corresponds to 100 μm.

markers similar to those obtained from human bone marrow, such as CD90, CD105, and CD73, as well as the absence of early hematopoietic markers. They also showed excellent proliferation and differentiation properties.

The proliferation and differentiation of hDPSC to odontoblast-like cells are known to depend on stimulation with specific differentiation media [18, 19]. Thus, we used TGF-β1 in addition to ascorbic acid, β-glycerophosphate, and dexamethasone. TGF-β is a multifunctional cytokine involved in cell growth, proliferation, and differentiation, as well as in extracellular matrix synthesis, indicating an important role for the signaling pathways involved in odontogenic

differentiation; this includes the upregulation of the expression of noncollagenous proteins such as DSPP and DMP-1 [20], as supported by our results. The differential profiles of genes expression during odontoblast differentiation suggest that further lineage determination of mineralizing cells is the result of differences in signaling pathways that control gene expression in odontogenic differentiation.

We found that OLC had a population doubling time that was higher than those of hDPSC or EXP-21 cells, which is consistent with the notion that synchronization between the proliferation and differentiation of stem cells to specific lineages is a key aspect to maintain stem cell function. The

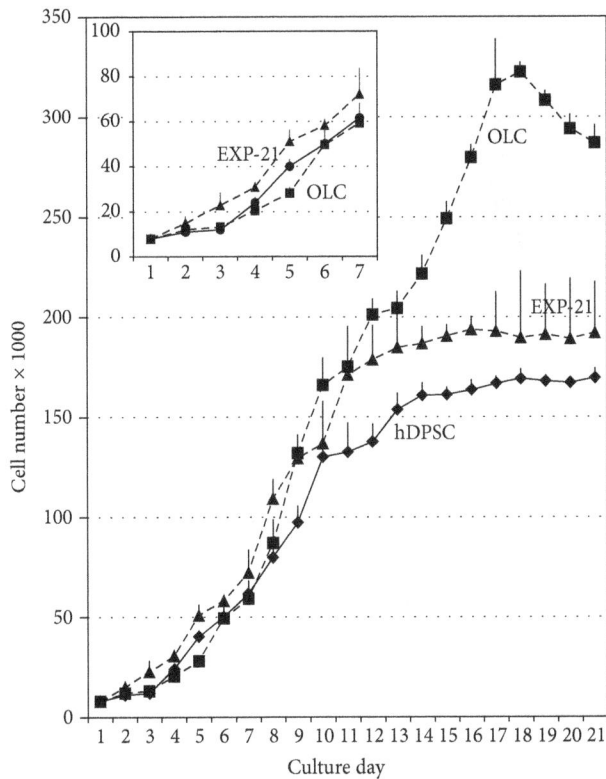

FIGURE 4: *Cell growth curves.* The hDPSC, OLC, and EXP-21 cells were detached daily for 21 days to determine the average number of cells in each cell population. hDPSC were maintained in culture medium without differentiation factors for 21 days, the OLC were stimulated with differentiation medium, and EXP-21 cells were maintained with differentiation medium for additional 21 days after detaching. Data are shown as mean ± SD and are based on three independent experiments and each was quantitated by triplicate ($n = 9$).

conditions for inducing differentiation are not the same as those for inducing cell proliferation; therefore, during the start of differentiation, large gene expression changes occur so that the cells can acquire the specific functions of the differentiated cells [21]. There were no changes in the initial proliferation rates in either the hDPSC or EXP-21 cells, which both reached the plateau phase around the tenth day, possibly because the monolayer ceased proliferation due to high confluence and contact inhibition.

DSPP and DMP-1 have been suggested to be the main phenotypic markers of odontoblasts [7]. There was a correspondence between DSPP transcription and protein synthesis, as evaluated by qPCR and immunocytochemistry, respectively. The dentin sialoprotein (DSP) has been described to have specific temporal and spatial expression patterns within the developing and mineralizing dentin [1], although this role appears to be more important during *in vivo* studies than those of *in vitro* biomineralization and hydroxyapatite crystal deposition studies. For example, the insertion of the DSP gene in DSPP-null mice induced the recovery of dentin mineralization, which facilitated hydroxyapatite formation along the collagen fibrils. This process describes

the predentin to dentin transition in the mineralization front [22].

Interestingly, we detected DSP located in the OLC nuclear compartment at early differentiation periods and in both fresh and cryopreserved EXP-21 cells, indicating a regulatory function during dentinogenesis. The cytoplasmic localization of DSP in differentiating OLC-14 and the extracellular staining in OLC-21 suggest active secretion during differentiation to promote the initial mineralization, connecting calcium ions to collagen fibers as is normal in dentin formation. The DSP protein and transcript expression of DSPP peaked after 14 days of differentiation, and this pattern was maintained in OLC-21 and in both fresh and thawed expanded cells (EXP-21). These results agree with those of Chen et al. (2008), who detected high DSP expression in mice MO6-G3 cells, indicating the differentiation of odontoblast cells; they also found staining in both the nuclei and cytoplasm [23]. However, the role of nuclear DSP remains to be elucidated.

Immunoperoxidase and immunofluorescence staining corroborated DMP-1 translocation from the nucleus to the cytoplasm and subsequent secretion to the ECM, a process clearly involved in the in vitro differentiation of odontoblasts. This protein plays an important role in odontogenesis, and its expression is required from early to late developmental stages. The role of DMP-1 signaling has been well studied; after endocytosis by preodontoblasts, DMP-1 induces calcium release from the endoplasmic reticulum, activating the p38 pathway and RUNX2 nuclear translocation, which are responsible for initiating the transcription of odontogenesis-related genes [24]. The nuclear/cytoplasmic localization of DSP and DMP-1 found in this work is compatible with the dual role regulator and nucleating proteins during ECM mineralization.

We described a substrate adherence pattern of OLC that may be important for optimizing and controlling the cellular microenvironment in surface engineering experiments [25]. Each cell type presented different affinities to ECM molecules depending on the expression of integrin α and $\beta1$ subunits [26]. EXP-21 cells had the best adhesion to collagen (I and IV), fibronectin, and fibrinogen but not to laminin, possibly due to the high expression of integrin $\alpha2$, $\alpha6$, and $\beta1$ induced by TGF-$\beta1$ treatment and the absence of $\alpha1$, $\alpha3$, and $\alpha7$ necessary for laminin adherence, as reported by Warstat et al. (2010) [26].

ECM protein secretion and mineralization are the main characteristics of mature secretory odontoblasts. Our findings demonstrated that differentiation media upregulated the proteins necessary to establish a functional ECM. COL-I was highly upregulated at 7 days of differentiation, as observed in extracellular deposits. After 21 days of differentiation, all of the extracellular space was filled with fibers, with a structural band pattern present around each cell. Differentiation media induced increases in the size and number of mineralized nodes due to the upregulation of gene and ALP activity in the early stages of differentiation. It therefore appears evident that, in our cultures, there was an orchestration of multiple events leading the cells to synthetize suitable proteins and matrix elements to guarantee the controlled growth of hydroxyapatite crystals and the formation of a

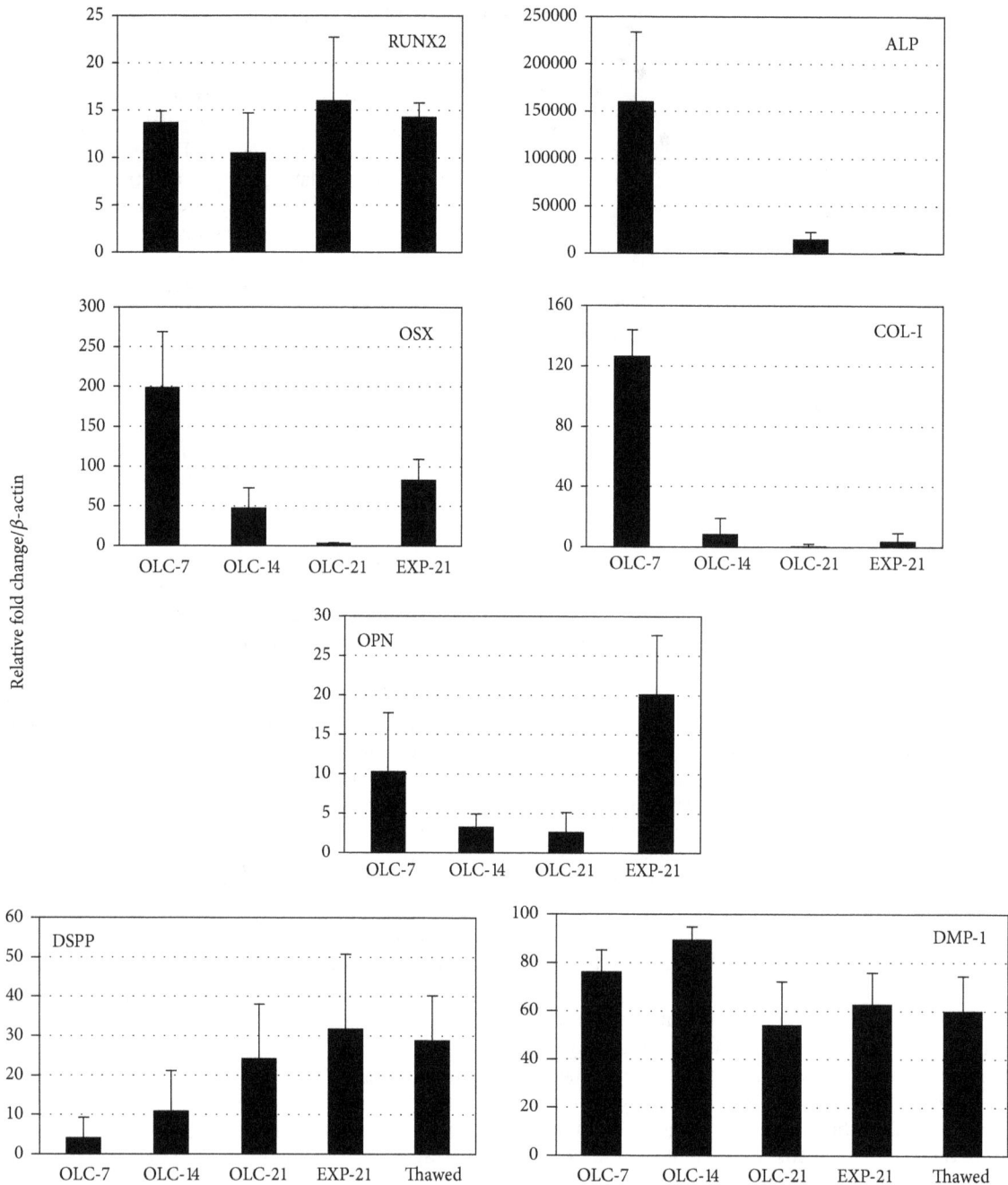

FIGURE 5: *Relative gene expression quantitation during differentiation.* Specific odontoblast and mineralizing transcripts were quantitated in differentiating OLC (7, 14, and 21 days) and in both fresh and thawed EXP-21 cells by real-time PCR considering a value of 1 in hDPSC. Data are shown as mean ± SD of three independent experiments in duplicate (n = 6). Runt-related transcription factor 2 (RUNX2), alkaline phosphatase (ALP), osterix (OSX), type I-collagen (COL-I), dentin sialophosphoprotein (DSPP), dentin matrix protein-1 (DMP-1), osteopontin (OPN).

dentin matrix by OLC and both fresh and thawed EXP-21 cells.

EXP-21 cells after subcultivation did not show significant changes in the proliferation rate, the phenotype, or cell clustering. Large impacts on survival or phenotype are expected be associated with trypsin activity during detachment [27] due to the loss of the cell-ECM interaction and actin-myosin contraction. However, an enzyme plus mild EDTA treatment,

such as that used in our study, could help maintain the viability and differentiation status of dissociated cells until medium with appropriate supplements restarts treatment, as has been reported for induced stem cells [28]. Interestingly, EXP-21 cells express 10-fold more OPN transcript than OLC-21. This ECM molecule is frequently induced by mechanical stress and is well recognized as a signal transduction activator when pulp repair is necessary [22, 29]. Therefore, EXP-21 cells

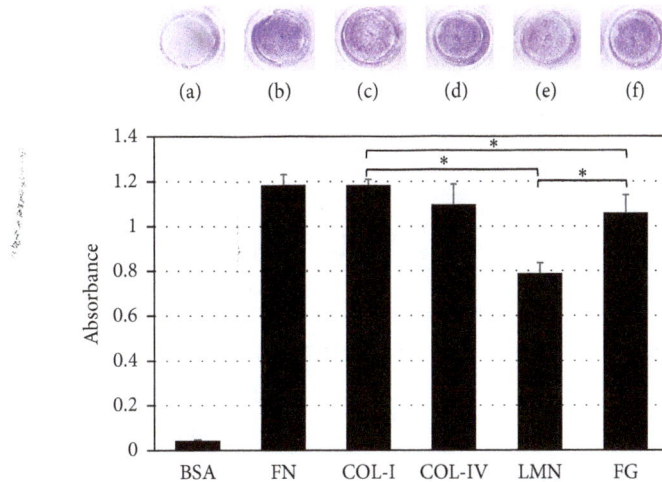

FIGURE 6: *EXP-21 cells extracellular matrix (ECM) protein adhesion assay.* Cell adhesion assay CytoSelect™ (Cell Biolabs) was employed to characterize cell attachment to ECM proteins like (b) fibronectin (FN), (c) type I-collagen (COL-I), (d) type IV-collagen (COL-IV), (e) laminin (LMN), (f) fibrinogen (FG). (a) Bovine serum albumin (BSA) is used as adhesion control. Attached cells after 1 h were stained with Coomassie blue dye which was subsequently extracted to compare the absorbances. The lowest adhesion was obtained on LMN. Data shown are mean ± SD of three independent experiments in duplicate ($n = 6$). Asterisks show significant differences ($p < 0.05$).

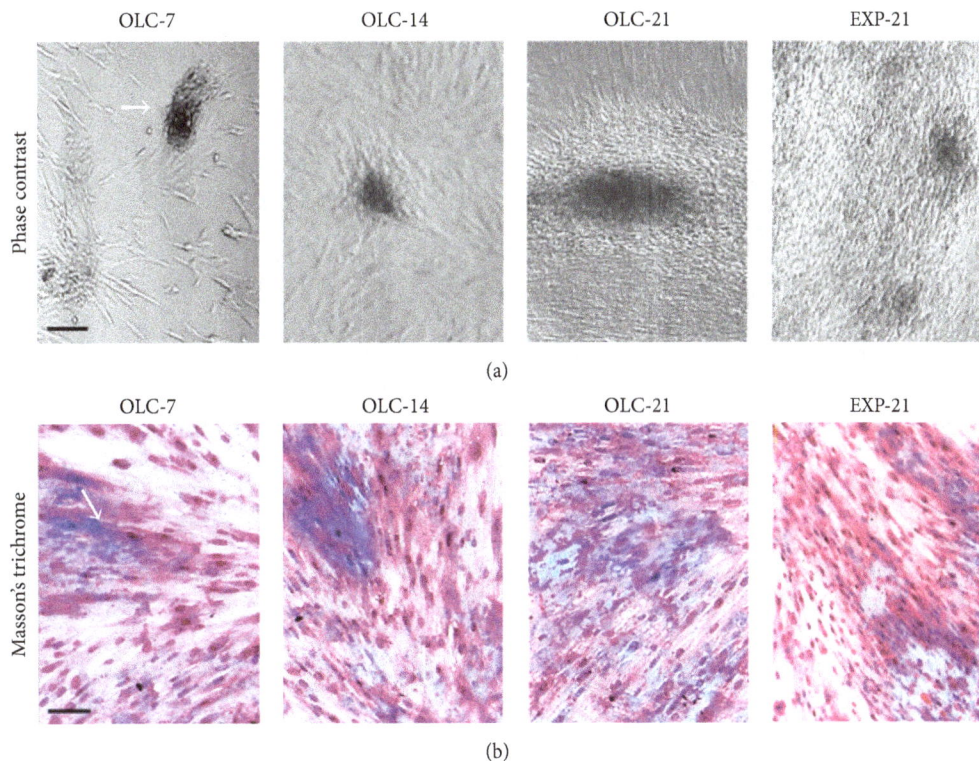

FIGURE 7: *Extracellular matrix deposition.* OLC and EXP-21 cultures were exposed to differentiation-inductive conditions. Cell aggregates (black arrow) became the centers where there was an extensive deposition of extracellular matrix components. Masson's trichrome staining (black arrow, (b)) shows the deposition of COL-I. Scale bar: 100 μm.

can be suggested to have a differentiated phenotype like that present in active cells during the dentin matrix repair process.

In this culture system, maintaining the local microenvironment with the appropriate differentiation inducer signal also helps preserve the differentiated phenotype in subcultures. Therefore, the addition of differentiation medium containing TGF-β1 guarantees that the phenotype reached will be permanent in subcultures, preserving the odontoblast differentiation markers and the maintenance of ECM mineralization capacity observed in EXP-21 cells.

Based on the described findings, the reported *in vitro* model preserves most of the physiological, biochemical, and

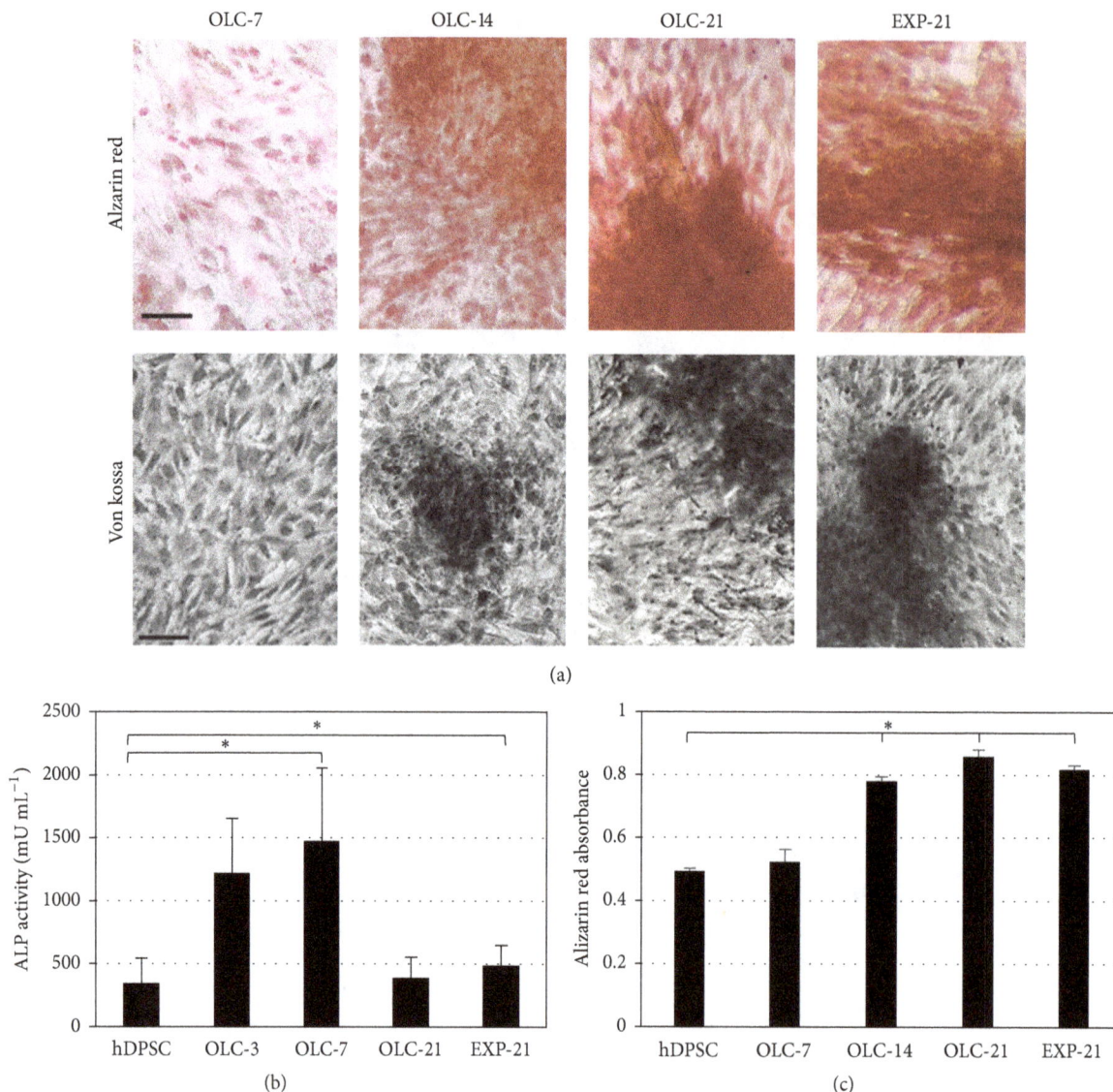

FIGURE 8: *Extracellular matrix mineralization.* (a) Mineralization was determined by alizarin red (upper panel) and von Kossa staining (lower panel). The strong staining of matrix indicates the apparent formation of calcification nodes. Scale bar: 100 μm. (b) The ALP activity as early marker of differentiation was associated with progressive mineralization of the extracellular matrix. Data shown are the mean ± SD of six replicas of three independent experiments ($n = 18$). Asterisks show significant differences with respect to negative control (hDPSC) ($p < 0.05$). (c) Absorbance measurement of alizarin red stain extracted at different stages of differentiation. Data shown are mean ± SD of six replicas of three independent experiments ($n = 18$).

genetic characteristics of odontoblasts. Consequently, the expanded cells may be a useful tool for studying odontoblast differentiation and the underlying cell signaling. Further, we demonstrated that subcultivation, as well as cryopreservation and thawing of differentiated odontoblasts, is possible. Only minor changes in proliferation or differentiated phenotype were observed, suggesting the utility of subcultured cells in tissue engineering and biocompatibility studies on dental materials.

Disclosure

The authors alone are responsible for the content and writing of the article.

Conflicts of Interest

The authors report no conflicts of interest.

Acknowledgments

This work was supported by grants from Universidad Nacional de Colombia, Banco de La República, Colombia, Project code: 3824, and Universidad El Bosque (UEB), project code: 398-2015.

References

[1] J. Hao, A. Ramachandran, and A. George, "Temporal and spatial localization of the dentin matrix proteins during dentin

biomineralization," *Journal of Histochemistry & Cytochemistry*, vol. 57, no. 3, pp. 227–237, 2009.

[2] J. Hao, K. Narayanan, A. Ramachandran et al., "Odontoblast cells immortalized by telomerase produce mineralized dentin-like tissue both in vitro and in vivo," *The Journal of Biological Chemistry*, vol. 277, no. 22, pp. 19976–19981, 2002.

[3] J. V. Ruch, H. Lesot, and C. Begue-Kirn, "Odontoblast differentiation," *The International Journal of Developmental Biology*, vol. 39, no. 1, pp. 51–68, 1995.

[4] J. V. Ruch, "Odontoblast commitment and differentiation," *The International Journal of Biochemistry & Cell Biology*, vol. 76, no. 6, pp. 923–938, 1998.

[5] G. Spagnuolo, V. D'Antò, C. Cosentino, G. Schmalz, H. Schweikl, and S. Rengo, "Effect of N-acetyl-L-cysteine on ROS production and cell death caused by HEMA in human primary gingival fibroblasts," *Biomaterials*, vol. 27, no. 9, pp. 1803–1809, 2006.

[6] C. T. Hanks, D. Fang, Z. Sun, C. A. Edwards, and W. T. Butler, "Dentin-specific proteins in MDPC-23 cell line," *European Journal of Oral Sciences*, vol. 106, supplement 1, pp. 260–266, 1998.

[7] H. S. Chinga, N. Luddin, I. A. Rahman, and K. T. Ponnuraj, "Expression of odontogenic and osteogenic markers in DPSCs and SHED: A review," *Current Stem Cell Research & Therapy*, vol. 12, no. 1, pp. 71–79, 2017.

[8] R. Wilson, N. Urraca, C. Skobowiat et al., "Assessment of the tumorigenic potential of spontaneously immortalized and hTERT immortalized cultured dental pulp stem cells," *Stem Cells Translational Medicine*, vol. 4, no. 8, pp. 905–912, 2015.

[9] S. G. Kim, Y. Zheng, J. Zhou et al., "Dentin and dental pulp regeneration by the patient's endogenous cells," *Endodontic Topics*, vol. 28, no. 1, pp. 106–117, 2013.

[10] M. J. Coelho, A. Trigo Cabral, and M. H. Fernandes, "Human bone cell cultures in biocompatibility testing. Part I: Osteoblastic differentiation of serially passaged human bone marrow cells cultured in α-MEM and in DMEM," *Biomaterials*, vol. 21, no. 11, pp. 1087–1094, 2000.

[11] S. Gronthos, M. Mankani, J. Brahim, P. G. Robey, and S. Shi, "Postnatal human dental pulp stem cells (DPSCs) in vitro and in vivo," *Proceedings of the National Acadamy of Sciences of the United States of America*, vol. 97, no. 25, pp. 13625–13630, 2000.

[12] M. Dominici, K. Le Blanc, I. Mueller et al., "Minimal criteria for defining multipotent mesenchymal stromal cells. The International Society for Cellular Therapy position statement," *Cytotherapy*, vol. 8, no. 4, pp. 315–317, 2006.

[13] G. Teti, V. Salvatore, A. Ruggeri et al., "In vitro reparative dentin: A biochemical and morphological study," *European Journal of Histochemistry*, vol. 57, no. 3, pp. 151–158, 2013.

[14] J. H. Schefe, K. E. Lehmann, I. R. Buschmann, T. Unger, and H. Funke-Kaiser, "Quantitative real-time RT-PCR data analysis: current concepts and the novel 'gene expression's CT difference' formula," *Journal of Molecular Medicine*, vol. 84, no. 11, pp. 901–910, 2006.

[15] V. Roth, *Doubling Time Computing*, 2006, http://www.doubling-time.com/compute.php.

[16] M. L. Velandia, R. Pérez-Castro, H. Hurtado, and J. E. Castellanos, "Ultrastructural description of rabies virus infection in cultured sensory neurons," *Memórias do Instituto Oswaldo Cruz*, vol. 102, no. 4, pp. 441–447, 2007.

[17] G. T.-J. Huang, S. Gronthos, and S. Shi, "Mesenchymal stem cells derived from dental tissues *vs.* those from other sources: their

biology and role in regenerative medicine," *Journal of Dental Research*, vol. 88, no. 9, pp. 792–806, 2009.

[18] P. R. Arany, A. Cho, T. D. Hunt et al., "Photoactivation of endogenous latent transforming growth factor-β1 directs dental stem cell differentiation for regeneration," *Science Translational Medicine*, vol. 6, no. 238, Article ID 238ra69, 2014.

[19] A. Bakopoulou, G. Leyhausen, J. Volk et al., "Comparative analysis of in vitro osteo/odontogenic differentiation potential of human dental pulp stem cells (DPSCs) and stem cells from the apical papilla (SCAP)," *Archives of Oral Biolog*, vol. 56, no. 7, pp. 709–721, 2011.

[20] J. A. Jadlowiec, X. Zhang, J. Li, P. G. Campbell, and C. Sfeir, "Extracellular matrix-mediated signaling by dentin phosphophoryn involves activation of the Smad pathway independent of bone morphogenetic protein," *The Journal of Biological Chemistry*, vol. 281, no. 9, pp. 5341–5347, 2006.

[21] L. J. A. Hardwick, F. R. Ali, R. Azzarelli, and A. Philpott, "Cell cycle regulation of proliferation versus differentiation in the central nervous system," *Cell and Tissue Research*, vol. 359, no. 1, pp. 187–200, 2015.

[22] G. Yuan, G. Yang, G. Song, Z. Chen, and S. Chen, "Immunohistochemical localization of the NH2-terminal and COOH-terminal fragments of dentin sialoprotein in mouse teeth," *Cell and Tissue Research*, vol. 349, no. 2, pp. 605–614, 2012.

[23] S. Chen, L. Chen, A. Jahangiri et al., "Expression and processing of small integrin-binding ligand N-linked glycoproteins in mouse odontoblastic cells," *Archives of Oral Biolog*, vol. 53, no. 9, pp. 879–889, 2008.

[24] K. Narayanan, A. Ramachandran, J. Hao et al., "Dual functional roles of dentin matrix protein 1. Implications in biomineralization and gene transcription by activation of intracellular Ca2+ store," *The Journal of Biological Chemistry*, vol. 278, no. 19, pp. 17500–17508, 2003.

[25] F. Paduano, M. Marrelli, L. J. White, K. M. Shakesheff, M. Tatullo, and X. Liu, "Odontogenic differentiation of human dental pulp stem cells on hydrogel scaffolds derived from decellularized bone extracellular matrix and collagen type I," *PLoS ONE*, vol. 11, no. 2, Article ID e0148225, 2016.

[26] K. Warstat, D. Meckbach, M. Weis-Klemm et al., "TGF-beta enhances the integrin alpha2 beta1-mediated attachment of mesenchymal stem cells to type I collagen," *Stem Cells & Development*, vol. 19, no. 5, pp. 645–656, 2010.

[27] H.-L. Huang, H.-W. Hsing, T.-C. Lai et al., "Trypsin-induced proteome alteration during cell subculture in mammalian cells," *Journal of Biomedical Science*, vol. 17, no. 1, article 36, 2010.

[28] J. Beers, D. R. Gulbranson, N. George et al., "Passaging and colony expansion of human pluripotent stem cells by enzyme-free dissociation in chemically defined culture conditions," *Nature Protocols*, vol. 7, no. 11, pp. 2029–2040, 2012.

[29] R. C. D'Alonzo, A. J. Kowalski, D. T. Denhardt, G. Allen Nickols, and N. C. Partridge, "Regulation of collagenase-3 and osteocalcin gene expression by collagen and osteopontin in differentiating MC3T3-E1 cells," *The Journal of Biological Chemistry*, vol. 277, no. 27, pp. 24788–24798, 2002.

11

Microwave-Assisted Tissue Preparation for Rapid Fixation, Decalcification, Antigen Retrieval, Cryosectioning, and Immunostaining

Kazuo Katoh

Laboratory of Human Anatomy and Cell Biology, Faculty of Health Sciences, Tsukuba University of Technology,
4-12-7 Kasuga, Tsukuba, Ibaraki 305-8521, Japan

Correspondence should be addressed to Kazuo Katoh; katoichi@k.tsukuba-tech.ac.jp

Academic Editor: Richard Tucker

Microwave irradiation of tissue during fixation and subsequent histochemical staining procedures significantly reduces the time required for incubation in fixation and staining solutions. Minimizing the incubation time in fixative reduces disruption of tissue morphology, and reducing the incubation time in staining solution or antibody solution decreases nonspecific labeling. Reduction of incubation time in staining solution also decreases the level of background noise. Microwave-assisted tissue preparation is applicable for tissue fixation, decalcification of bone tissues, treatment of adipose tissues, antigen retrieval, and other special staining of tissues. Microwave-assisted tissue fixation and staining are useful tools for histological analyses. This review describes the protocols using microwave irradiation for several essential procedures in histochemical studies, and these techniques are applicable to other protocols for tissue fixation and immunostaining in the field of cell biology.

1. Introduction

Microwave irradiation during tissue processing markedly reduces the time required for fixation, decalcification, staining with chemical reagents, and incubation with antibodies.

Since the mid-1980s, microwave irradiation has been increasingly used in histological preparation. Microwave irradiation induces rapid oscillation of water molecules (2.45 GHz) and thus increases tissue temperature. Conventional microwave devices irradiate tissues both rapidly and uniformly, and microwave irradiation protocols differ according to the specific microwave devices used.

Microwave irradiation is routinely applied for special staining [1–12]. Microwave irradiation has also been applied during fixation [13] and subsequent staining procedures, such as enzyme-based staining and immunofluorescence staining.

During preparation of tissues for immunohistological studies, many artifacts that disrupt the original signals may occur, most of which are commonly associated with late fixation or low fixative volume. Late preparation of tissues causes decomposition of proteins, resulting in a lack of certain epitopes. Disruption of proteins during fixation adversely affects the epitope-antibody reaction during immunohistochemistry. Moreover, morphological changes also occur during fixation of cryosections and/or samples for electron microscopy. Conventional fixation may also result in shrinkage of tissues, such as skeletal or smooth muscle cells, or of cultured cells due to insufficient penetration of fixative (e.g., formalin solution) to completely fix tissues, and a long time is needed for fixation.

Microwave irradiation can be used to achieve more rapid fixation, solution processing, and immunostaining [13–38]. Microwave irradiation is also applied for fluorescence in situ hybridization (FISH) analysis of paraffin-embedded tissues [39–41]. Recently, the author described microwave-irradiated blood vessel fixation and immunofluorescence microscopy [42]. In this case, microwave irradiation was used to increase penetration of fixatives. The use of microwave irradiation also reduced nonspecific binding of fluorescently labeled antibodies when fixed samples were immunostained. Rapid tissue fixation and immunofluorescence staining of cultured cells using microwave irradiation have also been described [43].

Microwave irradiation was shown to significantly reduce the required incubation times with primary and secondary antibodies in immunofluorescence microscopy. We utilized a technique involving exposure of cultured cells to intermittent microwave irradiation during fixation, which resulted in good preservation of tissue immunoreactivity compared with conventional fixation, along with reduced fixation time [43].

Another issue affecting histological analysis is the effect of pretreating hard tissues, such as bone, which requires decalcification after fixation to soften the tissue and allow it to be cut using a microtome. A long time is also required to remove fat from some tissues. Conventional decalcification requires a period of about 1-2 weeks, which prevents early diagnosis in histological research [44, 45]. Tissue preparation for electron microscopy, which involves fixation and subsequent solution treatment, is also problematic. Fixation using formalin-based fixatives causes tissue shrinkage. Solution treatment, such as dehydration by passage through an alcohol series, requires a relatively long time in conventional protocols.

Conventional antigen retrieval was generally performed using an autoclave chamber at high temperature (~121°) and high pressure and always caused tissue disruption and removal from the slides. Microwave irradiation is also highly applicable for antigen retrieval on paraffin-embedded tissue sections [46–49].

Microwave tissue processing markedly reduces the processing time required for enzyme reaction, peroxidase processing, and blocking procedures. Microwave irradiation reduces the processing time to 1/3–1/10 compared to that of conventional procedures. Moreover, microwave irradiation yields low-background, high-contrast images due to the reduced nonspecific binding of staining solution or antibodies for immunofluorescence staining.

Several microwaves that allow user-selectable control of irradiation power from 150 to 400 W are available. It is also possible to precisely control the temperature using two independent systems, for example, infrared and thermocouple temperature measurement systems.

This review describes a microwave-assisted tissue preparation protocol for tissue fixation, decalcification of bone tissue, fixation of fatty (adipose) tissues, antigen retrieval of paraffin-embedded tissues, and other techniques for which microwave irradiation is applicable. In addition, application of microwave irradiation for electron microscopy of blood vessel cells in situ is also discussed.

2. Application for Tissue Fixation

2.1. Fixation of Blood Vessels In Situ [42, 50]. Due to the difficulties associated with fixation of blood vessels, because of the shrinkage of smooth muscle tissues, there have been only a few studies using blood vessels in situ. It is very difficult to obtain good fixation of blood vessels in animals, especially endothelial cells, compared to those obtained from other organs. Perfusion of paraformaldehyde causes smooth muscle contraction according to the penetration of formalin-based fixatives. During fixation, blood vessels shrink rapidly during perfusion of paraformaldehyde solution. However, we have used microwave irradiation during fixation of blood

vessels and achieved good preservation of both tissue morphology and immunoreactivity.

2.2. Protocol. Aortae are obtained from normal adult guinea pigs 400–600 g in body weight.

(1) Perfusion with 0.85% NaCl containing heparin sodium (1 U/mL) is performed via the left ventricle, and the descending thoracic aorta, abdominal aorta, and inferior vena cava are excised.

(2) Vessels are cut open along the dorsal wall and pinned onto a dental wax plate, exposing the luminal surface.

(3) For light microscopy, the aorta is placed in a 100-mL beaker containing 50 mL of 2% paraformaldehyde in phosphate-buffered saline (PBS).

For both scanning and electron microscopy, it is recommended to use 1/2 Karnovsky's fixative (2.5% glutaraldehyde and 2% paraformaldehyde in 0.1 M sodium cacodylate buffer, pH 7.2).

(4) The beaker is placed on the turntable of a microwave oven and subjected to intermittent microwave irradiation at 200 W for 5 minutes (4 s on/3 s off).

After microwave irradiation, the blood vessels are rinsed with two or three changes of PBS for 10 minutes each time without microwave irradiation. They are then cut crosswise into small segments (about 5–10 mm in length) and processed for either paraffin embedding or immunofluorescence microscopy (see Application for Immunofluorescence Microscopy).

3. Fixation of Cultured Cells [43]

Microwave irradiation is applicable for fixation of cultured cells. Conventional fixation of cultured cells requires at least 30–60 minutes with 1% paraformaldehyde. Microwave irradiation during fixation significantly reduces the times required for both fixation and staining for immunofluorescence microscopy. All procedures, including fixation and antibody staining, are completed within 30–45 minutes without any loss of cell morphology and without nonspecific binding of dyes.

3.1. Protocol

(1) Cells are cultured on coverslips according to standard procedures.

(2) Cells are washed quickly with three changes of PBS.

(3) Cells are fixed with 1% paraformaldehyde with intermittent microwave irradiation (total 5 minutes; 4 s on/3 s off at 200 W).

(4) Fixed cells are quickly rinsed three times in PBS for a total of 5 minutes without microwave irradiation followed by immunofluorescence microscopy (see Application for Immunofluorescence Microscopy).

3.2. Notes. Two coverslips (18 × 18 mm) are placed in plastic culture dishes 50 mm in diameter. About 4 mL of fixative is added to the culture dish.

If using culture dishes 100 mm in diameter, more than 10 coverslips (18 × 18 mm) can be fixed with 10 mL of fixative.

For microwave irradiation, the authors' laboratory uses a MI-77 type microwave irradiation device (Azumaya, Tokyo, Japan) with controllable microwave irradiation power and temperature control.

4. Application for Decalcification [26, 27, 33, 44, 45]

Decalcification is essential after bone fixation to obtain good paraffin sections. Decalcification of bone tissues for histological research requires a very long time, that is, 2–4 days with 10% formic acid or 1-2 weeks with 10% EDTA.

With microwave irradiation, however, the processing time can be reduced to 1/5–1/10 of the original preparation time.

4.1. Decalcification Solution. 10% formic acid in distilled water or 10% neutral EDTA.

4.2. Comments

Formic Acid Decalcification

(1) For all procedures, use intermittent microwave irradiation at 400 W (5 s on/5 s off).

(2) Preparation time is reduced to 1/10 of the original procedure. For formic acid, irradiation can be performed overnight.

(3) The formic acid solution temperature should not exceed 45°C.

EDTA Decalcification

(1) Use intermittent microwave irradiation at 400 W (5 s on/5 s off).

(2) Preparation time is reduced to 1/10 of the original procedure. For EDTA decalcification, irradiation can be performed for 2 or 3 days.

(3) The EDTA solution temperature should not exceed 45°C.

(4) Decalcification solution should be changed every day.

Procedure time should be determined by each researcher. Processing time should be modified according to the size and hardness of bones. For microwave irradiation, the authors' laboratory uses a MI-77 type microwave irradiation device (Azumaya, Tokyo, Japan) with controllable microwave irradiation power and temperature control.

5. Application for Immunohistochemistry [7, 17, 21, 27, 33, 36, 38, 51–53]

For immunohistochemistry experiments, avidin-biotin complex interaction (the ABC method) is generally used for detection of certain types of proteins using specific primary antibodies (Vectastain ABC Kit; Vector Laboratories, Burlingame, CA). Although it is a well-established procedure for histochemical investigations, conventional protocols for ABC method require 2-3 hours. Microwave irradiation reduces both the procedure time and background noise [51].

Deparaffinization should be performed according to standard methods. After deparaffinization, microwave irradiation can be applied. A protocol for immunohistochemistry using the ABC complex is presented below.

5.1. Protocol

(1) Deparaffinize tissues using standard procedures without microwave treatment.

(2) H_2O_2 treatment is performed to block endogenous peroxidase for 5 minutes with microwave treatment (intermittent irradiation; 5 s on/5 s off at 200 W).

(3) Antigen retrieval should be performed in this step (see Application for Antigen Retrieval of Paraffin-Embedded Samples).

(4) Wash briefly with PBS.

(5) Block with blocking solution for 5 minutes with microwave treatment (5 s on/5 s off at 200 W).

(6) Incubate with 1st antibody for 5 minutes with microwave treatment (5 s on/5 s off at 200 W).

(7) Wash samples briefly with PBS for 10 s–1 minute without microwave treatment.

(8) Incubate with biotinylated anti-mouse or anti-rabbit IgG with microwave treatment. (Secondary antibody varies according to the origin of the 1st antibody.)

(9) Wash samples briefly with PBS for 1 minute without microwave treatment.

(10) Incubate in Vectastain ABC solution for 3–5 minutes with microwave treatment (5 s on/5 s off at 200 W).

(11) Wash samples briefly with PBS twice for 1 minute each time without microwave treatment.

(12) Development of ABC complex with diaminobenzidine until staining develops without microwave treatment.

All reagents should be prepared according to the manual supplied with the Vectastain ABC Kit.

6. Application for Immunofluorescence Microscopy [35, 37, 42, 50, 54]

For immunofluorescence microscopy, microwave irradiation reduces incubation time to about 1/5–1/10 of the original time. Our laboratory protocol for staining of guinea pig aorta and vena cava is presented below. This is an example of immunofluorescence staining of blood vessels. The following protocol should be applicable for other tissues, although the exact irradiation power and time of microwave treatment should be determined by each researcher. The detailed cultured cell protocol was reported previously [43].

For fixation using microwave irradiation (see also Application for Tissue Fixation), aortae are rinsed several times with PBS and cut into small pieces, and en face preparations are made.

6.1. Protocol

(1) Permeabilize the tissue with 0.5% Triton X-100 in PBS for 5 minutes without microwave irradiation.

(2) The specimens are then incubated with 10% normal goat serum for 5 minutes with intermittent microwave irradiation (4 s on/3 s off at 200 W).

(3) The specimens are rinsed several times with PBS without microwave irradiation.

(4) Incubation should then be performed with one of the primary antibodies (in this case, antipaxillin as a marker of adhesion plaques located at sites of cell-substrate adhesion) for 5 minutes with intermittent microwave irradiation (4 s on/3 s off at 200 W).

(5) Rinse specimens with PBS several times without microwave treatment.

(6) Incubate with FITC-labeled secondary antibody for 5 minutes with microwave irradiation (4 s on/3 s off at 200 W).

(7) [Optional] Further staining with dyes can be performed, such as rhodamine-labeled phalloidin for F-actin or propidium iodide for nuclear staining with microwave irradiation (4 s on/3 s off at 200 W).

(8) Mount samples on slide glasses, followed by immunofluorescence or confocal laser scanning microscopy.

The temperature of samples should not exceed 40°C for immunofluorescence microscopy.

An example of typical double staining with anti-phosphotyrosine (PY-20) antibody (BD Biosciences, Franklin Lakes, NJ) as a tyrosine-phosphorylated protein marker and rhodamine-labeled phalloidin for actin filament staining is shown in Figure 1. The above protocol is applicable for staining of cryosections (see also [42]). See also immunofluorescence microscopy of paraffin-embedded samples by other authors [35, 37].

7. Application for Antigen Retrieval of Paraffin-Embedded Samples [23, 37, 46–49]

In general, paraffin-embedded samples that have been fixed with paraformaldehyde are not suitable for immunohistochemical staining or in situ hybridization due to the masking of antigenic sites by protein-protein cross-linking with paraformaldehyde. In some formalin-fixed samples, antigens of certain proteins are masked according to protein cross-linking by formaldehyde. Therefore, formalin-fixed samples require antigen retrieval using special treatment. Heat-induced antigen retrieval is often used to retrieve the epitopes

of proteins by autoclaving or using a commercially available pressure cooker. Heat-induced antigen retrieval sometimes causes damage to fixed samples.

Conventional procedures involved using an autoclave (121°C with pressure). In some cases, a household kitchen microwave oven (nearly 100°C) was used. In many cases, tissues on slide glasses became detached, and high levels of background noise were observed because of tissue damage by high temperature.

Microwave irradiation using a purpose-built apparatus can yield stable experimental results for research [46, 49].

7.1. Comments

(1) Beaker with antigen retrieval solution 400 mL (e.g., Target Retrieval solution; DAKO, Produktionsvej, Denmark).

(2) Maximum solution temperature: 70°C–99°C.

(3) Continuous irradiation for 20–30 minutes (intermittent irradiation: 5 s on/5 s off at 400 W).

(4) Precise reaction time determined by each researcher.

8. Application for Processing Fatty Tissues

Thymus gland, breast, and lymph node specimens are typical fatty (adipose) tissues. Fixation of fatty tissues is difficult because of the low penetration of paraformaldehyde solutions. Poor tissue fixation causes tissue destruction and "bubbles" in paraffin-embedded sections. Due to the poor penetration of paraformaldehyde solution, fatty tissues require treatment with a mixture of xylene and methanol after fixation. In general, treatment of fatty tissues requires about 10–30 hours. With microwave irradiation, however, the fixation time can be significantly reduced to about 1/20–1/30 with good fixation.

8.1. Procedure. Take thymus gland as an example.

Refer to Application for Tissue Fixation for tissue fixation protocol.

(1) Incubate fixed fatty tissues in beaker with treatment solution: xylene : methanol = 1 : 1 (500 mL).

(2) Intermittent microwave irradiation for 30–60 minutes at 400 W (5 s on/5 s off): replace treatment solution if there is sedimentation of fat.

(3) During microwave irradiation, the temperature should not exceed 50°C.

Irradiation times should change according to the volume of the fatty tissues. Total irradiation time should be determined by each researcher.

9. Other Applications of Microwave Irradiation

9.1. Application for Electron Microscopy. Tissues to be analyzed by electron microscopy are usually embedded in resin for ultramicrotomy. Although the embedding procedures

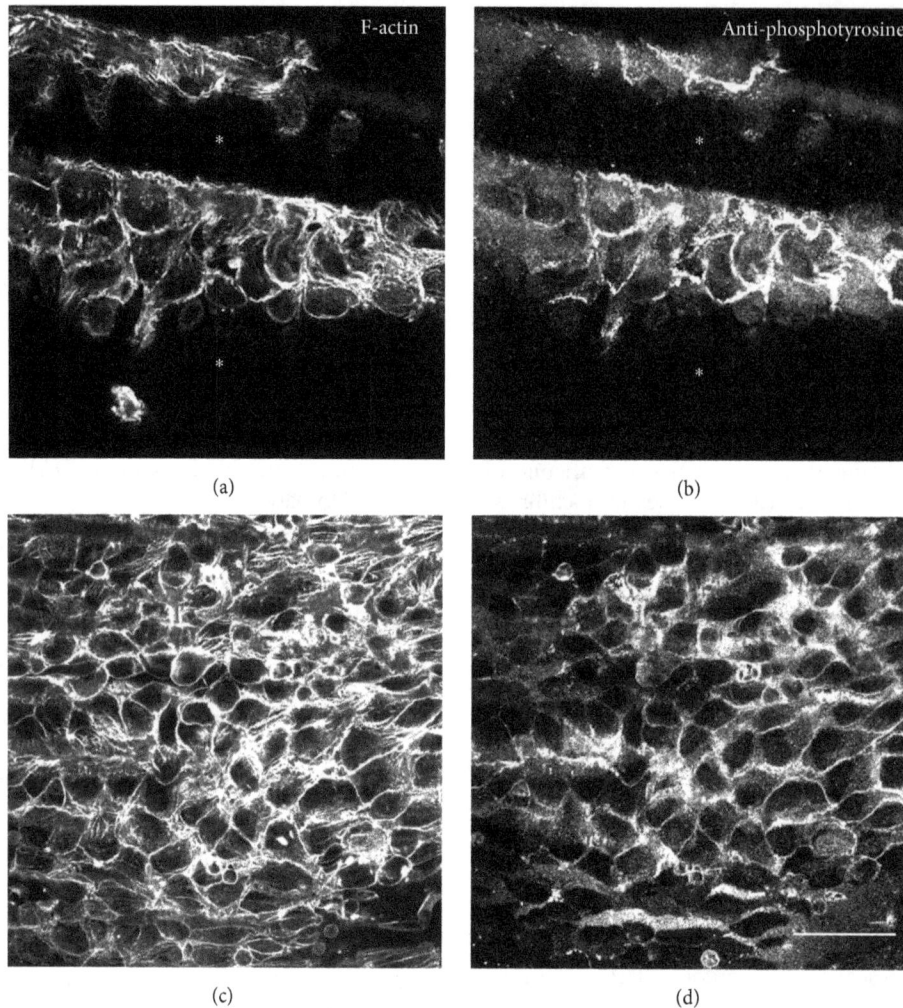

FIGURE 1: Comparison between conventional fixation (a and b) and microwave irradiation (c and d) of guinea pig aortic endothelial cells in whole-mount preparations. Conventionally fixed specimens showed shrinkage in the smooth muscle cell layer. Asterisks in (a) and (b) indicate gutters caused by shrinkage of the smooth muscle cell layer. Fixation with microwave irradiation showed well-preserved and flattened morphology of the endothelium (c and d). Samples were stained with both anti-phosphotyrosine antibody to reveal tyrosine-phosphorylated proteins (b and d) and rhodamine-labeled phalloidin for F-actin staining (a and c). Samples were observed by confocal laser scanning microscopy, and focus was adjusted at the endothelial cell layer. Bar = 10 μm. See also [42].

vary with respect to chemical composition, they generally require about 1 week for processing. During sample preparation for electron microscopy, use of microwave irradiation reduces the procedure time to only 2 days without any loss of fine structure in tissue samples. Our microwave irradiation technique for fixation seems to be applicable to both transmission and scanning electron microscopy (see Figure 2 for SEM image).

Refer to Application for Tissue Fixation for tissue fixation protocol.

9.2. Protocols

(1) Postfixation with 1% osmium tetroxide in distilled water with microwave treatment for 15 minutes* (all procedures in this protocol use intermittent irradiation, 5 s on/5 s off at 200 W).

(1% uranyl acetate incubation for 10 minutes with intermittent microwave irradiation if required.)

(2) Alcohol 50% with microwave irradiation for 5 minutes.

(3) Alcohol 75% with microwave irradiation for 5 minutes.

(4) Alcohol 90% with microwave irradiation for 5 minutes.

(5) Alcohol 100% with microwave irradiation for 5 minutes × 2 times.

(6) Propylene oxide 100% with microwave irradiation for 5 minutes × 2 times.

For Transmission Electron Microscopy [56, 57]

(7) Propylene oxide : epoxy resin = 1 : 1 with microwave irradiation for 30–60 minutes.

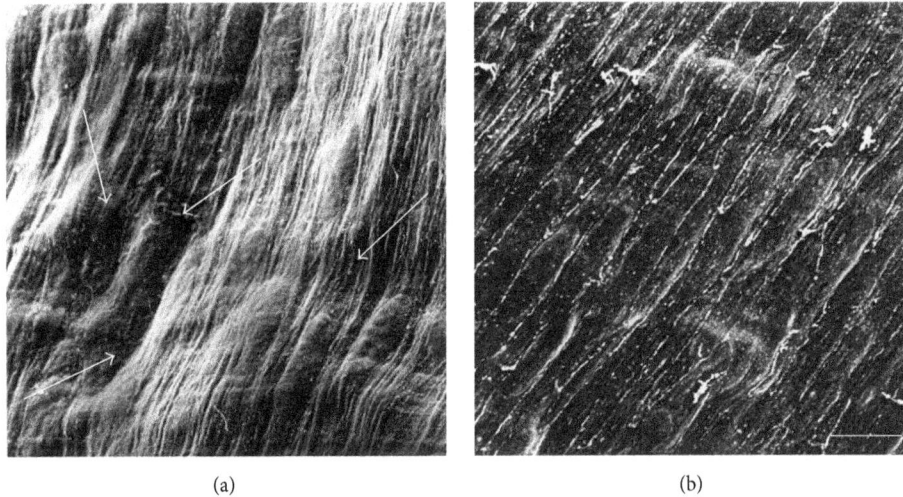

(a)   (b)

FIGURE 2: Scanning electron micrographs of venous endothelial cells without (a) or with (b) microwave irradiation. Guinea pig venous blood vessels were fixed conventionally with 1/2 Karnovsky's solution (a). Shrinkage of the smooth muscle cell layer occurred (a: arrows). After fixation with microwave irradiation, the flattened endothelial cell layer located in the inner surface of blood vessels was well-preserved (b). Arrows in (a) indicate the wavy artifacts caused by shrinkage of the smooth muscle cell layer. Compare (a) without microwave irradiation and (b) with microwave irradiation; the microwave irradiation showed good preservation of the endothelial cell layer morphology (b). Bar = 10 μm. See also [55].

(8) Epoxy resin 100% and microwave irradiation for 30–60 minutes.

(9) Embedding in epoxy resin according to standard procedures (60°C) for 15–20 hours, followed by thin sectioning.

Notes. For transmission electron microscopy, microwave irradiation also can reduce staining time with uranyl acetate and lead citrate [24, 58].

For Scanning Electron Microscopy [55]

(7) Transfer to 100% *t*-butyl alcohol with microwave irradiation for 5 minutes × 2 times.

(8) Samples are freeze-dried under high vacuum.

(9) Dehydrated samples are coated using an ion-sputtering device (gold-platinum) and observed by scanning electron microscopy.

Samples with/without microwave treatment are shown in Figure 2.

9.3. Notes. Sample size should not exceed 2 mm (increase time for large-sized specimens >2 mm in length). Do not exceed the maximum temperature of 50°C. The exact time should be determined by each researcher. Do not exceed the maximum temperature of 50°C during the dehydration procedure using an alcohol series (50%–100%).

For postfixation in osmium tetroxide, sample temperature should not exceed 37°C when using microwave irradiation. To prevent exposure to osmium tetroxide gas, the procedure should be performed in a laboratory fume hood. Alternatively, postfixation can be performed conventionally

for 1 hour at room temperature without microwave irradiation in a tightly closed screw-topped vial (1 mL of 1% osmium tetroxide solution in 10-mL vial).

10. Application for Cryosectioning [42]

Frozen sections are used for pathological diagnosis, enzyme detection, and immunofluorescent microscopy using antibodies. During freezing, crystallized water rapidly damages the tissues. Microwave irradiation reduces the formation of water crystals in tissues and can maintain the fine structure. Microwave irradiation mixes well with generally used Tissue-Tek OCT compound (Sakura Finetek, Tokyo, Japan) and can be sliced very easily. Compared with nonirradiated specimens, microwave-irradiated samples have fewer bubbles around the tissues and structures are well-preserved.

Refer to Application for Tissue Fixation for tissue fixation protocol.

10.1. Protocols

(1) For cryoprotection, fixed tissues are immersed in 30% sucrose solution with microwave irradiation until the sample sinks to the bottom of the beaker (all procedures in this protocol use intermittent irradiation, 5 s on/5 s off at 200 W). Change sucrose solution 2-3 times to prevent reduction of sucrose concentration.

(2) Dissect samples into small blocks (5–10 mm) using a sharp knife.

(3) Immerse trimmed tissues in 10 mL of Tissue-Tek/sucrose solution (1 : 1) in a small beaker or plastic

bottle with microwave irradiation for 10–15 minutes (5 s on/5 s off at 200 W).

(4) Place the above tissues on parafilm (Bemis Flexible Packaging, Chicago, IL) and drop Tissue-Tek over the sample.

(5) [Optional] If bubbles are observed around the tissue, the sample could be treated with microwave irradiation for 5 minutes (5 s on/5 s off at 200 W), which can remove some bubbles.

(6) Place Tissue-Tek on precooled cryostat head, mount above sample, and drop Tissue-Tek over the samples as soon as possible.

(7) Quickly freeze Tissue-Tek and samples with HFC134a aerosol cold spray (CRYON: Oken-Syoji, Tokyo, Japan).

(8) Cut cryosections according to standard procedures.

Comments. When samples are mounted on the precooled cryostat head, freeze as quickly as possible to prevent formation of ice crystals in the samples.

11. Application for Special Staining

For special staining, such as that using periodic acid-methenamine-silver stain (PAM), Azan staining, Grimelius' method, Fontana-Masson stain, methenamine silver-nitrate Gomori-Grocott's variation, and Congo red stain, microwave irradiation allows good results without background staining within a very short time. Conventionally, the above staining procedures take 1-2 hours, while microwave irradiation allows completion of staining within 5–10 minutes [1, 3, 5, 7, 8, 59].

12. Conclusions

Microwave irradiation can be applied during paraformaldehyde fixation and several types of staining procedure. It also significantly reduces the time required for staining procedures, such as immunohistochemistry, immunofluorescence microscopy, and special staining of tissues. Moreover, it reduces the time required for decalcification of bone. Conventional microwave ovens are unsuitable for laboratory use because the irradiation power is too high or they do not allow precise control of the power and sample temperature. Use of a conventional microwave oven requires calibration [60]. Modern microwave devices built specifically for laboratory use allow precise control over the power of microwave irradiation and sample temperature. Microwave irradiation should be highly applicable for many histological and cell biological techniques without any loss of morphology or immunoreactivity of tissues.

Competing Interests

The author declares that there is no conflict of interests regarding the publication of this paper.

Acknowledgments

The work reported here was supported by grants-in-aid for Promotional Projects for Advanced Education and Research, National University Corporation Tsukuba University of Technology.

References

[1] S. Valle, "Special stains in the microwave oven," *Journal of Histotechnology*, vol. 9, no. 4, pp. 237–239, 1986.

[2] M. E. Boon, L. P. Kok, H. E. Moorlag, P. O. Gerrits, and A. J. H. Suurmeijer, "Microwave-stimulated staining of plastic embedded bone marrow sections with the romanowsky-giemsa stain: improved staining patterns," *Biotechnic and Histochemistry*, vol. 62, no. 4, pp. 257–266, 1987.

[3] K. Matthews and J. K. Kelly, "A microwave oven method for the combined alcian blue-periodic acid-Schiff stain," *Journal of Histotechnology*, vol. 12, no. 4, pp. 295–297, 1989.

[4] A. S.-Y. Leong and P. Gilham, "A new, rapid, microwave-stimulated method of staining melanocytic lesions," *Stain Technology*, vol. 64, no. 2, pp. 81–85, 1989.

[5] C. F. Danielson, T. Bloch, G. G. Brown, and D.-J. Summerlin, "The effect of microwave processing on histochemical staining reactions," *Journal of Histotechnology*, vol. 13, no. 3, pp. 181–183, 1990.

[6] J. Bonner and P. J. Armati, "Microwave assisted staining of nerve and muscle biopsy tissue," *Biotechnic and Histochemistry*, vol. 66, no. 5, pp. 236–238, 1991.

[7] H. J. G. van de Kant, M. E. Boon, and D. G. de Rooij, "Microwave applications before and during immunogold-silver staining," *Journal of Histotechnology*, vol. 16, no. 3, pp. 209–215, 1993.

[8] C. J. Churukian, "Microwave Giemsa technique for paraffin embedded tissue sections," *Journal of Histotechnology*, vol. 18, no. 4, pp. 319–322, 1995.

[9] C. Ilgaz, H. Kocabiyik, D. Erdogan, C. Özogul, and T. Peker, "Double staining of skeleton using microwave irradiation," *Biotechnic & Histochemistry*, vol. 74, no. 2, pp. 57–63, 1999.

[10] Z. Kahveci, F. Z. Minday, and I. Cavusoglu, "Safranin O staining using a microwave oven," *Biotechnic & Histochemistry*, vol. 75, no. 6, pp. 264–268, 2000.

[11] S. G. Temel, S. Noyan, I. Cavusoglu, and Z. Kahveci, "A simple and rapid microwave-assisted hematoxylin and eosin staining method using 1,1,1 trichloroethane as a dewaxing and a clearing agent," *Biotechnic and Histochemistry*, vol. 80, no. 3-4, pp. 123–132, 2005.

[12] B. Avci, N. Kahveci, Z. Kahveci, and S. A. Sirmali, "Using microwave irradiation in Marchi's method for demonstrating degenerated myelin," *Biotechnic and Histochemistry*, vol. 81, no. 2-3, pp. 63–69, 2006.

[13] Z. Kahveci, I. Çavuşoğlu, and Ş. A. Sirmali, "Microwave fixation of whole fetal specimens," *Biotechnic and Histochemistry*, vol. 72, no. 3, pp. 144–147, 1997.

[14] A. S.-Y. Leong and J. Milios, "Rapid immunoperoxidase staining of lymphocyte antigens using microwave irradiation," *Journal of Pathology*, vol. 148, no. 2, pp. 183–187, 1986.

[15] K. Y. Chiu and K. W. Chan, "Rapid immunofluorescence staining of human renal biopsy specimens using microwave irradiation," *Journal of Clinical Pathology*, vol. 40, no. 6, pp. 689–692, 1987.

[16] J. Milios and A. S. Y. Leong, "The application of the avidin-biotin technique to previously stained histological sections," *Stain Technology*, vol. 62, no. 6, pp. 411–416, 1987.

[17] R. A. Moran, F. Nelson, J. Jagirdar, and F. Paronetto, "Application of microwave irradiation to immunohistochemistry: preservation of antigens of the extracellular matrix," *Stain Technology*, vol. 63, no. 5, pp. 263–269, 1988.

[18] P. A. Takes, J. Kohrs, R. Krug, and S. Kewley, "Microwave technology in immunohistochemistry: application to avidin-biotin staining of diverse antigens," *Journal of Histotechnology*, vol. 12, no. 2, pp. 95–98, 1989.

[19] M. Werner, R. von Wasielewski, and A. Georgii, "Immunodetection of a tumor associated antigen (TAG-12): comparison of microwave accelerated and conventional method," *Biotechnic and Histochemistry*, vol. 66, no. 2, pp. 79–81, 1991.

[20] L. van Vlijmen-Willems and P. van Erp, "Microwave irradiation for rapid and enhanced lmmunohistochemical staining: application to skin antigens," *Biotechnic & Histochemistry*, vol. 68, no. 2, pp. 67–74, 1993.

[21] M. E. Boon and L. P. Kok, "Microwaves for immunohistochemistry," *Micron*, vol. 25, no. 2, pp. 151–170, 1994.

[22] A. S.-Y. Leong, "Microwaves in diagnostic immunohistochemistry," *European Journal of Morphology*, vol. 34, no. 5, pp. 381–383, 1996.

[23] A. K. Katoh, N. Stemmler, S. Specht, and F. D'Amico, "Immunoperoxidase staining for estrogen and progesterone receptors in archival formalin fixed, paraffin embedded breast carcinomas after microwave antigen retrieval," *Biotechnic and Histochemistry*, vol. 72, no. 6, pp. 291–298, 1997.

[24] I. Cavusoglu, Z. Kahveci, and S. A. Sirmali, "Rapid staining of ultrathin sections with the use of a microwave oven," *Journal of Microscopy*, vol. 192, no. 2, pp. 212–216, 1998.

[25] S. Noyan, Z. Kahveci, I. Cavusoglu, F. Z. Minbay, F. B. Sunay, and S. A. Sirmali, "Effects of microwave irradiation and chemical fixation on the localization of perisinusoidal cells in rat liver by gold impregnation," *Journal of Microscopy*, vol. 197, no. 1, pp. 101–106, 2000.

[26] M. Kaneko, T. Tomita, T. Nakase et al., "Rapid decalcification using microwaves for in situ hybridization in skeletal tissues," *Biotechnic & Histochemistry*, vol. 74, no. 1, pp. 49–54, 1999.

[27] E. M. Keithley, T. Truong, B. Chandronait, and P. B. Billings, "Immunohistochemistry and microwave decalcification of human temporal bones," *Hearing Research*, vol. 148, no. 1-2, pp. 192–196, 2000.

[28] I. N. Sheriffs, D. Rampling, and V. V. Smith, "Paraffin wax embedded muscle is suitable for the diagnosis of muscular dystrophy," *Journal of Clinical Pathology*, vol. 54, no. 7, pp. 517–520, 2001.

[29] F. Z. Minbay, Z. Kahveci, and I. Cavusoglu, "Rapid Bielschowsky silver impregnation method using microwave heating," *Biotechnic & Histochemistry*, vol. 76, no. 5-6, pp. 233–237, 2001.

[30] T. Kumada, K. Tsuneyama, H. Hatta, S. Ishizawa, and Y. Takano, "Improved 1-h rapid immunostaining method using intermittent microwave irradiation: practicability on 5 years application in Toyama Medical and Pharmaceutical University Hospital," *Modern Pathology*, vol. 17, no. 9, pp. 1141–1149, 2004.

[31] S. G. Temel, F. Z. Minbay, Z. Kahveci, and L. Jennes, "Microwave-assisted antigen retrieval and incubation with cox-2 antibody of archival paraffin-embedded human oligodendroglioma and astrocytomas," *Journal of Neuroscience Methods*, vol. 156, no. 1-2, pp. 154–160, 2006.

[32] H. Hatta, K. Tsuneyama, T. Kumada et al., "Freshly prepared immune complexes with intermittent microwave irradiation result in rapid and high-quality immunostaining," *Pathology Research and Practice*, vol. 202, no. 6, pp. 439–445, 2006.

[33] L. L. Emerson, S. R. Tripp, B. C. Baird, L. J. Layfield, and L. R. Rohr, "A comparison of immunohistochemical stain quality in conventional and rapid microwave processed tissues," *American Journal of Clinical Pathology*, vol. 125, no. 2, pp. 176–183, 2006.

[34] R. J. Buesa, "Microwave-assisted tissue processing: real impact on the histology workflow," *Annals of Diagnostic Pathology*, vol. 11, no. 3, pp. 206–211, 2007.

[35] D. J. Long II and C. Buggs, "Microwave oven-based technique for immunofluorescent staining of paraffin-embedded tissues," *Journal of Molecular Histology*, vol. 39, no. 1, pp. 1–4, 2008.

[36] C. C. Abreu, P. A. Nakayama, C. I. Nogueira et al., "Domestic microwave processing for rapid immunohistochemical diagnosis of bovine rabies," *Histology and Histopathology*, vol. 27, no. 9, pp. 1227–1230, 2012.

[37] S. Shi, Q. Cheng, P. Zhang et al., "Immunofluorescence with dual microwave retrieval of paraffin-embedded sections in the assessment of human renal biopsy specimens," *American Journal of Clinical Pathology*, vol. 139, no. 1, pp. 71–78, 2013.

[38] A. Bond and J. C. Kinnamon, "Microwave processing of gustatory tissues for immunohistochemistry," *Journal of Neuroscience Methods*, vol. 215, no. 1, pp. 132–138, 2013.

[39] Y. Kitayama, H. Igarashi, and H. Sugimura, "Initial intermittent microwave irradiation for fluorescence in situ hybridization analysis in paraffin-embedded tissue sections of gastrointestinal neoplasia," *Laboratory Investigation*, vol. 80, no. 5, pp. 779–781, 2000.

[40] Y. Kitayama, H. Igarashi, and H. Sugimura, "Different vulnerability among chromosomes to numerical instability in gastric carcinogenesis: stage-dependent analysis by FISH with the use of microwave irradiation," *Clinical Cancer Research*, vol. 6, no. 8, pp. 3139–3146, 2000.

[41] K. Kobayashi, Y. Kitayama, H. Igarashi et al., "Intratumor heterogeneity of centromere numerical abnormality in multiple primary gastric cancers: application of fluorescence in situ hybridization with intermittent microwave irradiation on paraffin-embedded tissue," *Japanese Journal of Cancer Research*, vol. 91, no. 11, pp. 1134–1141, 2000.

[42] K. Katoh, Y. Kano, and S. Ookawara, "Microwave irradiation for fixation and immunostaining of endothelial cells in situ," *Biotechnic and Histochemistry*, vol. 84, no. 3, pp. 101–108, 2009.

[43] K. Katoh, "Rapid fixation and immunofluorescent staining of cultured cells using microwave irradiation," *Journal of Histotechnology*, vol. 34, no. 1, pp. 29–34, 2011.

[44] G. Tornero, L. L. Latta, and G. Godoy, "Use of microwave radiation for the histological study of bone canaliculi," *Journal of Histotechnology*, vol. 14, no. 1, pp. 27–30, 1991.

[45] A. L. Marr and A. Wong, "Effects of microwave fixation and decalcification on rodent tissue," *Journal of Histotechnology*, vol. 32, no. 4, pp. 190–192, 2009.

[46] H. Utsunomiya, L. Shan, I. Kawano et al., "Immunolocalization of parathyroid hormone in human parathyroid glands with special references to microwave antigen retrieval," *Endocrine Pathology*, vol. 6, no. 3, pp. 223–227, 1995.

[47] R. W. M. Hoetelmans, H.-J. van Slooten, R. Keijzer, C. J. H. van de Velde, and J. H. van Dierendonck, "Comparison of the effects of microwave heating and high pressure cooking for antigen retrieval of human and rat Bcl-2 protein in formaldehyde-fixed,

paraffin-embedded sections," *Biotechnic & Histochemistry*, vol. 77, no. 3, pp. 137–144, 2002.

[48] Z. Kahveci, F. Z. Minbay, S. Noyan, and I. Çavusoglu, "A comparison of microwave heating and proteolytic pretreatment antigen retrieval techniques in formalin fixed, paraffin embedded tissues," *Biotechnic and Histochemistry*, vol. 78, no. 2, pp. 119–128, 2003.

[49] L. Gu, J. Cong, J. Zhang, Y. Tian, and X. Zhai, "A microwave antigen retrieval method using two heating steps for enhanced immunostaining on aldehyde-fixed paraffin-embedded tissue sections," *Histochemistry and Cell Biology*, vol. 145, no. 6, pp. 675–680, 2016.

[50] Y. Kano, K. Katoh, and K. Fujiwara, "Lateral zone of cell-cell adhesion as the major fluid shear stress-related signal transduction site," *Circulation Research*, vol. 86, no. 4, pp. 425–433, 2000.

[51] R. D. Krug and P. A. Takes, "Microwave technology in immunohistochemistry: II. A universal staining protocol using diaminobenzidine as the chromogen," *Journal of Histotechnology*, vol. 14, no. 1, pp. 31–34, 1991.

[52] R. A. Shiurba, E. T. Spooner, K. Ishiguro et al., "Immunocytochemistry of formalin-fixed human brain tissues: microwave irradiation of free-floating sections," *Brain Research Protocols*, vol. 2, no. 2, pp. 109–119, 1998.

[53] T. Kumada, K. Tsuneyama, H. Hatta, S. Ishizawa, and Y. Takano, "Improved 1-h rapid immunostaining method using intermittent microwave irradiation: practicability based on 5 years application in Toyama Medical and Pharmaceutical University Hospital," *Modern Pathology*, vol. 17, no. 9, pp. 1141–1149, 2004.

[54] K. Katoh and Y. Noda, "Distribution of cytoskeletal components in endothelial cells in the guinea pig renal artery," *International Journal of Cell Biology*, vol. 2012, Article ID 439349, 10 pages, 2012.

[55] K. Katoh, Y. Kano, and S. Ookawara, "Morphological differences between guinea pig aortic and venous endothelial cells in situ," *Cell Biology International*, vol. 31, no. 6, pp. 554–564, 2007.

[56] A. S.-Y. Leong and R. T. Sormunen, "Microwave procedures for electron microscopy and resin- embedded sections," *Micron*, vol. 29, no. 5, pp. 397–409, 1998.

[57] P. Webster, "Microwave-assisted processing and embedding for transmission electron microscopy," *Methods in Molecular Biology*, vol. 1117, pp. 21–37, 2014.

[58] F. Hernandez-Chavarria and M. Vargas-Montero, "Rapid contrasting of ultrathin sections using microwave irradiation with heat dissipation," *Journal of Microscopy*, vol. 203, no. 2, pp. 227–230, 2001.

[59] N. T. Brinn, "Rapid metallic histological staining using the microwave oven," *Journal of Histotechnology*, vol. 6, no. 3, pp. 125–129, 1983.

[60] R. J. Buesa, "Haven't you calibrated your microwave oven yet?" *Journal of Histotechnology*, vol. 25, no. 1, pp. 39–43, 2002.

12

The Human IL-23 Decoy Receptor Inhibits T-Cells Producing IL-17 by Genetically Engineered Mesenchymal Stem Cells

Masoumeh Rostami,[1] Kamran Haidari,[2] and Majid Shahbazi ⓘ[1]

[1]Medical Cellular and Molecular Research Center, Golestan University of Medical Sciences, Gorgan, Iran
[2]Department of Anatomy, Faculty of Medical Sciences, Golestan University of Medical Sciences, Gorgan, Iran

Correspondence should be addressed to Majid Shahbazi; shahbazimajid@yahoo.co.uk

Academic Editor: Maria Roubelakis

The immunomodulatory and self-renewable features of human adipose mesenchymal stem cells (hAD-MSCs) mark their importance in regenerative medicine. Interleukin 23 (IL- 23) as a proinflammatory cytokine suppresses T regulatory cells (Treg) and promotes the response of T helper 17 (Th17) and T helper 1 (Th1) cells. This pathway starts inflammation and immunosuppression in several autoimmune diseases. The current study for producing recombinant IL- 23 decoy receptor (RIL- 23R) using hAD-MSCs as a good candidate for ex vivo cell-based gene therapy purposes reducing inflammation in autoimmune diseases. hAD-MSCs was isolated from lipoaspirate and then characterized by differentiation. RIL- 23R was designed and cloned into a pCDH-813A-1 lentiviral vector. The transduction of hAD-MSCs was performed at MOI (multiplicity of infection) = 50 with pCDH- EFI α- RIL- 23R- PGK copGFP. Expressions of RIL- 23R and octamer-binding transcription factor 4 (OCT- 4) were determined by real-time polymerase chain reaction (real time-PCR). Self-renewing properties were assayed with OCT- 4. Bioactivity of the designed RIL- 23R was evaluated by IL- 17 and IL- 10 expression of mouse splenocytes. Cell differentiation confirmed the true isolation of hAD-MSCs from lipoaspirate. Restriction of the enzyme digestion and sequencing verified the successful cloning of RIL- 23R in the CD813A-1 lentiviral vector. The green fluorescent protein (GFP) positive transduction rate was up to 90%, and real-time PCR showed the expression level of RIL-23R. Oct-4 had a similar expression pattern with nontransduced hAD-MSCs and transduced hAD-MSCs/ RIL-23R indicating that lentiviral vector did not affect hAD-MSCs characteristics. Downregulation of IL-17 and upregulation of IL-10 showed the correct activity of the engineered hAD-MSCs. The results showed that the transduced hAD-MSCs/ RIL- 23R, expressing IL-23 decoy receptor, can give a useful approach for a basic research on cell-based gene therapy for autoimmune disorders.

1. Introduction

The mesenchymal stem cells (MSCs) are one of the adult stem cells capable of being differentiating into mesoderm-lineage cells, including osteoblasts, chondrocytes, and adipocytes. They are characterized by attachment to plastic culture vessels and the ability to express the CD44, CD73, CD90, and CD105 but not CD45, CD34, and CD14 cell surface markers [1]. Many studies have found that MSCs have both hypoimmunogenic and immunomodulatory properties that allow them to home to damaged tissues and initiate healing through repair processes [2–4]. These characteristics of MSCs support the idea that the creation of genetically modified

MSCs could be the best option for combining cell and gene therapy to treat diverse forms of autoimmune diseases [5–10].

Among which, the Adipose-MSCs (AD-MSCs) are a rich source for MSCs in the therapeutic purposes as adipose tissue is easily accessible and large amounts of AD-MSCs are easily obtainable [11–13]. Currently, AD-MSCs are clinically applied for regenerative treatments and wound healing [14, 15]. Thus, the ability of human adipose-derived MSCs to serve as vehicles for a cell-based gene therapy is promising [16–19].

Autoimmune diseases are multi-factorial disorders with complicated immune system dysregulation mediated by immune cytokines and immune cells [20]. The IL- 23 belonging to the family IL-12 plays an active role to proliferate the memory T helper 1 cells. The hetero dimerized IL-23 receptor constituted of specific (IL23A) and common (IL12Rβ1) subunits [21]. The transforming growth factor beta (TGF-β) and IL-6 in most of the autoimmune diseases can induce Th17 cells to secrete increasingly IL-23 and IL-17 [22]. The IL-23 is able to suppress the Treg cells and promote the response of Th17 and Th1 cells, initiating inflammation and immunosuppression in several autoimmune and inflammatory diseases [23]. Based on the strong evidence, the IL-23/IL-17 axis is important for the development of chronic inflammation [24]. Recent studies determined that the suppression of IL-23, the IL-23R, or the IL-23/IL-17 axes potentially can be therapeutic targets for the autoimmune diseases [25]. However, there are no specific treatments for inhibition of the IL-23 proinflammatory responses.

Targeting IL-23p19, but not IL-12p40, in gene knockout studies showed that the decrease of proinflammatory responses and resistance to different autoimmune diseases are due to the absence of IL-17-producing T-cells (i.e., Th17 cells) [26].

Based on earlier studies, extensive alternative splicing exerted on the IL-23R gene transcript [27]. Generation of IL-23RΔ9 form (GenBank AM990318), which encodes a soluble version of the entire external domain of the specific receptor chain (IL23A), is an example of this splicing [28]. After binding human IL-23 in solution, this soluble decoy receptor dependently inhibits STAT3 phosphorylation and functional maturation of human Th17 cells in vitro [29].

Based on the immunosuppressive functions of hAD-MSCs and RIL-23R, they can cooperate to improve the immunomodulatory and prevent the initiation of inflammation and latter autoimmune diseases. The current report was the first case of RIL-23R gene transduction into hAD-MSCs. This study successfully transduced hAD-MSCs by recombinant Interleukin 23R-harbouring lentiviral particles and evaluated the expression of RIL-23R in transduced hAD-MSCs by real time-PCR. It also analyzed the RIL-23R bioactivity and the effect of this cytokine on the T-cells.

2. Materials and Methods

2.1. Isolation of Mesenchymal Stem Cell from Human Adipose Tissue. Lipoaspirate samples were washed with phosphate-buffered saline (PBS) containing 3 X penicillin/ streptomycin and amphotericin three times. The adipose tissue was added dispase (50 u/ml)/ collagenase I (250 u/ml) (Sigma-Aldrich, St. Louis, MO), followed by shaking for 30 min at 37°C and then centrifugation at 1500 rpm. The plated cells after suspending were distributed in the flasks with α-MEM (minimum essential medium eagle-alpha modification) containing 10% fetal bovine serum (FBS) for three days.

2.2. Cloning of RIL-23R cDNA into a Lentiviral Vector. IL-23 soluble form, IL-23RΔ9 (GenBank AM990318), which encodes a version of the entire external domain of the receptor chain, (alternative splicing product) was designed and amplified in the pORF plasmid (Biomatik, Cambridge, Ontario, Canada) using the Forward primer of 5′

ACAACAGCTCGGCTTTGGTAT -3′, and Reverse primer of 5′ TACTGGCAGCCTTGGAGTTC -3′.

RIL-23R-pORF-For synthesizing the digestion of GFP-harboring lentiviral particles (LvGFP), GFP-carrying vector was digested by EcorV and Sall, cDNA subcloned into pCDH813A-1 (System Bioscience, Mountain View, CA, United States) to make a recombinant vector (pCDH-EF-1α–RIL-23R-PGK-copGFP).

2.3. Generation, Concentration, and Titration of Recombinant Lentivirus. The recombinant lentivirus was produced according to the previous protocol [30, 31]. In summary, the transfection of plasmid DNA into 1×10^6 HEK-293T cells was based on calcium–phosphate method (21 μg of lentivirus involving pCDH- RIL-23R-copGFP, 21 μg of packaging plasmid psPAX2, and 10.5 μg of envelope plasmid per 10-cm plate). The lentivirus particles contained in the supernatant of HEK-293 T cells were collected at 36, 48, 60, and 72 hours posttransfection and then passed through a 0.45 μm filter. The concentration process of lentivirus particles was performed according to the precipitation method using 50% PEG-8000 (Sigma-Aldrich, St. Louis, MO, USA) to reach a final concentration of 5%. The concentration was continued with NaCl 5 M (Sigma-Aldrich, St. Louis, MO, USA) overnight to achieve a final concentration of 0.15 M. Then, 6×10^4 HEK-293T cells were plated into plates for titration, followed by transduction with 1, 4 and 16 μl of the virus. After three days of incubation, the cells were detached and flow cytometry was used to analyze cell fluorescence.

2.4. Transduction of hAD-MSCs. The second passage hAD-MSCs were trypsinized and were seeded in 6- well plates at a density of 1×105 cells in 1 ml DMEM-F12. The virus particles (MOI = 50) were added along with 6 μg/ml of Polybrene (Sigma-Aldrich, St. Louis, MO), followed by shaking on a rotator at 5 rpm for 18 hours. The medium was renewed by DMEM-F12. The flow cytometry was applied to calculate the GFP-positive transduced hAD-MSCs at 72 hours after transduction and data were analyzed by BD Accuri c6 software.

2.5. Charactherization of Normal/Transduced hAD-MSCs by Differentiation, Flow Cytometry, and CFU-F Assay. In this step, 2×10^4 cells/ml of third passage RIL-23R-transduced hAD-MSCs and hAD-MSCs were poured in a 6-well plate. The incubation was performed with α-MEM containing 10% FBS until reaching confluence. The medium was renewed by osteogenic medium consisting of DMEM having 10 nM of dexamethasone (Sigma-Aldrich, St. Louis, MO, USA), 50 μg/ml of ascorbic acid 2-phosphate (Sigma-Aldrich, St. Louis, MO, USA), and 10 mM of β-glycerol phosphate (Sigma-Aldrich, St. Louis, MO, USA). The action was done twice a week for three weeks. Next, the cells were fixed with 10% formalin for 10 min and stained with alizarin red (Sigma-Aldrich, St. Louis, MO, USA) at room temperature for 2 min. In the adipogenesis, the third passage RIL-23R-transduced hAD-MSCs and hAD-MSCs cells were incubated in the presence of differentiation medium containing DMEM with

50 μg/ml of indomethacin (Sigma-Aldrich, St. Louis, MO, USA) and 100 nM of dexamethasone (Sigma-Aldrich, St. Louis, MO, USA). The medium was renewed twice a week for three weeks. The cells staining was performed with 0.5% oil red O (Sigma-Aldrich, St. Louis, MO, USA) in methanol for 2 min at room temperature.

The trypsinization of cells after third passage was carried out by 0.25% trypsin and 0.02% EDTA. The washing was continued twice using PBS, and staining was based on the manufacturer's instruction for running the flow cytometry. The antibody-labeled cells had the antibodies (e-Bioscience) of [1] : PE-conjugated mouse anti-human CD11b, PE-conjugated mouse anti-human CD34, PE-conjugated mouse anti-human CD45, PE-conjugated mouse anti-human CD105, and PE-conjugated mouse anti-human CD73. PE-conjugated mouse IgG1, FITC-conjugated mouse IgG2, and PE-conjugated rat IgG2 were considered to be isotype controls. The incubated cell suspensions were rinsed with PBS to discard any unlabeled antibodies, followed by suspension of labeled cells in PBS. The analysis was done by FACScan flow cytometer (Becton Dickinson, San Diego, CA, USA). The data were analyzed by BD Accuri c6 software.

The self-renewing feature of the cells can be evaluated by Colony-forming unit-fibroblast (CFU-F) assay. $1x10^3$ cells at third passage were seeded in 10 cm dishes. Following cultivation for 2 weeks, the cells were washed with PBS and stained with 0.5% crystal violet (Sigma-Aldrich, St. Louis, MO, USA) for 5 min at room temperature. Stained colonies were counted.

2.6. Stemness Markers Expression in Normal/Transduced hAD-MSCs.
The expression of OCT-4, sox2, Nanog, and c-Myc was examined in hAD-MSCs and transduced hAD-MSCs by RT-PCR. The GAPDH expression was the endogenous reference gene.

2.7. RIL- 23R Expression in Normal/Transduced hAD-MSCs.
The RIL- 23R expression was studied in hAD-MSCs and transduced hAD-MSCs. The GAPDH expression was the endogenous reference gene. Qiazol lysis reagent (Alameda, CA, United States) was used to isolate total RNA from the cells. The reactions of standard reverse transcription were done using 5 μg of total RNA by oligo (dT) 18 as a primer according to the manufacturer's instructions for the cDNA synthesis kit (Fermentas). Additional PCR components were 2.5 μl of cDNA, 1 X PCR buffer (AMS), 200 μM of dNTPs, 0.5 μM of each primer pair, and 1 unit per 25 μL of reaction mix Taq DNA polymerase (Fermentas). The above-mentioned pairs of primers detected the expression of RIL- 23R gene.

Whole cell lysate of normal/transduced hAD-MSC was generated from confluent cell layers in 6-well plates using lysis buffer made of 50 mM Tris-Cl, pH 7.4; 150 mM NaCl, 1% NP-40, 1% Na-deoxycholate, 1 mM EDTA, 0.1% SDS (all from Sigma), and Mini Complete, EDTA-free protease inhibitor cocktail (Roche). Proteins were separated by 12%

SDS-PAGE, blotted to nitrocellulose membranes (GE Healthcare, Munich, Germany), blocked with TBS containing 5% nonfat dry milk, and analyzed for RIL-23R with the anti-RIL-23R antibody (R&D systems) followed by anti-goat HRP conjugate (Santa Cruz, Santa Cruz, CA, USA).

2.8. RIL- 23R Function Assay by Co-Culturing.
The IL- 23R secretion and function were investigated by bioassays of naïve T cells of C57BL/ 6 mouse co-cultured with normal / transduced hAD-MSCs and HT1080 cells. The mice were killed in accordance with the laboratory animal protocol to dissect the spleen for digesting with 100 u /ml of dispase/collagenase. The spleen cells were gathered via centrifugation at 1200 rpm and the RBCs lysis via an RBC lysis buffer. The cells were then suspended in RPMI culture medium supplemented with 10% fetal calf serum at a concentration of 1×10^6 cells/ml. To generate short-term cultures of activated T cells, splenocytes were activated by the addition of the lectin phytohaemagglutinin (PHA) at a concentration of 1 mg/ml in RPMI supplemented with 10% fetal calf serum for 4 days [32]. The cells were then washed and maintained in interleukin-2 (IL-2) supplemented media at a concentration of 2 ng/ml for at least 7 days before use.

2×10^4 cells/well were seeded in 96-well plates. The next day, cells were washed and preincubated with a mitomycin C (0.4 mg/ml for 5 minutes) [33]. Following treatment, the cells were washed immediately at least three times and incubated in RPMI (Sigma) growth medium. To decide whether transduced cells induced Th2 cells proliferation in results from the upregulated RIL-23R expression, IL-2 dependent PHA activated T cells were added to the culture medium with an effector: target cell ratio of 20:1 [34]. T cell-adherent cell contact was inhibited in some experiments by seeding T cells into cell culture inserts (Becton Dickinson). These inserts incorporate membranes that are transparent and contain 0.4 mm pores which allow free passage of soluble factors but are too small to let cell migration [32]. Supernatants were harvested after 24 or 48 h and concentration of IL-10 and IL-17 determined by RT-PCR [35]. Total RNA extraction was carried out from the T cells cocultured with normal/transduced cells. The expression of glyceraldehyde-3-phosphate dehydrogenase (GAPDH) was considered as an endogenous reference gene.

2.9. Ethical Considerations.
The present study was conducted with the recommendations in the guide for the Care and Use of Laboratory Animals and was approved by the ethics committee of The Golestan University of Medical Sciences. The human adipose tissue was collected after obtaining informed consent considering the Declarations of the Golestan University of Medical Science, Gorgan, Iran

2.10. Statistical Analysis.
All statistical analyses were carried out by Graph Pad Prism (version 6) using one-way ANOVA and independent t-test to check the differences between groups at the statistical significant level of $P < 0.05$.

3. Results

3.1. hAD-MSCs Were Isolated and Differentiated into Adipocytes and Osteoblasts. After the cells were passaged, the adipocytes were separated from lipoaspirate tissue using mechanical and enzymatic digestion. The cells displayed a strong proliferative ability (Figure 1). The confirmation of the cell multilineage capacity was done following their differentiation into adipocyte and osteoblast cells after three weeks. Visualization of many lipid vacuoles after oil red staining revealed the adipocyte features of hAD-MSCs (Figure 1(a)). We identified the presence of cell surface markers CD73, CD105 and the absence of hematopoietic (CD45, CD11b) and endothelial (CD34) antigens in isolated adipose MSC cells. Our data show that this type of cell has similar expression profiles for the selected markers (Figure 1(b)).

3.2. RIL- 23R Was Designed and Successfully Cloned into a Lentiviral Vector. The designed recombinant open reading frame of the IL- 23R gene (RIL- 23R) was amplified using PCR and then subcloned into a lentiviral vector (Figure 2(a)). The insertion of RIL-23R cDNA into the pCDH813A-1 vector was confirmed by digestion of the shuttle vector by EcorV and Sal1. The 1400 bp fragment corresponding to RIL-23R was successfully separated from the pCDH813A- 1 vector (Figure 2(b)). Moreover, the colony-PCR was used to detect RIL-23R cDNA in this vector. The cDNA then was sequenced, and no mutations were found in the cDNA sequence.

3.3. The Recombinant Viral Particle Was Produced by HEK-293T. HEK- 293T cells with 70% confluency (Figure 3(a)) were cotransfected with this construct and successfully produced RIL- 23R- lentiviral particles (Figure 3(b)). The transfection efficiency was over 90% (Figure 3(c)).

3.4. hAD-MSCs Were Transduced with Recombinant Lentiviral Particles. AD-MSCs were about 50% confluent when transduced with lentiviral particles (Figure 4(a)). GFP-positive transduced hAD-MSCs measured at 72 h after transduction (Figure 4(b)) and were selected for with 2.5 μg of puromycin (Figure 4(c)). Flow cytometry was used to get about 95% pure transduced cells 14 days post-transduction (Figure 4(d)). The transduced hAD-MSCs at MOI = 50 was conducted after doing virus concentration and titration.

3.5. Normal/Transduced hAD-MSCs Were Characterized by Differentiation, Flow Cytometry, and CFU-F Assay. We identified the presence of cell surface markers CD73 and CD105 and the absence of hematopoietic (CD45, CD11b) and endothelial (CD34) antigens in both hAD-MSCs and transduced hAD-MSCs (Figure 5(a)). Our data show that these two types of cells have similar expression profiles for the selected markers. But the stem cell marker CD105 was less abundant in transduced hAD-MSCs (55.6% ±4.39 SEM, n=5 for hAD-MSCs, and 27.1% ±2.3% SEM, n=5 for transduced-hAD-MSCs).

Although mesenchymal stromal cells have been defined by the positive expression of CD105 [36], several groups have also observed considerable phenotypic drift within ASCs during in vitro expansion [37–39]. We found that CD105 expression on hMSCs is heterogeneous in agreement with previous studies [36, 40–42]. The differences observed in CD105 expression could be a consequence of culture conditions (passage number, culture time, cell confluence [43, 44], oxygen pressure, TNF-α [45], IFN-γ [4], and Serum-Free Medium) [46] and MSC source [47]. Also, Absence or low expression of Endoline (CD105) correlated with a subgroup of adipose-derived cells with increased osteogenic gene expression [36, 48], while the selection of CD105 positive (CD105+) MSCs favors chondrogenesis [49]. These results also showed that the MSCs were not derived from endothelial or hematopoietic cells.

The normal/transduced hAD-MSCs displayed a strong proliferative ability. The cell multilineage capacity was done following their differentiation into adipocyte and osteoblast cells (Figure 5(b)). Visualization of massive calcium depositions around differentiated cells after alizarin Red staining confirmed the presence of osteoblasts in hAD-MSCs and transduced hAD-MSCs. Moreover, observation of many lipid vacuoles after oil red staining revealed the adipocyte features of hAD-MSCs and transduced hAD-MSCs. Therefore, lentiviral particles containing RIL-23R keep the mesodermal properties of hAD-MSCs.

The CFU-F assay was performed to check the self-renewing properties of the cells. There were no significant differences in the number of CFU-Fs following seeding cells in 10 cm dishes after 2 weeks (Figure 5(c)). The normal and transduced hAD-MSCs displayed a higher self-renewal feature regardless of growth rate, although the differences were not significant.

3.6. Self-Renewing Features of hAD-MSCs Were Analyzed by RT-PCR. We examined the expression of molecular markers in the hAD-MSCs and transduced hAD-MSCs (Figure 6). The OCT4, c-Myc, SOX2, and NANOG were detected in the hAD-MSCs and transduced hAD-MSCs. the expression of SOX2 was low in both cell lines. These results suggest that hAD-MSCs and transduced hAD-MSCs have the highest capacity for self-renewal and differentiation potential. Primer sets used for RT-PCR was shown in Table 1.

3.7. RIL- 23R Expression Were Analyzed by RT-PCR and Western Blot. For RIL-23R expression, total RNA of non-transduced and transduced hAD-MSCs were isolated, and RT- PCR showed expression of RIL-23R in transduced hAD-MSCs but not in control hAD-MSCs (Figures 7(a) and 7(b)). Subsequently, the expression of RIL-23R was determined on the protein level. By immunoblot analysis, RIL-23R was almost detectable in whole cell lysate of transduced hAD-MSCs (Figure 7(c)).

3.8. RIL- 23R Function Was Assayed by Co-Culture with CD4+ T. To prove that the inhibitory effect of hMSCs on murine T cells is specific to the transduced h-ADMSCs, we also used a

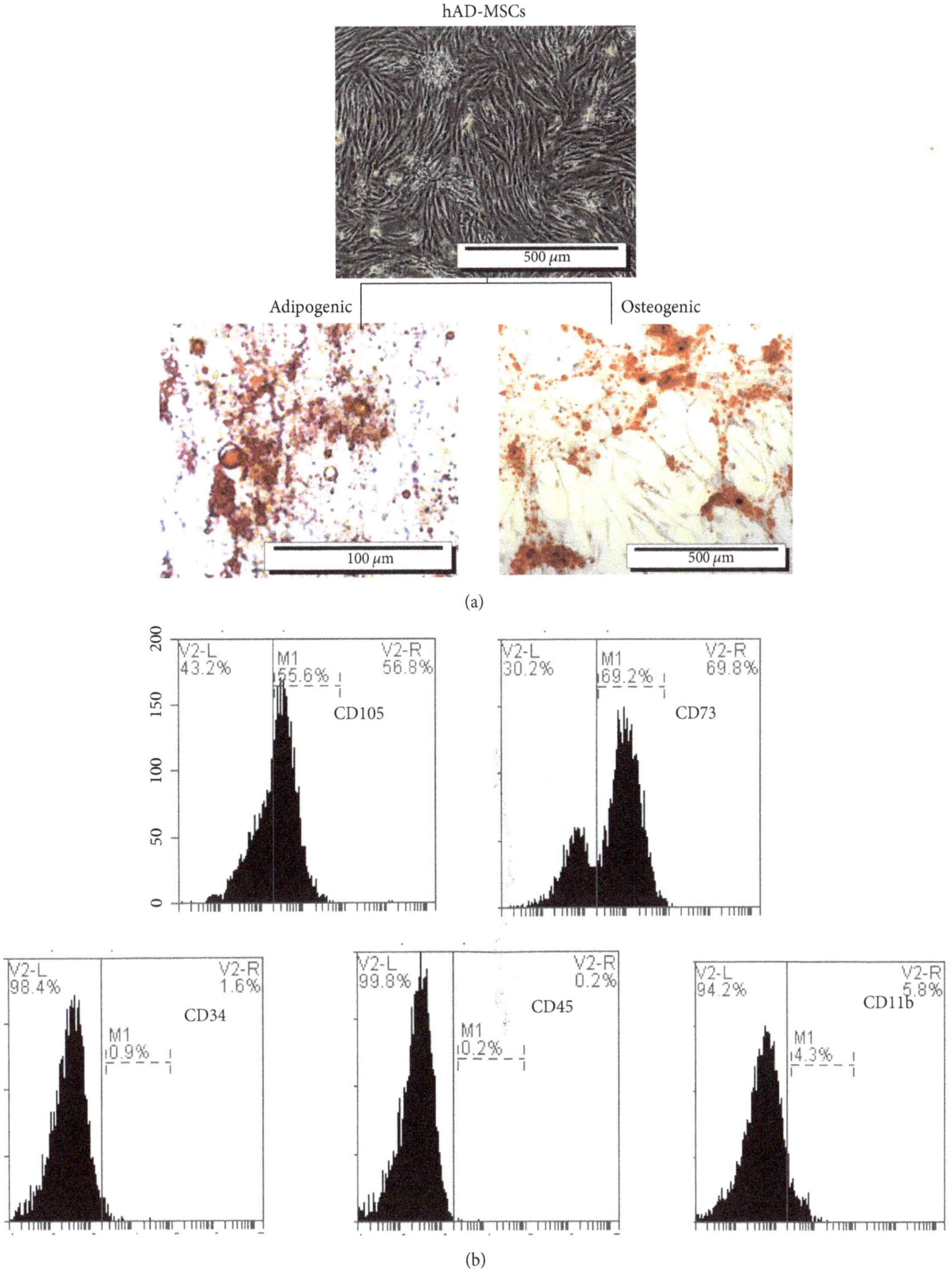

FIGURE 1: The isolated mesenchymal stem cells (MSCs) characterization by flow cytometry and differentiation. (a) The second passage of isolated hAD- MSCs. Adipogenic: staining of isolated hAD- MSCs after 2nd passage by Oil Red O reagent. Osteogenic: staining of isolated hAD-MSCs by Alizarin Red S reagent. (b) Expression of surface antigens in isolated hAD-MSCs as determined using flow cytometry.

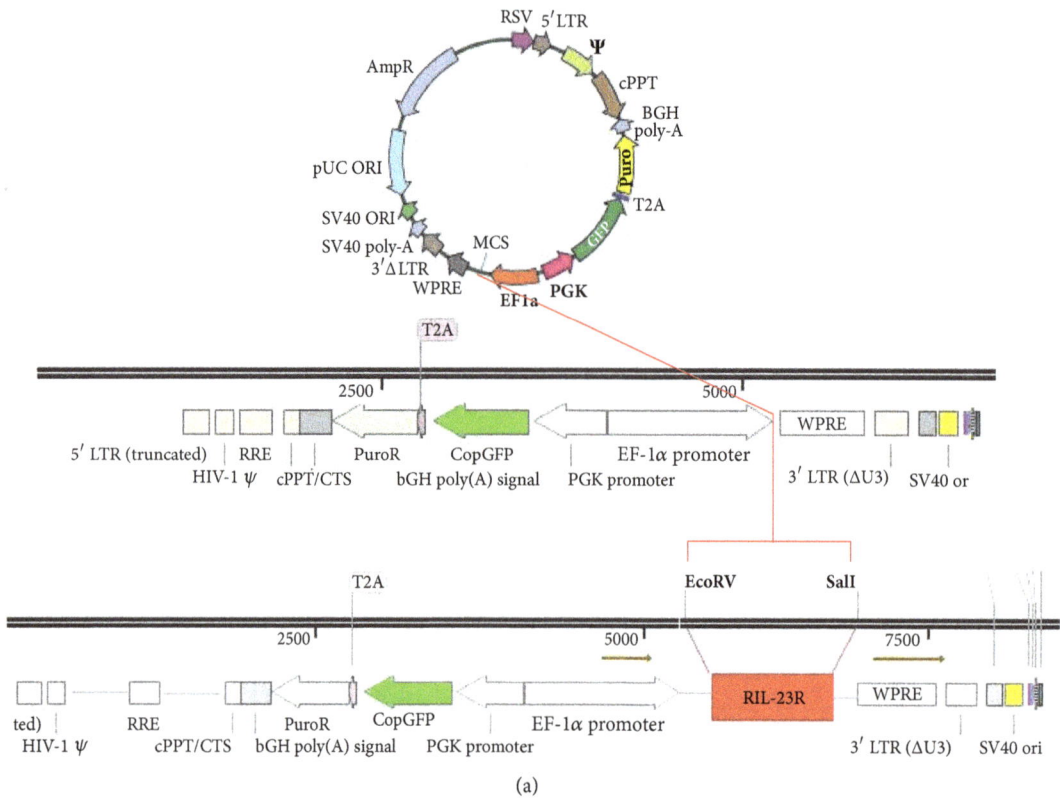

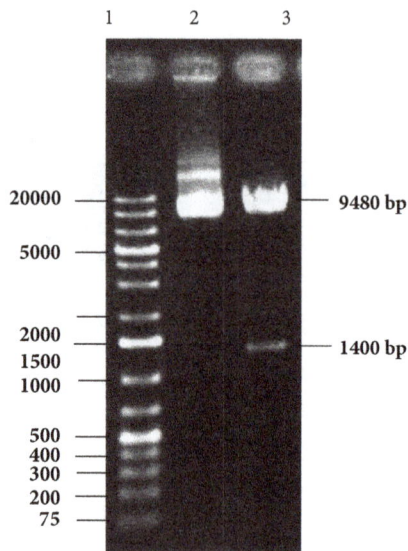

FIGURE 2: Recombinant IL-23R gene construct. (a) Recombinant IL-23R cDNA inserted into the pCDH-813A-1 lentiviral vector. (b) The presence of a 1400 bp segment related to RIL- 23R and a segment of a 9480 bp related to pCDH-813A-1 confirmed that the cloning established. A genetic map confirmed these data.

(a)   (b)

(c)

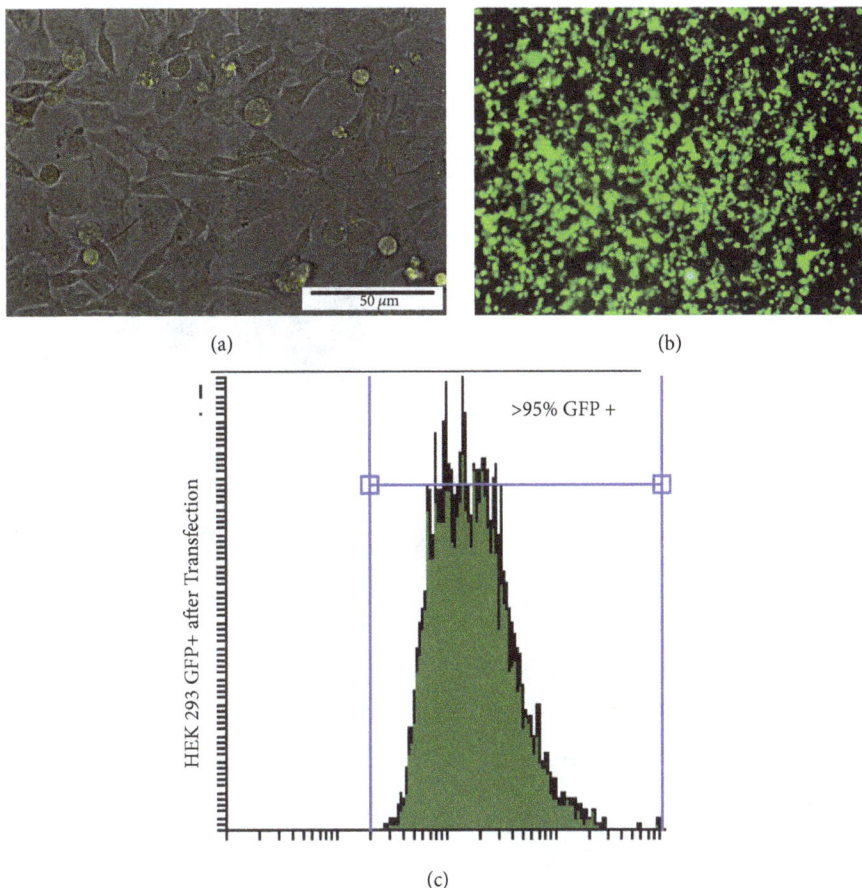

FIGURE 3: Transfection of HEK-293T for recombinant viral particles production. (a) The nontransfected culture of HEK-293T. (b) Transfected HEK- 293T at 24 hours after transfection by pCDH- EF1α- RIL- 23R- PGK-copGFP. (c) Flow cytometry analysis shows High expression of GFP in HEK- 293T and the high rate of transfection.

TABLE 1: Primer sets used for RT-PCR.

Gene	Primer sequence (5′ ⟶ 3′)	Product size (bp)
c-Myc	Forward: TCGGATTCTCTGCTCTCCTC	413
	Reverse: CGCCTCTTGACATTCTCCTC	
4-Oct	Forward: GACAACAATGAGAACCTTCAGGAGA	218
	Reverse: TTCTGGCGCCGGTTACAGAACCA	
SOX2	Forward: AACCAAGACGCTCATGAAGAAG	341
	Reverse: GCGAGTAGGACATGCTGTAGGT	
NANOG	Forward: ATAGCAATGGTGTGACGCAG	219
	Reverse: GATTGTTCCAGGATTGGGTG	
GAPDH	Forward: GTGGTCTCCTCTGACTTCAACA	210
	Reverse: CTCTTCCTCTTGTGCTCTTGCT	

different adherent human cell line in our experiments, HT-1080 (human fibrosarcoma cell line) [50]. These cells have similar morphological features to hMSCs, without having any known MSC-like properties (the multi-potent ability to differentiate into osteocytes, chondrocytes and adipocytes); also the fibroblasts do not have any inhibitory effect on the T cells [50], therefore, they were used as controls in our experiments. We used these cell lines in the same ratios as the hAD-MSCs and the same experimental settings.

4. Discussion

Autoimmune disease occurs when body organs are attacked by autoimmune cells as a result of an unfit immune response directed to autoantigens [51]. The most immunosuppressive drugs for the treatment of autoimmune disease belong to the corticosteroids family [52]. Many medical investigators are seeking new immunotherapeutic strategies with fewer side effects. These strategies have included a gene or recombinant

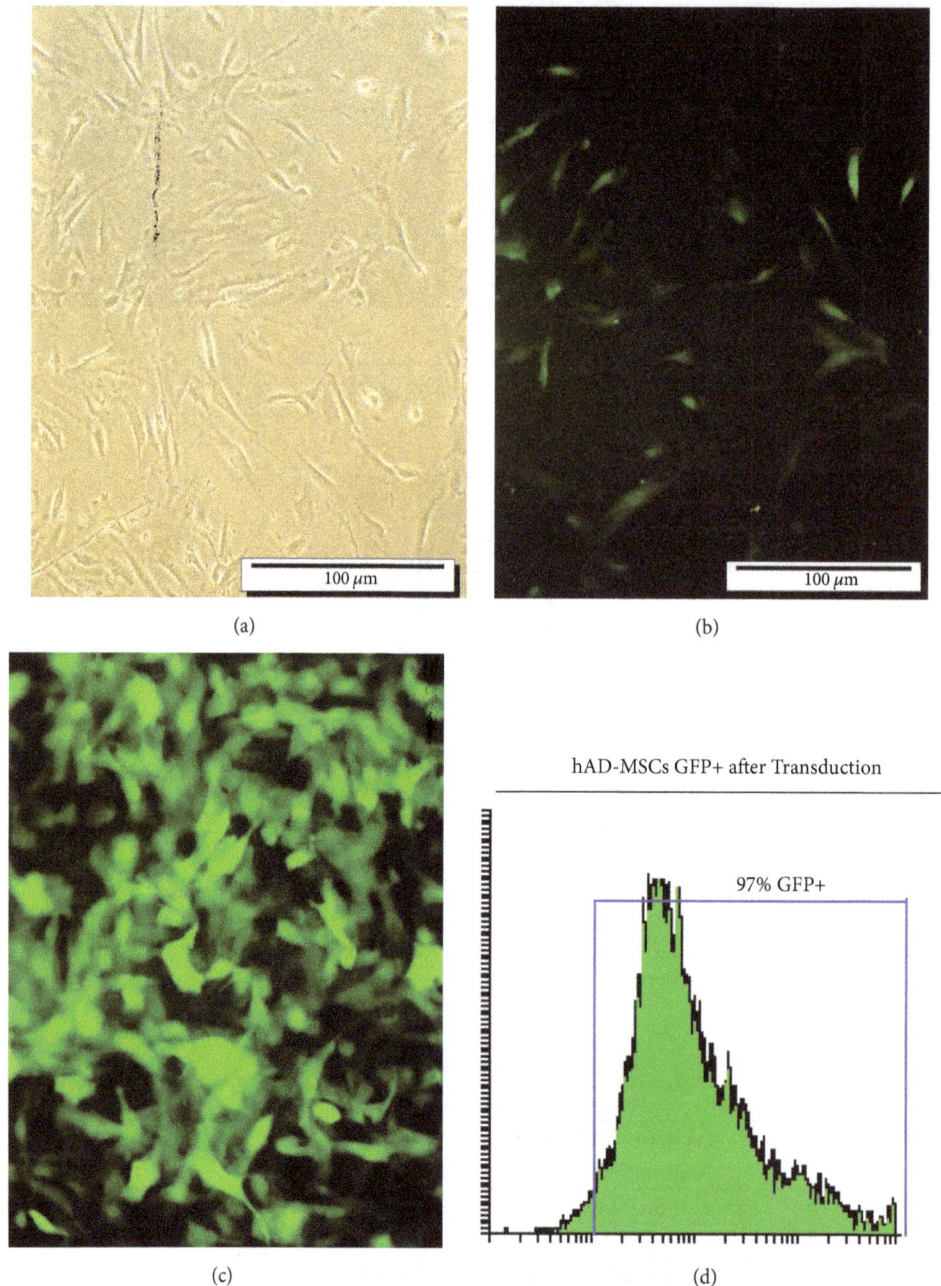

(a)

(b)

(c)

hAD-MSCs GFP+ after Transduction

97% GFP+

(d)

FIGURE 4: Adipocyte-derived mesenchymal stem cells (hAD-MSCs) transduction with lentiviral particles. (a) hAD-MSCs before transduction. (b) hAD-MSCs transduced with pCDH-RIL-23R-PGK-copGFP lentiviral vector. (c) Transduced hAD-MSCs after puromycin selection; abundant green cells and GFP expression exhibit high transduction level (d) FACS analysis of the transduction rate after puromycin selection indicated that more than 90% of the cells were GFP positive.

protein [53] therapies for affecting specific immune cells [54] and molecules such as cytokines [55], chemokines [56], and costimulatory molecules [57].

Mesenchymal stem cells (MSCs) are known to have immunomodulatory, self-renewing, and multilineage differentiation properties [58]. These characteristics have led to recognition of the true capacity of MSC-based cell therapy. Due to the efficiency with which MSCs can be transduced with different genes [59], genetic modifications of these cells have been carried out to enhance MSC efficacy in tissue repair/regeneration [9].

The high amount of MSCs found in adipose tissues and the relative ease with which they can be isolated makes MSCs a good source of adult stem cells in regenerative medicine [19]. Some studies [12, 13] have shown that manifestly more stem cells can be obtained from adipose tissue compared to the same volume of bone marrow [60].

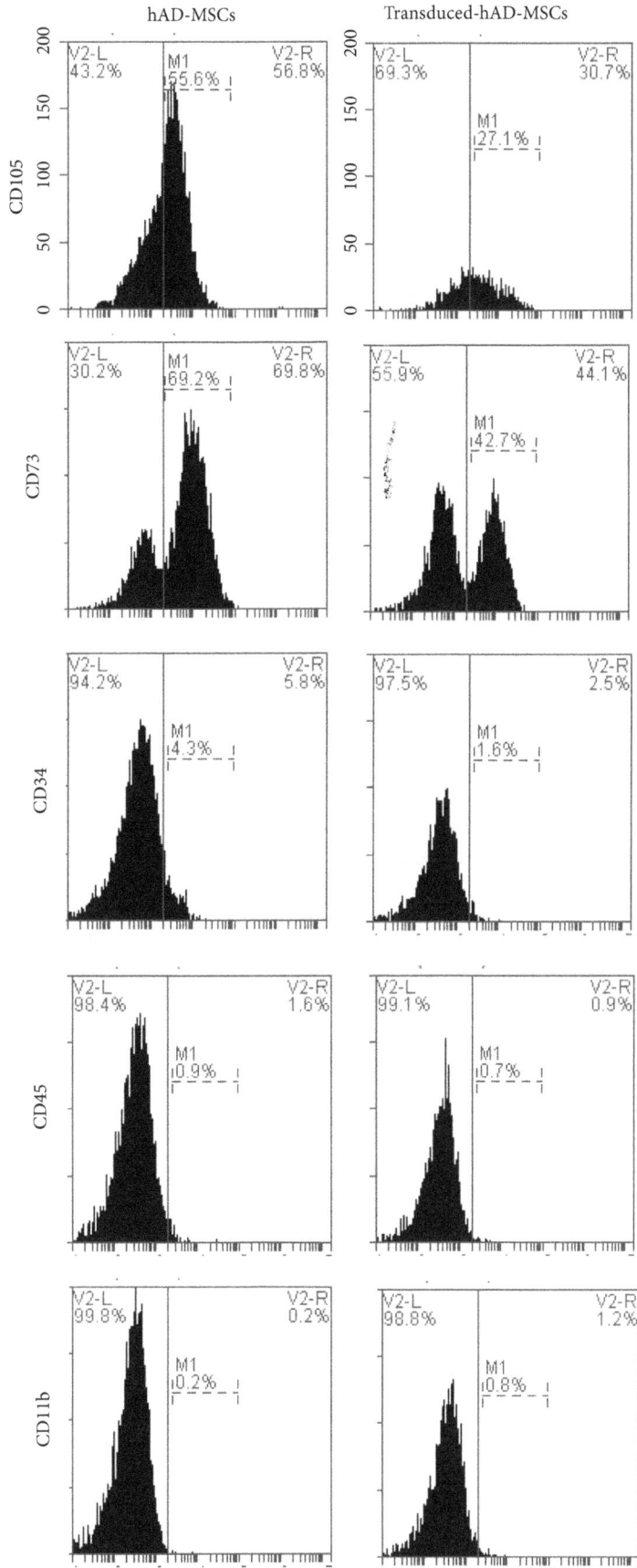

(a)

FIGURE 5: Continued.

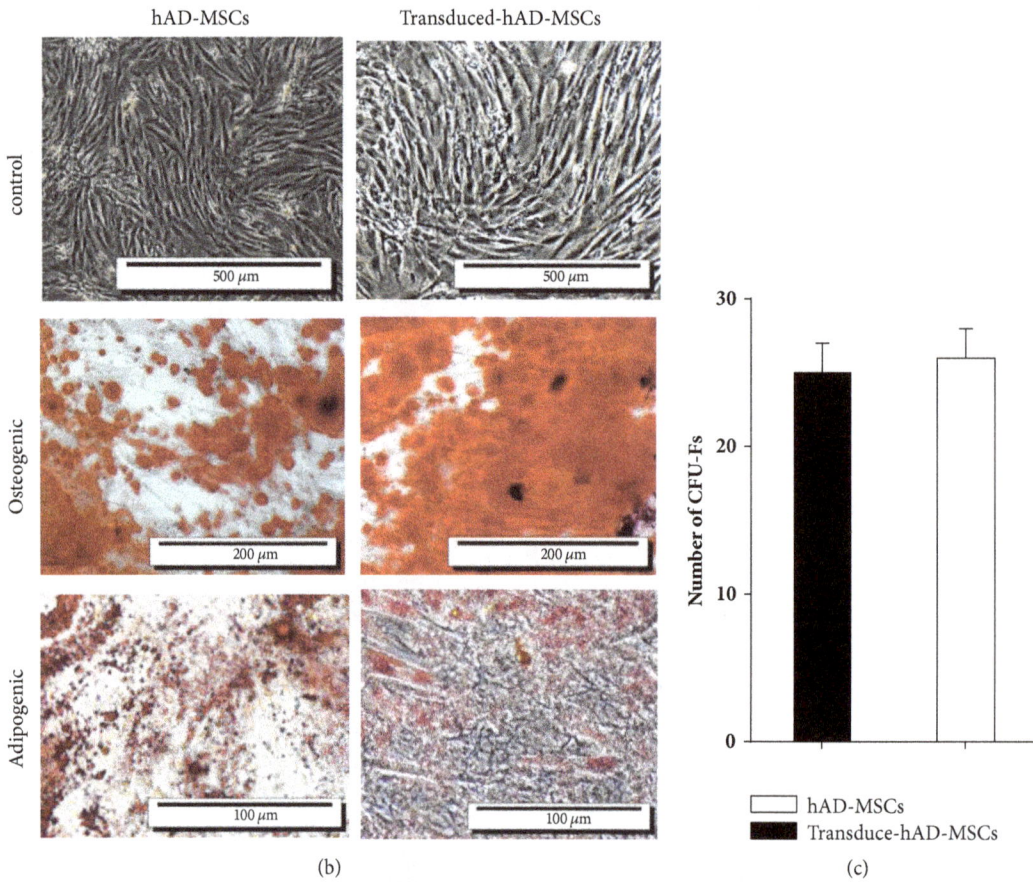

FIGURE 5: The mesenchymal stem cells (MSCs) characterization by flow cytometry, differentiation, and CFU-F assay. (a) Expression of surface antigens in hAD-MSCs, and transduced hAD-MSCs as determined using flow cytometry. (b) Control: the second passage of hAD- MSCs and transduced hAD-MSCs. Osteogenic: staining of hAD-MSCs and transduced hAD-MSCs by Alizarin Red S reagent; calcium deposits stained bright orange-red around hAD-MSCs differentiated into osteocytes on the 21th day. Adipogenic: staining of hAD-MSCs and transduced hAD-MSCs after 2nd passage by Oil Red O reagent; intracellular vesicles of oil accumulated and stained bright red within hAD-MSCs differentiated into adipocytes on the 21th day. (c) Clonogenic capacity was measured by colony forming unit-fibroblast (CFU-F) assay. The results are represented as the means ± SD.

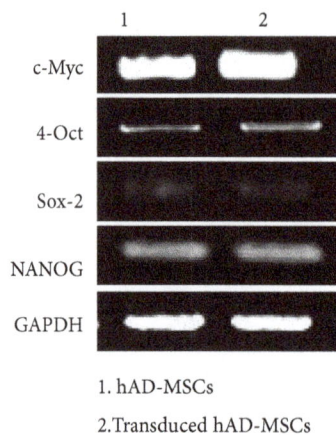

1. hAD-MSCs

2.Transduced hAD-MSCs

FIGURE 6: Stemness markers expression in hAD-MSCs and transduced hAD-MSCs.

(a)

(b)

(c)

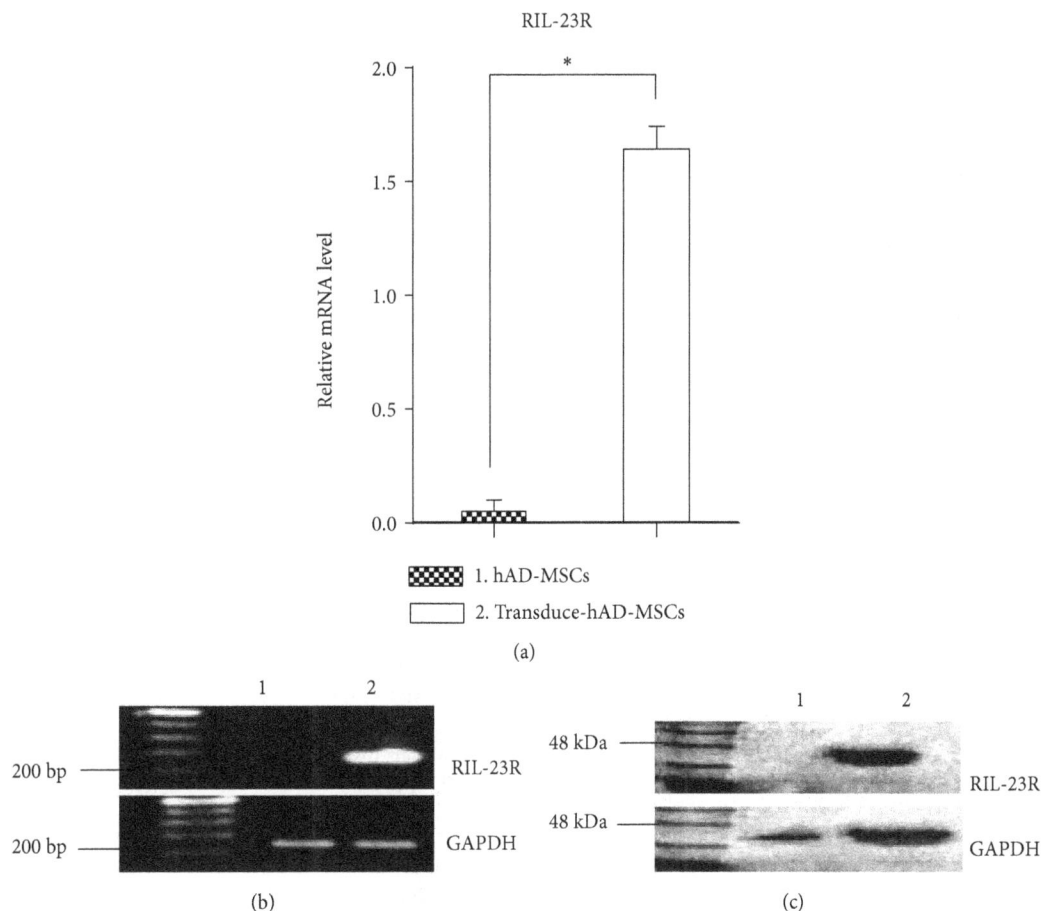

FIGURE 7: Expression of RIL- 23R in hAD-MSCs and transduced hAD-MSCs. (a-b) qPCR analysis of the RIL-23R in hAD-MSCs and transduced hAD-MSCs. the relative copy numbers of RIL-23R in transduced hAD-MSCs were compared with the copy number in hAD-MSC. The data are shown as means ± SD of 5 independent experiments. (c) Western blot analysis of the RIL-23R in whole cell lysates of hAD-MSCs (20 μg protein was loaded per lane).

In the present research, the hAD-MSCs were successfully isolated from lipoaspirate tissue samples and characterized by flow cytometry. As well, their osteogenic and adipogenic differentiation were analyzed. All transduced and non-transduced cultures were able to differentiate into adipogenic and osteogenic lineages, to express stem cell markers such as OCT- 4, SOX2, NANOG and c-Myc retain the nature of hAD-MSCs. There were no significant variations in proliferation capacity or cell surface marker expression between transduced and non-transduced cultures. There was a decrease in CD105, an MSC-specific marker. The transduced and non-transduced cells showed the same morphologies and differentiation [61].

Endoline (CD105), the TGF-b receptor III, is generally considered an important marker for MSCs [62]. We have reported that the stem cell marker CD105 was less abundant in transduced hAD-MSCs. several reports have shown that its expression varies depending upon MSC source, culture time in vitro, and differentiation state [40, 63].

Anderson et al. [40] demonstrated that both CD105- and CD105+ mASCs had a similar proliferative capacity, colony-forming unitfibroblast (CFU-F) potential, and expression of differentiation-related genes and shared all other MSC markers analyzed. CD105-mASCs had a greater capacity to differentiate into adipocytes and osteocytes also were more efficient at inhibiting T cell proliferation in vitro compared to CD105+ mASCs. When the mASC cultures reached confluency, membrane type 1 matrix metalloproteinase, which is a membrane-tethered MMP, can cleave CD105 from the cell surface [64]. Levi et al. [36] found that expression patterns for CD105 to be closely associated with the osteogenic potential of ASCs. Although they cannot necessarily predict the nature of these differences, with respect to CD105 and its role as a coreceptor for TGF-β1, they would expect that CD105 low cells would show reduced TGF-β1 signaling.

The natural form of IL-23R is encoded by at least 12 exons. Here, we designed recombinant IL-23R (RIL-23R) that can produce a secreted form of the receptor with an antagonistic function against IL-23 [29]. Some studies have shown that the IL-12Rβ2 chain or IL-23R exhibited tumor suppressor functions [65]. In addition, evidence indicates that soluble receptors play roles as agonists and antagonists during disease and normal homeostasis [66]. Therefore, we propose that generation of this soluble receptor might give a method of immune maintenance in autoimmune diseases.

The MSCs can be used as a gene delivery system because of their ability to be easily transformed and home in on injured tissues and their lack of immunogenic properties. Additionally, they have been transduced with different vectors to optimize transgene expression [9]. In some cases, such as therapy for long-life illnesses, permanent transduction of MSCs may be needed and can be attained by using retroviruses to change MSCs with high efficiency to get long-term expression [67].

We obtained more than 95% lentiviral transduction efficiency of AD-MSCs according to the selection of puromycin as a beneficial agent for ideal therapeutic purposes [68].

The RT-PCR verified expression of RIL- 23R. The present findings and other those from similar studies found no negative effects on multipotency properties after transgene expression by lentiviral transduction [30].

The IL-23 is effective in the function of Th17 cells but unable to induce the differentiation of these cells *in vitro*. The ability of the soluble isoform of IL-23R to inhibit IL-23 signaling [69] led us to hypothesize that our RIL-23R, acting as an IL-23 inhibitor, may bind the p19 chain of IL-23 and inhibit the functional maturation of Th17 cells, resulting in T cell production leading to large amount secretion of IL-10 in a shift to the Th2 cell type [70]. Then, we expected overexpression of IL-10 and suppression of IL-17. The RT-PCR is capable of exhibiting the expression level of the target genes; however, bioassays exactly confirmed the potency of gene expression. The stem cells have some features of cancer cells including long lifespan, relative apoptosis resistance, and ability to replicate for extended periods of time [71]. In addition, similar growth regulators and control mechanisms are involved in both cancer and stem cell maintenance [72]. STS cell lines harbor differentiation capacity similar to MSCs [73]. Recent evidence strongly suggests that sarcomas originate from mesenchymal stem cells [74–77], which represent a plastic-adherent spectrum of different bone marrow-derived cells until recently. Identically treated cultures of the highly metastatic fibrosarcoma cell line HT1080 were included as positive controls. We isolated mouse CD4+T-cells from splenocytes and cocultured them for three days in the presence of the transduced hAD-MSCs and HT1080 cells and hAD-MSCs and HT1080 cells as control cells. The ability of our recombinant RIL-23R was also evaluated to express IL-17 and IL-10 by RT-PCR. Our RIL-23R induced the Th2 cell proliferation and improved their function by IL-10 overexpression [74].

5. Conclusions

It was the first report of RIL-23R transduction into hAD-MSCs, which may be an effective approach to use such cells as a good vehicle for cell-based gene therapy in autoimmune disorders.

Conflicts of Interest

The authors declare that there are no conflicts of interest.

Acknowledgments

This study was supported by grants from the Research and Technology Council of Golestan University of Medical Sciences (Grant no. 940805182).

References

[1] A. A. Nery, I. C. Nascimento, T. Glaser, V. Bassaneze, J. E. Krieger, and H. Ulrich, "Human mesenchymal stem cells: From immunophenotyping by flow cytometry to clinical applications," *Cytometry Part A*, vol. 83, no. 1, pp. 48–61, 2013.

[2] M. E. Bernardo and W. E. Fibbe, "Mesenchymal stromal cells: sensors and switchers of inflammation," *Cell Stem Cell*, vol. 13, no. 4, pp. 392–402, 2013.

[3] A. Gebler, O. Zabel, and B. Seliger, "The immunomodulatory capacity of mesenchymal stem cells," *Trends in Molecular Medicine*, vol. 18, no. 2, pp. 128–134, 2012.

[4] G. Ren, L. Zhang, X. Zhao et al., "Mesenchymal stem cell-mediated immunosuppression occurs via concerted action of chemokines and nitric oxide," *Cell Stem Cell*, vol. 2, no. 2, pp. 141–150, 2008.

[5] T. J. Myers, F. Granero-Molto, L. Longobardi, T. Li, Y. Yan, and A. Spagnoli, "Mesenchymal stem cells at the intersection of cell and gene therapy," *Expert Opinion on Biological Therapy*, vol. 10, no. 12, pp. 1663–1679, 2010.

[6] Y. Y. Lipsitz, P. Bedford, A. H. Davies, N. E. Timmins, and P. W. Zandstra, "Achieving efficient manufacturing and quality assurance through synthetic cell therapy design," *Cell Stem Cell*, vol. 20, no. 1, pp. 13–17, 2017.

[7] D. M. Smith, E. J. Culme-Seymour, and C. Mason, "Evolving industry partnerships and investments in cell and gene therapies," *Cell Stem Cell*, vol. 22, no. 5, p. 779, 2018.

[8] S. Kumar, D. Chanda, and S. Ponnazhagan, "Therapeutic potential of genetically modified mesenchymal stem cells," *Gene Therapy*, vol. 15, no. 10, pp. 711–715, 2008.

[9] V. Mundra, I. C. Gerling, and R. I. Mahato, "Mesenchymal stem cell-based therapy," *Molecular Pharmaceutics*, vol. 10, no. 1, pp. 77–89, 2013.

[10] J. S. Park, S. Suryaprakash, Y. H. Lao, and K. W. Leong, "Engineering mesenchymal stem cells for regenerative medicine and drug delivery," *Methods*, vol. 84, pp. 3–16, 2015.

[11] A. El-Badawy, M. Amer, and R. Abdelbaset, "Adipose stem cells display higher regenerative capacities and more adaptable electro-kinetic properties compared to bone marrow-derived mesenchymal stromal cells," *Scientific Reports*, vol. 24, no. 6, Article ID 37801, 2016.

[12] M. J. Oedayrajsingh-Varma, S. M. van Ham, M. Knippenberg et al., "Adipose tissue-derived mesenchymal stem cell yield and growth characteristics are affected by the tissue-harvesting procedure," *Cytotherapy*, vol. 8, no. 2, pp. 166–177, 2006.

[13] L. Aust, B. Devlin, S. J. Foster et al., "Yield of human adipose-derived adult stem cells from liposuction aspirates," *Cytotherapy*, vol. 6, no. 1, pp. 7–14, 2004.

[14] S. Kanji and H. Das, "Advances of stem cell therapeutics in cutaneous wound healing and regeneration," *Mediators of Inflammation*, vol. 2017, Article ID 5217967, 14 pages, 2017.

[15] G. Muhammad, J. Xu, J. W. Bulte, A. Jablonska, P. Walczak, and M. Janowski, "Transplanted adipose-derived stem cells can be short-lived yet accelerate healing of acid-burn skin wounds: a multimodal imaging study," *Scientific Reports*, vol. 7, no. 1, 2017.

[16] C. Barranco, "Stem cells: Mesenchymal stem cells from adipose tissue could be used to deliver gene therapy to the liver," *Nature Reviews Gastroenterology & Hepatology*, vol. 8, no. 2, p. 64, 2011.

[17] S. A. Choi, Y. E. Lee, P. A. Kwak et al., "Clinically applicable human adipose tissue-derived mesenchymal stem cells delivering therapeutic genes to brainstem gliomas," *Cancer Gene Therapy*, vol. 22, no. 6, pp. 302–311, 2015.

[18] S. Bougioukli, O. Sugiyama, W. Pannell et al., "Gene therapy for bone repair using human cells: superior osteogenic potential of bone morphogenetic protein 2-transduced mesenchymal stem cells derived from adipose tissue compared to bone marrow," *Human Gene Therapy*, vol. 29, no. 4, pp. 507–519, 2018.

[19] A. Bajek, N. Gurtowska, J. Olkowska, L. Kazmierski, M. Maj, and T. Drewa, "Adipose-Derived Stem Cells as a Tool in Cell-Based Therapies," *Archivum Immunologiae et Therapia Experimentalis*, vol. 64, no. 6, pp. 443–454, 2016.

[20] S. Onuora, "Cytokines in disease: Genetic variation affects IL-6 response in synovial fibroblasts," *Nature Reviews Rheumatology*, vol. 12, no. 1, p. 2, 2016.

[21] B. D. Wines et al., "Distinctive expression of interleukin-23 receptor subunits on human Th17 and gammadelta T cells," *Immunol Cell Biol*, vol. 95, no. 3, pp. 272–279, 2017.

[22] L. Zhou, I.I. Ivanov, and R. Spolski, "IL-6 programs T(H)-17 cell differentiation by promoting sequential engagement of the IL-21 and IL-23 pathways," *Nature Immunology*, vol. 8, no. 9, pp. 967–974, 2007.

[23] M. T. Villanueva, "IL-23 assists the transition from autoimmunity to inflammatory disease," *Nature Reviews Rheumatology*, vol. 13, no. 1, p. 1, 2017.

[24] S. L. Gaffen, R. Jain, A. V. Garg, and D. J. Cua, "The IL-23-IL-17 immune axis: from mechanisms to therapeutic testing," *Nature Reviews Immunology*, vol. 14, no. 9, pp. 585–600, 2014.

[25] C. Tang, S. Chen, H. Qian, and W. Huang, "Interleukin-23: As a drug target for autoimmune inflammatory diseases," *The Journal of Immunology*, vol. 135, no. 2, pp. 112–124, 2012.

[26] D. J. Cua, J. Sherlock, Y. Chen et al., "Interleukin-23 rather than interleukin-12 is the critical cytokine for autoimmune inflammation of the brain," *Nature*, vol. 421, no. 6924, pp. 744–748, 2003.

[27] S.-H. Kan, G. Mancini, and G. Gallagher, "Identification and characterization of multiple splice forms of the human interleukin-23 receptor α chain in mitogen-activated leukocytes," *Genes & Immunity*, vol. 9, no. 7, pp. 631–639, 2008.

[28] R. Y. Yu and G. Gallagher, "A naturally occurring, soluble antagonist of human IL-23 inhibits the development and in vitro function of human Th17 cells," *The Journal of Immunology*, vol. 185, no. 12, pp. 7302–7308, 2010.

[29] X.-Y. Zhang, H.-J. Zhang, Y. Zhang et al., "Identification and expression analysis of alternatively spliced isoforms of human interleukin-23 receptor gene in normal lymphoid cells and selected tumor cells," *Immunogenetics*, vol. 57, no. 12, pp. 934–943, 2006.

[30] S. Hajizadeh-Sikaroodi et al., "Lentiviral mediating genetic engineered mesenchymal stem cells for releasing IL-27 as a gene therapy approach for autoimmune diseases," *Cell Journal*, vol. 16, no. 3, p. 255, 2014.

[31] N. Klages, R. Zufferey, and D. Trono, "A stable system for the high-titer production of multiply attenuated lentiviral vectors," *Molecular Therapy*, vol. 2, no. 2, pp. 170–176, 2000.

[32] J. G. Crowston, "T lymphocyte mediated lysis of mitomycin C treated Tenon's capsule fibroblasts," *British Journal of Ophthalmology*, vol. 88, no. 3, pp. 399–405, 2004.

[33] P. T. Khaw et al., "Five-minute treatments with fluorouracil, floxuridine, and mitomycin have long-term effects on human Tenon's capsule fibroblasts," *Archives of Ophthalmology*, vol. 110, pp. 1150–1154, 1992.

[34] W. Gombert, N. J. Borthwick, D. L. Wallace et al., "Fibroblasts prevent apoptosis of IL-2-deprived T cells without inducing proliferation: A selective effect on Bcl-x(L) expression," *The Journal of Immunology*, vol. 89, no. 3, pp. 397–404, 1996.

[35] S. Sapski, N. Beha, R. Kontermann, and D. Müller, "Tumor-targeted costimulation with antibody-fusion proteins improves bispecific antibody-mediated immune response in presence of immunosuppressive factors," *OncoImmunology*, vol. 6, no. 12, Article ID e1361594, 2017.

[36] B. Levi, D. C. Wan, J. P. Glotzbach et al., "CD105 protein depletion enhances human adipose-derived stromal cell osteogenesis through reduction of transforming growth factor β1 (TGF-β1) signaling," *The Journal of Biological Chemistry*, vol. 286, no. 45, pp. 39497–39509, 2011.

[37] T. Rada, R. L. Reis, and M. E. Gomes, "Distinct stem cells subpopulations isolated from human adipose tissue exhibit different chondrogenic and osteogenic differentiation potential," *Stem Cell Reviews and Reports*, vol. 7, no. 1, pp. 64–76, 2011.

[38] T. Rada, M. E. Gomes, and R. L. Reis, "A novel method for the isolation of subpopulations of rat adipose stem cells with different proliferation and osteogenic differentiation potentials," *Journal of Tissue Engineering and Regenerative Medicine*, vol. 5, no. 8, pp. 655–664, 2011.

[39] J. B. Mitchell, K. McIntosh, S. Zvonic et al., "Immunophenotype of human adipose-derived cells: temporal changes in stromal-associated and stem cell-associated markers," *Stem Cells*, vol. 24, no. 2, pp. 376–385, 2006.

[40] P. Anderson, A. B. Carrillo-Gálvez, A. García-Pérez, M. Cobo, and F. Martín, "CD105 (Endoglin)-negative murine mesenchymal stromal cells define a new multipotent subpopulation with distinct differentiation and immunomodulatory capacities," *PLoS ONE*, vol. 8, no. 10, Article ID e76979, 2013.

[41] T. Jiang, W. Liu, X. Lv et al., "Potent in vitro chondrogenesis of CD105 enriched human adipose-derived stem cells," *Biomaterials*, vol. 31, no. 13, pp. 3564–3571, 2010.

[42] M. Leyva-Leyva, L. Barrera, C. López-Camarillo et al., "Characterization of mesenchymal stem cell subpopulations from human amniotic membrane with dissimilar osteoblastic potential," *Stem Cells and Development*, vol. 22, no. 8, pp. 1275–1287, 2013.

[43] E. Fonsatti, A. P. Jekunen, K. J. A. Kairemo et al., "Endoglin is a suitable target for efficient imaging of solid tumors: In vivo evidence in a canine mammary carcinoma model," *Clinical Cancer Research*, vol. 6, no. 5, pp. 2037–2043, 2000.

[44] K. C. Russell, H. A. Tucker, B. A. Bunnell et al., "Cell-surface expression of neuron-glial antigen 2 (NG2) and melanoma cell adhesion molecule (CD146) in heterogeneous cultures of marrow-derived mesenchymal stem cells," *Tissue Engineering Part A*, vol. 19, no. 19-20, pp. 2253–2266, 2013.

[45] C. Li et al., "TNFα down-regulates CD105 expression in vascular endothelial cells: a comparative study with TGFß1," 2003.

[46] P. Mark, M. Kleinsorge, R. Gaebel et al., "Human mesenchymal stem cells display reduced expression of CD105 after culture in

serum-free medium," *Stem Cells International*, vol. 2013, Article ID 698076, 8 pages, 2013.

[47] H. J. Jin, S. K. Park, W. Oh, Y. S. Yang, S. W. Kim, and S. J. Choi, "Down-regulation of CD105 is associated with multi-lineage differentiation in human umbilical cord blood-derived mesenchymal stem cells," *Biochemical and Biophysical Research Communications*, vol. 381, no. 4, pp. 676–681, 2009.

[48] M. Rosu-Myles, J. Fair, N. Pearce, and J. Mehic, "Non-multipotent stroma inhibit the proliferation and differentiation of mesenchymal stromal cells in vitro," *Cytotherapy*, vol. 12, no. 6, pp. 818–830, 2010.

[49] C. B. Chang, S. A. Han, E. M. Kim, S. Lee, S. C. Seong, and M. C. Lee, "Chondrogenic potentials of human synovium-derived cells sorted by specific surface markers," *Osteoarthritis and Cartilage*, vol. 21, no. 1, pp. 190–199, 2013.

[50] C. Nazarov, J. L. Surdo, S. R. Bauer, and C.-H. Wei, "Assessment of immunosuppressive activity of human mesenchymal stem cells using murine antigen specific CD4 and CD8 T cells *in vitro*," *Stem Cell Research & Therapy*, vol. 4, no. 5, article 128, 2013.

[51] E. W. Choi, "Adult stem cell therapy for autoimmune disease," *International Journal of Stem Cells*, vol. 2, no. 2, pp. 122–128, 2009.

[52] C. Hartono, T. Muthukumar, and M. Suthanthiran, "Immuno-suppressive drug therapy," *Cold Spring Harbor Perspectives in Medicine*, vol. 3, no. 9, 2013.

[53] P. Buckel, "Recombinant proteins for therapy," *Trends in Pharmacological Sciences*, vol. 17, no. 12, pp. 450–456, 1996.

[54] S. Khan, M. W. Ullah, R. Siddique et al., "Role of Recombinant DNA Technology to Improve Life," *International Journal of Genomics*, vol. 2016, Article ID 2405954, 14 pages, 2016.

[55] C. Qian, Y. L. Xin, and J. Prieto, "Therapy of cancer by cytokines mediated by gene therapy approach," *Cell Research*, vol. 16, no. 2, pp. 182–188, 2006.

[56] S. I. Grivennikov, F. R. Greten, and M. Karin, "Immunity, inflammation, and cancer," *Cell*, vol. 140, no. 6, pp. 883–899, 2010.

[57] J. R. Podojil and S. D. Miller, "Molecular mechanisms of T-cell receptor and costimulatory molecule ligation/blockade in autoimmune disease therapy," *Immunological Reviews*, vol. 229, no. 1, pp. 337–355, 2009.

[58] M. B. Murphy, K. Moncivais, and A. I. Caplan, "Mesenchymal stem cells: environmentally responsive therapeutics for regenerative medicine," *Experimental & Molecular Medicine*, vol. 45, no. 11, article e54, 2013.

[59] J. Reiser, X.-Y. Zhang, C. S. Hemenway, D. Mondal, L. Pradhan, and V. F. La Russa, "Potential of mesenchymal stem cells in gene therapy approaches for inherited and acquired diseases," *Expert Opinion on Biological Therapy*, vol. 5, no. 12, pp. 1571–1584, 2005.

[60] C. R. Fellows, C. Matta, R. Zakany, I. M. Khan, and A. Mobasheri, "Adipose, bone marrow and synovial joint-derived mesenchymal stem cells for cartilage repair," *Frontiers in Genetics*, vol. 7, 2016.

[61] F. A. van Vollenstee, C. Jackson, D. Hoffmann, M. Potgieter, C. Durandt, and M. S. Pepper, "Human adipose derived mesenchymal stromal cells transduced with GFP lentiviral vectors: assessment of immunophenotype and differentiation capacity in vitro," *Cytotechnology*, vol. 68, no. 5, pp. 2049–2060, 2016.

[62] H. Akbulut, G. Cüce, T. M. Aktan, and S. Duman, "Expression of mesenchymal stem cell markers of human adipose tissue surrounding the vas deferens," *Journal of Biomedical Research*, vol. 23, no. 2, pp. 166–169, 2012.

[63] S. Kern, H. Eichler, J. Stoeve, H. Klüter, and K. Bieback, "Comparative analysis of mesenchymal stem cells from bone marrow, umbilical cord blood, or adipose tissue," *Stem Cells*, vol. 24, no. 5, pp. 1294–1301, 2006.

[64] L. J. A. C. Hawinkels, P. Kuiper, E. Wiercinska et al., "Matrix metalloproteinase-14 (MT1-MMP)-mediated endoglin shedding inhibits tumor angiogenesis," *Cancer Research*, vol. 70, no. 10, pp. 4141–4150, 2010.

[65] I. Airoldi, E. Di Carlo, B. Banelli et al., "The IL-12Rβ2 gene functions as a tumor suppressor in human B cell malignancies," *The Journal of Clinical Investigation*, vol. 113, no. 11, pp. 1651–1659, 2004.

[66] S. A. Jones, "Directing transition from innate to acquired immunity: defining a role for IL-6," *The Journal of Immunology*, vol. 175, no. 6, pp. 3463–3468, 2005.

[67] M. Miura, Y. Miura, H. M. Padilla-Nash et al., "Accumulated chromosomal instability in murine bone marrow mesenchymal stem cells leads to malignant transformation," *Stem Cells*, vol. 24, no. 4, pp. 1095–1103, 2006.

[68] I. Ben-Dor, P. Itsykson, D. Goldenberg, E. Galun, and B. E. Reubinoff, "Lentiviral vectors harboring a dual-gene system allow high and homogeneous transgene expression in selected polyclonal human embryonic stem cells," *Molecular Therapy*, vol. 14, no. 2, pp. 255–267, 2006.

[69] R. Y. Yu and G. Gallagher, "A naturally-occurring, soluble antagonist of human IL-23 inhibits the development and in vitro function of human TH17 cells," *Gastroenterology*, vol. 138, no. 5, 2010.

[70] N. L. Payne, G. Sun, C. McDonald et al., "Human adipose-derived mesenchymal stem cells engineered to secrete IL-10 inhibit APC function and limit CNS autoimmunity," *Brain, Behavior, and Immunity*, vol. 30, pp. 103–114, 2013.

[71] Y. Wang, Z. Han, Y. Song, and Z. C. Han, "Safety of mesenchymal stem cells for clinical application," *Stem Cells International*, vol. 2012, Article ID 652034, 4 pages, 2012.

[72] C. A. Herberts, M. S. G. Kwa, and H. P. H. Hermsen, "Risk factors in the development of stem cell therapy," *Journal of Translational Medicine*, vol. 9, article 29, 2011.

[73] S. Wirths, E. Malenke, T. Kluba et al., "Shared cell surface marker expression in mesenchymal stem cells and adult sarcomas," *Stem Cells Translational Medicine*, vol. 2, no. 1, pp. 53–60, 2013.

[74] J. Yang, "The role of mesenchymal stem/progenitor cells in sarcoma: update and dispute," *Stem Cell Investig*, vol. 1, p. 18, 2014.

[75] Y. Li, C. Zhong, D. Liu et al., "Evidence for Kaposi sarcoma originating from mesenchymal stem cell through KSHV-induced mesenchymal-to-endothelial transition," *Cancer Research*, vol. 78, no. 1, pp. 230–245, 2018.

[76] M. Gaebler, A. Silvestri, J. Haybaeck et al., "Three-dimensional patient-derived in vitro sarcoma models: promising tools for improving clinical tumor management," *Frontiers in Oncology*, vol. 7, 2017.

[77] A. B. Mohseny and P. C. W. Hogendoorn, "Concise review: Mesenchymal tumors: When stem cells go mad," *Stem Cells*, vol. 29, no. 3, pp. 397–403, 2011.

13

Carbohydrate Moieties and Cytoenzymatic Characterization of Hemocytes in Whiteleg Shrimp *Litopenaeus vannamei*

Norma Estrada,[1] **Edwin Velázquez,**[2] **Carmen Rodríguez-Jaramillo,**[3] **and Felipe Ascencio**[2]

[1]*Programa Cátedras CONACyT, Centro de Investigaciones Biológicas del Noroeste, S.C. (CIBNOR), 23090 La Paz, BCS, Mexico*
[2]*Laboratorio de Patogénesis Microbiana, Centro de Investigaciones Biológicas del Noroeste, S.C., 23090 La Paz, BCS, Mexico*
[3]*Laboratorio de Histología e Histoquímica, Centro de Investigaciones Biológicas del Noroeste, S.C., 23090 La Paz, BCS, Mexico*

Correspondence should be addressed to Norma Estrada; nestrada@cibnor.mx

Academic Editor: Salvatore Desantis

Hemocytes represent one of the most important defense mechanisms against foreign material in Crustacea and are also involved in a variety of other physiological responses. Fluorescent lectin-binding assays and cytochemical reactions were used to identify specificity and distribution of carbohydrate moieties and presence of several hydrolytic enzymes, in hemocytes of whiteleg shrimp *Litopenaeus vannamei*. Two general classes of circulating hemocytes (granular and agranular) exist in *L. vannamei*, which express carbohydrates residues for FITC-conjugated lectins WGA, LEA, and PNA; UEA and Con-A were not observed. Enzymatic studies indicated that acid phosphatase, nonspecific esterase, and specific esterases were present; alkaline phosphatase was not observed. The enzymes and carbohydrates are useful tools in hemocyte classification and cellular defense mechanism studies.

1. Introduction

In crustacean decapods, the defense system relies on humoral and cellular mechanisms, with cellular defense coordinated by circulating hemocytes [1]. Hemocytes in crustaceans also are known to be involved in rapid sealing of wounds to prevent loss of hemolymph and prevent infection [2, 3]. Many studies of morphology, structure, function, and classification of hemocytes in decapods indicate three fundamental types, according to the number and size of granules present: hyaline cells, small-granule cells, and large-granule cells [4]. These types are described for penaeid shrimp [5, 6] and freshwater crayfish [7, 8]. Hemocytes can recognize and eliminate or sequester invading pathogens through phagocytosis, encapsulation, and secretion of lysosomal enzymes and bacteriostatic substances [9, 10]. Principally, granular hemocytes carry out phagocytosis by engulfing small foreign particles. Granules are known to contain many enzymes, such as lysozymes, esterases, phosphatases, phospholipases, peroxidases, and proteases and also oxidative enzymes that help to eliminate the foreign material [11–14]. Hemocytes

play an important role in producing and discharging agglutinins, such as lectins [15] and antibacterial peptides [16]. Invertebrate hemocytes also have several carbohydrate moieties that act as receptors for invading pathogens, where binding of lectins and carbohydrates leads to structural changes of the complex that induce activation of hemocytes [17, 18].

The Pacific whiteleg shrimp, *Litopenaeus vannamei* (Boone, 1931), is one of the most important farmed species in the world. Economic losses due to disease in shrimp aquaculture have made it necessary to increase our knowledge of the invertebrate immune system. Enzyme cytochemistry has often been used in functional characterization of hemocytes [9]. Differentiating hemocytes with lectins to study specificity and distribution of carbohydrate moieties has been useful in classifying hemocytes and defining their function [18]. To study hemocytes in *L. vannamei* in a basal state for a comprehensive description for the structure and function of the hemocytes, we used microscopic analysis, enzymatic activity by cytochemistry, and glycoconjugates with lectin probes in cells.

TABLE 1: Lectins used to investigate cell surface oligosaccharides domains in hemocytes and their competitive sugars used as controls for lectin-binding.

Lectin	Latin name	Source	Resuspension solution	Specificity sugar	Inhibition*
Con-A	*Canavalia ensiformis* (Sigma, C7642)	Jack bean	NaCl 0.9% pH 6.5 with 5 mM $CaCl_2$	a-D-Mannose, a-D-glucose	Glucose
WGA	*Triticum vulgaris* (Sigma, L4895)	Wheat germ	PBS pH 7.4	(D-GlcNAc), NeuNAc	N-Acetylglucosamine
PNA	*Arachis hypogaea* (Sigma, L7381)	Peanut	PBS pH 7.4	β-Galactose(1-3) galNAc	Galactosamine
UEA-I	*Ulex europaeus* (Sigma, L9006)	Gorse, furze	NaCl 0.9%	a-L-Fucose	Fucose
LEA	*Lycopersicon esculentum* (Sigma, L0401)	Tomato	**	$(glcNAc)_3$	N-Acetylglucosamine

GalNAc = N-acetyl-galactosamine, GlcNAc = N-acetyl-glucosamine, and NeuNAc = N-acetylneuraminic acid.

*Competing sugar used in the study at 0.2 M.

**10 mM HEPES, 0.15 M NaCl, 0.1 mM Ca^{2+}, 0.08% sodium azide, and 5 mg mL^{-1} β-cyclodextrin.

2. Material and Methods

2.1. Extraction of Hemolymph. Juvenile shrimp (average length = 10.3 ± 2.0 cm). Before experiments, the shrimp were kept in laboratory under controlled conditions in a recirculating system in 1500 L fiberglass tanks with 1 μm filtered seawater at 35 PSU and 25°C at pH 8, with constant aeration, and fed commercial pellet feed daily. Hemolymph was extracted with a 27-gauge syringe at the junction between the basis and ischium of the fifth walking leg. Prior to bleeding, the sample area was wiped with 70% ethanol. The syringe contained an equal volume of citrate/EDTA anticoagulant composed of 0.45 M NaCl, 0.1 M glucose, 30 mM sodium citrate, 26 mM citric acid, and 10 mM EDTA at pH 5.4 and put on ice [19].

2.2. Type of Hemocytes and Total and Differential Count. Total hemocytes counts were measured with an electronic particle counter (Multisizer, Beckman Coulter, Brea, CA). Cellular viability of each sample was estimated in Neubauer chambers with a fluorescence microscope, after adding a solution of propidium iodide, which is not membrane permeable and usually excluded from viable cells. It is commonly used to label dead cells and as a counterstain in multicolor fluorescent techniques. Hemocytes were more than 90% viable. To determine the type of hemocytes, we counted types of hemocytes in histological sections. For this procedure, we pooled hemolymph mixed at 1 : 1 with Karnovsky's fixative [20] for 24 h at 4°C, washed in several changes of Karnovsky's buffer, dehydrated through an ethanol series, and embedded in catalyzed acrylic monomer (JB-4 plus embedding kit, Polysciences, Warrington, PA). Histological samples were sectioned to 1 μm and stained with hematoxylin-eosin and May-Grünwald Giemsa. Permanent slides were examined under a microscope, and digital photographs were taken. Digital images were used to analyze at least 500 hemocytes at 40x (10 pictures in 10 random fields per slide) (Image-Pro Plus v4, Media Cybernetics, Bethesda, MD, USA), identifying small-granule cells (SGC), large-granule cells (LGC), and hyaline

cells (HC), according to Heng and Lei [21]. The percentage of each type of cell was calculated, based on the total number of cells. Data were tested with ANOVA with the *post hoc* Tukey test to test for differences among hemocyte types and staining methods. Numerical data are represented as mean ± SD, using the SPSS 16.0 software (IBM SPSS, Armonk, NY). Results were considered significant at $P < 0.05$.

2.3. Lectin-Binding Assays. Differences in the distribution of hemocyte oligosaccharides were studied using five fluorescein isothiocyanate- (FITC-) labeled lectins. The lectins are listed in Table 1. Aliquots of 100 μL of pooled hemolymph (1×10^6 cells mL^{-1}) were placed in 0.7 μL polypropylene tubes in triplicate and 5 μL formaldehyde was added to each tube for 30 min. Solutions of the lectins were prepared in their respective buffer to concentrations of 100, 50, 25, and 10 mg lectin mL^{-1}. To each tube 100 μL of a lectin solution was added and incubated in the dark for 1.5 h at room temperature. The same procedure was performed for all lectins. After incubation with labeled lectins, the hemocytes were washed and resuspended in 100 μL filtered PBS at pH 7.2 for analysis in a Neubauer chamber with an epifluorescence microscope, using the appropriate filters (Olympus BX41, Tokyo, Japan) at 40x. For the controls, each lectin was incubated in its appropriate competing sugar at a 0.2 M final concentration (see Table 1) for 30 min at room temperature. This solution was then centrifuged for 15 min at 16,000 ×g and hemocytes were incubated in the supernatant for 1.5 h at room temperature. Then, tissues were rinsed in PBS containing the competing sugar and prepared for fluorescence microscopy, as described below. Other control slides were incubated with PBS without lectins. For each treatment, digital images were analyzed (at least 500 hemocytes, 10 pictures in 10 random small Neubauer chamber squares per sample), counting fluorescent and nonfluorescent hemocytes. Intensity and area covered by fluorescence were determined with an image analyzer (Image-Pro Plus v4, Media Cybernetics, Bethesda, MD) at 40x. The percentage of positive (fluorescing) cells per sample was calculated, as defined in reference to total hemocyte number. Multiple independent samples were analyzed to ensure reproducibility. Data were tested with ANOVA with the *post hoc* Tukey test to test differences among different lectins. Numerical data are represented as mean ± SD, using the SPSS 16.0 software. Results were considered significant at $P < 0.05$.

2.4. Enzyme Cytochemistry. Hemocytes were prepared by cell adhesion for histochemistry of enzymes. After withdrawing hemolymph, 150 μL was placed on glass slides at densities of 5×10^6 cells mL^{-1} and allowed to adhere for 1 h at 25°C under sterilized moist chamber conditions. The slides were then carefully washed with PBS at pH 7.2 and stained with reagents for a range of lysosomal enzymes. Adhered cells were fixed according to each enzyme tested. Diagnostic kits were used to test for four hydrolytic enzymes, carried out according to the manufacturer's instructions: (1) acid phosphatase (naphthol AS-BI phosphoric acid substrate, #386-A, Sigma, St. Louis, MO), (2) alkaline phosphatase (naphthol AS-BI alkaline substrate, #86-C, Sigma), (3) nonspecific esterase (α-naphthyl acetate substrate, #91-A, Sigma), and (4) specific esterase (naphthol AS-D chloroacetate, #91-C, Sigma). For acid phosphatase, duplicate films are treated with L(+)-tartrate-containing substrate; cells containing tartaric acid-sensitive acid phosphatase are devoid of activity. Slides were rinsed in deionized water, allowed to air-dry, and studied microscopically, using an oil immersion lens. For each treatment, digital images were used to analyze at least 500 hemocytes (10 pictures in 10 random fields per slide), identifying positive and nonpositive cells. For all enzymes, the percentage of positive cells was calculated, based on the total number of cells. Control slides were incubated only in buffer; all solutions were made immediately before use. Following this, observations were made with a light microscope. Multiple independent samples were analyzed to ensure reproducibility ($n = 9$ slides from three independent, pooled samples for each technique). Data were tested with ANOVA with the *post hoc* Tukey test to test differences among different enzymes. Numerical data are represented as mean ± SD, using the SPSS 16.0 software. Results were considered significant at $P < 0.05$.

3. Results

Hemocytes stained in histological sections, with hematoxylin-eosin and May-Grünwald Giemsa dyes, showed many cytoplasmic granules of different sizes, as well as cells without granules (Figures 1(a) and 1(b)). Description of hemocyte types using morphological criteria previously developed for the shrimp [21] shows three basic cell types: small-granule cells (SGC), large-granule cells (LGC), and hyaline cells (HC). Figure 1(c) shows the percentage of different hemocyte types per sample, calculated with reference to the total number of hemocytes.

Binding of five lectins with different carbohydrate specificities to hemocytes of *L. vannamei* was studied using FITC-labeled lectins (Table 1). Circulating hemocytes isolated from *L. vannamei* were able to bind to WGA, PNA, and LEA

TABLE 2: Fluorescent lectin-binding.

Lectin	Litopenaeus vannamei	
	Lectin-binding	Optimal concentration* (μg mL^{-1})
Con-A	—	100
WGA	C, S	25
PNA	S	50
UEA-I	—	100
LEA	C, S	25

* Optimal concentration for binding assays. C = clumps. S = single cells.

(Figures 2(a), 2(b), and 2(c)); carbohydrate moieties for UEA and Con-A were not observed. Table 2 shows whether fluorescence was present in clump of cells or individual cells and shows optimal concentration of lectin for labeling assays. The percentage of positive cells per sample was calculated, defined with reference to the total number of hemocytes, along with percentages of hyaline, small-granule, and large-granule hemocytes (Figure 2(d)). There were no statistical differences in the expression of carbohydrates on the hemocytes. Figures 2(a) and 2(c) and Table 2 show that WGA and LEA reacted with hemocytes of *L. vannamei* in agglutinated and individual cells, with clearly dotted staining of different sizes with WGA, and with LEA also dotted structures were observed, along with uniform label pattern in many cells which seems to label surfaces. PNA was present in high percentages in *L. vannamei* hemocytes also in dotted structures (Figure 2(b)). Known inhibiting carbohydrate residues were used to inhibit labeling of lectin-treated hemocytes. Controls with the inhibitory saccharides demonstrated that the binding of FITC-lectin was specific, since the presence in the incubation medium of the appropriate haptenic sugar (Table 1) abolished or markedly decreased fluorescence.

Hemocytes from *L. vannamei* were analyzed by cell adhesion for several hydrolytic enzymes. Of the four enzymes that we studied, only three were present: acid phosphatase, α-naphthyl acetate (nonspecific esterase), and naphthol AS-D chloroacetate (specific esterase) (Figures 3(a), 3(b), and 3(c)). Distribution and location of these enzymes were not homogeneous in the hemocytes; however, strong reactions were observed when positive. Acid phosphatase is clearly localized as patchy structures of different sizes, and nonspecific and specific esterase activities are also found in dotted structures, similar to those found with acid phosphatase activity. Figure 3(d) shows shrimp hemocytes with positive enzyme activities, along with percentages of hyaline, small-granule, and large-granule hemocytes. A high percentage of hemocytes were positive for acid phosphatase; many hemocytes were resistant to tartrate. A high percentage of *L. vannamei* hemocytes were positive for nonspecific esterase, and also a significant percentage of hemocytes were positive for specific esterases. Reaction sites demonstrating hydrolytic enzymes were less abundant in hyaline cells and more abundant in granulocytes.

(a)

(b)

(c)

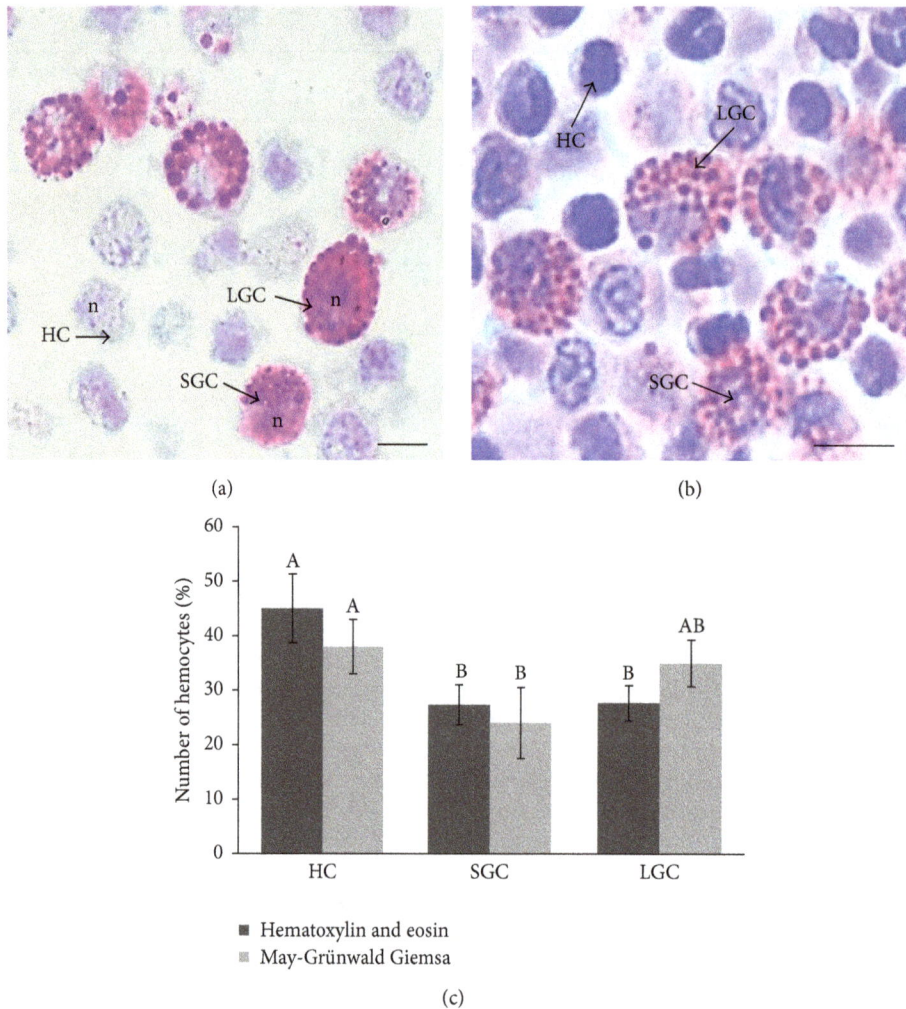

FIGURE 1: Hemocytes of *Litopenaeus vannamei*. (a) Micrographs of hemocytes stained with hematoxylin-eosin in resin histological sections and (b) hemocytes stained with May-Grünwald Giemsa, showing three types of hemocytes: small-granule cells (SGC), large-granule cells (LGC), and hyaline cells (HC). Scale bar = 5 μm. (c) Differential hemocyte count (DHC) in histological resin sections stained with hematoxylin-eosin and May-Grünwald Giemsa. The relative percentage of DHC was calculated by observing digital images analyzing at least 500 hemocytes. Data represent mean ± SD. Letters indicate significant differences ($P < 0.05$) between hemocyte types. n = nucleus.

4. Discussion

There are many studies of morphology, structure, function, and classification of hemocytes in arthropod decapods that show two fundamental types of hemocytes in hemolymph: granular and hyaline types [5, 9, 21, 22]. Hose et al. [23] and Kondo et al. [24] show that hemocytes of the Penaeidae shrimp comprise hyaline cells (HC), small-granule cells (SGC), and large-granule cells (LGC). In our results with hematoxylin-eosin, 45% were HC, 27% were SGC, and 28% were LGC, whereas with May-Grünwald Giemsa 38% were HC, 24% were SGC, and 35% were LGC. These results are different compared to studies of *L. vannamei* and other shrimp species, where the most abundant cells are the SGC, ~40–60% [5, 21, 25]. Small-granule cells are the only hemocyte type involved in all the four known biological functions (phagocytosis, encapsulation, cytotoxicity, and prophenoloxidase activity), while hyaline and large-granule cells are only

involved in some of these functions [1]. It is common to use other classification schemes, such as size and shape of the cells. These schemes have inadequacies. Separation of hemocytes in this study using resin histological sections is an adequate method because we can count more than 500 hemocytes and clearly identify them with classical hematoxylin-eosin or May-Grünwald Giemsa staining, an easy way to avoid tedious techniques due to fragility of the cells. Different counts were recorded by relative percentage of the hemocytes in different decapod species having many variations. Hose et al. [26] found that *Sicyonia ingentis* had 50–60% HC, 30% SGC, and 10% LGC. Kakoolaki et al. [27] found that *F. indicus* had 10–15% HC, 20–25% LGC, and 60–65% SGC. In addition, in our study total hemocyte count (THC) is about 6–10 × 10^6 cells mL^{-1}, similar to other works of Penaeidae shrimp [5, 27, 28]. However variation of THC also has been usually reported by researchers. Total and differential hemocyte counts can vary greatly in response

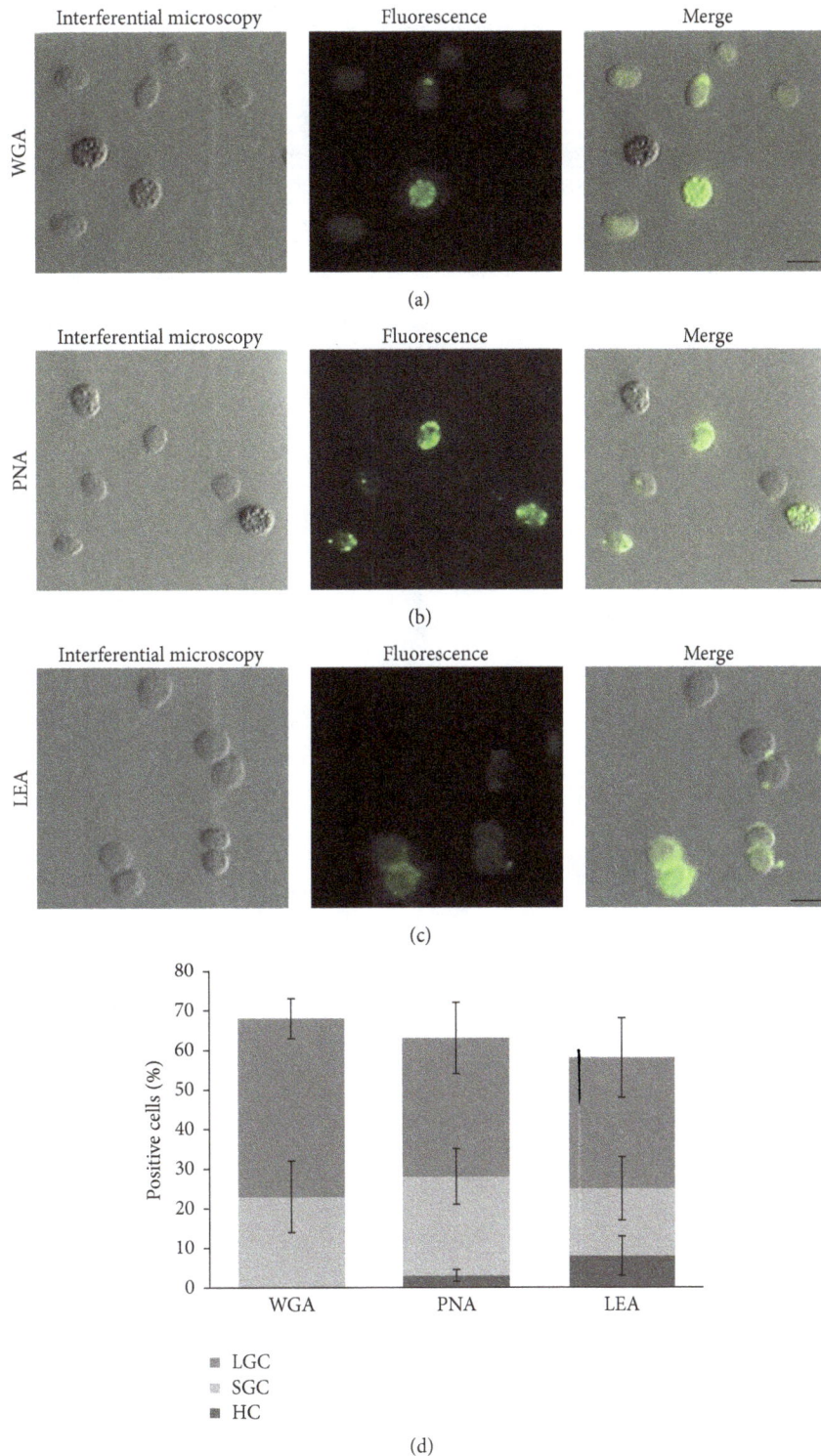

FIGURE 2: Labeling pattern of fixed hemocytes isolated from *Litopenaeus vannamei* after incubation with FITC-conjugated lectins. Many hemocytes express ligands differentially: (a) wheat germ agglutinin (WGA), (b) peanut agglutinin (PNA), and (c) *Lycopersicon esculentum* agglutinin (LEA). Hemocytes were evaluated by interference contrast microscopy and epifluorescence microscopy. Scale bar = 5 μm. (d) Percentage of total hemocytes and cell types of positive cells labeled by FITC-conjugated lectins. For each treatment, digital images were used to analyze at least 500 hemocytes. The percentage of positive (i.e., fluorescing) cells per sample was calculated, as defined with reference to the total hemocyte number. Data represent mean ± SD.

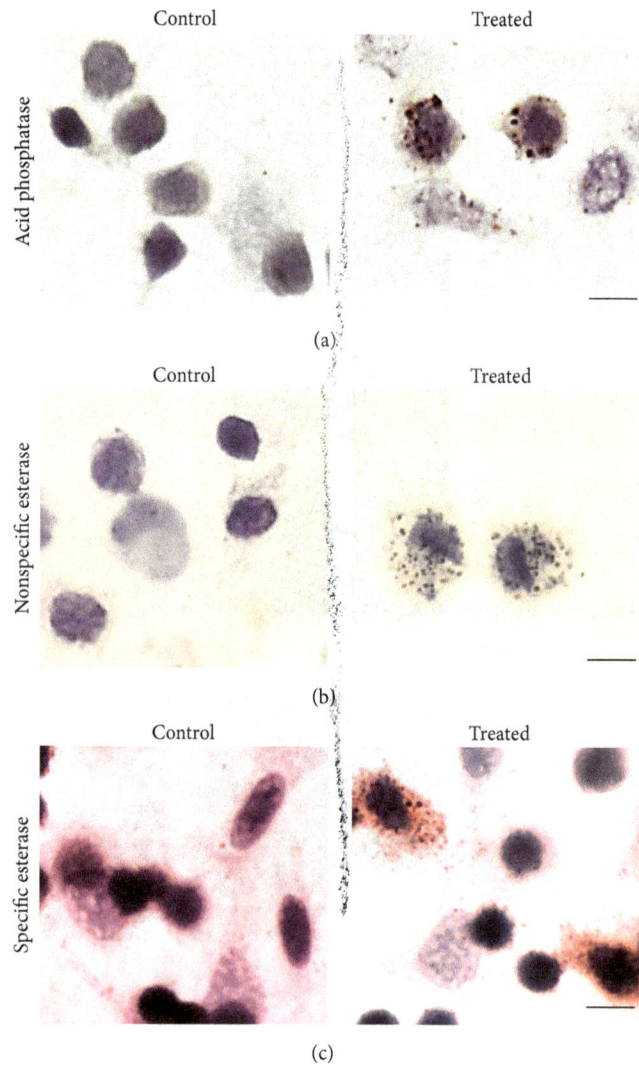

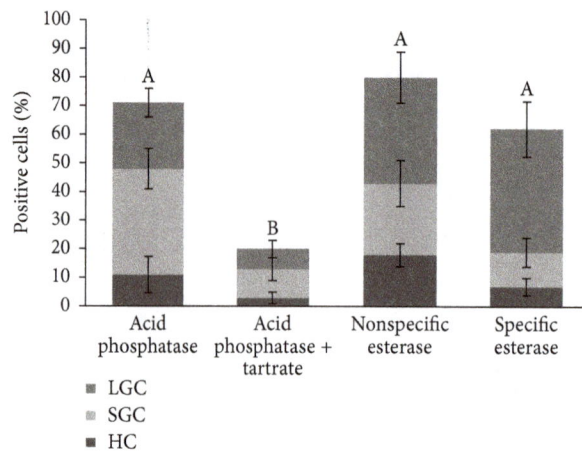

FIGURE 3: Enzyme cytochemistry of adhered hemocytes isolated from *Litopenaeus vannamei*. The first column of micrographs corresponds to the control cells incubated without the enzyme substrate and the second column shows enzymatic activities. (a) Acid phosphatase, (b) nonspecific esterase (alpha-naphthyl esterase), and (c) specific esterase (AS-D chloroacetate esterase). Scale bar = 5 μm. (d) Percentage of adhered hemocytes of *L. vannamei*, which react with the substrates for different enzymes, along with cell types of positive cells. Data represent mean ± SD. Letters indicate significant differences ($P < 0.05$) in percentage of positive cells.

Carbohydrate Moieties and Cytoenzymatic Characterization of Hemocytes in Whiteleg Shrimp...

173

to infection, environmental stress, and endocrine activity during moulting cycle [29–31].

We characterized hemocytes of *L. vannamei* with cytochemical tests. The combination of many cytochemical tests is suggested for classification of shrimp hemocytes [9]. Differentiating hemocytes using lectins to recognize glycoconjugates and the activity of diverse hydrolytic enzymes can be useful for classification of different cell types. The method provides information on the function and relationships between cell types. Hemocytes of crustaceans have binding sites for a number of lectins as constituents of glycoproteins and glycolipids in the cell [18, 32]. Our results confirm that circulating hemocytes express ligands (carbohydrates) for lectin-binding in a very heterogeneous manner. The percentage of lectin-labeled hemocytes in *L. vannamei* is similar to WGA and LEA. The WGA lectin selectively binds to N-acetyl-glucosamine (GlcNAc) and to N-acetylneuraminic acid (sialic acid) residues of glycoproteins and glycolipids. The labeling pattern with WGA appears as dotted cytoplasmic structures which correspond to granules present in cytoplasm of many hemocytes of decapods and these are granular hemocytes [18, 32] and are distinct from hyaline hemocytes with fewer or without granules. LEA showed a homogeneous distribution on the cell surface; although this tomato lectin is specific for oligomers of β-(1,4)-linked N-acetyl-D-glucosamine, with the binding site being able to accommodate up to 4 carbohydrate units, also LEA binds well to sialoglycoproteins of the membrane. Comparisons have been made between several lectins that share a similar specificity, including WGA, showing that their reactivity with glycoproteins varies. The tomato lectin does not seem to require that the GlcNAc residues be consecutive, a finding that has not been noted for WGA.

Lectin-labeled hemocytes in *L. vannamei* with PNA bound the carbohydrate sequence galactosamine-β(1-3)-N-acetyl-D-glucosamine and showed a similar dotted pattern as WGA, but PNA and WGA recognize different carbohydrate determinants; thus the similarity in fluorescence patterns indicates that carbohydrate moieties that are recognized by these two lectins are shared by the same glycoproteins or occur as different glycoproteins that are closely associated. This suggests that the granule hemocytes in *L. vannamei* possess a cocktail of glycoconjugates, whereas LEA recognizes surface membrane receptors of hemocytes. Lectins are widely used to study recognition of carbohydrates by proteins in model systems to understand the molecular basis of how proteins recognize specific terminal sugars or groups of sugars in glycoproteins and glycolipids because they are relatively easy to obtain and have a wide variety of sugar specificities. Lectins assume a particular significance because they act as pattern-recognition proteins and recognize specific carbohydrate moieties on surfaces of pathogen cells that cause agglutination and facilitate binding of foreign particles to and promote ingestion by phagocytes [15, 33, 34]. Lectin-carbohydrate binding leads to a structural change of the complex that induces hemocyte activation [17].

To identify and characterize subpopulations of hemocytes, we used enzyme cytochemistry. Specific esterase and nonspecific esterase were detected in hemocytes from *L. vannamei*, as in similar organisms [9]. *L. vannamei* hemocytes were not positive for alkaline phosphatase. The lack of alkaline phosphatase is reported in other crustaceans [9] but had been detected in mud crab, *Scylla serrata* [35]. Since we tested hemocytes in the basal state, production of this enzyme could be induced by external agents. However, hemocytes of *L. vannamei* had a high percentage of cells that were positive for acid phosphatase. The presence of lysosomes is commonly corroborated with acid phosphatase staining in hemocytes and serum of crustaceans [9]. Many cells were tartrate resistant, which indicates that there are cells that have a similar enzyme to human 5-acid phosphatase (Acp5) that is present intracellularly in many cell types, is secreted *in vitro* by macrophages, and participates in phagocytosis of bacteria [36, 37]. In *L. vannamei*, the percentage of cells that are positive for esterases and acid phosphatase was almost the same, 60–80% of the total hemocyte population. This probably reflects the number of granular cells obtained using other methods of detection, such as WGA-binding to granules in hemocytes. The dominance of granular hemocytes in many crustaceans suggests that they are capable of active defense reactions and might be phagocytic cells that contain abundant hydrolytic enzymes and other proteins [4, 23, 38, 39]. Hyalinocytes show limited phagocytic ability and lower levels of hydrolytic enzymes, probably having other functions that are different than phagocytosis, such as use of nutrients or coagulation [27, 40].

In summary, it is interesting to note that most invertebrates possess subpopulations of granular and hyaline hemocytes that can be distinguished with classical staining and microscopic methods. Although recent research has expanded the physiological roles played by decapod hemocytes, extension of this information from one species to another is difficult because there is no unified classification of hemocytes of either group, principally because different techniques are used and the objectives of investigation are different. The immune function of *L. vannamei* is partly based on a sugar code, matching glycan diversity with the presence of lectins and various glycan epitopes identified as ligands to destroy foreign organisms. Lectin-binding staining properties and the presence of lysosomal enzymes seem to be good markers to provide valuable information on hemocyte morphofunctional characteristics. Understanding the immune systems of crustacean decapods is necessary to assess the relative contribution of environmental, anthropogenic, and pathological stresses. Morphological and functional characterization of hemocytes in *L. vannamei* provide some insights into the response of the immune system, yet more work is necessary to identify markers that will facilitate our understanding of hemocyte characteristics.

Competing Interests

The authors declare that there is no conflict of interests regarding the publication of this paper.

Acknowledgments

The authors thank Eulalia Meza Chávez of the Laboratorio de Histología e Histoquímica of CIBNOR for technical assistance. Ira Fogel of CIBNOR provided valuable editorial services. Financial support was provided by CIBNOR (Grant AC 3.0).

References

[1] K. Söderhäll and L. Cerenius, "Crustacean immunity," *Annual Review of Fish Diseases*, vol. 2, pp. 3–23, 1992.

[2] J. Levin, "The role of amebocytes in the blood coagulation mechanism of the horseshoe crab *Limulus polyphemus*," in *Blood Cells of Marine Invertebrates: Experimental Systems in Cell Biology and Comparative Physiology*, W. D. Cohen, Ed., pp. 145–163, Alan R. Liss, New York, NY, USA, 1985.

[3] G. G. Martin and J. E. Hose, "Vascular elements and blood (hemolymph)," in *Microscopic Anatomy of Invertebrates*, F. W. Harrison and A. G. Humes, Eds., pp. 117–146, Wiley-Liss, New York, NY, USA, 1992.

[4] A. G. Bauchau, "Crustaceans," in *Invertebrate Blood Cells*, N. A. Ratcliffe and A. F. Rowley, Eds., vol. 2, pp. 385–420, Academic Press, New York, NY, USA, 1981.

[5] G. G. Martin and B. L. Graves, "Fine structure and classification of shrimp hemocytes," *Journal of Morphology*, vol. 185, no. 3, pp. 339–348, 1985.

[6] J. Rodriguez, V. Boulo, E. Mialhe, and E. Bachère, "Characterization of shrimp haemocytes and plasma components by monoclonal antibodies," *Journal of Cell Science*, vol. 108, no. 3, pp. 1043–1050, 1995.

[7] H. Lanz, V. Tsutsumi, and H. Aréchiga, "Morphological and biochemical characterization of *Procambarus clarki* blood cells," *Developmental and Comparative Immunology*, vol. 17, no. 5, pp. 389–397, 1993.

[8] S. Söderhäll, M. W. Johansson, and V. J. Smith, "Internal defence mechanisms," in *Freshwater Crayfish: Biology, Management and Exploitation*, D. M. Holdich and R. S. Lowery, Eds., pp. 213–235, Croom Helm, London, UK, 1988.

[9] J. E. Hose, G. G. Martin, V. A. Nguyen, J. Lucas, and T. Rosenstein, "Cytochemical features of shrimp hemocytes," *The Biological Bulletin*, vol. 173, no. 1, pp. 178–187, 1987.

[10] K. Söderhäll and V. J. Smith, "The prophenoloxidase activating cascade as a recognition and defense system in arthropods," in *Humoral and Cellular Immunity in Arthropods*, A. P. Gupta, Ed., pp. 251–285, Wiley-Liss, New York, NY, USA, 1986.

[11] J. E. Hose, D. V. Lightner, R. M. Redman, and D. A. Danald, "Observations on the pathogenesis of the imperfect fungus, *Fusarium solani*, in the California brown shrimp, *Penaeus californiensis*," *Journal of Invertebrate Pathology*, vol. 44, no. 3, pp. 292–303, 1984.

[12] D. A. Millar and N. A. Ratcliffe, "Invertebrates," in *Immunology: A Comparative Approach*, R. J. Turner, Ed., pp. 29–68, John Wiley & Sons, Oxford, UK, 1994.

[13] M. Muñoz, R. Cedeño, J. Rodríguez, W. P. W. Van der Knaap, E. Mialhe, and E. Bachère, "Measurement of reactive oxygen intermediate production in haemocytes of the penaeid shrimp, *Penaeus vannamei*," *Aquaculture*, vol. 191, no. 1–3, pp. 89–107, 2000.

[14] Y.-L. Song and Y.-T. Hsieh, "Immunostimulation of tiger shrimp (*Penaeus monodon*) hemocytes for generation of microbicidal substances: analysis of reactive oxygen species," *Developmental and Comparative Immunology*, vol. 18, no. 3, pp. 201–209, 1994.

[15] M. R. F. Marques and M. A. Barracco, "Lectins, as non-self-recognition factors, in crustaceans," *Aquaculture*, vol. 191, no. 1–3, pp. 23–44, 2000.

[16] D. Destoumieux, M. Muñoz, C. Cosseau et al., "Penaeidins, antimicrobial peptides with chitin-binding activity, are produced and stored in shrimp granulocytes and released after microbial challenge," *Journal of Cell Science*, vol. 113, no. 3, pp. 461–469, 2000.

[17] C. J. Bayne, "Phagocytosis and non-self recognition in invertebrates," *BioScience*, vol. 40, no. 10, pp. 723–731, 1990.

[18] G. G. Martin, C. Castro, N. Moy, and N. Rubin, "N-acetyl-D-glucosamine in crustacean hemocytes; possible functions and usefulness in hemocyte classification," *Invertebrate Biology*, vol. 122, no. 3, pp. 265–270, 2003.

[19] K. Söderhäll and V. J. Smith, "Separation of the haemocyte populations of *Carcinus Maenas* and other marine decapods, and prophenoloxidase distribution," *Developmental and Comparative Immunology*, vol. 7, no. 2, pp. 229–239, 1983.

[20] M. J. Karnovsky, "A formaldehyde-glutaraldehyde fixative of high osmolarity for use in electron microscopy," *The Journal of Cell Biology*, vol. 27, p. 137A, 1964.

[21] L. Heng and W. Lei, "On the ultrastructure and classification of the hemocytes of penaeid shrimp, *Penaeus vannamei* (Crustacea, Decapoda)," *Chinese Journal of Oceanology and Limnology*, vol. 16, no. 4, pp. 333–338, 1998.

[22] M. W. Johansson, P. Keyser, K. Sritunyalucksana, and K. Söderhäll, "Crustacean haemocytes and haematopoiesis," *Aquaculture*, vol. 191, no. 1–3, pp. 45–52, 2000.

[23] J. E. Hose, G. G. Martin, and A. S. Gerard, "A decapod hemocyte classification scheme integrating morphology, cytochemistry, and function," *The Biological Bulletin*, vol. 178, no. 1, pp. 33–45, 1990.

[24] M. Kondo, T. Itami, Y. Takahashi, R. Fujii, and S. Tomonaga, "Ultrastructural and cytochemical characteristics of phagocytes in kuruma prawn," *Fish Pathology*, vol. 33, no. 4, pp. 421–427, 1998.

[25] Y. Ye and K. Cheng, "Studies on the circulating hemocytes of *Penaeus chinensis*," *Journal of Ocean University of Qingdao*, vol. 23, pp. 35–42, 1993.

[26] J. E. Hose, G. G. Martin, S. Tiu, and N. McKrell, "Patterns of hemocyte production and release throughout the molt cycle in the penaeid shrimp *Sicyonia ingentis*," *The Biological Bulletin*, vol. 183, no. 2, pp. 185–199, 1992.

[27] S. Kakoolaki, I. Sharifpour, M. Soltani, H. A. Ebrahimzadeh Mousavi, S. Mirzargar, and M. Rostami, "Selected morphochemical features of hemocytes in farmed shrimp, *Fenneropenaeus indicus* in Iran," *Iranian Journal of Fisheries Sciences*, vol. 9, no. 2, pp. 219–232, 2010.

[28] R. Chotikachinda, W. Lapjatupon, S. Chaisilapasung, D. Sangsue, and C. Tantikitti, "Effect of inactive yeast cell wall on growth performance, survival rate and immune parameters in pacific white shrimp (*Litopenaeus vannamei*)," *Songklanakarin Journal of Science and Technology*, vol. 30, no. 6, pp. 687–692, 2008.

[29] V. J. Smith and N. A. Ratcliffe, "Host defence reactions of the shore crab, *Carcinus maenas* (L.); clearance and distribution of injected particles," *Journal of Marine Biology Association of the United Kingdom*, vol. 60, no. 1, pp. 89–102, 1980.

[30] V. J. Smith and P. A. Johnston, "Differential haemotoxic effect of PCB congeners in the common shrimp, *Crangon crangon*,"

Comparative Biochemistry and Physiology Part C: Comparative Pharmacology, vol. 101, no. 3, pp. 641–649, 1992.

[31] T. Sequeira, D. Tavares, and M. Arala-Chaves, "Evidence for circulating hemocyte proliferation in the shrimp *Penaeus japonicus*," *Developmental and Comparative Immunology*, vol. 20, no. 2, pp. 97–104, 1996.

[32] D. R. Coombe, P. L. Ey, and C. R. Jenkin, "Self/non-self recognition in invertebrates," *The Quarterly Review of Biology*, vol. 59, no. 3, pp. 231–255, 1984.

[33] N. Estrada, E. Velázquez, C. Rodríguez-Jaramillo, and F. Ascencio, "Morphofunctional study of hemocytes from lions-paw scallop *Nodipecten subnodosus*," *Immunobiology*, vol. 218, no. 8, pp. 1093–1103, 2013.

[34] L. Renwrantz, "Lectins in molluscs and arthropods: their occurrence, origin and roles in immunity," in *vol 56*, vol. 56, pp. 81–93, 1986.

[35] S. Saha, M. Ray, and S. Ray, "Activity of phosphatases in the hemocytes of estuarine edible mudcrab, *Scylla serrata* exposed to arsenic," *Journal of Environmental Biology*, vol. 30, no. 5, pp. 655–658, 2009.

[36] S. R. Räisänen, J. Halleen, V. Parikka, and H. K. Väänänen, "Tartrate-resistant acid phosphatase facilitates hydroxyl radical formation and colocalizes with phagocytosed *Staphylococcus aureus* in alveolar macrophages," *Biochemical and Biophysical Research Communications*, vol. 288, no. 1, pp. 142–150, 2001.

[37] A. J. Janckila, R. N. Parthasarathy, L. K. Parthasarathy et al., "Properties and expression of human tartrate-resistant acid phosphatase isoform 5a by monocyte-derived cells," *Journal of Leukocyte Biology*, vol. 77, no. 2, pp. 209–218, 2005.

[38] J. E. Hose and G. G. Martin, "Defense functions of granulocytes in the ridgeback prawn *Sicyonia ingentis*," *Journal of Invertebrate Pathology*, vol. 53, no. 3, pp. 335–346, 1989.

[39] M. A. Barracco and G. A. Amirante, "Morphological and cytochemical studies of the hemocytes of *Squilla mantis* (Stomatopoda)," *Journal of Crustacean Biology*, vol. 12, no. 3, pp. 372–382, 1992.

[40] S. A. Omori, G. G. Martin, and J. E. Hose, "Morphology of hemocyte lysis and clotting in the ridgeback prawn, *Sicyonia ingentis*," *Cell and Tissue Research*, vol. 255, no. 1, pp. 117–123, 1989.

14

Grammatophyllum speciosum Ethanolic Extract Promotes Wound Healing in Human Primary Fibroblast Cells

Saraporn Harikarnpakdee ⓘ[1,2] **and Verisa Chowjarean** ⓘ[1,3]

[1]*Cosmeceutical Research, Development and Testing Center, College of Pharmacy, Rangsit University, Pathum Thani 12000, Thailand*
[2]*Department of Industrial Pharmacy, College of Pharmacy, Rangsit University, Pathum Thani 12000, Thailand*
[3]*Department of Pharmaceutical Technology, College of Pharmacy, Rangsit University, Pathum Thani 12000, Thailand*

Correspondence should be addressed to Verisa Chowjarean; verisa.b@rsu.ac.th

Academic Editor: Paul J. Higgins

Grammatophyllum speciosum is a plant in Orchidaceae family which contains a variety of phytochemical compounds that might be beneficial for medicinal use. This study aimed to evaluate the activity of pseudobulb of *G. speciosum* extract (GSE) in wound healing processes in human primary fibroblast cells along with *in vitro* antioxidant activity and total phenolic content of GSE. Scratch wound healing assay indicated that GSE was capable of increasing migration rate after 6 and 9 hours of treatment. Besides, the extract was able to scavenge DPPH, ABTS, and superoxide anion radicals indicating the antioxidative property of GSE. This study suggested a novel role of the of pseudobulb extract of *G. speciosum* as a wound healing enhancer. The results from this study might be beneficial for the development of further novel active compounds for skin wound healing.

1. Introduction

Wound healing process is divided into four phases classified as vasoconstriction and coagulation, acute inflammation, cellular proliferation, and wound remodeling [1]. Briefly, coagulation causes the development of platelet thrombosis and fibrin clot leading to the recruitment of neutrophils and macrophages which leads to inflammatory response. Growth factors and proinflammatory cytokines are then released to activate cells which involve antimicrobial process such as keratinocyte, endothelial, and fibroblast. During inflammatory response, reactive oxygen species (ROS) are produced to defend cells against bacteria and microorganism invasion. Histamine and other factors leading to an increase in vasodilatation are also released. Cellular proliferation phase is started by the accumulation of extracellular matrix (ECM), including collagen and fibronectin which are produced from fibroblasts. The proliferation of endothelial cells stimulated by vascular-promoting growth factors leads to an increasing amount of blood vessels. Finally, reepithelializing mechanism and tissue remodeling are then started. At this process,

fibroblasts are activated by keratinocytes to synthesize growth factors to regulate their proliferation. A noncellular scar, and cross-linked collagen matrix will substitute fibroblast-rich granulation tissue and provide scar tensile strength or intact skin. During the tissue remodeling phase, a decrease in cellularity from the apoptosis of myofibroblasts, endothelial cells, and inflammatory cells will lead to the process of impaired wound healing. This impairment may be caused by uncontrolled inflammatory, infection, and overproduction of ROS. Excessive ROS production could damage ECM protein and cellular function of fibroblasts and keratinocytes, resulting in slow wound healing process [1–4]. The plant extracts with potential for skin protection from its antioxidant property might be useful for the acceleration of tissue wound healing [5].

Grammatophyllum speciosum Blume is a plant in Orchidaceae family mostly found in Southeast Asia. The pseudobulb extract of *G. speciosum* was used for relieving pains from scorpion venom (*Heterometrus laoticus*). In addition, *G. speciosum* ethanolic extract (GSE) was reported to have potential for an increase in stem cell phenotypes of human keratinocytes [6]. Moreover, the extract also had an ability to protect the cells against superoxide anion-induced cell death [6]. *G. speciosum* contains various phytochemical compounds such as glucosyloxybenzyl derivatives, grammatophyllosides, cronupapine, vandateroside II, gastodin, vanilloloside, orcinol glucoside, and isovitexin [7]. Due to their benefits, these phytochemicals in this plant may be used as medicine; however, the effect of *G. speciosum* on tissue wound healing has never been investigated.

This study examined the potential wound healing effects of GSE in human primary skin fibroblast cells. Also, the in vitro antioxidant activity of GSE was evaluated for its ability to scavenge DPPH, ABTS, and superoxide anion radicals. The results from this study might be beneficial for the development of further novel active compounds for tissue wound healing treatment.

2. Materials and Methods

2.1. Plant Material Collection and Extraction. Fresh pseudobulbs of *G. speciosum* Blume were collected from the area of Khao Hin Sorn Royal Development Study Center, Chachoengsao Province, Thailand. Dry pseudobulbs of *G. speciosum* were ground up and macerated for 3 days in ethanol (1:9 w/v) at 25°C for a total of 3 times. The extract was then filtered and evaporated under vacuum pressure at a temperature below 40°C.

2.2. Quality Control of G. speciosum Bulb Extract. For reproducibility data of the extract, determination of gastrodin content which is a major compound of *G. speciosum* extract from three different batches was performed using a rapid high-performance liquid chromatography method under the conditions using gradient elution with a mobile phase of acetonitrile–water, as previous described by our group [8].

2.3. Total Phenolic Content Determination. A total phenolic content was determined using Folin-Ciocalteu assay [9]. In a 96-well plate, the reagent was mixed with the samples to which Na_2CO_3 was consequently added. The mixture was incubated for 1 hour and the absorbance was measured using the UV-spectrophotometer at 765 nm. The total phenolic content was compared with the epigallocatechin gallate (EGCG) and expressed as g of EGCG equivalents (gEGCG) per 100 g of GSE extract. The experiment was repeated in triplicate.

2.4. Assessment of Antioxidant Activity

2.4.1. DPPH Assay. The 2,2-diphenyl-2-picrylhydrazyl (DPPH) radical scavenging activity of GSE was determined by using DPPH radical. The samples were added to 0.15 mM DPPH solution in ethanol in a 96-well plate. The mixture was incubated for 30 minutes in the dark at room temperature. The absorbance was measured using the UV-spectrophotometer at 517 nm. The results were expressed as % inhibition of DPPH. DPPH solution with the vehicle was used as a negative control whereas the vehicle without DPPH solution was used as a blank for background subtraction.

2.4.2. ABTS Radical Scavenging Activity. Twenty micrometers of the samples were added to a 96-well plate before adding ABTS$^{•+}$ solution 180 μl. The solutions were mixed using a shaker and incubated in the dark at room temperature for 5 minutes. The absorbance of the ABTS•$^+$ was measured using the UV-spectrophotometer at 750 nm. The results were expressed as %inhibition of ABTS•$^+$ by GSE compared to the ABTS•$^+$ solution with the vehicle used as the negative control. A 50% reduction of the ABTS•$^+$ was calculated and presented in an IC50 value.

2.4.3. Superoxide Anion Radical Scavenging (SOSA) Determination. Twenty micrometers of the samples was added to a 96-well plate before adding 20 μl phosphate buffer, 80 μl NADH, 80 μl NBT, and 20 μl PMS solutions. The solutions were incubated in the dark at room temperature for 15 minutes. The absorbance of the mixture was measured using the UV-spectrophotometer at 560 nm. The results were expressed as % inhibition of SOSA by GSE compared to the ABTS•$^+$ solution with the vehicle used as the negative control. A 50% reduction of the superoxide anion radical was calculated and presented in an IC50 value.

2.5. Cell Viability Assay. Human primary skin fibroblast cells (ATCC® CRL 2097, USA) were cultivated in a 96-well plate for approximately 5×10^3 cells/well density using 10% FBS-supplemented DMEM medium and were incubated in 37°C with 5% humidity and CO_2 condition overnight. After the incubation, DMEM was replaced by 10% FBS DMEM containing 5–100 μg/mL of GSE, except in control. The cells were incubated in 37°C, with humidified 5% and CO_2 atmosphere for 24 h, respectively. Afterwards, 100 μL of 3-(4,5-dimethylthiazol-2-yl)-2,5-diphenyltetrazolium bromide (MTT) solution (5 mg/mL in PBS) was used to replace culture medium in each well, and the plate was then incubated at 37°C, dark, in 5% humidity and CO_2 condition for 3 h. After removing MTT solution, 100 μL of DMSO was instead added to each well and a microplate reader was then used to read the absorbance at a wavelength of 540 nm. Cell viability was calculated into percentage and compared to the control sample.

2.6. Scratch Wound Healing Assay. The stimulatory effect of GSE on migration of human primary fibroblast cells was determined by the scratch wound healing assay. The cells were seeded at a density of 3×10^4 cells/well in a DMEM culture medium supplemented with 10% FBS in 96-well plates and incubated for overnight. After the incubation, DMEM was completely removed and the adherent cell layer was scratched with a sterile yellow pipette tip. Cellular debris was removed by PBS rinsing. The complete medium with GSE or without GSE was then added, and the cells were incubated for 9 h. At 0, 6, and 9 h, the image of scratch area was recorded under bright field microscopy (10×). The wound area was measured using the Olympus DP controller software. The results were expressed as relative cell migrations by dividing the percentage change in the space of the GSE-treated cells at 6 and 9 h compared to 0 h in each experiment.

FIGURE 1: The ethanolic extract from *G. speciosum* pseudobulbs.

TABLE 1: Gastrodin contents of GSE at different batches.

Batch	No. 1	No. 2	No. 3
Gastrodin (mg/g)	63.62±0.2	54.96±0.2	56.65±0.3

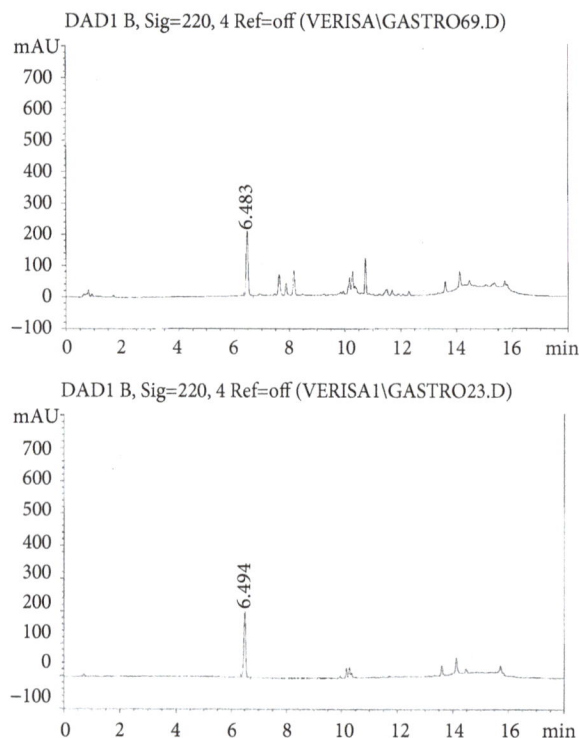

FIGURE 2: Typical HPLC chromatogram of the *G. speciosum* extract (top) and gastrodin standard (bottom).

2.7. Statistical Analysis. Data from independent experiments were presented as mean ± SD. The statistical differences among the multiple groups were analyzed using the analysis of variance ANOVA), and the individual comparisons were performed by Scheffe's post hoc test. Statistical significance was accepted at $p < 0.05$.

3. Results

3.1. Percent Yield and Appearance of G. speciosum Extract. The crude extract of *G. speciosum* Blume pseudobulbs was 194.6 g accounted as 6.49% w/w yield compared to the initial dry weight. Crude herb of *G. speciosum* were ground and extracted according to the extraction process to obtain herb-to-extract ratio (HER) of 1:9. The *G. speciosum* extract obtained was highly viscous with a dark-brown color. The appearance of *G. speciosum* extract is shown in Figure 1.

3.2. Determination of Gastrodin Content by HPLC. The determination of gastrodin content which is a major compound of *G. speciosum* extract was performed using a rapid high-performance liquid chromatography method under the conditions using gradient elution with a mobile phase of acetonitrile–water. The retention time of *G. speciosum* extract samples and standard solutions were found to elute at 6.483 and 6.494 min, respectively (Figure 2).

The amount of gastrodin in 3 different batches of GSE was analyzed. The results showed that the amount of gastrodin from batch 1, batch 2, and batch 3 contained 63.62±0.2, 54.96±0.2, and 56.65±0.3 mg/g, respectively (Table 1), which was considered to be not significantly different from one another ($p > 0.05$). The result thus suggested that gastrodin levels were not significantly different among batches, indicating uniformity among batches collected and extracted from different harvest periods.

3.3. Total Phenolic Contents and Antioxidant Activity of the GSE. Phenolic compounds may directly play a role in the antioxidant effect. The GSE was determined for the total phenolic content using Folin-Ciocalteu assay compared to the standard compound EGCG. The total phenolic content of GSE was 48.2±0.4 mg EGCG equivalent/g.

The present study demonstrated the antioxidant capacities of GSE against the DPPH radical, in which DPPH is a free radical compound normally used for screening the radical scavenging effect. Table 2 shows the percentage of the scavenging activity of GSE ranging from 24.8 to 55.7% with IC50 at 0.1 mg/mL. The percentage of the scavenging activity of ascorbic acid, the standard compound, was 21.7 to 76.3% with IC50 at 2.3 mg/mL. The IC50 of GSE was lower than that of ascorbic acid indicating that GSE has high antioxidant capacity than ascorbic acid.

The calculation of the total antioxidant activity was performed based on the decolorization of $ABTS\bullet^{+}$. Trolox was used as a standard compound. The results were expressed as %inhibition of $ABTS\bullet^{+}$ radical. GSE (Figure 3(a)) or trolox (Figure 3(b)) suppressed the absorbance of the $ABTS\bullet^{+}$ radical in a dose dependent manner. The IC50 of %inhibition of $ABTS\bullet^{+}$ of GSE and trolox were 0.12 ± 0.01 and 0.06 ± 0.01 mg/mL, respectively. Moreover, a further investigation was made to discover the antioxidant capacities of GSE against superoxide anion by determining SOSA. The IC50 of SOSA inhibition of GSE was 1.37 mg/mL (Figure 3(c)). However, the IC50 of GSE was higher than ascorbic acid

TABLE 2: Antioxidant activity of GSE and ascorbic acid determined by the DPPH assay.

| GSE | | Ascorbic acid | |
mg/mL	%Inhibition of DPPH radical	mg/mL	%Inhibition of DPPH radical
0.030	24.8 ± 0.5	1.0	21.70 ± 2.3
0.045	32.2 ± 2.2	1.5	32.50 ± 5.6
0.060	38.8 ± 2.1	2.0	45.52 ± 5.6
0.075	45.2 ± 1.2	2.5	56.52 ± 5.1
0.090	50.6 ± 2.0	3.0	67.59 ± 5.3
0.105	55.7 ± 1.1	3.5	76.28 ± 3.9

Values are presented in mean ± SD (n = 3).

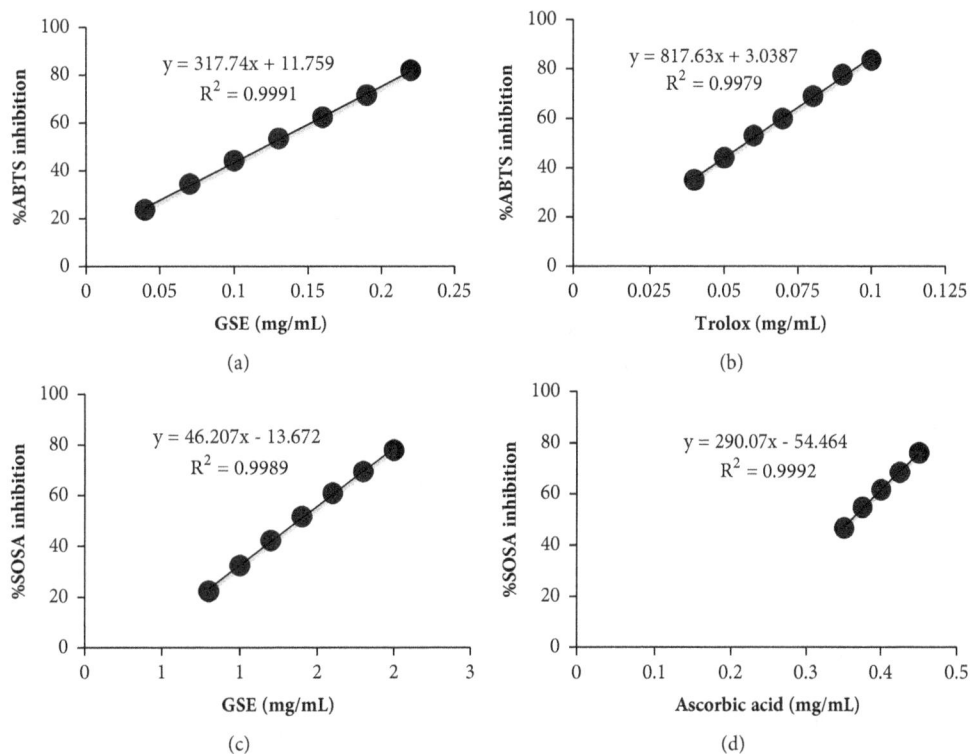

(a)

(b)

(c)

(d)

FIGURE 3: Total antioxidant activity. Total antioxidant activity of GSE and trolox. Effects of (a) GSE and (b) reference compound trolox on the decolorization of ABTS radical cation. Effects of (c) GSE and (d) ascorbic acid on SOSA inhibition were investigated. The percentage inhibition was plotted against the concentrations of the samples. Data are expressed as mean ± SD (n = 3).

(IC50 of ascorbic acid was 0.36 mg/mL) as presented in Figure 3(d).

3.4. Effects of GSE on Human Primary Fibroblasts Cells Viability. An effect of GSE on cell viability was evaluated in human primary fibroblasts cells in different concentrations after 24 h, using MTT assay. The concentration of GSE 5–100 μg/mL was shown to be noncytotoxic in primary fibroblast cells, and the cell viability was higher than 80% (Figure 4). Nontoxic concentrations were selected to study scratch wound healing assay.

3.5. GSE in the Enhancement of Wound Healing in Human Primary Fibroblasts Cells. Human primary fibroblasts cell line was tested through the wound healing assay to determine the capacity of these cells to migrate after GSE treatment. Cell migration was determined at 6 and 9 h after exposure to GSE. GSE at a concentration of 5–25 μg/mL increased

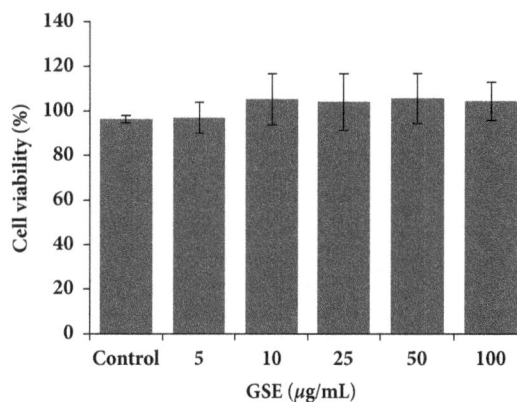

FIGURE 4: Effects of GSE on human primary fibroblast viability. Human primary fibroblast cells were treated with GSE (0–100 μg/mL) for 24 h and tested for their viability using a MTT assay. The cell viability (%) was calculated to those of the untreated controls. Data are presented in mean ± SD (n = 3). $*$ $p < 0.05$ versus untreated control.

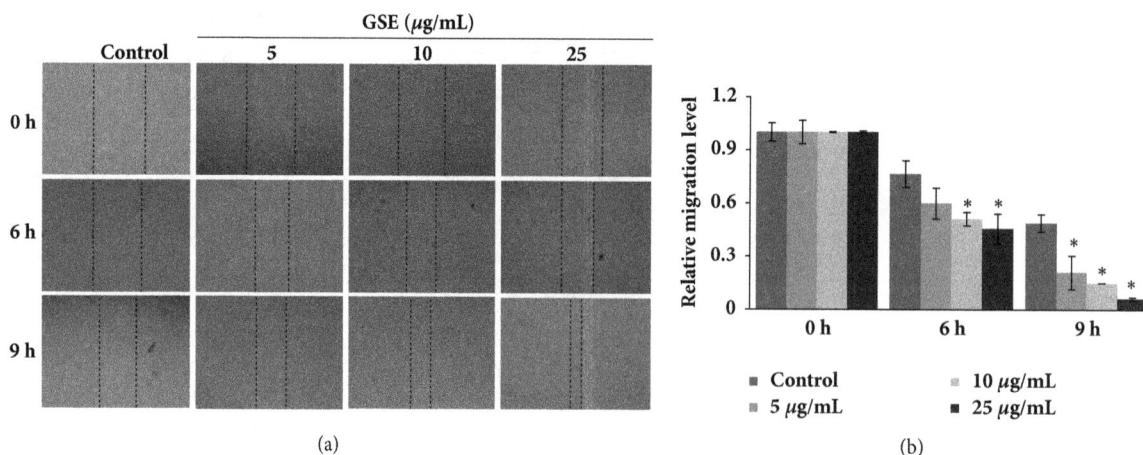

GSE (μg/mL)

(a)

(b)

FIGURE 5: Effects of GSE on human primary fibroblast cell migration. Human primary fibroblast cells were treated with GSE (0–25 μg/mL) for 9 h and subsequently tested for their migratory abilities by scratch wound healing assay. (a) Phase-contrast images (10×) captured at 0, 6, and 9 h and (b) the relative cell migration compared to that of the untreated controls. Data are presented in mean ± SD (n = 3). ∗ $p < 0.05$ versus untreated control.

the cell migration rates at 6 and 9 h when compared to 0 h (Figure 5(a)). The relative migration levels of human primary fibroblasts cells were presented in Figure 5(b), supporting that GSE significantly enhanced the migration rates after 6 and 9 h of GSE treatment.

4. Discussion

Natural plant extracts can be beneficial for wound healing if their phytochemical has the antioxidant activities and free radical scavengers, as reported by many previous studies [5]. *Caesalpinia mimosoides* extract contains phenolic contents and has a remarkable antioxidant activity leading to the enhancement of the wound healing property [10]. In line with that, this phenomenon was studied by treating GSE on human primary fibroblast cells to analyze its potential on wound healing.

Cell migration is an important process in the tissue formation phase of wound healing [1–3]. The wound healing assay is the method imitating a wound in vitro and investigating a cell migration rate. After a cell monolayer is destroyed by scratching, a loss of cell-cell interaction occurs, resulting in the initiation of cell migration and proliferation. In this study, GSE is able to increase a cell migration rate, indicating the wound healing enhancement.

The formation of reactive oxygen species (ROS), including hydrogen peroxide (H_2O_2), superoxide anion ($\cdot O2-$), hydroxyl radical ($\cdot OH$), lipid peroxides (LOOH), and their radicals (LOO·) occurred and accumulated in skin cells that expose to UVA and UVB (290–320 nm and 320–400 nm respectively). These ROS could break down cell membranes and induce skin inflammation, aging, and phototoxicity. Excessive amounts of these ROS might impair the skin wound healing process. In this sense, the plant extracts with antioxidant activity might be beneficial for the enhancement of the wound healing process [11, 12]. This study found that GSE has antioxidant activity, as determined by its ability to scavenge DPPH, ABTS and superoxide anion radicals.

Gastrodin, a phenolic glycoside [p-hydroxymethylphenyl-β-D-glucopyranoside], is found as a major active component in *G. Speciosum*. This study selected gastrodin as an active marker to control the quality of GSE due to its presence in significant amount, thereby facilitating detection and analysis. It was also used as an analytical marker for quantitative evaluation of *Rhizoma Gastrodiae* (Tianma) extracts using HPLC [13, 14]. Gastrodin is a potent antioxidant, which showed protective effects against osteoporosis linking to a reduction in ROS [15]. The antioxidant effect of GSE might be exerted from the ability of gastrodin in scavenging of ROS.

Bhat et al. and Pitz et al. found the enhancement of wound healing effects of *Caesalpinia mimosoides* extract and Jaboticaba fruit peel extract, respectively, probably caused by their phenolic contents [10, 16]. Likewise, the wound healing property of GSE might be from its phenolic compound.

5. Conclusion

GSE has a potential to enhance the wound healing process of human primary fibroblast cells. Besides, in relation to the increase in wound healing property, the extract is able to scavenge DPPH, ABTS, and superoxide anion radicals. Hence, this study provides interesting effects of GSE for the first time on tissue wound healing enhancement that warrants further investigations and may provide useful information for further therapeutic applications.

Conflicts of Interest

The authors declare that there are no competing or conflicts of interest regarding this paper publication.

Acknowledgments

This research was supported by the Cosmeceutical Research, Development and Testing Centre, Faculty of Pharmacy

(Grant no. CDRT-1600-01-2017), and the Research Institute of Rangsit University (Grant no. 24/2560), Rangsit university.

References

[1] B. M. Borena, A. Martens, S. Y. Broeckx et al., "Regenerative skin wound healing in mammals: state-of-the-art on growth factor and stem cell based treatments," *Cellular Physiology and Biochemistry*, vol. 36, no. 1, pp. 1–23, 2015.

[2] C. Dunnill, T. Patton, J. Brennan et al., "Reactive oxygen species (ROS) and wound healing: the functional role of ROS and emerging ROS-modulating technologies for augmentation of the healing process," *International Wound Journal*, vol. 12, no. 6, pp. 1–8, 2015.

[3] N. Lebonvallet, B. Laverdet, L. Misery, A. Desmoulière, and D. Girard, "New insights into the roles of myofibroblasts and innervation during skin healing and innovative therapies to improve scar innervation," *Experimental Dermatology*, vol. 27, no. 9, pp. 950–958, 2018.

[4] M. Takeo, W. Lee, and M. Ito, "Wound healing and skin regeneration," *Cold Spring Harbor Perspectives in Medicine*, vol. 5, no. 1, 2015.

[5] J. Artem Ataide, L. Caramori Cefali, F. Machado Croisfelt, A. Arruda Martins Shimojo, L. Oliveira-Nascimento, and P. Gava Mazzola, "Natural actives for wound healing: A review," *Phytotherapy Research*, vol. 32, no. 9, pp. 1664–1674, 2018.

[6] V. Chowjarean, U. Nimmannit, C. Chaotham et al., "Grammatophyllum speciosum extract potentiates stemness in keratinocyte cells," *Chiang Mai Journal of Science*, vol. 45, no. 1, pp. 237–248, 2018.

[7] P. Sahakitpichan, C. Mahidol, W. Disadee, N. Chimnoi, S. Ruchirawat, and T. Kanchanapoom, "Glucopyranosyloxybenzyl derivatives of (R)-2-benzylmalic acid and (R)-eucomic acid, and an aromatic glucoside from the pseudobulbs of Grammatophyllum speciosum," *Tetrahedron*, vol. 69, no. 3, pp. 1031–1037, 2013.

[8] V. Chowjarean, A. Sucontphunt, S. Vchirawongkwin, T. Charoonratana, T. Songsak, S. Harikarnpakdee et al., "Validated rp-hplc method for quantification of gastrodin in ethanolic extract from the pseudobulbs of Grammatophyllum speciosum blume," *Malaysian Journal of Analytical Science*, vol. 22, no. 2, pp. 219–226, 2018.

[9] N. Lourith and M. Kanlayavattanakul, "Antioxidant activities and phenolics of Passiflora edulis seed recovered from juice production residue," *Journal of Oleo Science*, vol. 62, no. 4, pp. 235–240, 2013.

[10] P. B. Bhat, S. Hegde, V. Upadhya, G. R. Hegde, P. V. Habbu, and G. S. Mulgund, "Evaluation of wound healing property of Caesalpinia mimosoides Lam," *Journal of Ethnopharmacology*, vol. 193, pp. 712–724, 2016.

[11] P. H. Hart, M. A. Grimbaldeston, and J. J. Finlay-Jones, "Sunlight, immunosuppression and skin cancer: Role of histamine and mast cells," *Clinical and Experimental Pharmacology and Physiology*, vol. 28, no. 1-2, pp. 1–8, 2001.

[12] U. Wölfle, G. Seelinger, G. Bauer, M. C. Meinke, J. Lademann, and C. M. Schempp, "Reactive molecule species and antioxidative mechanisms in normal skin and skin aging," *Skin Pharmacology and Physiology*, vol. 27, no. 6, pp. 316–332, 2014.

[13] W.-C. Chen, Y.-S. Lai, K.-H. Lu et al., "Method development and validation for the high-performance liquid chromatography assay of gastrodin in water extracts from different sources of Gastrodia elata Blume," *Journal of Food and Drug Analysis*, vol. 23, no. 4, pp. 803–810, 2015.

[14] J.-G. Lee, S.-O. Moon, S.-Y. Kim et al., "Rapid HPLC determination of gastrodin in Gastrodiae Rhizoma," *Journal of the Korean Society for Applied Biological Chemistry*, vol. 58, no. 3, pp. 409–413, 2015.

[15] Q. Huang, J. Shi, B. Gao et al., "Gastrodin: An ancient Chinese herbal medicine as a source for anti-osteoporosis agents via reducing reactive oxygen species," *Bone*, vol. 73, pp. 132–144, 2015.

[16] H. D. S. Pitz, A. Pereira, M. B. Blasius et al., "In Vitro Evaluation of the Antioxidant Activity and Wound Healing Properties of Jaboticaba (Plinia peruviana) Fruit Peel Hydroalcoholic Extract," *Oxidative Medicine and Cellular Longevity*, vol. 2016, 2016.

15

Biologics for Targeting Inflammatory Cytokines, Clinical uses, and Limitations

Peleg Rider,[1] Yaron Carmi,[1] and Idan Cohen[2]

[1]*The Department of Pathology, Sackler Faculty of Medicine, Tel-Aviv University, 6997801 Tel-Aviv, Israel*
[2]*Galilee Medical Center, 22100 Nahariya, Israel*

Correspondence should be addressed to Idan Cohen; idan5161@gmail.com

Academic Editor: Paul J. Higgins

Proinflammatory cytokines are potent mediators of numerous biological processes and are tightly regulated in the body. Chronic uncontrolled levels of such cytokines can initiate and derive many pathologies, including incidences of autoimmunity and cancer. Therefore, therapies that regulate the activity of inflammatory cytokines, either by supplementation of anti-inflammatory recombinant cytokines or by neutralizing them by using blocking antibodies, have been extensively used over the past decades. Over the past few years, new innovative biological agents for blocking and regulating cytokine activities have emerged. Here, we review some of the most recent approaches of cytokine targeting, focusing on anti-TNF antibodies or recombinant TNF decoy receptor, recombinant IL-1 receptor antagonist (IL-1Ra) and anti-IL-1 antibodies, anti-IL-6 receptor antibodies, and TH17 targeting antibodies. We discuss their effects as biologic drugs, as evaluated in numerous clinical trials, and highlight their therapeutic potential as well as emphasize their inherent limitations and clinical risks. We suggest that while systemic blocking of proinflammatory cytokines using biological agents can ameliorate disease pathogenesis and progression, it may also abrogate the hosts defense against infections. Moreover, we outline the rational need to develop new therapies, which block inflammatory cytokines only at sites of inflammation, while enabling their function systemically.

1. Introduction

The use of recombinant proteins as biological drugs has been known for the past three decades; however, this field is continuously emerging and in the last decade an increasing number of new biologic entities (biologics) in the area of cytokines were developed. Biologics can be an antibody which neutralizes an inflammatory cytokine or blocks its receptors, decoy receptors targeting the cytokine, or a recombinant protein, which can either be receptor agonist or, alternatively, an antagonist that occupies and prevents receptor binding.

The benefits of cytokines as therapeutic targets are as follows: (i) unlike in chemical drugs, specific protein which mediate the inflammatory process can be inhibited; (ii) cytokines are well studied in animal models using neutralizing antibodies or genetic models like knockout mice; thus the process in which these cytokines are involved can be thoroughly researched; (iii) with the advancement of biotechnology techniques, the expression and isolation of highly purified recombinant proteins becomes a relatively easier and cheaper process than in the past years.

The drawbacks of cytokine therapy come due to the basic properties of cytokines: (i) cytokines are pleiotropic, meaning that they affect several processes in parallel; (ii) cytokines are also known to have redundancy, meaning that the effects achieved by blocking one specific cytokine activity can be compensated by others (although this can be also beneficial, since a biological agent can be replaced to different cytokine blocker when incomplete remission or in case of intolerance); (iii) the cytokine network is a regulated and balanced system and its alteration may lead to impaired immune response. For example, inhibiting proinflammatory cytokines can result in compromised host defense against infections. On the other hand, inhibition of regulatory cytokines can result in autoimmunity or tissue damage; (iv) the production and manufacturing of biologics is still an expensive process, since their production requires sterile conditions (i.e., GMP conditions) and multiple stages of purification; (v) compared to chemical drugs, recombinant cytokines and antibodies have limited shelf half-life, require special/controlled storage conditions, and are typically administrated by a physician.

In this review we discuss some of the key approaches of anticytokine blockers focusing on approved anti-inflammatory biologics. In particular, we highlight their beneficial effects and present their possible side effects and risk factors. Most importantly, we suggest several potential solutions for the anticytokine adverse effects and propose new approaches to this emerging field.

2. Therapeutic Use of Proinflammatory Cytokines

Cytokine therapy emerged from the need to increase immunity against tumors using the lymphocyte activator and proliferative factor, interleukin-2 (IL-2). Based on its remarked efficacy in mice, cancer patients bearing renal cell carcinoma (RCC) and melanoma were administered high doses of IL-2 in order to increase antitumor immunity [1, 2]. Unfortunately, systemic administration of IL-2 has been related to severe toxicity, mainly capillary leak syndrome, associated with edema and hypotension, damage to the kidneys, heart, and brain (as well as tachycardia, atrial fibrillation, fever and chills, muscle and joint pain, and catheter related urinary tract infections) [2, 3]. In spite of numerous restrictions and warnings, a recombinant modified version of IL-2 (aldesleukin) was approved in 1992 for metastatic RCC and in 1998 for metastatic melanoma patients [4].

As early as the 1990s, several years following the discovery of IL-1 by Auron et al. [5], IL-1 was used to treat cancer patients undergoing chemotherapy or patients suffering from anemia. It was assumed that since IL-1 has neutrophilic effects, it could restore neutrophil counts back to normal numbers in neutropenic patients [6–8]. However, IL-1 is a potent proinflammatory cytokine; thus the treatment resulted in toxicity with side effects such as fever, rigors, fatigue, joint aches, headache, and nausea.

Another example is IFNα, a cytokine involved in the response to viral infections. IFNα in a PEGylated form is given in order to increase antiviral immunity by elevated CD8+ cell response in cases of chronic hepatitis-B virus (HBV) and hepatitis-C virus (HCV) [9, 10] or in the case of immediate treatment for acute HCV. The IFNα can be given alone [11, 12] or together with the nucleoside analog, ribavirin [13]. This treatment facilitates the clearance of the HCV virus and can prevent the chronic disease which can result in cirrhosis and hepatocellular carcinoma [14]. However, this type I IFN cytokine can cause serious adverse effects that can result in limitation of the doses given or even in discontinuation of the treatment. Among these adverse effects are decreased granulocytes and thrombocytes production in the bonemarrow, flu-like symptoms, neuropsychiatric disorders, and autoimmunity syndromes, mainly thyroiditis [15].

3. Anti-Inflammatory Cytokine Biologics

3.1. Anti-TNF-α Biologics. TNF-α is a proinflammatory cytokine; it appears early during the response to trauma or

bacterial infections and was first cloned in 1985 by several groups [16–19]. Initially it was described as a soluble factor with two important abilities, inducing hemorrhagic necrosis of tumors in vivo, combined with the ability to kill tumor cells in vitro [20]. TNF-α is a central alarm cytokine, which is mainly secreted from activated macrophages or dendritic cells in response to ligation of pattern-recognition receptors. Both TNF and IL-1 are attractive therapeutic targets, since they are the upstream factors of the inflammatory cascade. The role of TNF receptor signaling has been correlated with several diseases including rheumatoid arthritis (RA), Crohn's disease, atherosclerosis, psoriasis, sepsis, diabetes, and obesity [21]. TNF-α is expressed as a precursor, anchored to the cell membrane and further cleaved to its soluble form. TNF-α binds the inflammatory TNFR1 and regulatory TNFR2 and in addition to the inflammatory cascade affects cell death, proliferation, and differentiation [21].

The TNF-α inhibitor etanercept was the first biologic on the market for the treatment of RA. Etanercept is a FC fused recombinant form of a natural TNF inhibitor that was first described in 1988 [22] and later was found to be a soluble TNF receptor [23, 24]. Infliximab is a monoclonal chimeric human-mouse anti-TNF antibody and was approved by the FDA together with etanercept in 1998. Later on, by 2002, a fully human monoclonal antibody against TNF-α (adalimumab) was approved as well. Etanercept and anti-TNF antibodies carry differences in their abilities to bind TNF. While infliximab binds both monomeric and trimeric forms of TNF (the inactive and active forms), etanercept binds mainly the active trimeric form in a less stable manner [25], as well as binding TNF-β [26]. The anti-TNF antibodies are capable of lysing cells they bind by recruiting the complement system [27]. These differences appear in the molecules effectiveness against different inflammatory diseases and might be related to the antibodies binding to membranal TNF-α on T cells [28]. Infliximab was first approved for the treatment of severe Crohn's disease and later also for RA, where etanercept was first approved only for the treatment of RA. Infliximab was further approved for ulcerative colitis, psoriatic arthritis, ankylosing spondylitis, and chronic plaque psoriasis, and etanercept was further approved for psoriatic arthritis, ankylosing spondylitis, chronic plaque psoriasis, and juvenile idiopathic arthritis in children. Adalimumab is approved for rheumatoid arthritis, psoriatic arthritis, ankylosing spondylitis, Crohn's disease, ulcerative colitis, moderate-to-severe chronic psoriasis, moderate-to-severe hidradenitis suppurativa, juvenile idiopathic arthritis, and noninfectious uveitis. These days, certolizumab and golimumab are the newer, less studied anti-TNF-α antibodies most recently approved by the FDA for the treatment of RA, psoriatic arthritis, ankylosing spondylitis, Crohn's disease unresponsive to regular medications (certolizumab), and ulcerative colitis (golimumab).

Although the TNF inhibitors were shown effective for the treatment of skin and joint inflammation [29], they carry the risk of several adverse effects, mainly concerning infections. TNF-α is a fundamental factor for fighting intracellular bacteria and is therefore not surprising that TNF-α inhibition was shown to increase the risk for reactivation of tuberculosis [30]. In a 3-year French study of 69 newly diagnosed tuberculosis patients undergoing anti-TNF therapy, it was concluded that anti-TNF antibodies (infliximab and adalimumab) have a high risk for tuberculosis. Etanercept, the soluble TNF receptor, also carries such a risk but at a lower level [31]. It can be assumed that the differences in the anti-TNF strategies, which allow antibodies to be more effective against IBD, affect also the ability to inhibit the immune system to fight tuberculosis. Similarly, reactivation of HBV is higher during TNF inhibition [32]. Additionally, RA patients who were treated with anti-TNF antibodies experienced a higher rate of outbreaks of herpes zoster virus (HZV) compared to etanercept or disease-modifying antirheumatic drug (DMARD) treatments [33]. Blocking TNF plays an opposing role regarding the development of malignancies. On the one hand, TNF is an inflammatory mediator and the inflammatory process itself can lead to cancerous diseases [34]; hence, inhibiting TNF, like other proinflammatory molecules, can be beneficial in the aspect of cancer initiation and progression. On the other hand, TNF plays a role in cell proliferation, differentiation, and apoptosis [35], and, therefore, its inhibition can be a result and indeed was correlated with hematological malignances, like increased hepatosplenic T cell lymphoma in young IBD patients treated with infliximab [36]. In addition, since TNF inhibitors are immunosuppressive drugs, they carry the risk for development of malignancies. Indeed, TNF inhibitors carry warnings for increased risk of hematological malignancies in children, adolescents, and young adults, primarily treated for ulcerative colitis or Crohn's disease also treated with immunosuppressant (azathioprine and/or mercaptopurine). The fact that TNF inhibitors are often combined with methotrexate which also increases the risk for malignancy [37] and, in addition, the association of diseases treated with TNF inhibitors, for example, IBD or RA, with increased risk for cancer [38–41] is making the direct link between TNF inhibitors and malignancies harder to determine. TNF inhibition, using infliximab or etanercept, was trialed for the treatment of congestive heart disease and not only were they shown to be inefficient but also increased the chance of hospitalization or death due to heart failure [42, 43]. Patients treated with anti-TNF therapy were also reported for increased risk for demyelinating disorders, like multiple sclerosis, optic neuritis, and acute transverse myelitis [44, 45]; paradoxical psoriasis consisting of severe skin lesions was observed in IBD patients treated with anti-TNF agents [46]. In addition, unlike anti-TNF antibodies, etanercept, which is not effective for the treatment of IBD, was correlated with the development of newly diagnosed ulcerative colitis and Crohn's disease in treated patients [47, 48].

3.2. Anti-IL-1 Therapy. Following the failure to use IL-1 as a therapeutic agent in order to treat neutropenic patients and the increasing data demonstrating the potency of this cytokine to induce inflammation, it was comprehended that IL-1 inhibition rather than IL-1 administration could be beneficial. Following inflammatory stimuli, like bacterial products, the proinflammatory cytokines, IL-1α and IL-1β, are elevated. However, an additional inhibitory protein that reduces these IL-1 molecules is secreted [49, 50]. The anti-inflammatory mediator was isolated in 1990, and the sequence of the IL-1 receptor antagonist (IL-1Ra) was published [51]. It is a cytokine, which belongs to the IL-1 family with about 40% similarity to IL-1β that binds the same IL-1 receptor type 1 (IL-1R1) albeit occupying it without inducing the signal transduction. The significance of the IL-1Ra as a natural anti-inflammatory cytokine is demonstrated by the genetic loss of function of the IL1RN gene. This results in a lethal systemic inflammatory disease with severe skin and bone involvement, termed deficiency of interleukin-1 receptor antagonist (DIRA) [52].

Anakinra is a recombinant nonglycosylated form of IL-1Ra that was approved in 2001 for the treatment of RA in adult patients that did not respond to other antirheumatoid drugs, like DMARD. Anakinra was shown beneficial for the treatment of RA by reducing symptoms and joint damage; however it is recommended to use when other biologics, like anti-IL-6 or anti-TNF therapies which are preferable, are refractory or contraindicated [53–55].

Anakinra competes with IL-1β for the receptor binding. The inflammasome-caspase-1 pathway mediates IL-1β activation and secretion. Mutations in the inflammasome related genes can result in autoinflammatory syndromes due to excess IL-1 [56]. Anakinra is therefore approved for the treatment of patients suffering from a form of Cryopyrin-Associated Periodic Syndromes (CAPS) called Neonatal-Onset Multisystem Inflammatory Disease (NOMID). CAPS is a common name for three autoinflammatory syndromes (familial cold autoinflammatory syndrome, Muckle-Wells syndrome, and NOMID), in which dysregulated inflammasome results in IL-1β activation and secretion and a broad inflammation occurs. Since IL-1 is the major mediator of these autoinflammatory diseases, it is obvious why anakinra, which blocks IL-1 activity, is preferable for therapy [57–60]. Anakinra is also given to other inflammatory or autoinflammatory diseases off-label. Familial Mediterranean Fever (FMF) is a hereditary chronic inflammatory disease which IL-1 plays a major role in, and blocking IL-1 reduces the symptoms [61, 62]. Anakinra was also shown to be effective in the case of nonhereditary chronic systemic inflammatory diseases like the adult-onset Still disease [63, 64], which involves arthritis, fever, and systemic inflammation or the childhood version—systemic-onset juvenile idiopathic arthritis (SJIA) [65–67]. In addition, there are more common inflammatory diseases like gout [68], hemodialysis patients [69], post-myocardial infarction cardiac remodeling [70], and type 2 diabetes, in which the glycaemia and beta-cell secretory function are improved [71], in addition to vast types of other inflammatory disorders responding to anakinra (reviewed n [72–75]).

Anakinra has a short half-life of about 6 h; treatment therefore requires frequent subcutaneous injections and the most common side effect of anakinra is injection site reaction. The short half-life of anakinra allows immediate withdrawal of the treatment if needed. During the administration of anakinra, the immune systems ability to fight infections is reduced. Meta-analysis of four RA trials using anakinra showed increased risk of infections, mainly pneumonia but also osteomyelitis, cellulitis, bursitis, herpes zoster, infected bunion, and gangrene [76]. Gouty arthritis patients treated with anakinra were also in increased risk for infections, mostly by *S. aureus* [68]. Since IL-1 is a neutrophil attractant and growth factor, the risk for neutropenia in patients treated with anakinra increased as well [77, 78], and during administration of anakinra neutrophil numbers must be followed. Anakinra is forbidden to patients receiving TNF blockers or patients getting live vaccines. The combination of anakinra together with corticosteroids or other immunosuppressive drugs increases the risk of infections. Combining anakinra with prednisolone was shown to risk RA patient with serious infections of *S. aureus*, hemolytic streptococci, and *E. coli* [79]. Patients with a history of tuberculosis are not recommended for anakinra treatment or for those participating in clinical trials, since the chance for reactivation of tuberculosis during administration of anakinra is high [76].

Rilonacept (also termed IL-1 trap), a dimer of IL-1R and IL-1R accessory protein (IL-1RacP) extracellular chains fused to the Fc fragments of IgG, was trialed and found effective for the treatment of CAPS [80]. Rilonacept was approved as biological drug in 2008, and canakinumab, a monoclonal anti-IL-1β antibody that was also shown beneficial for the treatment of CAPS [81–86], was approved in 2009. Like anakinra, both were shown to reduce symptoms in additional inflammatory diseases, such as gout [87, 88], and canakinumab was also shown to be effective for SJIA. Side effects associated with canakinumab resemble those of anakinra, such as increased risk of infections [89], neutropenia, and low platelet count [90]; therefore it is not recommended for patients with a high risk for infections. Canakinumab is administered once every four to eight weeks, dependent on disease severity, due to its extended half-life. Nonetheless, withdrawal will not terminate the effects of the drug immediately, like in the case of anakinra. Hyper-IgD syndrome (HIDS) is a genetic autoinflammatory syndrome associated with high IgD blood levels, caused by a mutation in the gene encoding mevalonate kinase (MK) [91]. TNFR1-associated periodic syndrome (TRAPS) is caused by intracellular accumulation of misfolded mutated TNFR1 and an elevated IL-1 production [92]. HIDS and TRAPS were shown to respond to anakinra [93–100]. Canakinumab is currently tested in a phase III trial in colchicine resistant FMF, HIDS/MK deficiency, and TRAPS patients (ClinicalTrials.gov identifier: NCT02059291). In addition, the effect of canakinumab on cardiovascular events and type 2 diabetes is currently held by the Canakinumab Anti-Inflammatory Thrombosis Outcome Study (CANTOS) trial [101].

3.3. Anti-IL-6.

IL-6 is another major proinflammatory cytokine with pleotropic effects on the immune system. IL-6 is the ligand for IL-6 receptor (IL-6R). Following its binding, gp130, a transmembranal glycoprotein forms a homodimer and transmits the signaling. Unlike IL-1R1 or TNFR1, which are ubiquitous, IL-6R is restricted to hepatocytes, monocytes, macrophages, and lymphocytes. Another difference from the IL-1 and TNF cytokines is that the soluble form of IL-6R facilitates and induces the signal rather than serving as an inhibitor. Soluble IL-6R binds IL-6 and this complex further binds membranal gp130, which, unlike IL-6R, is expressed in all cell types. This kind of signaling is termed trans-signaling [102], a process which allows IL-6 to mediate its response on cells that lack IL-6R; among these are embryonic stem cells, endothelial cells, hematopoietic progenitor cells, osteoclasts, and neuronal cells [102]. The proinflammatory cytokines, IL-1 and TNF-α, were assumed to be responsible for the acute phase response of liver cells in vivo. Nevertheless, when hepatocytes response to stimulation by crude macrophage cytokines was compared to isolated cytokines IL-6, IL-1, and TNF, only IL-6 could induce fully comparable response [103]. Among the many IL-6 effects, it was found that it induces immunologic and metabolic responses. IL-6 can alter the T helper cell phenotype programming [104]; it can stimulate B cells, NK cells, osteoclasts, and cancer cells [105] and is secreted by a variety of cells; among these are lymphocytes, macrophages, endothelial cells, epithelial cells, and fibroblasts; these then play a major role in autoimmune diseases, especially RA, in which increased levels of IL-6 are found in synovial fluid [106]. The myeloma receptor antibody (MRA), a humanized antibody against IL-6R, was first trialed in 2003. It was then demonstrated to decrease serum acute phase protein in RA patients, which were not responsive to DMARD or other immunosuppressive drugs [107]. The MRA antibody was renamed tocilizumab and its efficiency for RA was demonstrated in a large trial consisting of 633 patients. The trial showed reduced disease activity [108] and the FDA approved tocilizumab in 2010 for the treatment of RA patients refractory to TNF inhibitors; additionally it was also shown efficient in another trial for the treatment of SJIA [109] where in 2011 the FDA expanded the use of the antibody to include the treatment of SJIA patients. Unfortunately, together with the benefits of IL-6 inhibition came adverse effects. Data pooled from five clinical trials, two ongoing extension trials, and one clinical pharmacology study summarized the following adverse effects among trial participants: serious infections mainly pneumonia, gastroenteritis, and urinary tract infections, opportunistic infections (such as tuberculosis, candidiasis), gastrointestinal perforation, and anaphylactic reactions. Other side effects were neutropenia and increased lipid levels, which are assumed to induce cardiovascular events [110]. Since IL-6 elevates CRP levels, its inhibition by tocilizumab results in milder elevation of CRP during infections. This can put the patients at risk since it is harder to diagnose an infection in patients undergoing treatment [111]. One of SJIA complications is Macrophage-Activating Syndrome (MAS), a life threatening disease, associated with impaired bone-marrow and liver functions. Tocilizumab treatment does not prevent or worsen MAS [112]; however, it does mask the clinical symptoms, again by reducing the CRP levels, which allow diagnosing the outbreak of this syndrome [113]. Blocking IL-1 with anakinra, on the other hand, was shown to reduce MAS severity in SJIA patients [114–116].

Siltuximab is a human-mouse chimeric anti-IL-6 antibody approved in 2014 for HIV-negative and herpes virus-8 negative patients for the treatment of multicentric Castleman's disease, a lymphoproliferative disorder associated with increased IL-6 in the enlarged hyperplastic lymph nodes

[117]. Siltuximab was further studied for its beneficial anti-IL-6 effects in other malignancies, like multiple myeloma, myelodysplastic syndrome, prostate cancer, ovarian cancer and lung cancer, and cancer-associated cachexia and anorexia [118–122]. However, the treatment with siltuximab increases the risk of upper respiratory tract infections and other adverse effects including nausea, fatigue pruritus, increased weight gain, rash, hyperuricemia, thrombocytopenia, dyspnea, leukopenia, and neutropenia [123, 124].

3.4. Biologics Targeting TH17 Cytokines. Ustekinumab is a human monoclonal antibody against IL-12 and IL-23, which share the same IL-12p40 subunit. The antibody recognition of this cytokine reduces the differentiation of naïve CD4+ T helper cells into effector T cells, TH1, and TH17. Previously termed "IL-23-derived autoreactive CD4 T cells," TH17 cells were named after IL-17 cytokine (which they produce) and are correlated with autoimmunity disorders including RA, lupus, colitis, and EAE [125, 126]. IL-12 and IL-23 and their associated T helper cells are correlated to psoriasis which is an immune-mediated chronic inflammatory skin disease, and psoriasis patients have an increased risk to develop psoriatic arthritis [127]. Ustekinumab was shown to be more effective compared to etanercept [128] and was approved in 2009 for plaque psoriasis and in 2013 for psoriatic arthritis. However, ustekinumab treated patients are recommended to receive prophylactic treatment due to increased risk of tuberculosis reactivation [129], as well as the issue of reduced CD4+ lymphocytes during this treatment, that should be taken into account [130, 131].

Secukinumab is a human anti-IL-17A antibody that was trialed and shown ineffective in clinical trials for the treatment of Crohn's disease, as the treatment aggravated the disease severity in addition to increased adverse effects, like upper respiratory tract infections and local fungal infections [132]. However, much like ustekinumab, IL-17 inhibition using secukinumab reduced symptoms and improved physiological functioning compared to placebo or etanercept in plaque psoriasis and was approved by the FDA in 2015. Secukinumab was also reported for its efficiency for psoriatic arthritis [133] and ankylosing spondylitis [134] and was approved for these indications. In March 2016 an additional monoclonal anti-IL-17 antibody—ixekizumab—was approved for patients with plaque psoriasis [135, 136]. Long-term data from experiences of these antibodies targeting effector helper T cells cytokines is required for further evaluation of the adverse effects and safety of these biologics.

4. Reducing Infections in Anti-Inflammatory Biologics

Anticytokine therapy is a powerful tool to fight autoimmune and autoinflammatory diseases in addition to many other diseases in which the inflammatory process enhances the disease activity. For example, IL-1Ra, anakinra, was shown beneficial in vast types of diseases, among which are autoimmune RA [54], autoinflammatory diseases like CAPS [57–60], hereditary inflammatory FMF, improved beta cells function

in type 2 diabetes [71], remodeling following myocardial infarction [70], smoldering myeloma [137], and a variety of other disorders [72]. Other inflammatory mediators like TNF-α and IL-6 have also great potential as targets in anti-inflammatory treatment. However, there is always a major cause for concern when systemically reducing inflammation by biologics that can compromise the patient ability to overcome infections. For example, the ability to reduce the rheumatoid process in patient's joints without inhibiting the neutrophils migration into the lungs in order to fight pneumonia is the objective for new biologics. One strategy to do so is to inject the patient with inactive biologic that would be activated whenever it meets the inflammatory site; the chimeric-IL-1Ra recently published carries such an approach [138]. This molecule is composed of the N-terminal peptide of IL-1β fused to IL-1Ra in its C-terminal side that mimics the structure of the precursor of IL-1β; thus it is expressed as an inactive procytokine. In the inflammatory sites, increased levels of neutrophil serine proteases (like elastase, cathepsin G, or chymotrypsin) [139], macrophages-derived PR3 and caspase-1 [140, 141], or granzymes from NK cells [142] are released from activated or dying cells and these enzymes cleave the N-terminal peptide of IL-1β and release the active free C-terminal cytokine part (Figure 1). For example in the inflamed joint of gouty arthritis patients, IL-1β is active due to the increased activity of neutrophils where the short-lived neutrophils are rich in serine proteases and are released to the site of inflammation [140]. As for the chimeric-IL-1Ra, the active IL-1Ra part is released in the same manner as IL-1β (i.e., an inactive precursor that transforms into an active cytokine due to the inflamed environment). At the same time, the patient's unaffected tissues are spared from the excessive systemic IL-1R1 blockade. Chronic inflammation in the microenvironment of tumors facilitates the tumors mechanisms of invasion and growth [143]. The tumor is surrounded with myeloid cells, rich with inflammatory enzymes and cytokines when IL-1 facilitates tumor growth, angiogenesis, and metastases [144, 145]. It was shown that the inflammatory tumorigenic microenvironment is derived from IL-1β secreted from the myeloid cells around the tumor, and the IL-1α secreted from the tumor cells accompanied with hypoxia, necrosis, or DNA damage [146–149]. It was therefore why anti-IL-1 therapy was suggested for trials in cancer patients [150, 151]. Recently, it was shown that IL-1α neutralization using a monoclonal antibody would be beneficial in cancer patients in prolonging their survival [152, 153]. Cancer patients are often treated with immunosuppressive and bone-marrow suppressing drugs; therefore, they are exposed to increased risk of infections. Thus, biologics like the chimeric-IL-1Ra that might reduce the inflammatory process in the tumor site without reducing the patient's ability to fight infection is a desirable approach. In order to inhibit cell surface TNF, in a cell-type restricted manner, Efimov et al. constructed bispecific antibody that recognizes both the F4/80 macrophage marker and the membranal TNF-α [154]. In this manner, the antibody favors binding of TNF-α on myeloid cells rather than free TNF-α or T lymphocytes derived TNF-α. The aim was to reduce anti-TNF side effects by blocking macrophage-derived inflammation,

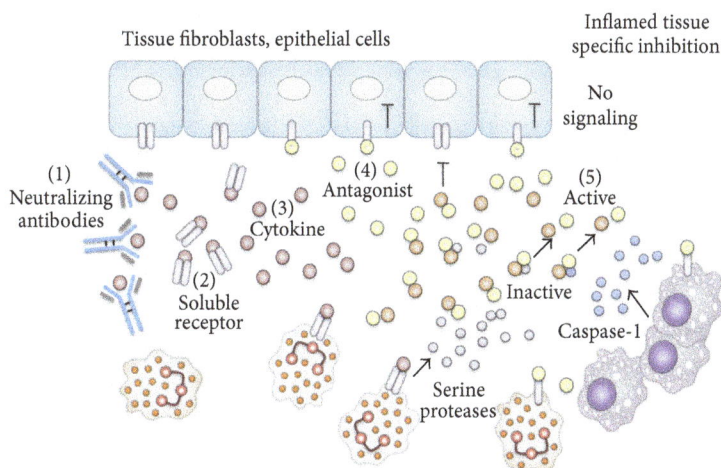

FIGURE 1: Biological drugs strategies for targeting inflammatory cytokines. The biologics can be composed of anticytokine or antireceptor neutralizing antibodies (1) or a soluble receptor that binds the cytokine (2). An inflammatory cytokine, like IL-1β, binds the IL-1R1 and the coreceptor IL-1R accessory protein (3) and transmits cell signaling, while an antagonist, like IL-1Ra, binds the receptor without recruiting the coreceptor (4), thus inhibiting signaling from the receptor and reducing the inflammation. Inflammation-dependent anticytokine strategy: enzymes such as neutrophil serine proteases or macrophage caspase-1 are released into the environment and cleave the two parts of the chimeric-IL-1Ra inactive precursor into an active antagonist (5), which blocks the receptors of tissue cells and the inflammatory cells.

while maintaining T cell activity. The authors claim that this antibody can prevent reactivation of latent tuberculosis and reduce anti-TNF liver toxicity.

5. Concluding Remarks

Unregulated levels of cytokines are central mediators of many inflammatory diseases. Targeting these cytokines using recombinant anti-inflammatory cytokines, recombinant soluble receptors, or antibodies against cytokines has demonstrated preferable clinical outcomes in patients with autoimmune diseases, which are refractory to glucocorticoids treatments. However, systemic cytokine blocking suffers from a number of serious limitations. For one, the lack of danger signals, which is crucial for adequate immune cell activation as well as hematopoiesis alterations, common features in all biologics, expose the host to increased risks of infections. In addition, the pleiotropic nature of most cytokines and their necessity to the function of multiple cell types across different organs make it almost impossible to inhibit their signaling cascade in a long-term therapy without severe complications. Therefore, new approaches based on site-restricted biologics, which maintain the cytokine activity in other sites, are highly advised.

Competing Interests

The authors declare that there is no conflict of interests regarding the publication of this paper.

Acknowledgments

Dr. Idan Cohen is supported by Ministry of Aliyah and Immigrant Absorption, The Center for Absorption in Science, Israel.

References

[1] S. A. Rosenberg, J. C. Yang, S. L. Topalian et al., "Treatment of 283 consecutive patients with metastatic melanoma or renal cell cancer using high-dose bolus interleukin," *Journal of the American Medical Association*, vol. 271, no. 12, pp. 907–913, 1994.

[2] M. B. Atkins, M. T. Lotze, J. P. Dutcher et al., "High-dose recombinant interleukin 2 therapy for patients with metastatic melanoma: analysis of 270 patients treated between 1985 and 1993," *Journal of Clinical Oncology*, vol. 17, no. 7, pp. 2105–2116, 1999.

[3] L. C. Hartmann, W. J. Urba, R. G. Steis et al., "Use of prophylactic antibiotics for prevention of intravascular catheter-related infections in interleukin-2-treated patients," *Journal of the National Cancer Institute*, vol. 81, no. 15, pp. 1190–1193, 1989.

[4] M. V. Doyle, M. T. Lee, and S. Fong, "Comparison of the biological activities of human recombinant interleukin-2125 and native interleukin-2," *Journal of Biological Response Modifiers*, vol. 4, no. 1, pp. 96–109, 1985.

[5] P. E. Auron, A. C. Webb, L. J. Rosenwasser et al., "Nucleotide sequence of human monocyte interleukin 1 precursor cDNA," *Proceedings of the National Academy of Sciences of the United States of America*, vol. 81, no. 24, pp. 7907–7911, 1984.

[6] T. Iizumi, S. Sato, T. Iiyama et al., "Recombinant human interleukin-1 beta analogue as a regulator of hematopoiesis

in patients receiving chemotherapy for urogenital cancers," *Cancer*, vol. 68, no. 7, pp. 1520–1523, 1991.

[7] J. Crown, A. Jakubowski, N. Kemeny et al., "A phase I trial of recombinant human interleukin-1β alone and in combination with myelosuppressive doses of 5-fluorouracil in patients with gastrointestinal cancer," *Blood*, vol. 78, no. 6, pp. 1420–1427, 1991.

[8] C. E. Walsh, J. M. Liu, S. M. Anderson, J. L. Rossio, A. W. Nienhuis, and N. S. Young, "A trial of recombinant human interleukin-1 in patients with severe refractory aplastic anaemia," *British Journal of Haematology*, vol. 80, no. 1, pp. 106–110, 1992.

[9] F. Zoulim, F. Lebossé, and M. Levrero, "Current treatments for chronic hepatitis B virus infections," *Current Opinion in Virology*, vol. 18, pp. 109–116, 2016.

[10] J. L. Dienstag and J. G. McHutchison, "American gastroenterological association technical review on the management of hepatitis C," *Gastroenterology*, vol. 130, no. 1, pp. 231–264, 2006.

[11] E. Jaeckel, M. Cornberg, H. Wedemeyer et al., "Treatment of acute hepatitis C with interferon alfa-2b," *New England Journal of Medicine*, vol. 345, no. 20, pp. 1452–1457, 2001.

[12] T. Santantonio, M. Fasano, E. Sinisi et al., "Efficacy of a 24-week course of PEG-interferon α-2b monotherapy in patients with acute hepatitis C after failure of spontaneous clearance," *Journal of Hepatology*, vol. 42, no. 3, pp. 329–333, 2005.

[13] J. W. Choi, J. S. Lee, W. H. Paik et al., "Acute pancreatitis associated with pegylated interferon-alpha-2a therapy in chronic hepatitis C," *Clinical and Molecular Hepatology*, vol. 22, no. 1, pp. 168–171, 2016.

[14] J. H. Hoofnagle, "Course and outcome of hepatitis C," *Hepatology*, vol. 36, no. 5, supplement 1, pp. S21–S29, 2002.

[15] M. P. Manns, H. Wedemeyer, and M. Cornberg, "Treating viral hepatitis C: efficacy, side effects, and complications," *Gut*, vol. 55, no. 9, pp. 1350–1359, 2006.

[16] A. M. Wang, A. A. Creasey, M. B. Ladner et al., "Molecular cloning of the complementary DNA for human tumor necrosis factor," *Science*, vol. 228, no. 4696, pp. 149–154, 1985.

[17] T. Shirai, H. Yamaguchi, H. Ito, C. W. Todd, and R. B. Wallace, "Cloning and expression in Escherichia coli of the gene for human tumour necrosis factor," *Nature*, vol. 313, no. 6005, pp. 803–806, 1985.

[18] L. Fransen, R. Muller, A. Marmenout et al., "Molecular cloning of mouse tumour necrosis factor cDNA and its eukaryotic expression," *Nucleic Acids Research*, vol. 13, no. 12, pp. 4417–4429, 1985.

[19] D. Pennica, J. S. Hayflick, T. S. Bringman, M. A. Palladino, and D. V. Goeddel, "Cloning and expression in Escherichia coli of the cDNA for murine tumor necrosis factor," *Proceedings of the National Academy of Sciences of the United States of America*, vol. 82, no. 18, pp. 6060–6064, 1985.

[20] P. Ghezzi and A. Cerami, "Tumor necrosis factor as a pharmacological target," *Molecular Biotechnology*, vol. 31, no. 3, pp. 239–244, 2005.

[21] N. Parameswaran and S. Patial, "Tumor necrosis factor-α signaling in macrophages," *Critical Reviews in Eukaryotic Gene Expression*, vol. 20, no. 2, pp. 87–103, 2010.

[22] P. Seckinger, S. Isaaz, and J.-M. Dayer, "A human inhibitor of tumor necrosis factor α," *Journal of Experimental Medicine*, vol. 167, no. 4, pp. 1511–1516, 1988.

[23] P. Seckinger, J.-H. Zhang, B. Hauptmann, and J.-M. Dayer, "Characterization of a tumor necrosis factor α (TNF-α) inhibitor: evidence of immunological cross-reactivity with the TNF receptor," *Proceedings of the National Academy of Sciences of the United States of America*, vol. 87, no. 13, pp. 5188–5192, 1990.

[24] H. Engelmann, D. Novick, and D. Wallach, "Two tumor necrosis factor-binding proteins purified from human urine. Evidence for immunological cross-reactivity with cell surface tumor necrosis factor receptors," *The Journal of Biological Chemistry*, vol. 265, no. 3, pp. 1531–1536, 1990.

[25] B. Scallon, A. Cai, N. Solowski et al., "Binding and functional comparisons of two types of tumor necrosis factor antagonists," *Journal of Pharmacology and Experimental Therapeutics*, vol. 301, no. 2, pp. 418–426, 2002.

[26] K. M. Mohler, D. S. Torrance, C. A. Smith et al., "Soluble Tumor Necrosis Factor (TNF) receptors are effective therapeutic agents in lethal endotoxemia and function simultaneously as both TNF carriers and TNF antagonists," *The Journal of Immunology*, vol. 151, no. 3, pp. 1548–1561, 1993.

[27] B. J. Scallon, M. A. Moore, H. Trinh, D. M. Knight, and J. Ghrayeb, "Chimeric anti-TNF-α monoclonal antibody cA2 binds recombinant transmembrane TNF-α and activates immune effector functions," *Cytokine*, vol. 7, no. 3, pp. 251–259, 1995.

[28] R. Atreya, M. Zimmer, B. Bartsch et al., "Antibodies against tumor necrosis factor (TNF) induce T-cell apoptosis in patients with inflammatory bowel diseases via TNF receptor 2 and intestinal CD14+ macrophages," *Gastroenterology*, vol. 141, no. 6, pp. 2026–2038, 2011.

[29] M. Papoutsaki and A. Costanzo, "Treatment of psoriasis and psoriatic arthritis," *BioDrugs*, vol. 27, supplement 1, pp. 3–12, 2013.

[30] P. L. Lin, H. L. Plessner, N. N. Voitenok, and J. L. Flynn, "Tumor necrosis factor and tuberculosis," *Journal of Investigative Dermatology Symposium Proceedings*, vol. 12, no. 1, pp. 22–25, 2007.

[31] F. Tubach, D. Salmon, P. Ravaud et al., "Risk of tuberculosis is higher with anti-tumor necrosis factor monoclonal antibody therapy than with soluble tumor necrosis factor receptor therapy: the three-year prospective French research axed on tolerance of biotherapies registry," *Arthritis and Rheumatism*, vol. 60, no. 7, pp. 1884–1894, 2009.

[32] D. Vassilopoulos and L. H. Calabrese, "Management of rheumatic disease with comorbid HBV or HCV infection," *Nature Reviews Rheumatology*, vol. 8, no. 6, pp. 348–357, 2012.

[33] A. Strangfeld, J. Listing, P. Herzer et al., "Risk of herpes zoster in patients with rheumatoid arthritis treated with anti-TNF-α agents," *JAMA - Journal of the American Medical Association*, vol. 301, no. 7, pp. 737–744, 2009.

[34] A. Mantovani, "Molecular pathways linking inflammation and cancer," *Current Molecular Medicine*, vol. 10, no. 4, pp. 369–373, 2010.

[35] V. Baud and M. Karin, "Signal transduction by tumor necrosis factor and its relatives," *Trends in Cell Biology*, vol. 11, no. 9, pp. 372–377, 2001.

[36] A. C. Mackey, L. Green, C. Leptak, and M. Avigan, "Hepatosplenic T cell lymphoma associated with infliximab use in young patients treated for inflammatory bowel disease: update," *Journal of Pediatric Gastroenterology and Nutrition*, vol. 48, no. 3, pp. 386–388, 2009.

[37] F. Wolfe and K. Michaud, "Lymphoma in rheumatoid arthritis: the effect of methotrexate and anti-tumor necrosis factor therapy in 18,572 patients," *Arthritis & Rheumatism*, vol. 50, no. 6, pp. 1740–1751, 2004.

[38] C. N. Bernstein, J. F. Blanchard, E. Kliewer, and A. Wajda, "Cancer risk in patients with inflammatory bowel disease: A

Population-based Study," *Cancer*, vol. 91, no. 4, pp. 854–862, 2001.

[39] E. Thomas, D. H. Brewster, R. J. Black, and G. J. Macfarlane, "Risk of malignancy among patients with rheumatic conditions," *International Journal of Cancer*, vol. 88, no. 3, pp. 497–502, 2000.

[40] S. Raheel, C. S. Crowson, K. Wright, and E. L. Matteson, "Risk of malignant neoplasm in patients with incident rheumatoid arthritis 1980-2007 in relation to a comparator cohort: A Population-Based Study," *International Journal of Rheumatology*, vol. 2016, Article ID 10.1155/2016/4609486, 6 pages, 2016.

[41] E. Baecklund, A. Iliadou, J. Askling et al., "Association of chronic inflammation, not its treatment, with increased lymphoma risk in rheumatoid arthritis," *Arthritis & Rheumatism*, vol. 54, no. 3, pp. 692–701, 2006.

[42] E. S. Chung, M. Packer, K. H. Lo, A. A. Fasanmade, and J. T. Willerson, "Randomized, double-blind, placebo-controlled, pilot trial of infliximab, a chimeric monoclonal antibody to tumor necrosis factor-α, in patients with moderate-to-severe heart failure: results of the anti-TNF therapy against congestive heart failure (ATTACH) trial," *Circulation*, vol. 107, no. 25, pp. 3133–3140, 2003.

[43] D. L. Mann, J. J. V. McMurray, M. Packer et al., "Targeted anticytokine therapy in patients with chronic heart failure: results of the Randomized Etanercept Worldwide Evaluation (RENEWAL)," *Circulation*, vol. 109, no. 13, pp. 1594–1602, 2004.

[44] S. Bernatsky, C. Renoux, and S. Suissa, "Demyelinating events in rheumatoid arthritis after drug exposures," *Annals of the Rheumatic Diseases*, vol. 69, no. 9, pp. 1691–1693, 2010.

[45] E. Kaltsonoudis, P. V. Voulgari, S. Konitsiotis, and A. A. Drosos, "Demyelination and other neurological adverse events after anti-TNF therapy," *Autoimmunity Reviews*, vol. 13, no. 1, pp. 54–58, 2014.

[46] D. Pugliese, L. Guidi, P. M. Ferraro et al., "Paradoxical psoriasis in a large cohort of patients with inflammatory bowel disease receiving treatment with anti-TNF alpha: 5-year follow-up study," *Alimentary Pharmacology and Therapeutics*, vol. 42, no. 7, pp. 880–888, 2015.

[47] A. O'Toole, M. Lucci, and J. Korzenik, "Inflammatory bowel disease provoked by etanercept: report of 443 possible cases combined from an IBD Referral center and the FDA," *Digestive Diseases and Sciences*, vol. 61, no. 6, pp. 1772–1774, 2016.

[48] A. Dallocchio, D. Canioni, F. Ruemmele et al., "Occurrence of inflammatory bowel disease during treatment of juvenile idiopathic arthritis with etanercept: a French retrospective study," *Rheumatology*, vol. 49, no. 9, pp. 1694–1698, 2010.

[49] C. A. Dinarello, L. J. Rosenwasser, and S. M. Wolff, "Demonstration of a circulating suppressor factor of thymocyte proliferation during endotoxin fever in humans," *Journal of Immunology*, vol. 127, no. 6, pp. 2517–2519, 1981.

[50] W. P. Arend, F. G. Joslin, and R. J. Massoni, "Effects of immune complexes on production by human monocytes of interleukin 1 or an interleukin 1 inhibitor," *Journal of Immunology*, vol. 134, no. 6, pp. 3868–3875, 1985.

[51] S. P. Eisenberg, R. J. Evans, W. P. Arend et al., "Primary structure and functional expression from complementary DNA of a human interleukin-1 receptor antagonist," *Nature*, vol. 343, no. 6256, pp. 341–346, 1990.

[52] I. Aksentijevich, S. L. Masters, P. J. Ferguson et al., "An autoinflammatory disease with deficiency of the interleukin-1-receptor antagonist," *The New England Journal of Medicine*, vol. 360, no. 23, pp. 2426–2437, 2009.

[53] M. Mertens and J. A. Singh, "Anakinra for rheumatoid arthritis: a systematic review," *Journal of Rheumatology*, vol. 36, no. 6, pp. 1118–1125, 2009.

[54] D. E. Furst, "Anakinra: review of recombinant human interleukin-I receptor antagonist in the treatment of rheumatoid arthritis," *Clinical Therapeutics*, vol. 26, no. 12, pp. 1960–1975, 2004.

[55] M. Mertens and J. A. Singh, "Anakinra for rheumatoid arthritis," *Cochrane Database of Systematic Reviews (Online)*, no. 1, p. CD005121, 2009.

[56] L. Broderick, D. De Nardo, B. S. Franklin, H. M. Hoffman, and E. Latz, "The inflammasomes and autoinflammatory syndromes," *Annual Review of Pathology: Mechanisms of Disease*, vol. 10, pp. 395–424, 2015.

[57] H. M. Hoffman, S. Rosengren, D. L. Boyle et al., "Prevention of cold-associated acute inflammation in familial cold autoinflammatory syndrome by interleukin-1 receptor antagonist," *The Lancet*, vol. 364, no. 9447, pp. 1779–1785, 2004.

[58] S. K. Metyas and H. M. Hoffman, "Anakinra prevents symptoms of familial cold autoinflammatory syndrome and Raynaud's disease," *Journal of Rheumatology*, vol. 33, no. 10, pp. 2085–2087, 2006.

[59] P. N. Hawkins, H. J. Lachmann, E. Aganna, and M. F. McDermott, "Spectrum of clinical features in muckle-wells syndrome and response to anakinra," *Arthritis and Rheumatism*, vol. 50, no. 2, pp. 607–612, 2004.

[60] R. Goldbach-Mansky, N. J. Dailey, S. W. Canna et al., "Neonatal-onset multisystem inflammatory disease responsive to interleukin-1β inhibition," *The New England Journal of Medicine*, vol. 355, no. 6, pp. 581–592, 2006.

[61] K. Stankovic Stojanovic, Y. Delmas, P. U. Torres et al., "Dramatic beneficial effect of interleukin-1 inhibitor treatment in patients with familial Mediterranean fever complicated with amyloidosis and renal failure," *Nephrology Dialysis Transplantation*, vol. 27, no. 5, pp. 1898–1901, 2012.

[62] U. Meinzer, P. Quartier, J.-F. Alexandra, V. Hentgen, F. Retornaz, and I. Koné-Paut, "Interleukin-1 targeting drugs in familial mediterranean fever: a case series and a review of the literature," *Seminars in Arthritis and Rheumatism*, vol. 41, no. 2, pp. 265–271, 2011.

[63] K. Laskari, A. G. Tzioufas, and H. M. Moutsopoulos, "Efficacy and long-term follow-up of IL-1R inhibitor anakinra in adults with Still's disease: A Case-series Study," *Arthritis Research and Therapy*, vol. 13, no. 3, article R91, 2011.

[64] A. A. Fitzgerald, S. A. LeClercq, A. Yan, J. E. Homik, and C. A. Dinarello, "Rapid responses to anakinra in patients with refractory adult-onset Still's disease," *Arthritis & Rheumatism*, vol. 52, no. 6, pp. 1794–1803, 2005.

[65] V. Pascual, F. Allantaz, E. Arce, M. Punaro, and J. Banchereau, "Role of interleukin-1 (IL-1) in the pathogenesis of systemic onset juvenile idiopathic arthritis and clinical response to IL-1 blockade," *The Journal of Experimental Medicine*, vol. 201, no. 9, pp. 1479–1486, 2005.

[66] V. Ohlsson, E. Baildam, H. Foster et al., "Anakinra treatment for systemic onset juvenile idiopathic arthritis (SOJIA)," *Rheumatology*, vol. 47, no. 4, pp. 555–556, 2008.

[67] P. Quartier, F. Allantaz, R. Cimaz et al., "A multicentre, randomised, double-blind, placebo-controlled trial with the interleukin-1 receptor antagonist anakinra in patients with systemic-onset juvenile idiopathic arthritis (ANAJIS trial)," *Annals of the Rheumatic Diseases*, vol. 70, no. 5, pp. 747–754, 2011.

[68] S. Ottaviani, A. Moltó, H.-K. Ea et al., "Efficacy of anakinra in gouty arthritis: a retrospective study of 40 cases," *Arthritis Research and Therapy*, vol. 15, no. 5, article R123, 2013.

[69] A. M. Hung, C. D. Ellis, A. Shintani, C. Booker, and T. A. Ikizler, "IL-1β receptor antagonist reduces inflammation in hemodialysis patients," *Journal of the American Society of Nephrology*, vol. 22, no. 3, pp. 437–442, 2011.

[70] A. Abbate, M. C. Kontos, J. D. Grizzard et al., "Interleukin-1 Blockade With Anakinra to Prevent Adverse Cardiac Remodeling After Acute Myocardial Infarction (Virginia Commonwealth University Anakinra Remodeling Trial [VCU-ART] Pilot Study)," *American Journal of Cardiology*, vol. 105, no. 10, pp. 1371–1377.e1, 2010.

[71] C. M. Larsen, M. Faulenbach, A. Vaag et al., "Interleukin-1-receptor antagonist in type 2 diabetes mellitus," *New England Journal of Medicine*, vol. 356, no. 15, pp. 1517–1526, 2007.

[72] C. A. Dinarello, A. Simon, and J. W. M. van der Meer, "Treating inflammation by blocking interleukin-1 in a broad spectrum of diseases," *Nature Reviews Drug Discovery*, vol. 11, no. 8, pp. 633–652, 2012.

[73] A. A. Jesus and R. Goldbach-Mansky, "IL-1 blockade in autoinflammatory syndromes," *Annual Review of Medicine*, vol. 65, pp. 223–244, 2014.

[74] S. Federici, A. Martini, and M. Gattorno, "The central role of anti-IL-1 blockade in the treatment of monogenic and multifactorial autoinflammatory diseases," *Frontiers in Immunology*, vol. 4, article 351, 2013.

[75] G. Cavalli and C. A. Dinarello, "Treating rheumatological diseases and co-morbidities with interleukin-1 blocking therapies," *Rheumatology*, vol. 54, no. 12, pp. 2134–2144, 2015.

[76] L. D. Settas, G. Tsimirikas, G. Vosvotekas, E. Triantafyllidou, and P. Nicolaides, "Reactivation of pulmonary tuberculosis in a patient with rheumatoid arthritis during treatment with IL-1 receptor antagonists (anakinra)," *Journal of Clinical Rheumatology*, vol. 13, no. 4, pp. 219–220, 2007.

[77] G. Direz, N. Noël, C. Guyot, O. Toupance, J.-H. Salmon, and J.-P. Eschard, "Efficacy but side effects of anakinra therapy for chronic refractory gout in a renal transplant recipient with preterminal chronic renal failure," *Joint Bone Spine*, vol. 79, no. 6, p. 631, 2012.

[78] F. Perrin, A. Néel, J. Graveleau, A.-L. Ruellan, A. Masseau, and M. Hamidou, "Two cases of anakinra-induced neutropenia during auto-inflammatory diseases: drug reintroduction can be successful," *La Presse Médicale*, vol. 43, no. 3, pp. 319–321, 2014.

[79] C. Turesson and K. Riesbeck, "Septicemia with *Staphylococcus aureus*, beta-hemolytic streptococci group B and G, and *Escherichia coli* in a patient with rheumatoid arthritis treated with a recombinant human interleukin 1 receptor antagonist (Anakinra)," *Journal of Rheumatology*, vol. 31, no. 9, p. 1876, 2004.

[80] H. M. Hoffman, M. L. Throne, N. J. Amar et al., "Efficacy and safety of rilonacept (Interleukin-1 Trap) in patients with cryopyrin-associated periodic syndromes: results from two sequential placebo-controlled studies," *Arthritis and Rheumatism*, vol. 58, no. 8, pp. 2443–2452, 2008.

[81] G. M. Walsh, "Canakinumab for the treatment of cryopyrin-associated periodic syndromes," *Drugs of Today*, vol. 45, no. 10, pp. 731–735, 2009.

[82] S. Savic and M. F. McDermott, "Inflammation: canakinumab for the cryopyrin-associated periodic syndromes," *Nature Reviews Rheumatology*, vol. 5, no. 10, pp. 529–530, 2009.

[83] H. J. Lachmann, I. Kone-Paut, J. B. Kuemmerle-Deschner et al., "Use of canakinumab in the cryopyrin-associated periodic syndrome," *The New England Journal of Medicine*, vol. 360, no. 23, pp. 2416–2425, 2009.

[84] L. D. Church and M. F. McDermott, "Canakinumab, a fully-human mAb against IL-1β for the potential treatment of inflammatory disorders," *Current Opinion in Molecular Therapeutics*, vol. 11, no. 1, pp. 81–89, 2009.

[85] J. B. Kuemmerle-Deschner, E. Ramos, N. Blank et al., "Canakinumab (ACZ885, a fully human IgG1 anti-IL-1β mAb) induces sustained remission in pediatric patients with cryopyrin-associated periodic syndrome (CAPS)," *Arthritis Research & Therapy*, vol. 13, no. 1, article R34, 2011.

[86] I. Koné-Paut, H. J. Lachmann, J. B. Kuemmerle-Deschner et al., "Sustained remission of symptoms and improved health-related quality of life in patients with cryopyrin-associated periodic syndrome treated with canakinumab: results of a double-blind placebo-controlled randomized withdrawal study," *Arthritis Research & Therapy*, vol. 13, no. 6, p. R202, 2011.

[87] A. So, T. De Smedt, S. Revaz, and J. Tschopp, "A pilot study of IL-1 inhibition by anakinra in acute gout," *Arthritis Research and Therapy*, vol. 9, article R28, 2007.

[88] H. R. Schumacher Jr., R. R. Evans, K. G. Saag et al., "Rilonacept (interleukin-1 trap) for prevention of gout flares during initiation of uric acid-lowering therapy: results from a phase III randomized, double-blind, placebo-controlled, confirmatory efficacy study," *Arthritis Care & Research*, vol. 64, no. 10, pp. 1462–1470, 2012.

[89] N. Ruperto, H. I. Brunner, P. Quartier et al., "Two randomized trials of canakinumab in systemic juvenile idiopathic arthritis," *The New England Journal of Medicine*, vol. 367, no. 25, pp. 2396–2406, 2012.

[90] N. Schlesinger, R. E. Alten, T. Bardin et al., "Canakinumab for acute gouty arthritis in patients with limited treatment options: results from two randomised, multicentre, active-controlled, double-blind trials and their initial extensions," *Annals of the Rheumatic Diseases*, vol. 71, no. 11, pp. 1839–1848, 2012.

[91] M. Stoffels and A. Simon, "Hyper-IgD syndrome or mevalonate kinase deficiency," *Current Opinion in Rheumatology*, vol. 23, no. 5, pp. 419–423, 2011.

[92] F. C. Kimberley, A. A. Lobito, R. M. Siegel, and G. R. Screaton, "Falling into TRAPS—receptor misfolding in the TNF receptor 1-associated periodic fever syndrome," *Arthritis Research and Therapy*, vol. 9, no. 4, article 217, 2007.

[93] S. Peciuliene, B. Burnyte, R. Gudaitiene et al., "Perinatal manifestation of mevalonate kinase deficiency and efficacy of anakinra," *Pediatric Rheumatology*, vol. 14, no. 1, p. 19, 2016.

[94] R. Campanilho-Marques and P. A. Brogan, "Mevalonate kinase deficiency in two sisters with therapeutic response to anakinra: case report and review of the literature," *Clinical Rheumatology*, vol. 33, no. 11, pp. 1681–1684, 2014.

[95] E. J. Bodar, L. M. Kuijk, J. P. H. Drenth, J. W. M. Van Der Meer, A. Simon, and J. Frenkel, "On-demand anakinra treatment is effective in mevalonate kinase deficiency," *Annals of the Rheumatic Diseases*, vol. 70, no. 12, pp. 2155–2158, 2011.

[96] T. Lequerré, O. Vittecoq, S. Pouplin et al., "Mevalonate kinase deficiency syndrome with structural damage responsive to anakinra," *Rheumatology*, vol. 46, no. 12, pp. 1860–1862, 2007.

[97] C. Grimwood, V. Despert, I. Jeru, and V. Hentgen, "On-demand treatment with anakinra: a treatment option for selected TRAPS patients," *Rheumatology*, vol. 54, no. 9, pp. 1749–1751, 2015.

[98] M. Cattalini, A. Meini, P. Monari et al., "Recurrent migratory angioedema as cutaneous manifestation in a familiar case of TRAPS: dramatic response to Anakinra," *Dermatology Online Journal*, vol. 19, no. 11, Article ID 20405, 2013.

[99] M. Andrés and E. Pascual, "Anakinra for a refractory case of intermittent hydrarthrosis with a TRAPS-related gene mutation," *Annals of the Rheumatic Diseases*, vol. 72, no. 1, p. 155, 2013.

[100] L. Obici, A. Meini, M. Cattalini et al., "Favourable and sustained response to anakinra in tumour necrosis factor receptor-associated periodic syndrome (TRAPS) with or without AA amyloidosis," *Annals of the Rheumatic Diseases*, vol. 70, no. 8, pp. 1511–1512, 2011.

[101] P. M. Ridker, T. Thuren, A. Zalewski, and P. Libby, "Interleukin-1β inhibition and the prevention of recurrent cardiovascular events: Rationale and Design of the Canakinumab Antiinflammatory Thrombosis Outcomes Study (CANTOS)," *American Heart Journal*, vol. 162, no. 4, pp. 597–605, 2011.

[102] S. Rose-John, "Interleukin-6 biology is coordinated by membrane bound and soluble receptors," *Acta Biochimica Polonica*, vol. 50, no. 3, pp. 603–611, 2003.

[103] "Regulation of the acute phase and immune responses: interleukin-6," *Annals of the New York Academy of Sciences*, vol. 557, pp. 1–583, 1989.

[104] B. M. Baranovski, G. S. Freixo-Lima, E. C. Lewis, and P. Rider, "T helper subsets, peripheral plasticity, and the acute phase protein, α1-antitrypsin," *BioMed Research International*, vol. 2015, Article ID 184574, 14 pages, 2015.

[105] E. T. Keller, J. Wanagat, and W. B. Ershler, "Molecular and cellular biology of interleukin-6 and its receptor," *Frontiers in Bioscience*, vol. 1, pp. d340–d357, 1996.

[106] F. A. Houssiau, J.-P. Devogelaer, J. van Damme, C. N. Deuxchaisnes, and J. van Snick, "Interleukin-6 in synovial fluid and serum of patients with rheumatoid arthritis and other inflammatory arthritides," *Arthritis & Rheumatology*, vol. 31, no. 6, pp. 784–788, 1988.

[107] N. Nishimoto, K. Yoshizaki, K. Maeda et al., "Toxicity, pharmacokinetics, and dose-finding study of repetitive treatment with the humanized anti-interleukin 6 receptor antibody MRA in rheumatoid arthritis. Phase I/II Clinical Study," *Journal of Rheumatology*, vol. 30, no. 7, pp. 1426–1435, 2003.

[108] J. S. Smolen, A. Beaulieu, A. Rubbert-Roth et al., "Effect of interleukin-6 receptor inhibition with tocilizumab in patients with rheumatoid arthritis (OPTION study): a double-blind, placebo-controlled, randomised trial," *The Lancet*, vol. 371, no. 9617, pp. 987–997, 2008.

[109] S. Yokota, T. Imagawa, M. Mori et al., "Efficacy and safety of tocilizumab in patients with systemic-onset juvenile idiopathic arthritis: a randomised, double-blind, placebo-controlled, withdrawal phase III trial," *The Lancet*, vol. 371, no. 9617, pp. 998–1006, 2008.

[110] M. H. Schiff, J. M. Kremer, A. Jahreis, E. Vernon, J. D. Isaacs, and R. F. van Vollenhoven, "Integrated safety in tocilizumab clinical trials," *Arthritis Research and Therapy*, vol. 13, no. 5, article R141, 2011.

[111] V. R. Lang, M. Englbrecht, J. Rech et al., "Risk of infections in rheumatoid arthritis patients treated with tocilizumab," *Rheumatology*, vol. 51, no. 5, Article ID ker223, pp. 852–857, 2012.

[112] S. Yokota, Y. Itoh, T. Morio, N. Sumitomo, K. Daimaru, and S. Minota, "Macrophage activation syndrome in patients with systemic juvenile idiopathic arthritis under treatment with tocilizumab," *Journal of Rheumatology*, vol. 42, no. 4, pp. 712–722, 2015.

[113] M. Shimizu, Y. Nakagishi, K. Kasai et al., "Tocilizumab masks the clinical symptoms of systemic juvenile idiopathic arthritis-associated macrophage activation syndrome: the diagnostic significance of interleukin-18 and interleukin-6," *Cytokine*, vol. 58, no. 2, pp. 287–294, 2012.

[114] M. Durand, Y. Troyanov, P. Laflamme, and G. Gregoire, "Macrophage activation syndrome treated with anakinra," *Journal of Rheumatology*, vol. 37, no. 4, pp. 879–880, 2010.

[115] A. Kelly and A. V. Ramanan, "A case of macrophage activation syndrome successfully treated with anakinra," *Nature Clinical Practice Rheumatology*, vol. 4, no. 11, pp. 615–620, 2008.

[116] N. Bruck, M. Suttorp, M. Kabus, G. Heubner, M. Gahr, and F. Pessler, "Rapid and sustained remission of systemic juvenile idiopathic arthritis-associated macrophage activation syndrome through treatment with anakinra and corticosteroids," *Journal of Clinical Rheumatology*, vol. 17, no. 1, pp. 23–27, 2011.

[117] H. E. El-Osta and R. Kurzrock, "Castleman's disease: from basic mechanisms to molecular therapeutics," *Oncologist*, vol. 16, no. 4, pp. 497–511, 2011.

[118] A. Markham and T. Patel, "Siltuximab: first global approval," *Drugs*, vol. 74, no. 10, pp. 1147–1152, 2014.

[119] S. Bagcchi, "Siltuximab in transplant-ineligible patients with myeloma," *The Lancet Oncology*, vol. 15, no. 8, article e309, 2014.

[120] R. L. Stone, A. M. Nick, I. A. McNeish et al., "Paraneoplastic thrombocytosis in ovarian cancer," *New England Journal of Medicine*, vol. 366, no. 7, pp. 610–618, 2012.

[121] I. T. Cavarretta, H. Neuwirt, M. H. Zaki et al., "Mcl-1 is regulated by IL-6 and mediates the survival activity of the cytokine in a model of late stage prostate carcinoma," *Advances in Experimental Medicine and Biology*, vol. 617, pp. 547–555, 2008.

[122] L. Song, M. A. Smith, P. Doshi et al., "Antitumor efficacy of the anti-interleukin-6 (IL-6) antibody siltuximab in mouse xenograft models of lung cancer," *Journal of Thoracic Oncology*, vol. 9, no. 7, pp. 974–982, 2014.

[123] A. Deisseroth, C.-W. Ko, L. Nie et al., "FDA approval: siltuximab for the treatment of patients with multicentric castleman disease," *Clinical Cancer Research*, vol. 21, no. 5, pp. 950–954, 2015.

[124] S. K. Thomas, A. Suvorov, L. Noens et al., "Evaluation of the QTc prolongation potential of a monoclonal antibody, siltuximab, in patients with monoclonal gammopathy of undetermined significance, smoldering multiple myeloma, or low-volume multiple myeloma," *Cancer Chemotherapy and Pharmacology*, vol. 73, no. 1, pp. 35–42, 2014.

[125] C. L. Langrish, Y. Chen, W. M. Blumenschein et al., "IL-23 drives a pathogenic T cell population that induces autoimmune inflammation," *The Journal of Experimental Medicine*, vol. 201, no. 2, pp. 233–240, 2005.

[126] S. Fujino, A. Andoh, S. Bamba et al., "Increased expression of interleukin 17 in inflammatory bowel disease," *Gut*, vol. 52, no. 1, pp. 65–70, 2003.

[127] N. Koutruba, J. Emer, and M. Lebwohl, "Review of ustekinumab, an interleukin-12 and interleukin-23 inhibitor used for the treatment of plaque psoriasis," *Therapeutics and Clinical Risk Management*, vol. 6, pp. 123–141, 2010.

[128] C. E. M. Griffiths, B. E. Strober, P. Van De Kerkhof et al., "Comparison of ustekinumab and etanercept for moderate-to-severe psoriasis," *New England Journal of Medicine*, vol. 362, no. 2, pp. 118–128, 2010.

[129] T.-F. Tsai, V. Ho, M. Song et al., "The safety of ustekinumab treatment in patients with moderate-to-severe psoriasis and latent tuberculosis infection," *British Journal of Dermatology*, vol. 167, no. 5, pp. 1145–1152, 2012.

[130] C. L. Kauffman, N. Aria, E. Toichi et al., "A phase I study evaluating the safety, pharmacokinetics, and clinical response of a human IL-12 p40 antibody in subjects with plaque psoriasis," *Journal of Investigative Dermatology*, vol. 123, no. 6, pp. 1037–1044, 2004.

[131] A. B. Gottlieb, K. D. Cooper, T. S. McCormick et al., "A phase 1, double-blind, placebo-controlled study evaluating single subcutaneous administrations of a human interleukin-12/23 monoclonal antibody in subjects with plaque psoriasis," *Current Medical Research and Opinion*, vol. 23, no. 5, pp. 1081–1092, 2007.

[132] W. Hueber, B. E. Sands, S. Lewitzky et al., "Secukinumab, a human anti-IL-17A monoclonal antibody, for moderate to severe Crohn's disease: unexpected results of a randomised, double-blind placebo-controlled trial," *Gut*, vol. 61, no. 12, pp. 1693–1700, 2012.

[133] A. B. Gottlieb, R. G. Langley, S. Philipp et al., "Secukinumab improves physical function in subjects with plaque psoriasis and psoriatic arthritis: results from two randomized, phase 3 trials," *Journal of Drugs in Dermatology*, vol. 14, no. 8, pp. 821–833, 2015.

[134] D. Baeten, J. Sieper, J. Braun et al., "Secukinumab, an interleukin-17A inhibitor, in ankylosing spondylitis," *The New England Journal of Medicine*, vol. 373, no. 26, pp. 2534–2548, 2015.

[135] C. Leonardi, R. Matheson, C. Zachariae et al., "Anti-interleukin-17 monoclonal antibody ixekizumab in chronic plaque psoriasis," *New England Journal of Medicine*, vol. 366, no. 13, pp. 1190–1191, 2012.

[136] V. Ren and H. Dao Jr., "Potential role of ixekizumab in the treatment of moderate-to-severe plaque psoriasis," *Clinical, Cosmetic and Investigational Dermatology*, vol. 6, pp. 75–80, 2013.

[137] J. A. Lust, M. Q. Lacy, S. R. Zeldenrust et al., "Induction of a chronic disease state in patients with smoldering or indolent multiple myeloma by targeting interleukin 1β-induced interleukin 6 production and the myeloma proliferative component," *Mayo Clinic Proceedings*, vol. 84, no. 2, pp. 114–122, 2009.

[138] P. Rider, Y. Carmi, R. Yossef et al., "IL-1 receptor antagonist chimeric protein: context-specific and inflammation-restricted activation," *The Journal of Immunology*, vol. 195, no. 4, pp. 1705–1712, 2015.

[139] M. G. Netea, F. L. Van De Veerdonk, J. W. M. Van Der Meer, C. A. Dinarello, and L. A. B. Joosten, "Inflammasome-independent regulation of IL-1-family cytokines," *Annual Review of Immunology*, vol. 33, pp. 49–77, 2015.

[140] L. A. B. Joosten, M. G. Netea, G. Fantuzzi et al., "Inflammatory arthritis in caspase 1 gene-deficient mice: contribution of proteinase 3 to caspase 1-independent production of bioactive interleukin-1β," *Arthritis and Rheumatism*, vol. 60, no. 12, pp. 3651–3662, 2009.

[141] F. L. van de Veerdonk, M. G. Netea, C. A. Dinarello, and L. A. B. Joosten, "Inflammasome activation and IL-1β and IL-18 processing during infection," *Trends in Immunology*, vol. 32, no. 3, pp. 110–116, 2011.

[142] M. Irmler, S. Hertig, H. R. MacDonald et al., "Granzyme A is an interleukin 1β-converting enzyme," *Journal of Experimental Medicine*, vol. 181, no. 5, pp. 1917–1922, 1995.

[143] A. Mantovani, "Cancer: inflammation by remote control," *Nature*, vol. 435, no. 7043, pp. 752–753, 2005.

[144] Y. Carmi, E. Voronov, S. Dotan et al., "The role of macrophage-derived IL-1 in induction and maintenance of angiogenesis," *The Journal of Immunology*, vol. 183, no. 7, pp. 4705–4714, 2009.

[145] Y. Carmi, S. Dotan, P. Rider et al., "The role of IL-1β in the early tumor cell-induced angiogenic response," *Journal of Immunology*, vol. 190, no. 7, pp. 3500–3509, 2013.

[146] I. Cohen, P. Rider, Y. Carmi et al., "Differential release of chromatin-bound IL-1α discriminates between necrotic and apoptotic cell death by the ability to induce sterile inflammation," *Proceedings of the National Academy of Sciences of the United States of America*, vol. 107, no. 6, pp. 2574–2579, 2010.

[147] I. Cohen, P. Rider, E. Vornov et al., "Corrigendum: IL-1α is a DNA damage sensor linking genotoxic stress signaling to sterile inflammation and innate immunity," *Scientific Reports*, vol. 6, Article ID 19100, 2016.

[148] P. Rider, Y. Carmi, O. Guttman et al., "IL-1α and IL-1β recruit different myeloid cells and promote different stages of sterile inflammation," *Journal of Immunology*, vol. 187, no. 9, pp. 4835–4843, 2011.

[149] P. Rider, I. Kaplanov, M. Romzova et al., "The transcription of the alarmin cytokine interleukin-1 alpha is controlled by hypoxia inducible factors 1 and 2 alpha in hypoxic cells," *Frontiers in Immunology*, vol. 3, article 290, 2012.

[150] C. A. Dinarello, "Why not treat human cancer with interleukin-1 blockade?" *Cancer and Metastasis Reviews*, vol. 29, no. 2, pp. 317–329, 2010.

[151] C. A. Dinarello, "An expanding role for interleukin-1 blockade from gout to cancer," *Molecular Medicine*, vol. 20, supplement 1, pp. S43–S58, 2014.

[152] C. A. Dinarello, "Interleukin-1α neutralisation in patients with cancer," *The Lancet Oncology*, vol. 15, no. 6, pp. 552–553, 2014.

[153] D. S. Hong, D. Hui, E. Bruera et al., "MABp1, a first-in-class true human antibody targeting interleukin-1α in refractory cancers: an open-label, phase 1 dose-escalation and expansion study," *The Lancet Oncology*, vol. 15, no. 6, pp. 656–666, 2014.

[154] G. A. Efimov, A. A. Kruglov, Z. V. Khlopchatnikova et al., "Cell-type-restricted anti-cytokine therapy: TNF inhibition from one pathogenic source," *Proceedings of the National Academy of Sciences of the United States of America*, vol. 113, no. 11, pp. 3006–3011, 2016.

16

Nesprin-2 Interacts with Condensin Component SMC2

Xin Xing,[1,2] **Carmen Mroß,**[1,2] **Linlin Hao,**[1,2] **Martina Munck,**[1,2]
Alexandra Herzog,[1,2] **Clara Mohr,**[1,2] **C. P. Unnikannan,**[1,2] **Pranav Kelkar,**[1,2]
Angelika A. Noegel,[1,2] **Ludwig Eichinger,**[1,2] **and Sascha Neumann**[1,2]

[1]*Institute of Biochemistry I, Medical Faculty, University Hospital Cologne, Joseph-Stelzmann-Str. 52, 50931 Cologne, Germany*
[2]*Center for Molecular Medicine Cologne (CMMC) and Cologne Cluster on Cellular Stress Responses in Aging-Associated Diseases (CECAD), Medical Faculty, University of Cologne, Cologne, Germany*

Correspondence should be addressed to Angelika A. Noegel; noegel@uni-koeln.de and
Ludwig Eichinger; ludwig.eichinger@uni-koeln.de

Academic Editor: Arnoud Sonnenberg

The nuclear envelope proteins, Nesprins, have been primarily studied during interphase where they function in maintaining nuclear shape, size, and positioning. We analyze here the function of Nesprin-2 in chromatin interactions in interphase and dividing cells. We characterize a region in the rod domain of Nesprin-2 that is predicted as SMC domain (aa 1436–1766). We show that this domain can interact with itself. It furthermore has the capacity to bind to SMC2 and SMC4, the core subunits of condensin. The interaction was observed during all phases of the cell cycle; it was particularly strong during S phase and persisted also during mitosis. Nesprin-2 knockdown did not affect condensin distribution; however we noticed significantly higher numbers of chromatin bridges in Nesprin-2 knockdown cells in anaphase. Thus, Nesprin-2 may have an impact on chromosomes which might be due to its interaction with condensins or to indirect mechanisms provided by its interactions at the nuclear envelope.

1. Introduction

The nucleus of a eukaryotic cell harbors the genetic material that is organized in long DNA polymers and is associated with numerous proteins to form chromatin. Chromatin is separated from the cytoplasm by the nuclear envelope (NE), a continuous membrane system consisting of an inner (INM) and an outer nuclear membrane (ONM) enclosing the perinuclear space (PNS). Both membranes are connected at the nuclear pore complexes, the ONM continues into the endoplasmic reticulum (ER). The NE is not a simple membrane barrier but is lined with and crossed by large protein assemblies that provide it with various cellular functions. Nesprins (nuclear envelope spectrin repeat proteins) together with SUN proteins are central components of the NE. Currently four Nesprins (Nesprins-1–4) are known in mammals. They reside at the INM and ONM, have different sizes, and exist in many isoforms [1]. Nesprins are characterized by a varying number of spectrin repeats followed by a C-terminal KASH (Klarsicht, ANC-1, Syne Homology) domain which anchors the proteins in the nuclear membrane and interacts with the SUN domain of SUN proteins in the perinuclear space [2]. Nesprin-1 and Nesprin-2 harbor at their N-terminus paired calponin homology domains that mediate the binding to F-actin [3, 4]. The N-terminus of Nesprin-3 binds to plectin, a cytoskeletal crosslinker that establishes the connection to the intermediate filament system [5]. Nesprin-4 interacts with kinesin-1, a motor protein that uses microtubules as cellular routes [6]. Microtubule interaction through kinesin-1 has also been described for Nesprin-2 [7].

Based on the nucleo-cytoskeletal interactions, Nesprins integrate the nucleus into the cytoskeleton of a cell and participate in the maintenance of nuclear shape and stability [8, 9]. The spectrin repeats (SRs) are platforms for protein-protein or self-interactions [10]. Furthermore, the number of SRs and therefore the length of the rod have been proposed to modulate the size of the nucleus [11]. In the central SR domain, an additional domain has been described in Nesprin-2, an SMC (structural maintenance of chromosomes) domain encompassing amino acid residues 1,464–1,771, which was identified by Dawe et al. [12] as an interaction site for meckelin, a protein with functions in the formation of primary cilia. Primary cilia are sensory organs that act as mechanoreceptors in various signaling pathways or sensors of chemical stimuli [13].

SMC proteins have core functions in regulating genome stability and the organization of the genetic material. They are present from bacteria to man [14]. Classical SMC proteins are composed of 1,000–1,300 amino acids. They have two coiled-coil regions interrupted by a central hinge. The coiled coils fold back on themselves and form an extended structure. At their ends, the N- and C-termini of a molecule interact with each other to form a globular ATP-binding domain [15]. The hinge regions are responsible for heterodimerization of SMC molecules [16]. Six SMC proteins have been described in man, SMC1-6. SMC1/3 form the core of the cohesin complex which mediates sister chromatid cohesion; SMC2/4 are present in the condensin complex that acts in chromosome assembly and segregation. They are present in two condensin complexes with distinct roles, condensins I and II, which contain SMC2 and SMC4 in combination with different non-SMC subunits. Condensins I and II are associated sequentially with chromosomes during the cell cycle and have different roles for chromosome architecture. Condensin I is not present in the nucleus in interphase. During mitosis, condensin I is required for removal of cohesin from chromosome arms and for chromosome shortening, whereas condensin II plays a role in chromosome condensation during early prophase [17]. Condensin I is, however, not completely excluded in interphase from the nucleus since a small pool was found in association with intergenic and intronic regions during interphase [18, 19]. By contrast, condensin II is always nuclear. It is associated with DNA throughout interphase and concentrates on chromosomes in prophase. Based on its interphase distribution, a role in nuclear architecture was proposed [20, 21]. Cohesin and condensin complexes have also roles in DNA repair and gene regulation throughout the cell cycle [20]. Moreover, condensin is involved in organizing the chromatin allowing intrachromosomal associations of gene loci as shown in fission yeast [22]. SMC5/6 is mainly implicated in DNA damage repair and DNA recombination and has specific roles in meiosis [23, 24].

We have carried out a biochemical and functional characterization of the Nesprin-2-SMC domain, hereafter referred to as Nesprin-2-SMC. We show that it can self-assemble to form dimers, trimers, and higher order structures and can interact with condensin proteins SMC2 and SMC4. Monoclonal antibodies directed against the SMC domain showed a distribution of the Nesprin-2 isoforms containing the SMC domain along the NE during interphase and a presence at the chromosomes during mitosis. We also uncovered an impact of Nesprin-2 on mitotic chromosomes that might be mediated by an interaction with the condensin core units SMC2/4.

2. Materials and Methods

2.1. Cell Culture, Transfection, and Cell Synchronization.
HaCaT (human keratinocyte cell line), COS7 (African green

monkey kidney fibroblasts), and HeLa (human cervical cancer cells) cells were grown in a humidified atmosphere containing 5% CO_2 at 37°C in DMEM (high glucose, Life Technologies) supplemented with 10% fetal bovine serum (FBS), 2 mM Glutamine (SIGMA), and 1% penicillin/streptomycin. Cells were transfected as described [11]. To knock down Nesprin-2, HaCaT cells were transfected twice at intervals of 72 h using the Amaxa Nucleofector Kit V Solution (Lonza). The plasmids used for knockdown of Nesprin-2 targeting the N-terminus and the C-terminus (Nesprin-2 N-term shRNA, Ne-2 N-term KD; Nesprin-2 C-term shRNA, Ne-2 C-term KD) as well as the control have been described previously [7]. The newly generated plasmids are described below. For cell cycle synchronization, HaCaT cells were treated with thymidine (2 mM) for 24 h and then with Nocodazole (100 ng/ml) for 12 h or alternatively first with 9 μM RO-3306 (Santa Cruz Biotechnology, sc-358700) for 20 to 22 h and then approximately 3 h release (depending on the desired mitotic phase) to obtain mitotic phases. RO-3306 is a CDK1 inhibitor and reversibly arrests proliferating cells at the G2/M phase of the cell cycle [26]. FACS analysis of cell cycle stages was performed with unsynchronized and synchronized cells. Staining was done with Nuclear-ID™ Red DNA (Enzo ENZ-52406).

Determination of cell proliferation was done by plating at one time point six wells each with the same number of cells and then counting two wells after 24 h, two after 48 h, and two after 72 h.

2.2. Cloning Strategies.
cDNAs from HaCaT cells encoding the SMC domain in Nesprin-2 (AAN60443, aa 1436–1766, and SRs 11–13) were used as PCR templates using the primers for the following: 5′ GAATTCAATGAACTC-CTTAAAAATATTCAAGATGTG 3′, rev: 5′ GAATTC-CTCGAGGGATTCAGTCATCCCGATCTGGGTCTTGG 3′ that contain *EcoRI* restriction sites for cloning into pGEX-4T1 (Amersham) yielding pGEX-4T1-Nesprin-2-SMC which encodes GST-Nesprin-2-SMC. GST is located at the amino terminus of the protein. Nesprin-2-SMC sequences were generated by PCR and cloned into pCMV-Myc (GE Healthcare) using pGEX-4T1 Nesprin-2-SMC as template and primers with *EcoRI* or *XhoI* restriction sites, SMC (SR11–13) for the following: 5′ GAATTCTGAATGAACTCC-TTAAAAATATTCAAGATGTG 3′, rev: 5′ CTCGAGCTA-GAGGGATTCAGTCATCCCGATCTGGGTCTT 3′. SR11 for: 5′ GAATTCTGAATGAACTCCTTAAAAATATTC-AAGATGTG 3′, rev: 5′ CTCGAGCTATCTCCCACATTG-TTCAAGACATTCGGTGAC 3′, SR12 for: 5′ CTCGAG-GTTTTGGAGCTCTTAAAACAATATCAGAAT 3′, rev: 5′ CTCGAGCTAACCAAGATTTTCATAGTAATCTTC-TGTCTT 3′, SR13 for: 5′ GAATTCTGCGAGCTCTAGCTT-TGTGGGACAAACTTTTTA 3′, rev: 5′ CTCGAGCTA-GAGGGATTCAGTCATCCCGATCTGGGTCTT 3′. Myc-SR53–56 corresponding to residues 6146–6799 of Nesprin-2 is described in Schneider et al. [7].

A Nesprin-2 SMC domain specific shRNA (Ne-2 SMC) was generated as described using the following oligonucleotides: sense 5′-ATTCTCCTGTTAAGC-ACTTCTGTACATGGAAGCTTGCATGTATAGGAG-TGCTTAGCAGGAGAATCCATTTTTT-3′, antisense 5′-GATCAAAAAATGGATTCTCCTGCTAAGCACTCC-TATACATGCAAGCTTCCATGTACAGAAGTGCTT-AACAGGAGAATCG-3′ and a random control using sense 5′-CCTTTCAGATACGTCTTGTACAGGTATTGAAGC-TTGAATGCCTGTACAGGATGTATCTGAAAGGCG-ATTTTTT-3′ and antisense 5′ GATCAAAAAATCGCC-TTTCAGATACATCCTGTACAGGCATTCAAGCTT-CAATACCTGTACAAGACGTATCTGAAAGGCG-3′ oligonucleotides [27]. The efficiency of the knockdown was evaluated by immunofluorescence and western blot analysis. Knockdown of SMC2 in COS7 cells was achieved with SMC2-specific siRNAs (E-006836-00-0005, Dharmacon, GE Healthcare). For control, corresponding scrambled shRNA was used. The cell line was recommended by the supplier in combination with the particular siRNAs. Transfection was carried out using Dharmafect transfection reagent according to the manufacturer's protocol. The cells were analyzed 96 h after the transfection. Successful knockdown was assessed by immunofluorescence analysis using SMC2 specific antibodies.

2.3. Expression and Purification of GST Proteins and GST Pulldown.

Plasmids encoding GST fusion proteins were transformed into *E. coli* XL-1 blue and grown overnight and diluted 1 : 50 into fresh LB media. The bacteria were grown to an OD_{600} of 0.6 to 0.8 when they were induced with 0.5 mM IPTG and the protein expression was continued overnight at 20°C. Bacteria were pelleted and washed with STE buffer (10 mM Tris-HCl, pH 8.0, 50 mM NaCl, and 1 mM EDTA). Lysis was achieved by the addition of 100 μg/ml lysozyme and mechanical shearing in a Dounce homogenizer followed by centrifugation. Fusion proteins were bound to Glutathione-Sepharose 4B (GE Healthcare). The GST-Nesprin-2-SMC polypeptide has a predicted molecular weight of 64.8 kDa. It was efficiently expressed in *E. coli* XL-1 blue and purified as soluble proteins. The protein was bound to Glutathione-Sepharose beads and Nesprin-2-SMC was released from the GST part by thrombin cleavage (Sigma-Aldrich). Alternatively, GST-Nesprin-2-SMC was eluted from the beads with reduced glutathione (20 mM) in 100 mM Tris-HCl, pH 8.0.

GST pulldown assays were performed by lysing HaCaT or COS7 cells in lysis buffer (50 mM Tris-HCl, pH 7.5, 150 mM NaCl, 1% Nonidet P-40, and 0.5% sodium deoxycholate) supplemented with protease inhibitor cocktail (Sigma-Aldrich) by pushing them through a 0.4 mm needle followed by sonication and centrifugation. Cell lysates were incubated with Glutathione-Sepharose beads overnight for binding to the GST fusion proteins or GST and washed 5 times with PBS or lysis buffer supplemented with protease inhibitors. Beads bound protein complexes were analyzed by SDS-PAGE and western blot (WB).

2.4. Antibodies and Immunofluorescence (IF) Microscopy.

The following antibodies were used: mouse monoclonal anti-Nesprin-2 mAb K20-478 raised against the actin binding domain (ABD) of Nesprin-2 (residues 1–285) [3] (IF, 1 : 200;

hybridoma supernatant, WB, 1 : 10), rabbit polyclonal antibodies pAbK1 raised against spectrin repeats in the C-terminal region of Nesprin-2 [28] (IF, 1 : 100; WB, 1 : 1,000), Nesprin-1 specific mAb K43-322-2 raised against N-terminal spectrin repeats 10 and 11 of Nesprin-1 [29] (hybridoma supernatant, undiluted), GFP-specific mAb K3-184-2 [30] (hybridoma supernatant, IF, 1 : 2; WB, 1 : 10), Myc-specific mAb 9E10 [31] (hybridoma supernatant, IF, undiluted; WB, 1 : 10), pAb against GST [32] (WB, 1 : 50,000), mAb K84-913 against GST (hybridoma supernatant, WB 1 : 10), pAb Lamin B1 (Abcam ab16048, IF, 1 : 200; WB, 1 : 4,000), pAb SMC2 (Novus Biologicals NB100-373, IF, 1 : 100; WB, 1 : 2,000), WB: mAb SMC4 (Abcam ab179803 1 : 2,000), IF: pAb SMC4 (Abcam ab17958, 1 : 500), pAb SMC1 (Abcam ab21583, WB 1 : 1000), goat SMC3 (Santa Cruz Biotechnology, sc-8135, WB 1 : 50), rabbit CAP-H (Biomol-Bethyl A300-603A-T, WB 1 : 1000), pAb CAP-H2 (Biomol-Bethyl A302-275A, WB 1 : 4000), mAb PDI (Abcam ab2792, 1 : 100), pAb calreticulin (Thermo Fisher PA3-900, IF 1 : 50–200), and rat mAb YL1/2 specific for α-tubulin (1 : 5). mAb K81-116-6 (hybridoma supernatant, undiluted) directed against the SMC domain in Nesprin-2 was generated in this study. The antibodies were used for immunofluorescence and western blot analysis. A polypeptide corresponding to Nesprin-2 aa 1436–1766 (calculated molecular weight 38.78 kDa) was produced as GST fusion polypeptide and bound to Glutathione-Sepharose beads as described above. The SMC polypeptide was liberated by thrombin cleavage and used for production of monoclonal antibodies by immunization of mice as described [33]. Alexa 568 or 488 fluorescently labeled and highly cross absorbed and affinity purified secondary antibodies were used (Thermo Fisher), and 4,6-diamino-2-phenylindole (DAPI, Sigma) was used to visualize DNA. For immunofluorescence, cells grown on cover slips were fixed in 3% paraformaldehyde (PFA) in phosphate-buffered saline (PBS) for 15 min followed by 4 min incubation with 0.5% Triton X-100/PBS. Alternatively, cells were fixed by 10 min incubation in ice cold methanol at −20°C. Blocking was done with PBG (0.5% BSA, 0.045% fish gelatine in PBS, pH 7.4) at room temperature (RT) for 30 min. Primary and secondary antibodies as well as DAPI were diluted in PBG and applied to the cells for 1 h at RT or overnight at 4°C. Microscopy was performed by using TCS-SP5 (Leica) or the Ångstrom Opti Grid confocal microscope (Leica). For control, cells were routinely labeled with secondary antibodies only. In no case was a signal obtained.

To test the specificity of the newly established mAb K81-116-6, the antibodies were removed from the hybridoma supernatant (depletion) and the supernatant was then used for immunofluorescence analysis. Depletion was performed in two ways. For one, the hybridoma supernatant was incubated with Glutathione-Sepharose beads carrying GST-Nesprin-2-SMC polypeptides. The beads were removed by centrifugation (2000 rpm, 2 min) and the supernatant was used for immunofluorescence analysis. Alternatively, GST-Nesprin-2-SMC was loaded onto a SDS-polyacrylamide gel, the protein was then transferred to a nitrocellulose membrane, detected by Ponceau S staining, and the part of the membrane carrying GST-Nesprin-2-SMC protein was cut out and incubated with mAb K81-116-6. After overnight incubation (4°C), the solution was removed from the membrane and applied for IF. For both approaches, an aliquot of the antibody solution before depletion was kept for control.

2.5. Immunoprecipitation. For immunoprecipitation (IP), HaCaT cells were harvested and lysed in lysis buffer (50 mM Tris-HCl, pH 7.5, 150 mM NaCl, 1% Nonidet P-40, 0.5% sodium deoxycholate, and protease inhibitor cocktail). Cells were lysed by pushing and pulling through a 0.4 mm needle and centrifuged (12.000 rpm, 20 min). Supernatants were incubated for 1 h with protein A Sepharose CL-4B beads (GE Healthcare) for preclearing. Subsequently, beads were removed by centrifugation (2000 rpm, 2 min) and cell lysates incubated with 5–8 μg of the antibody of interest for 2 h at RT. Protein A Sepharose CL-4B beads equilibrated with lysis buffer were then added to the cell lysates and incubation was continued overnight at 4°C. The beads were collected by centrifugation and washed five times with PBS and the bound proteins released from the beads by addition of SDS sample buffer and heating to 95°C for 5 min and analyzed by SDS-PAGE (3–12% acrylamide for gradient gels; 10% and 12% acrylamide as appropriate) and western blotting. Transfer of high molecular weight Nesprin-2 giant to nitrocellulose membranes (0.22 μm pore size) was done by wet blotting technique for two to three days.

2.6. Gel Filtration and Chemical Cross-Linking. To assess the oligomeric state of the native protein, the sample was applied to a gel filtration column (Sephadex G-200, GE Healthcare) as described [34]. For molecular weight determination, molecular weight standards (GE Healthcare) were separated under identical conditions. Chemical cross-linking of Nesprin-2-SMC (1 mg/ml) was performed with the zero-length cross-linking reagent EDC (1-ethyl-3-[3-dimethylaminopropyl]carbodiimide hydrochloride) (Thermo Fisher) together with sNHS (Sulfo-N-hydroxysuccinimide) in 0.1 M MES buffer (pH 6.5) [35].

3. Results

3.1. Nesprin-2 Contains an SMC Domain in Its Rod Domain. We investigate here a region in the SR containing rod domain of Nesprin-2 with homology to the SMC (Structural Maintenance of Chromosomes) domain (E value $9.34e - 0.3$). This domain encompasses amino acids 1436–1766 and extends over SR11–13 designated Nesprin-2-SMC (Figure 1(a)) [36]. In a comparison with mammalian SMC proteins, we found high degrees of homology with the coiled-coil regions of SMC2 and SMC4 (19.7% identity, 52.9% similarity and 21.5% identity, 53.9% similarity, resp.) (Figure 1(b)). To assess whether Nesprin-2-SMC can undergo self-interactions, we expressed it as GST fusion protein and analyzed the elution behavior of the 39 kDa polypeptide, which had been released from GST by thrombin cleavage, by size exclusion chromatography. The protein eluted in two peaks, one eluting at ~50 kDa and corresponding to the monomer and a broader and larger one eluting between 75 kDa and 158 kDa indicative of oligomers (Figure 1(c)). The proteins used for calibrating the column are

globular proteins, whereas Nesprin-2-SMC is expected to be a rod shaped molecule presumably affecting the elution behavior. The elution pattern was also confirmed by SDS-PAGE and staining with Coomassie Blue which showed that the protein eluted in fractions in front of ovalbumin indicating an oligomeric state (Figure S1(a)). Cross-linking experiments using varying concentrations of the zero-length cross-linking reagent EDC showed the presence of monomers, dimers, trimers, and even higher molecular weight complexes. With decreasing EDC concentration, the amount of higher molecular weight forms decreased whereas the monomeric form increased (Figure 1(d)). The oligomerization property of Nesprin-2-SMC was supported by data from pulldown experiments in which GST-Nesprin-2-SMC precipitated Nesprin-2 giant from HaCaT cell lysates (see Materials and Methods for experimental details). Human Nesprin-2 giant is a 6,885-amino-acid protein with a predicted molecular weight of 796 kDa. Mass spectrometric analysis identified peptides covering the entire Nesprin-2 giant molecule in the precipitate (Figure S1(b)). The high coverage of the sequence located between residues 1436 and 1766 was due to the polypeptide used for the pulldown. GST did not precipitate Nesprin-2.

We further expressed Myc-tagged Nesprin-2-SMC (Myc-Nesprin-2-SMC) corresponding to the full length SMC domain of Nesprin-2 and Myc-tagged polypeptides corresponding to its individual SR domains in COS7 cells and used the cell lysates for pulldown experiments with GST-Nesprin-2-SMC (Figure 1(e)). GST-Nesprin-2-SMC precipitated Myc-Nesprin-2-SMC and its individual SRs from COS7 cell lysates as shown in the immunoblot using Myc-specific antibody mAb 9E10 (Figure 1(f)). Taken together, the results suggest that the Nesprin-2-SMC domain has the potential to oligomerize. We then asked whether this interaction is specific to this Nesprin-2 domain and tested whether GST-Nesprin-2-SMC could interact with other spectrin repeats of Nesprin-2. We therefore expressed Myc-SR53–56 composed of the last four spectrin repeats of Nesprin-2 (SR53–SR56, aa 6116–6799, Figure 1(a)) in COS7 cells and carried out pulldown assays with GST for control and GST-Nesprin-2-SMC [7]. GST-Nesprin-2-SMC did not precipitate Myc-SR53–56 underlining the specificity of the interaction (Figure 1(g)).

3.2. Monoclonal Nesprin-2-SMC Domain Specific Antibodies Detect a High Molecular Weight Protein and Stain the Nuclear Envelope. To study Nesprin-2 isoforms harboring the SMC domain, we generated monoclonal antibodies by immunizing mice with Nesprin-2-SMC polypeptide that had been released from the GST part by thrombin cleavage. In western blots of HaCaT cell homogenates that had been separated in gradient gels (3–12% acrylamide) mAb K81-116-6 recognized primarily a high molecular weight protein which we presume corresponds to the ~800 kDa Nesprin-2 giant [3]. Faint bands below could be degradation products or N-terminal isoforms [1] (Figure 2(a)). In independent experiments, in which we immunoprecipitated Nesprin-2 from HaCaT cells and probed the precipitate with SMC2 and SMC4 antibodies, we excluded that any of the lower molecular weight bands corresponded to SMC proteins due to cross reactivity of the antibodies (data not shown). In mAb K20-478   mAb K81-116-6   pAbK1

ABD — 1 — // — 11 12 13 — // — 53 54 — 55 56 — KASH   Nesprin-2 giant

1   285   1431   1766   6146   6799   6885

(a)

```
           1470      1480      1490      1500      1510      1520      1530      1540      1550      1560      1570
Ne-2  KKSLIRLDKVLDEYEEEKRHLQEMANSLPHF--KDGREKTVNQQCQNTVVLWENTKALVTECLEQCGRVLELLKQYQNFKSILTTLIQKEESVISLQASYMGKENLKK
      . :...  : :..  .:   .    ::.. ..  . ::.. ..  .:.         .   :.:       .   . . :  . .  :::.. ..  :    : ::.
SMC2  EKNMVEDSKTLAAKEKEVKKITDGLHALQEASNKDAEALAAAQQHFNAV-----SAGLSSN---EDGAEATLAGQMMACKNDISK-AQTEAKQAQMKLKHAQQE-LKN
           340       350       360       370       380       390       400       410       420       430

           1580      1590      1600      1610      1620      1630      1640      1650      1660      1670      1680
      RIAEIEIVKEEFNEHLEVVDKINQVCKNLQFYLNKMKTFEEPPFEKEANIIVDRWL---DINEKTEDYYENLGRALALWDKLFNLKNVIDEWTEKALQKMELHQLTEEDRERLKE-ELQV
      :  :.   .... .  :  . .  ::.. ..  :  . ..     .  :::..  :   :   . . .:::.   .     .. . ...:     : .. .  ..      ..     : .
SMC2  KQAEVKKMDSGYRKDQEALEAVKRLKEKLEAEMKKLN-YEE---NKEESLLEKRRQLSRDIGRLKETYEALLARFPNL---RFAYKDPEKNWNRNCVKGLVASLISVKDTSATTALELVA
           440       450       460       470       480       490       500       510       520       530       540

           1690      1700      1710      1720      1730      1740
      HEQKTSEFSRRVAEIQFLLQSSE-------IPLELQVMESSILNKMEHVQKCLTGESNCH
      :. .: .     .   ::.   .:       :::.. ..      :  . : .:.: :  :
SMC2  GERLYNVVVDTEVTGKKLLERGELKRRYTIIPLN-KISARCIAPETLRVAQNLVGPDNVH
           550       560       570       580       590       600
```

```
           1470      1480      1490      1500      1510      1520      1530      1540      1550      1560      1570
Ne-2  KKSLIRLDKVLDEYEEEKRHLQEMANSLPHFKDGREKTVNQQCQN----TVVLWENTKALVTECLEQCGRVLELLKQYQNFKSILTTLIQKEESVISLQASYMGKEN-----LKKRIAEI
      :   ::.    .  :   .  .   .. .. .  ..     . :.    .:  :  .:  :   ..     . ...  . ..:   .    .   . . . . ..:         .  :.:
SMC4  QKRIAEMETQKEKIHEDTKEINEKSNILSNEMKAKNKDVKDTEKKLNKITKFIEENKEKFTQLDLEDV-QVREKLKHATSKAKKLEKQLQKDKEKVEEFKSIPAKSNNIINETTTRNNAL
           340       350       360       370       380       390       400       410       420       430       440

           1580      1590      1600      1610      1620      1630      1640      1650      1660      1670      1680
      EIVKEEFNEHL-EVVDKINQVCKNLQFYLNKMKTFEEPPFEKEANIIVDRWLDINEKTEDYYENLGRALALWDKLFNLKNVIDEWTEK------ALQKME--LHQLTEEDRERLKEELQV
      :  ::.      .  . .  ::.. ..  :  . ..     .  :::..  :  :   . . .: ..    .     .. . ...:           . .   :  ..      ..    : .
SMC4  EKEKEKEEKKLKEVMDSLKQETQGLQ-KEKESREKELMGFSKSVNEARSK-MDVAQSELDIY--LSRHNTAVSQLTKAKEALIAASETLKERKAAIRDIEGKLPQTEQELKEKEKE-LQK
      450       460       470       480       490       500       510       520       530       540       550       560

           1690      1700      1710      1720      1730
      HEQKTSEFSRRVAEIQFLLQSSEIPLELQVMESSILNKMEHVQK
      :. .: .     .  ::.   .:       :::.. ..      :
SMC4  LTQEETNFKSLVHDLFQKVEEAKSSLAMNRSRGKVLDAIIQEKK
           570       580       590       600
```

(b)

OD 280 nm

0,013
0,008
−0,003
−0,002
−0,007

158 kDa   75 kDa   43 kDa

1.500   2.000   2.500

(ml)

11 12 13

(c)

Higher
~120
~80
(kDa)
~40

Decreasing EDC concentration

(d)

Myc-Nesprin-2-SMC	1436	Myc-11 12 13	1766
Myc-SR11	1436	Myc-11	1531
Myc-SR12	1532	Myc-12	1641
Myc-SR13	1642	Myc-13	1766

(e)

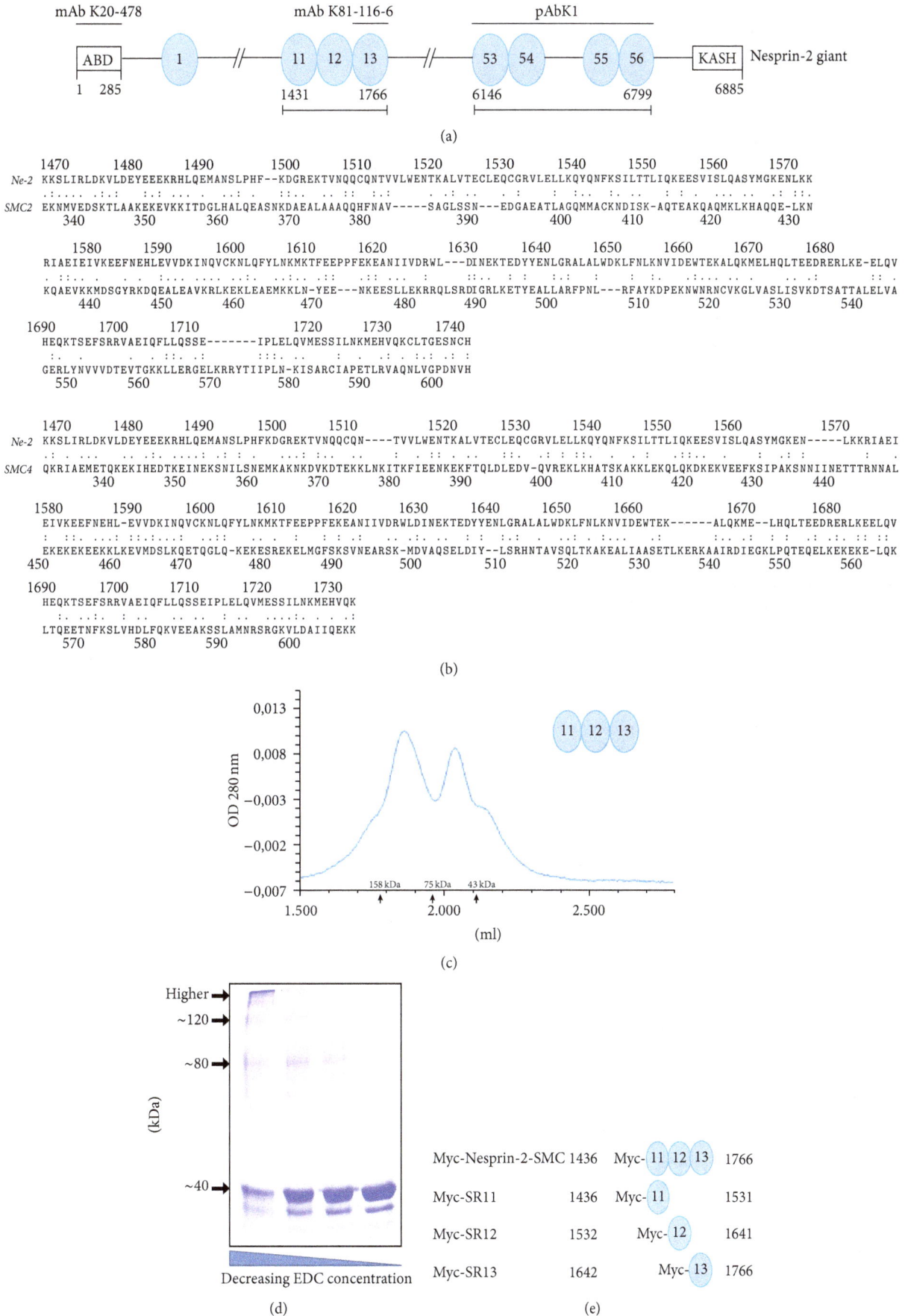

FIGURE 1: Continued.

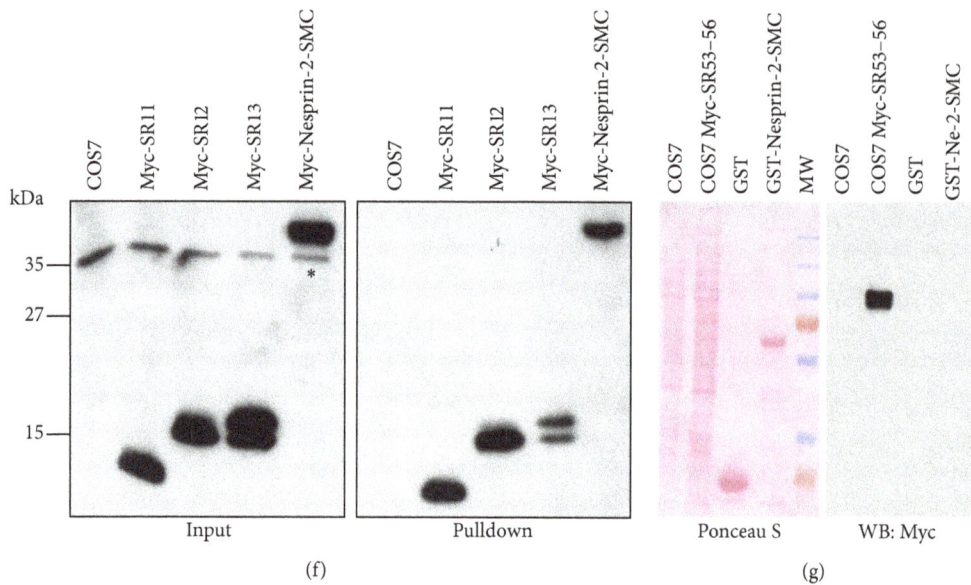

FIGURE 1: Characterization of the SMC domain of Nesprin-2. (a) Schematic of Nesprin-2 (not drawn to scale). The location of the SMC domain (spectrin repeats 11–13) and the C-terminal spectrin repeats (53–56) is shown. Epitopes of antibodies used are indicated above the schematic. ABD, actin binding domain; ovals, spectrin repeats. The spectrin repeat domain starts at position 308. (b) Sequence comparison of the Nesprin-2-SMC domain with coiled-coil regions of SMC2 and SMC4. The sequence comparison was performed using LALIGN, the Pairwise Sequence Alignment tool from EMBL-EBI (https://www.ebi.ac.uk/Tools/psa/lalign/). Nesprin-2 (NCBI GenBank accession number AF435011.1), SMC2 (NCBI GenBank accession number O95347.2), and SMC4 (NCBI GenBank accession number Q8WXH0.3) were used. :, identical amino acid; ., conservative substitution. (c) Analysis of Nesprin-2-SMC by gel filtration chromatography. UV traces of the elution profile are shown. Nesprin-2 SMC (calculated molecular weight 39 kDa). Molecular weight markers were ovalbumin (43 kDa), conalbumin (75 kDa), and aldolase (158 kDa). (d) Analysis of chemically crosslinked Nesprin-2-SMC. Zero-length cross-linking reagent EDC (1-ethyl-3-[3-dimethylaminopropyl] carbodiimide hydrochloride) was used at decreasing concentrations. The proteins were separated by SDS-PAGE (10% acrylamide) and stained with Coomassie Blue. (e) Schematic representation of Myc-tagged Nesprin-2-SMC polypeptides. Amino acid positions refer to human Nesprin-2 giant (accession number AF435011.1). (f) Interaction of GST-Nesprin-2-SMC with individual Myc-tagged spectrin repeats derived from Nesprin-2-SMC and expressed in COS7 cells. GST-Nesprin-2-SMC was used for pulldown (right panel). Western blots were probed with mAb 9E10 specific for Myc. Asterisk, endogenous Myc [25]. (g) Specificity of the Nesprin-2-SMC interaction. Myc-SR53–56 expressed in COS7 cells was used for pulldowns with GST for control and GST-Nesprin-2-SMC. COS7 and COS7 Myc-SR53–56 represent whole cell lysates. The Ponceau S stained blot and the corresponding blot probed with mAb 9E10 are shown. MW, molecular weight marker (from top to bottom: 200, 130, 100, 70, 55, 35, and 25 kDa).

immunofluorescence analysis, mAb K81-116-6 labeled the NE in HaCaT and HeLa cells overlapping with the pAbK1 staining (Figure 2(b)). The previously characterized pAbK1 polyclonal antibodies had been generated against the four C-terminal spectrin repeats of Nesprin-2 and are specific for Nesprin-2 (Figure 1(a)) [28]. In addition, mAb K81-116-6 stained structures in the cytoplasm in the vicinity of the nucleus which are possibly membranes of the endoplasmic reticulum (ER) as we observed colocalization with calreticulin, an ER protein (Figure 2(b), lower panel). The cytoplasmic staining was comparatively faint in HaCaT cells, whereas in HeLa cells it was more pronounced. pAbK1 also stained these structures; however the staining was less intense which might be due to different accessibility of the epitopes (Figure 2(b)). Nesprin-2 is a tail-anchored protein and its mRNA has been found anchored to the ER where it is translated. This might explain the observed localization [37].

To prove the specificity of mAb K81-116-6, we carried out antibody depletion studies. We found that the staining of the NE as well as the cytoplasmic staining was completely abrogated after depletion of mAb K81-116-6 from

the hybridoma supernatant by incubating the supernatant with nitrocellulose membrane strips carrying GST-Nesprin-2-SMC or with Glutathione-Sepharose 4B beads carrying GST-Nesprin-2-SMC. By contrast, the NE was still labeled by pAbK1 (Figure 2(c)). Furthermore, the protein was no longer detected in cell lysates after knocking down Nesprin-2 using shRNA directed against the SMC domain (Figure S2(a)) and no signals were detected when cells were analyzed by immunofluorescence (see below, Figures 4(b) and 4(c)).

3.3. SMC2 Is a Nesprin-2 Binding Partner. To identify binding partners for Nesprin-2, we performed immunoprecipitation experiments using mAb K20-478 directed against the N-terminus of Nesprin-2 and pAbK1 (Figure 1(a)). The proteins were separated by SDS-PAGE and stained with Coomassie Blue, bands were cut out, and the proteins were identified by mass spectrometry. For control GFP-specific antibody mAb K3-184-2 was used. Among the precipitated proteins were histones, SUN1, Lamin A/C, and SMC2 which were found in the immunoprecipitate of mAb K20-478. The SUN1 and Lamin A/C interactions have been previously described and

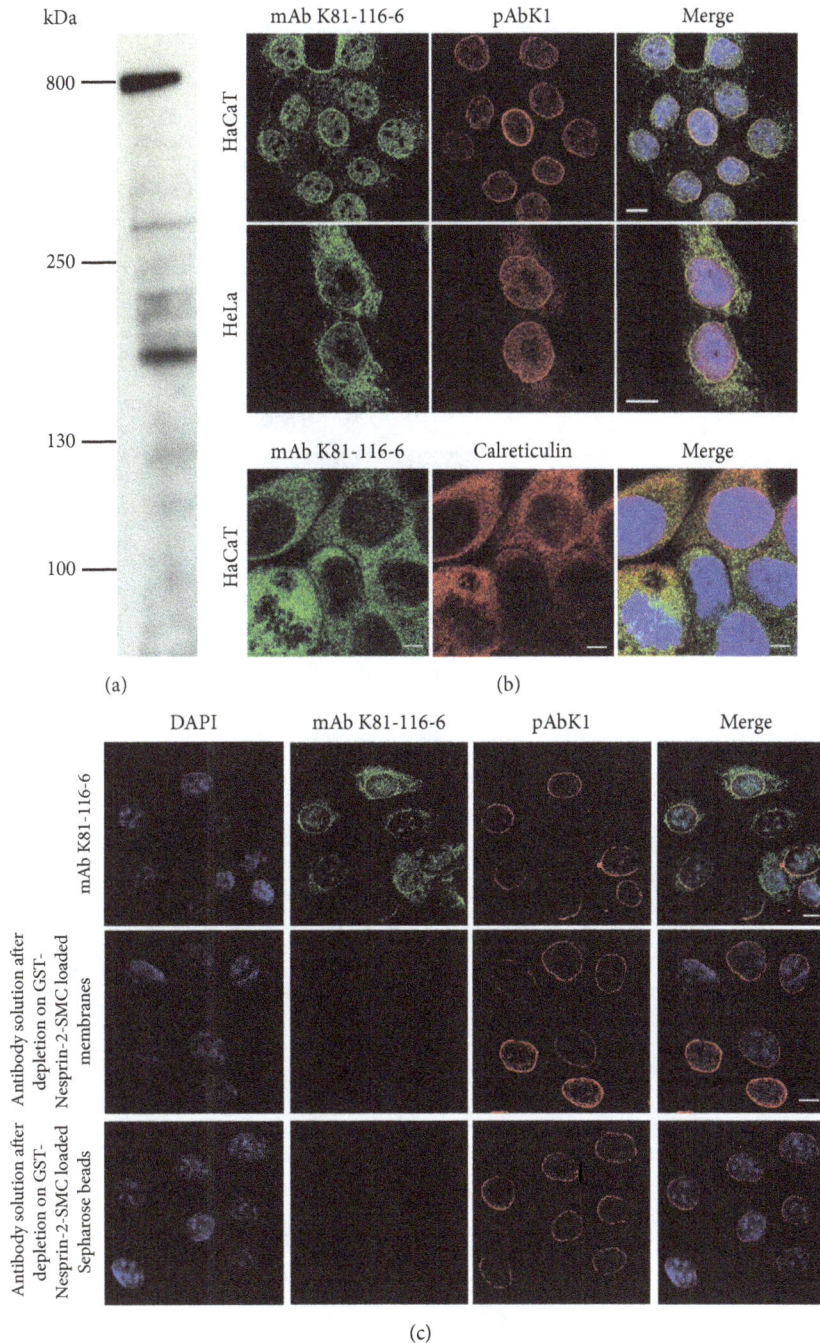

FIGURE 2: Characterization of monoclonal antibodies directed against the SMC domain. (a) Detection of Nesprin-2 with mAb K81-116-6 in HaCaT cell lysates. Proteins were separated by SDS-PAGE (3–12% acrylamide). (b) mAb K81-116-6 staining of HaCaT and HeLa cells. pAbK1 was used as bona fide Nesprin-2 antibody. DAPI stains the DNA (in Merge). Bar, 10 μm. Lower panel, colocalization of Nesprin-2 detected by mAb K81-116-6 with ER marker calreticulin in HaCaT cells. Bar, 5 μm. (c) Analysis of the specificity of mAb K81-116-6. Antibodies were depleted from the hybridoma supernatant by the indicated procedures. Antibody depleted supernatants were then used for immunofluorescence analysis. Bar, 10 μm.

are well characterized; the histone and SMC2 interactions are novel findings [2, 28, 38]. Here we followed up the SMC2 interaction. Because of the SMC homology in Nesprin-2, we speculated that this domain could interact with SMC2 and carried out pulldown assays with Glutathione-Sepharose 4B beads loaded with GST-Nesprin-2-SMC using HaCaT cell lysates as described in Materials and Methods and probed the pulldown for the presence of SMC2. GST loaded beads served as control. We could indeed detect SMC2 in the GST-Nesprin-2-SMC precipitate by SMC2 specific antibodies. SMC4 which forms a complex with SMC2 in condensin was also pulled down by GST-Nesprin-2-SMC. GST did not

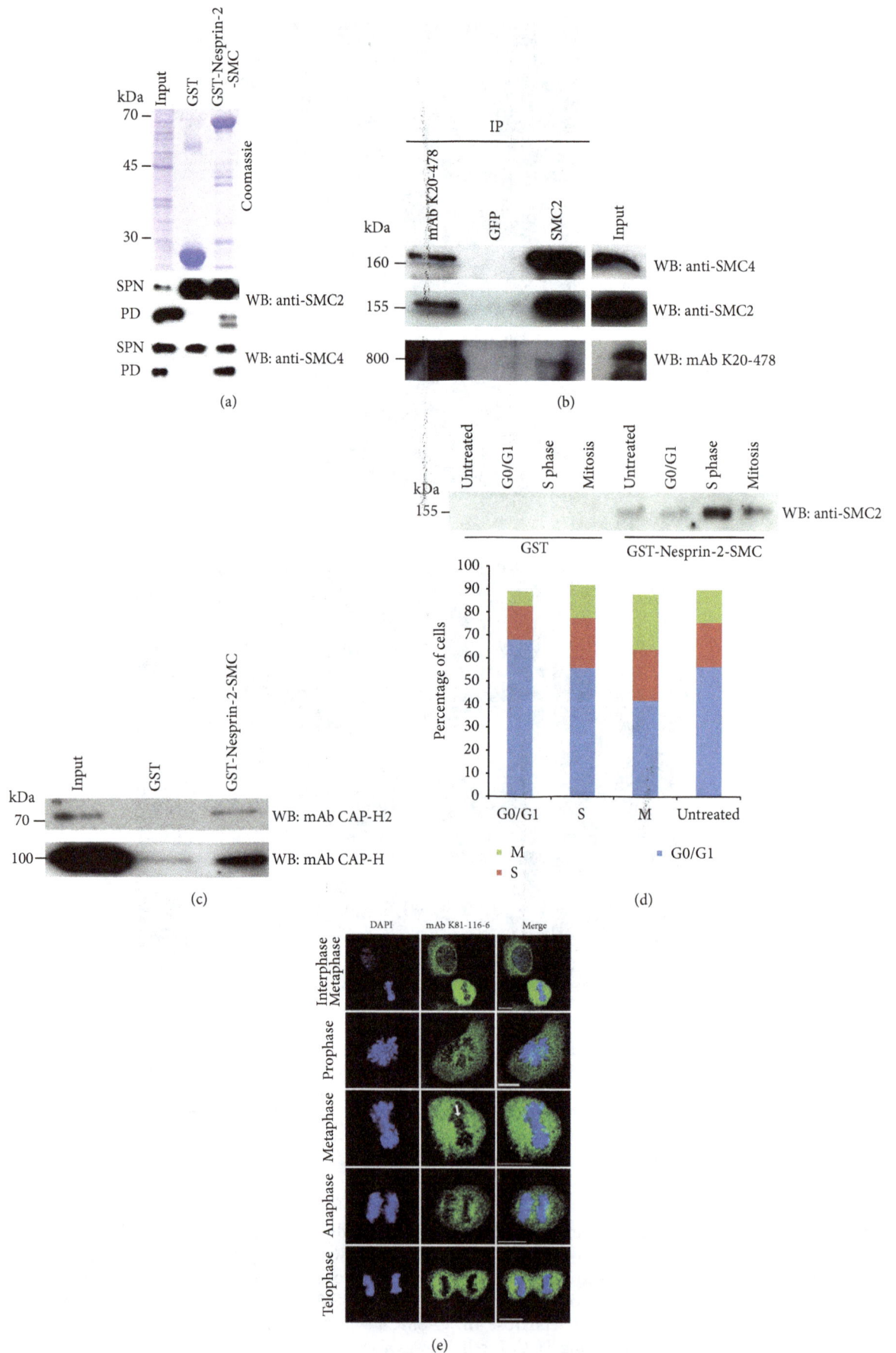

FIGURE 3: Continued.

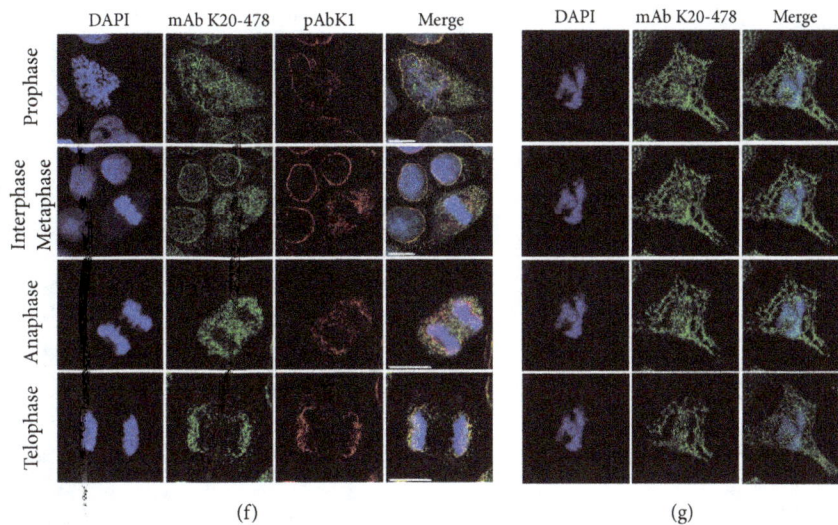

(f)   (g)

FIGURE 3: Interaction of Nesprin-2-SMC and Nesprin-2 with SMC2 and SMC4. (a) Precipitation of SMC2 and SMC4 with GST-Nesprin-2-SMC from HaCaT cell lysates. Precipitates were resolved on SDS-polyacrylamide gels (10% acrylamide) and probed with SMC2 and SMC4 specific antibodies. SPN, supernatant after pulldown; PD, pulldown. The lower molecular weight band in the SMC2 pulldown is presumably a breakdown product. (b) Immunoprecipitation of SMC2 from HaCaT cell lysates with Nesprin-2 specific mAbK20-478 and of Nesprin-2 with SMC2 specific antibodies. GFP-specific monoclonal antibodies were used for control. The antibodies used for immunoprecipitation are indicated above the panels (IP). The blots were probed with the antibodies listed on the right (WB). Immunoprecipitates were resolved on gradient gels (3–12% acrylamide) and 10% acrylamide gels as appropriate. The data are from one blot; however, the input was not directly adjacent to the SMC2 IP. (c) Interaction of CAP-H2 (condensin II) and CAP-H (condensin I) with Nesprin-2-SMC. Pulldowns were performed with HaCaT cell lysates and GST for control and GST-Nesprin-2-SMC as indicated. Unsynchronized cells were used for the experiments shown in (a)–(c). (d) Analysis of the Nesprin-2-SMC interaction with SMC2 during the cell cycle. HaCaT cells were synchronized with RO-3306 or other reagents as described in Materials and Methods in order to obtain the relevant cell cycle phases. Cell cycle phases were assessed by FACS analysis; the results are depicted in the accompanying diagram. Pulldown was carried out with GST-Nesprin-2-SMC bound to GST-Sepharose. GST was used for control. The blot was probed with SMC2 specific antibodies. (e) Localization of Nesprin-2 as detected with mAb K81-116-6 (green) during mitosis in HaCaT cells. DNA was stained with DAPI. Arrow points to filamentous staining across the chromosomes. (f) Nesprin-2 distribution in HaCaT cells during mitosis as detected with mAb K20-478 (green) and pAbK1 (red). DNA was detected with DAPI. Bar, 10 μm. (g) Nesprin-2 presence on chromosomes. Different Z-stacks (from top to bottom: 0 μm, 0.21 μm 0.42 μm, and 0.84 μm) from a COS7 cell in anaphase stained with mAb K20-478. DNA was stained with DAPI. Bar, 5 μm.

precipitate SMC2 or SMC4 (Figure 3(a)). Further proof for an interaction came from immunoprecipitation experiments from HaCaT cells with mAb K20-478 to precipitate Nesprin-2. In the Nesprin-2 pulldown, we detected SMC2 and SMC4. In the reverse experiment using SMC2 specific antibodies, Nesprin-2 was detected in the precipitate with mAb K20-478. GFP antibodies used for control did not bring down any of the proteins tested (Figure 3(b)).

As condensin exists in two complexes, condensin I and condensin II [18], we used CAP-H (kleisin γ, non-SMC condensin I complex subunit H) and CAP-H2 (kleisin β, non-SMC condensin II complex subunit H2) antibodies to probe the GST-Nesprin-2-SMC pulldown and identified CAP-H and CAP-H2 in the precipitate (Figure 3(c)). We also probed whether other SMC proteins interacted with Nesprin-2. However, the cohesin components SMC1 and SMC3 were not seen in the precipitate after carrying out a pulldown with GST-Nesprin-2-SMC (Figure S2(b)). These results make the interaction a specific one between condensin and Nesprin-2. Although SMC proteins are present in all phases of the cell cycle, they have specific roles in specific phases [17]. To find out whether the interaction is confined to a particular stage of the cell cycle, we used lysates from HaCaT cells

that had been treated with various reagents as described in Materials and Methods. This led to the enrichment of cells in particular cell cycle stages. Pulldown assays were carried out with GST-Nesprin-2-SMC and GST loaded Glutathione-Sepharose beads and the precipitates probed for the presence of SMC2. SMC2 was present in the precipitates obtained from lysates of untreated cells, cells in G0/G1, and from cell samples enriched for S and M phase. The signal was most prominent in lysates from S phase enriched cells followed by M phase cells. The GST-control did not bring down SMC2 (Figure 3(d)). The cell cycle stages were controlled by FACS analysis (Figure 3(d), bar graph).

A colocalization of SMC2 and SMC4 with Nesprin-2 was difficult to visualize at the immunofluorescence level because of the very strong signals for SMC2 and SMC4. However, some overlap indicating a colocalization could be seen particularly in telophase (see below, Figures 5(a) and 5(b), upper panels; see telophases of control cells for overlap).

3.4. Nesprin-2 Localization during Mitosis. For studying Nesprin-2 localization during mitosis, we performed immunofluorescence analysis using mAb K81-116-6, mAb K20-478, and pAbK1 (Figures 3(e), 3(f), and 3(g)). All

(a)

(b)

(c)

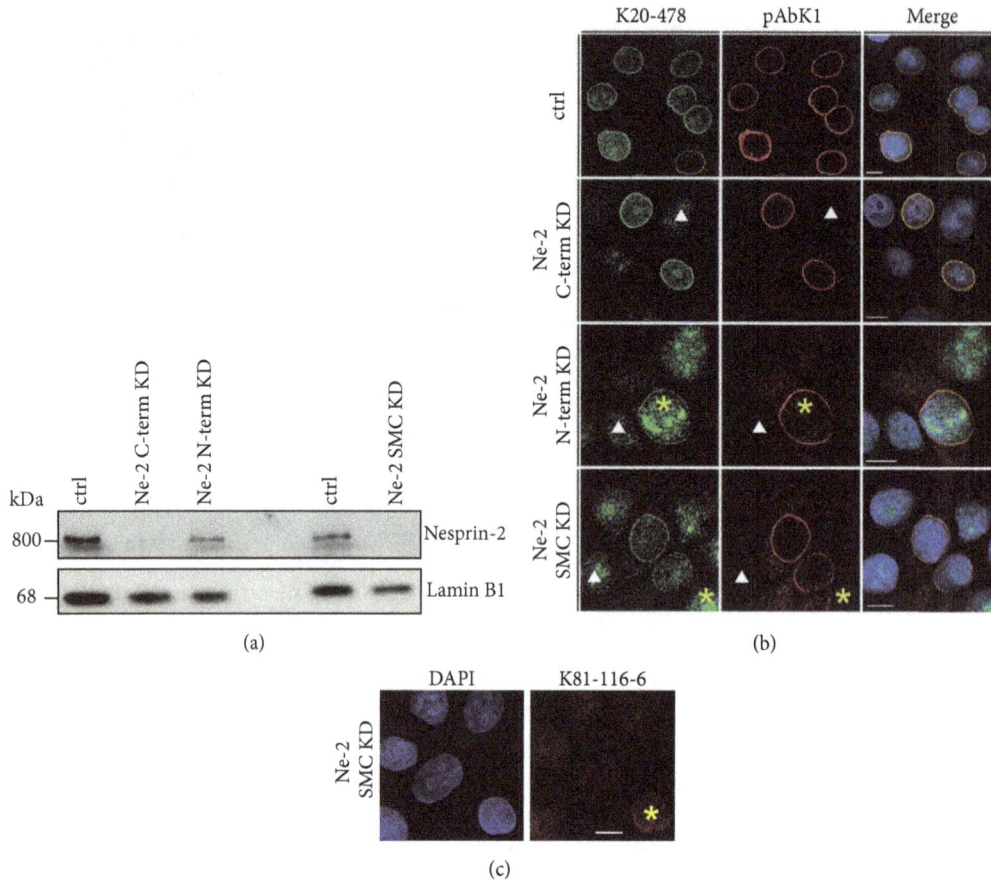

FIGURE 4: Knockdown of Nesprin-2 using shRNA directed against C-terminal, N-terminal, and SMC domain sequences. (a) Western blots showing the efficiency of the shRNA treatment at the protein level. HaCaT cells were transfected with shRNAs targeting the various regions and for control (ctrl) with the corresponding scrambled shRNAs. Nesprin-2 at ~800 kDa was detected by mAb K20-478. Lamin B1 was used for loading control. (b) Immunofluorescence analysis of HaCaT cells treated with shRNAs targeting the C-terminus (Ne-2 C-term KD), the N-terminus (Ne-2 N-term KD), or the SMC domain (Ne-2 SMC KD). Cells were stained with antibodies directed against the N-terminus (mAb K20-478, green) and the C-terminus (pAbK1, red) of Nesprin-2. DAPI was used to visualize DNA. Arrowhead indicates cells with successful knockdown; asterisk indicates cells which still express Nesprin-2. Bar, 10 μm. (c) Immunolabelling of Ne-2 SMC KD cells with mAb K81-116-6. Nuclei were labeled with DAPI. Asterisk indicates a cell which still expresses Nesprin-2. Bar, 10 μM.

antibodies showed that Nesprin-2 relocated to the cytoplasm upon nuclear envelope breakdown where it colocalized with the ER as revealed by costaining with an antibody specific for the ER marker PDI (protein disulfide isomerase) (Figure S3). It also still surrounded the condensed chromosomes, and Nesprin-2 positive structures extended across the chromosomes in all mitotic phases (Figures 3(e), 3(f), 3(g) and S4). Serial sections through the chromosomes of a mitotic cell confirmed the distribution of Nesprin-2 (Figure 3(g)). At the beginning of anaphase until telophase, we found signals at opposing ends of the dividing chromosome material presumably showing the reformation of the NE (Figure 3(f)). This localization was specific for Nesprin-2 as staining for Nesprin-1 with mAb K43-322-2 did not reveal an association with the chromosomes (Figure S5).

3.5. Nesprin-2 Knockdown Does Not Affect Condensin Distribution. To specifically explore the role of SMC domain containing Nesprin-2 isoforms, HaCaT cells were treated

with Nesprin-2-SMC shRNAs (Ne-2 SMC KD) and compared to cells treated with shRNAs targeting the Nesprin-2 N-terminus or the Nesprin-2 C-terminus (Ne-2 N-term KD; Ne-2 C-term KD) [7]. The sequences for the generation of the SMC-specific shRNAs were carefully chosen in order to exclude off-target effects due to homology to SMC sequences. In western blots labeling with mAb K20-478 revealed a strong reduction of Nesprin-2 giant at ~800 kDa in lysates from cells treated with Ne-2 C-term and Ne-2 SMC shRNAs (Figure 4(a)). Similar results were obtained with mAb K81-116-6 (see above and Figure S2(a)). The knockdown was confirmed at the immunofluorescence level with mAb K20-478, pAbK1, and mAb K81-116-6 (Figures 4(b) and 4(c)). Cell proliferation was not altered in the knockdown cells as compared to HaCaT control cells (two independent experiments, Figure S6(a)). Similarly, FACS analysis did not reveal changes in the progression through the cell cycle (three independent experiments, Figure S6(b)). Nesprin-2 depletion using Ne-2 SMC shRNA did not have an obvious effect on SMC2/4

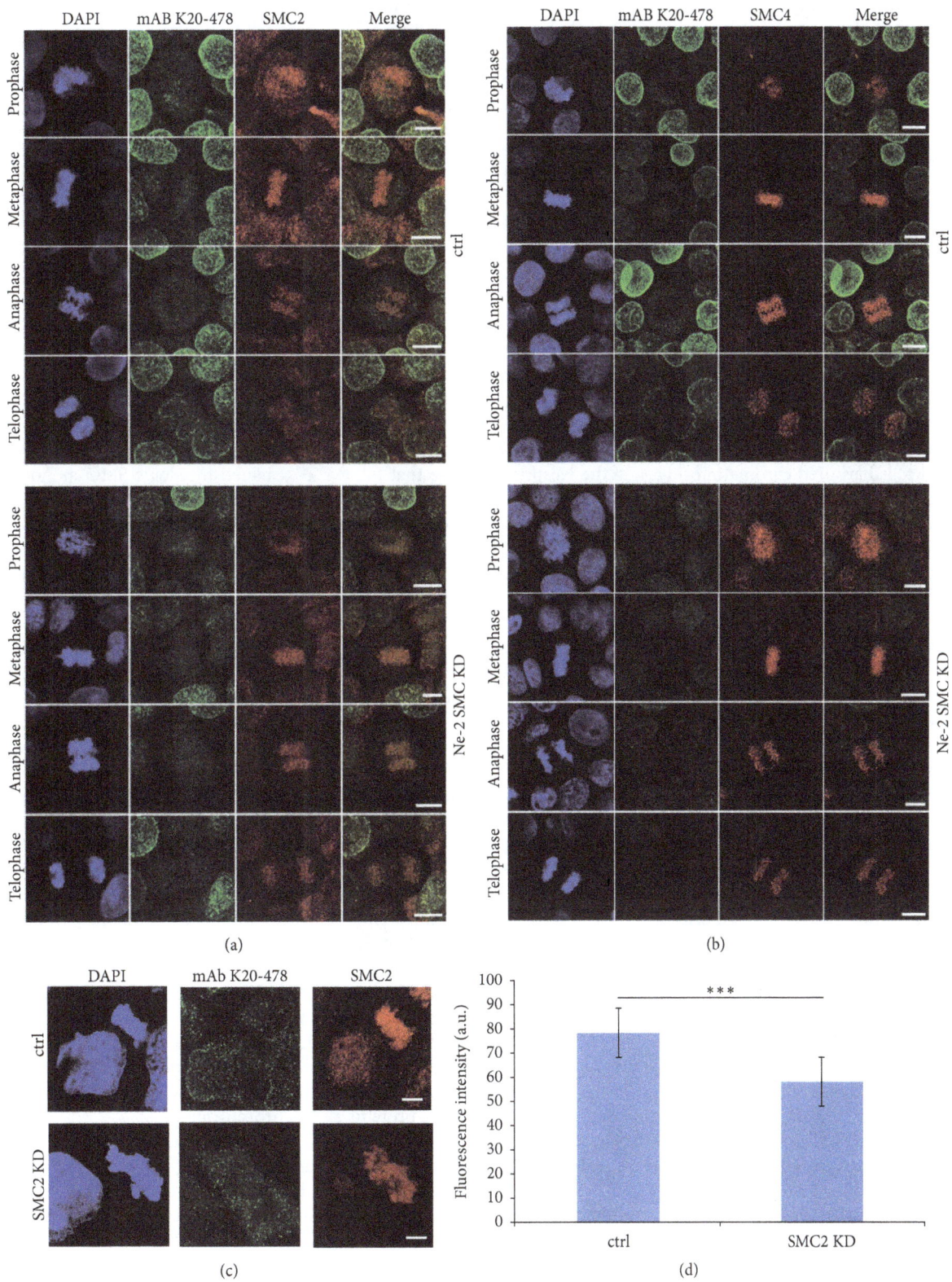

(a)

(b)

(c)

(d)

Figure 5: SMC2 (a) and SMC4 (b) in HaCaT keratinocytes treated with control shRNA (upper panels) and treated with Nesprin-2-SMC domain specific shRNA (lower panels). Nesprin-2 was detected with mAb K20-478. Bar, 10 μm. (c) Localization of Nesprin-2 after siRNA mediated knockdown of SMC2 in COS7 cells. Staining was with SMC2 specific antibodies and mAb K20-478 for Nesprin-2. Bar, 5 μm. (d) Evaluation of the SMC2 knockdown. SMC2 fluorescence intensity was measured in the center of mitotic chromosomes. 10 siRNA treated cells and 12 control cells (control treatment) were analyzed (*** P value = 0.0001).

location as the staining in immunofluorescence analysis was comparable to control cells. Also, SMC2/4 distribution during mitosis was not affected and the proteins had an apparently unaltered association with mitotic chromosomes at the level of analysis (Figures 5(a) and 5(b)). Furthermore, the protein levels appeared unaltered (Figure S2(c)).

We also performed the converse experiment by downregulating SMC2 in COS7 cells by transfection with a siRNA pool targeting SMC2. Since the knockdown was not complete, we searched for mitotic cells with reduced SMC2 staining and analyzed the Nesprin-2 distribution. We found that Nesprin-2 still surrounded the chromosomal mass indicating that Nesprin-2 localization is not strictly dependent on SMC2 (Figures 5(c) and 5(d)).

However, the analyses of the Nesprin-2 depleted cells revealed the presence of chromatin bridges during ana- and telophase. When we determined the chromatin bridges in cells transfected with SMC control and Ne-2 SMC shRNA at ana- and telophase, we observed that 4.4% (mean value) of control cells harbored chromatin bridges. In the Nesprin-2 knockdown cells, this number was increased to 10.3% (P value, 0.01; 440 and 544 ana- and telophases evaluated, resp.). This is a Nesprin-2 specific result as the Ne-2 N-term KD also led to enhanced chromatin bridge formation (15.25%, 445 ana- and telophases evaluated). Increased number of chromatin bridges in anaphase has been described for condensin II knockout cells as well as condensins I and II depleted cells [39, 40].

4. Discussion

Research on the Nesprins primarily focuses on the interphase nucleus and their role in nuclear positioning, maintaining mechanical and structural properties of the nucleus and the perinuclear cytoskeleton, and their role in signal transduction [1, 41, 42]. We found that during mitosis Nesprin-2 was present along mitotic condensed DNA. In previous studies, we reported that Nesprin-2 interacts with chromatin; in particular centromeric and other heterochromatic reads were enriched in the ChIP-seq data [9]. However, the nature of this interaction is unclear and it might well be an indirect one since Nesprin-2 interacts with proteins present in the chromatin such as histones or SMC proteins. We focused here specifically on the interaction with SMC proteins. In open mitosis, the NE breakdown (NEBD) starts during prophase resulting in a removal of the NE from chromatin. We found that Nesprin-2 was still associated with mitotic chromosomes and Nesprin-2 knockdown cells harbored increased numbers of chromatin bridges in anaphase cells.

In vertebrates, condensins I and II are both composed of the SMC2/4 heterodimer together with distinct additional non-SMC subunits, CAP-G/G2, CAP-D2/D3, and CAP-H/H2 [18]. A depletion of condensin I or II or a combination of both in HeLa cells led to delayed chromosome condensation and caused segregation problems resulting in cells with bridged or lagging chromosomes [17, 41]. In mouse embryonic stem cells, RNA interference studies revealed that condensins I and II are required for ES cell proliferation and that their loss leads to delayed initiation of anaphase

and formation of enlarged and misshapen interphase nuclei [43]. Altered nuclear architecture and size after condensin II knockdown were also described more recently [44].

Since we propose a role for Nesprin-2 on chromosomes and also on mitotic chromosomes, we searched publications reporting chromatin proteomes for the presence of Nesprin-2. Nesprin-2 was present in interphase chromatin [45] where it was listed in the category "non-expected chromatin function," and Nesprin-2 peptides were also identified in a report on nascent chromatin capture proteomics [46]. By contrast, in a publication describing the mitotic proteome, only Nesprin-1 was listed [47]. Taken together, data from independent proteomic approaches support our findings on the presence of Nesprin-2 on chromatin.

Based on the well-known structure and assembly of SMC monomers into pentameric ring complexes, it appears unlikely that the predicted SMC domain in Nesprin-2 fulfills the role of a classical SMC protein. SMC proteins form heterodimers and each dimer consists of a single polypeptide that follows a V-shaped topology. SMC monomers are connected along the hinge region and the terminal ends form catalytically active ATPases [16]. Currently, no Nesprin-2 isoform has been described that might exist as a separate isoform composed of the SMC domain only [48]. It might rather be that the SMC domain in Nesprin-2 interacts with SMC2/4 along their coiled coils. Alternatively, the interaction between condensin and Nesprin-2 is an indirect one. Interestingly, Nesprin-2 knockdown does not have an effect on mitotic progression but preliminary data indicate that the chromosomes in metaphase cells have a fuzzy appearance and a larger volume [49, 50]. Similar observations were made after SMC knockdown and this observation could place Nesprin-2 in this pathway [51]. In this context, Nesprin-2 might adopt a role similar to the one previously suggested for NE proteins in transcriptional regulation where they are thought to regulate the spatiotemporal accessibility of transcriptional regulators to their nuclear targets instead of directly acting as transcriptional regulators in the proximity of genes [52, 53]. Nesprin-2 might act on SMC2/4 in a similar way. Our data indicate that a loss of Nesprin did not prevent SMC2/4 proteins to assemble along mitotic chromosomes but an increased number of chromatin bridges were observed which hints at changes in the process of chromosome separation. It could therefore well be that Nesprin-2 affects directly or indirectly the spatiotemporal assembly or the function of SMC proteins along chromosomes.

In our analysis, we observed that the condensin Nesprin-2 interaction occurred throughout the cell cycle. Interestingly, condensins have roles not only during mitosis but also in interphase, where they are important particularly in gene regulation. For instance, a function in transcriptional regulation has been reported for condensins I and II by Li et al. [19] who found them on enhancers that had the estrogen receptor α bound. This led to full enhancer activation and efficient transcription of the respective genes [19]. Furthermore, Zhang et al. [54] reported that condensin I downregulation in chicken DT40 cells caused a misregulation of gene expression underlining its role in transcriptional regulation during interphase. Related findings were reported earlier for *C. elegans* where condensins were found at tRNA genes, promoters, and enhancers in interphase, and condensin II binding was associated with a repressive effect on transcription [55]. By contrast, in mouse embryonic stem cells, condensin II and cohesin were present at transcriptional elements of active genes during interphase and affected gene activity in a positive way [56].

In summary, we report a novel interaction partner of Nesprin-2 giant and show that the Nesprin-2 condensin interaction has an impact on mitotic chromosomes. The tight packaging of chromosomes during mitosis, to which the Nesprin-2 interaction might contribute, ensures their faithful segregation and allows them to withstand forces during segregation. Malfunctions in this process can cause DNA bridges which result in chromosome segregation errors and lead to micronucleus formation, and can make chromosomes more prone to DNA damage. It could well be that Nesprins and further NE proteins contribute to this chromosome phenotype. Therefore, mutations in these proteins have the potential to contribute to the formation of distinct clinical manifestations associated with condensin linked diseases [57]. Furthermore, since the Nesprin-2 condensin interaction also takes place during other phases of the cell cycle and since condensins have additional functions in interphase, the Nesprin-2 condensin complex could also affect these processes.

Disclosure

The present address of Linlin Hao is Animal Biotechnology Department, Jilin University, Changchun 130062, China. Part of this work was carried out as Ph.D. thesis (Xin Xing and Carmen Mroß).

Conflicts of Interest

The authors declare that they have no conflicts of interest with the contents of this article.

Authors' Contributions

Xin Xing and Carmen Mroß designed and carried out experiments, analyzed the data, prepared the figures, and wrote the manuscript. Linlin Hao, Martina Munck, Alexandra Herzog, Clara Mohr, C. P. Unnikannan, and Pranav Kelkar performed additional experiments and analyzed data. Sascha Neumann, Ludwig Eichinger, and Angelika A. Noegel conceived the study, reviewed all data, and prepared the final versions of the manuscript, text, and figures. Xin Xing and Carmen Mroß have equal contribution to this work.

Acknowledgments

Xin Xing was supported by a fellowship from the China Scholarship Council (CSC), Linlin Hao was supported by a fellowship from the Deutsche Akademische Austauschdienst (DAAD), and Carmen Mroß is a member of the International Graduate School in Development Health and Disease (IGS-DHD). The work was supported by the CMMC (C6) and CECAD (TPC05) and by a grant to Sascha Neumann from the Marga und Walter Boll-Stiftung. The authors thank Dr. M. Schleicher for help with chemical cross-linking, Berthold Gaßen for help with the generation of monoclonal Nesprin-2-SMC antibodies, Maria Stumpf for help with microscopy, and Rolf Müller for cloning, protein analysis, and invaluable help with figures. They thank Dr. Astrid Schauß and Nikolay Kladt from the CECAD imaging facility and Drs. S. Müller and G. Rappl for mass spectrometry and FACS cell sorting analysis, respectively, at the central facilities of the CMMC.

Supplementary Materials

Supplementary 1. Figure S1: (a) analysis of Nesprin-2 SMC by gel filtration chromatography followed by SDS-PAGE. The elution profile of the Nesprin-2-SMC polypeptide and ovalbumin is shown. (b) GST-Nesprin-2-SMC pulls down Nesprin-2 giant from HaCaT whole cell lysates. Several Nesprin-2 polypeptides derived from the 6885 amino acids protein were identified by mass spectrometry. Amino acid positions are given at the beginning and end of the identified sequences and refer to human Nesprin-2 giant (NCBI accession number: AF435011.1).

Supplementary 2. Figure S2: (a) Nesprin-2 giant is no longer detected by mAb K81-116-6 in lysates from HaCaT cells treated with shRNA directed against the SMC domain and the N-terminus of Nesprin-2. Whole cell lysates from cells treated with the indicated knockdown plasmids were separated in a gradient gel (3 to 12% acrylamide) and probed with mAb K81-116-6. Ne-2 ctrl KD corresponds to a scrambled SMC oligonucleotide. (b) SMC1 and SMC3 do not interact with GST-Nesprin-2-SMC. HaCaT cell lysates (input) were used for precipitation experiments employing GST, GST-Nesprin-2-SMC, and Glutathione-Sepharose beads, respectively, as indicated above the panels. Proteins were separated by SDS-PAGE (10% acrylamide) and the resulting western blots were probed with the antibodies indicated on the right. (c) SMC2 and SMC4 protein levels are not affected in Nesprin-2 knockdown cells. Whole cell lysates from cells treated with the indicated knockdown plasmids were separated by SDS-PAGE (10% acrylamide) and probed for SMC2 and SMC4. Lamin B1 served as control.

Supplementary 3. Figure S3: colocalization of Nesprin-2 and an ER marker in mitotic cells. HaCaT cells were stained with pAbK1 for Nesprin-2 and with protein disulfide isomerase (PDI) specific monoclonal antibodies as ER marker. DNA was stained with DAPI.

Supplementary 4. Figure S4: Nesprin-2 distribution during mitosis. HaCaT cells were labeled with pAbK1, mAb YL1/2 specific for α-tubulin, and DAPI for DNA. Bar, 5 μm.

Supplementary 5. Figure S5: specificity of Nesprin-2 association with chromosomes in mitosis. HaCaT cells were stained with pAbK1 for Nesprin-2 and mAb K43-322-2 for Nesprin-1. Bar, 5 μm.

Supplementary 6. Figure S6: (a) proliferation of Nesprin-2-SMC knockdown HaCaT cells. The mean of two independent experiments is shown. (b) Cell cycle progression is unaffected by the loss of Nesprin-2. The experiment was carried out for HaCaT control cells, Nesprin-2-SMC knockdown (Ne-2-SMC KD), and cells treated with a control plasmid containing scrambled sequences. The data show the mean of three independent experiments. No significant differences were noted. M, mitosis; S, S phase; G0/G1, G0, G1 phase.

References

[1] D. Rajgor and C. M. Shanahan, "Nesprins: from the nuclear envelope and beyond," *Expert Reviews in Molecular Medicine*, vol. 15, no. e5, 2013.

[2] V. C. Padmakumar, T. Libotte, W. Lu et al., "The inner nuclear membrane protein Sun1 mediates the anchorage of Nesprin-2 to the nuclear envelope," *Journal of Cell Science*, vol. 118, no. 15, pp. 3419–3430, 2005.

[3] Y.-Y. Zhen, T. Libotte, M. Munck, A. A. Noegel, and E. Korenbaum, "NUANCE, a giant protein connecting the nucleus and actin cytoskeleton," *Journal of Cell Science*, vol. 115, no. 15, pp. 3207–3222, 2002.

[4] V. C. Padmakumar, S. Abraham, S. Braune et al., "Enaptin, a giant actin-binding protein, is an element of the nuclear membrane and the actin cytoskeleton," *Experimental Cell Research*, vol. 295, no. 2, pp. 330–339, 2004.

[5] K. Wilhelmsen, S. H. M. Litjens, I. Kuikman et al., "Nesprin-3, a novel outer nuclear membrane protein, associates with the cytoskeletal linker protein plectin," *The Journal of Cell Biology*, vol. 171, no. 5, pp. 799–810, 2005.

[6] K. J. Roux, M. L. Crisp, Q. Liu et al., "Nesprin 4 is an outer nuclear membrane protein that can induce kinesin-mediated cell polarization," *Proceedings of the National Acadamy of Sciences of the United States of America*, vol. 106, no. 7, pp. 2194–2199, 2009.

[7] M. Schneider, W. Lu, S. Neumann et al., "Molecular mechanisms of centrosome and cytoskeleton anchorage at the nuclear envelope," *Cellular and Molecular Life Sciences*, vol. 68, no. 9, pp. 1593–1610, 2011.

[8] M. L. Lombardi, D. E. Jaalouk, C. M. Shanahan, B. Burke, K. J. Roux, and J. Lammerding, "The interaction between nesprins and sun proteins at the nuclear envelope is critical for force transmission between the nucleus and cytoskeleton," *The Journal of Biological Chemistry*, vol. 286, no. 30, pp. 26743–26753, 2011.

[9] R. N. Rashmi, B. Eckes, G. Glöckner et al., "The nuclear envelope protein Nesprin-2 has roles in cell proliferation and differentiation during wound healing.," *Nucleus (Austin, Tex.)*, vol. 3, no. 2, pp. 172–186, 2012.

[10] K. Djinovic-Carugo, M. Gautel, J. Ylänne, and P. Young, "The spectrin repeat: a structural platform for cytoskeletal protein assemblies," *FEBS Letters*, vol. 513, no. 1, pp. 119–123, 2002.

[11] W. Lu, M. Schneider, S. Neumann et al., "Nesprin interchain associations control nuclear size," *Cellular and Molecular Life Sciences*, vol. 69, no. 20, pp. 3493–3509, 2012.

[12] H. R. Dawe, M. Adams, G. Wheway et al., "Nesprin-2 interacts with meckelin and mediates ciliogenesis via remodelling of the actin cytoskeleton," *Journal of Cell Science*, vol. 122, no. 15, pp. 2716–2726, 2009.

[13] P. Satir, L. B. Pedersen, and S. T. Christensen, "The primary cilium at a glance," *Journal of Cell Science*, vol. 123, no. 4, pp. 499–503, 2010.

[14] T. Hirano, "SMC proteins and chromosome mechanics: from bacteria to humans," *Philosophical Transactions of the Royal Society B: Biological Sciences*, vol. 360, no. 1455, pp. 507–514, 2005.

[15] T. Hirano, "At the heart of the chromosome: SMC proteins in action," *Nature Reviews Molecular Cell Biology*, vol. 7, no. 5, pp. 311–322, 2006.

[16] C. H. Haering, J. Löwe, A. Hochwagen, and K. Nasmyth, "Molecular architecture of SMC proteins and the yeast cohesin complex," *Molecular Cell*, vol. 9, no. 4, pp. 773–788, 2002.

[17] T. Hirota, D. Gerlich, B. Koch, J. Ellenberg, and J.-M. Peters, "Distinct functions of condensin I and II in mitotic chromosome assembly," *Journal of Cell Science*, vol. 117, no. 26, pp. 6435–6445, 2004.

[18] I. Piazza, C. H. Haering, and A. Rutkowska, "Condensin: crafting the chromosome landscape," *Chromosoma*, vol. 122, no. 3, pp. 175–190, 2013.

[19] W. Li, Y. Hu, S. Oh et al., "Condensin I and II complexes license full estrogen receptor α-dependent enhancer activation," *Molecular Cell*, vol. 59, no. 2, pp. 188–202, 2015.

[20] A. J. Wood, A. F. Severson, and B. J. Meyer, "Condensin and cohesin complexity: the expanding repertoire of functions," *Nature Reviews Genetics*, vol. 11, no. 6, pp. 391–404, 2010.

[21] C. R. Bauer, T. A. Hartl, and G. Bosco, "Condensin II promotes the formation of chromosome territories by inducing axial compaction of polyploid interphase chromosomes," *PLoS Genetics*, vol. 8, no. 8, Article ID e1002873, 2012.

[22] O. Iwasaki, C. J. Corcoran, and K.-I. Noma, "Involvement of condensin-directed gene associations in the organization and regulation of chromosome territories during the cell cycle," *Nucleic Acids Research*, vol. 44, no. 8, pp. 3618–3628, 2016.

[23] E. Ampatzidou, A. Irmisch, M. J. O'Connell, and J. M. Murray, "Smc5/6 is required for repair at collapsed replication forks," *Molecular and Cellular Biology*, vol. 26, no. 24, pp. 9387–9401, 2006.

[24] D. E. Verver, G. H. Hwang, P. W. Jordan, and G. Hamer, "Resolving complex chromosome structures during meiosis: versatile deployment of Smc5/6," *Chromosoma*, vol. 125, no. 1, pp. 15–27, 2016.

[25] G. D. Spotts, S. V. Patel, Q. Xiao, and S. R. Hann, "Identification of downstream-initiated c-Myc proteins which are dominant-negative inhibitors of transactivation by full-length c-Myc proteins," *Molecular and Cellular Biology*, vol. 17, no. 3, pp. 1459–1468, 1997.

[26] L. T. Vassilev, C. Tovar, S. Chen et al., "Selective small-molecule inhibitor reveals critical mitotic functions of human CDK1," *Proceedings of the National Acadamy of Sciences of the United States of America*, vol. 103, no. 28, pp. 10660–10665, 2006.

[27] P. J. Paddison, A. A. Caudy, E. Bernstein, G. J. Hannon, and D. S. Conklin, "Short hairpin RNAs (shRNAs) induce sequence-specific silencing in mammalian cells," *Genes & Development*, vol. 16, no. 8, pp. 948–958, 2002.

[28] T. Libotte, H. Zaim, S. Abraham et al., "Lamin A/C-dependent localization of Nesprin-2, a giant scaffolder at the nuclear envelope," *Molecular Biology of the Cell (MBoC)*, vol. 16, no. 7, pp. 3411–3424, 2005.

[29] S. Taranum, I. Sur, R. Müller et al., "Cytoskeletal interactions at the nuclear envelope mediated by Nesprins," *International Journal of Cell Biology*, vol. 2012, Article ID 736524, 11 pages, 2012.

[30] A. A. Noegel, R. Blau-Wasser, H. Sultana et al., "The Cyclase-associated protein CAP as regulator of cell polarity and cAMP signaling in dictyostelium," *Molecular Biology of the Cell (MBoC)*, vol. 15, no. 2, pp. 934–945, 2004.

[31] G. I. Evan, G. K. Lewis, G. Ramsay, and J. M. Bishop, "Isolation of monoclonal antibodies specific for human c-myc proto-oncogene product," *Molecular and Cellular Biology*, vol. 5, no. 12, pp. 3610–3616, 1985.

[32] H. Xiong, F. Rivero, U. Euteneuer et al., "Dictyostelium Sun-1 connects the centrosome to chromatin and ensures genome stability," *Traffic*, vol. 9, no. 5, pp. 708–724, 2008.

[33] M. Schleicher, G. Gerisch, and G. Isenberg, "New actin-binding proteins from *Dictyostelium discoideum*," *EMBO Journal*, vol. 3, no. 9, pp. 2095–2100, 1984.

[34] P. Fucini, B. Köppel, M. Schleicher et al., "Molecular architecture of the rod domain of the Dictyostelium gelation factor (ABP120)," *Journal of Molecular Biology*, vol. 291, no. 5, pp. 1017–1023, 1999.

[35] Z. Grabarek and J. Gergely, "Zero-length crosslinking procedure with the use of active esters," *Analytical Biochemistry*, vol. 185, no. 1, pp. 131–135, 1990.

[36] J. G. Simpson and R. G. Roberts, "Patterns of evolutionary conservation in the nesprin genes highlight probable functionally important protein domains and isoforms," *Biochemical Society Transactions*, vol. 36, no. 6, pp. 1359–1367, 2008.

[37] X. A. Cui, H. Zhang, L. Ilan, A. X. Liu, I. Kharchuk, and A. F. Palazzo, "mRNA encoding Sec61β, a tail-anchored protein, is localized on the endoplasmic reticulum," *Journal of Cell Science*, vol. 128, no. 18, pp. 3398–3410, 2015.

[38] L. Yang, M. Munck, K. Swaminathan, L. E. Kapinos, A. A. Noegel, and S. Neumann, "Mutations in LMNA modulate the lamin A—Nesprin-2 interaction and cause LINC complex alterations," *PLoS ONE*, vol. 8, no. 8, Article ID e71850, 2013.

[39] L. C. Green, P. Kalitsis, T. M. Chang et al., "Contrasting roles of condensin I and condensin II in mitotic chromosome formation," *Journal of Cell Science*, vol. 125, no. 6, pp. 1591–1604, 2012.

[40] D. Gerlich, T. Hirota, B. Koch, J.-M. Peters, and J. Ellenberg, "Condensin I stabilizes chromosomes mechanically through a dynamic interaction in live cells," *Current Biology*, vol. 16, no. 4, pp. 333–344, 2006.

[41] D. T. Warren, T. Tajsic, J. A. Mellad, R. Searles, Q. Zhang, and C. M. Shanahan, "Novel nuclear nesprin-2 variants tether active extracellular signal-regulated MAPK1 and MAPK2 at promyelocytic leukemia protein nuclear bodies and act to regulate smooth muscle cell proliferation," *The Journal of Biological Chemistry*, vol. 285, no. 2, pp. 1311–1320, 2010.

[42] J. T. Morgan, E. R. Pfeiffer, T. L. Thirkill et al., "Nesprin-3 regulates endothelial cell morphology, perinuclear cytoskeletal architecture, and flow-induced polarization," *Molecular Biology of the Cell (MBoC)*, vol. 22, no. 22, pp. 4324–4334, 2011.

[43] T. G. Fazzio and B. Panning, "Condensin complexes regulate mitotic progression and interphase chromatin structure in embryonic stem cells," *The Journal of Cell Biology*, vol. 188, no. 4, pp. 491–503, 2010.

[44] C. George, J. Bozler, H. Nguyen, and G. Bosco, "Condensins are required for maintenance of nuclear architecture," *Cells*, vol. 3, no. 3, pp. 865–882, 2014.

[45] G. Kustatscher, N. Hégarat, K. L. H. Wills et al., "Proteomics of a fuzzy organelle: interphase chromatin," *EMBO Journal*, vol. 33, no. 6, pp. 648–664, 2014.

[46] C. Alabert, J.-C. Bukowski-Wills, S.-B. Lee et al., "Nascent chromatin capture proteomics determines chromatin dynamics during DNA replication and identifies unknown fork components," *Nature Cell Biology*, vol. 16, no. 3, pp. 281–291, 2014.

[47] S. Ohta, J.-C. Bukowski-Wills, L. Sanchez-Pulido et al., "The protein composition of mitotic chromosomes determined using multiclassifier combinatorial proteomics," *Cell*, vol. 142, no. 5, pp. 810–821, 2010.

[48] D. Rajgor, J. A. Mellad, F. Autore, Q. Zhang, and C. M. Shanahan, "Multiple novel nesprin-1 and nesprin-2 variants act as versatile tissue-specific intracellular scaffolds," *PLoS ONE*, vol. 7, no. 7, Article ID e40098, 2012.

[49] X. Xing, *Functional Characterization of The Predicted SMC Domain in Nesprin-2 [Ph.D. thesis]*, Math.-Nat. Fac, University of Cologne, Germany, 2013.

[50] C. Mroß, *Novel Functions of Nesprin-2 and Analysis of Its In Vivo Role [Ph.D. thesis]*, Math.-Nat. Fac, University of Cologne, Germany, 2017.

[51] T. Ono, A. Losada, M. Hirano, M. P. Myers, A. F. Neuwald, and T. Hirano, "Differential contributions of condensin I and condensin II to mitotic chromosome architecture in vertebrate cells," *Cell*, vol. 115, no. 1, pp. 109–121, 2003.

[52] S. Heessen and M. Fornerod, "The inner nuclear envelope as a transcription factor resting place," *EMBO Reports*, vol. 8, no. 10, pp. 914–919, 2007.

[53] C. Ivorra, M. Kubicek, J. M. González et al., "A mechanism of AP-1 suppression through interaction of c-Fos with lamin A/C," *Genes & Development*, vol. 20, no. 3, pp. 307–320, 2006.

[54] T. Zhang, J. R. Paulson, M. Bakhrebah et al., "Condensin I and II behaviour in interphase nuclei and cells undergoing premature chromosome condensation," *Chromosome Research*, vol. 24, no. 2, pp. 243–269, 2016.

[55] A.-L. Kranz, C.-Y. Jiao, L. H. Winterkorn, S. E. Albritton, M. Kramer, and S. Ercan, "Genome-wide analysis of condensin binding in Caenorhabditis elegans," *Genome Biology*, vol. 14, no. 10, article no. R112, 2013.

[56] J. M. Dowen, S. Bilodeau, D. A. Orlando et al., "Multiple structural maintenance of chromosome complexes at transcriptional regulatory elements," *Stem Cell Reports*, vol. 1, no. 5, pp. 371–378, 2013.

[57] C.-A. Martin, J. E. Murray, P. Carroll et al., "Mutations in genes encoding condensin complex proteins cause microcephaly through decatenation failure at mitosis," *Genes & Development*, vol. 30, no. 19, pp. 2158–2172, 2016.

PERMISSIONS

All chapters in this book were first published in IJCB, by Hindawi Publishing Corporation; hereby published with permission under the Creative Commons Attribution License or equivalent. Every chapter published in this book has been scrutinized by our experts. Their significance has been extensively debated. The topics covered herein carry significant findings which will fuel the growth of the discipline. They may even be implemented as practical applications or may be referred to as a beginning point for another development.

The contributors of this book come from diverse backgrounds, making this book a truly international effort. This book will bring forth new frontiers with its revolutionizing research information and detailed analysis of the nascent developments around the world.

We would like to thank all the contributing authors for lending their expertise to make the book truly unique. They have played a crucial role in the development of this book. Without their invaluable contributions this book wouldn't have been possible. They have made vital efforts to compile up to date information on the varied aspects of this subject to make this book a valuable addition to the collection of many professionals and students.

This book was conceptualized with the vision of imparting up-to-date information and advanced data in this field. To ensure the same, a matchless editorial board was set up. Every individual on the board went through rigorous rounds of assessment to prove their worth. After which they invested a large part of their time researching and compiling the most relevant data for our readers.

The editorial board has been involved in producing this book since its inception. They have spent rigorous hours researching and exploring the diverse topics which have resulted in the successful publishing of this book. They have passed on their knowledge of decades through this book. To expedite this challenging task, the publisher supported the team at every step. A small team of assistant editors was also appointed to further simplify the editing procedure and attain best results for the readers.

Apart from the editorial board, the designing team has also invested a significant amount of their time in understanding the subject and creating the most relevant covers. They scrutinized every image to scout for the most suitable representation of the subject and create an appropriate cover for the book.

The publishing team has been an ardent support to the editorial, designing and production team. Their endless efforts to recruit the best for this project, has resulted in the accomplishment of this book. They are a veteran in the field of academics and their pool of knowledge is as vast as their experience in printing. Their expertise and guidance has proved useful at every step. Their uncompromising quality standards have made this book an exceptional effort. Their encouragement from time to time has been an inspiration for everyone.

The publisher and the editorial board hope that this book will prove to be a valuable piece of knowledge for researchers, students, practitioners and scholars across the globe.

LIST OF CONTRIBUTORS

Alejandro Alvarez-Arce, Irene Lee-Rivera, Edith López, Arturo Hernández-Cruz and Ana María López-Colomé
Instituto de Fisiolog´ıa Celular, Universidad Nacional Aut´onoma de M´exico, M´exico City, Mexico

Charles S. Cox and Kevin P. Lally
Department of Pediatric Surgery, University of Texas McGovern Medical School, Houston, TX 77030, USA

Yohan Choi and Fanwei Meng
Department of Pediatric Surgery, University of Texas McGovern Medical School, Houston, TX 77030, USA
Center for Stem Cell and Regenerative Medicine, The Brown Foundation Institute of Molecular Medicine for the Prevention of Human Diseases (IMM),The University of Texas Health Science Center at Houston (UT Health), Houston, TX 77030, USA

Yong Li
Department of Pediatric Surgery, University of Texas McGovern Medical School, Houston, TX 77030, USA
Center for Stem Cell and Regenerative Medicine, The Brown Foundation Institute of Molecular Medicine for the Prevention of Human Diseases (IMM),The University of Texas Health Science Center at Houston (UT Health), Houston, TX 77030, USA
Center for Tissue Engineering and Aging Research,The IMM,The University of Texas Health Science Center at Houston (UT Health), Houston, TX 77030, USA

Johnny Huard
Department of Orthopaedic Surgery, University of Texas McGovern Medical School, Houston, TX 77030, USA
Center for Tissue Engineering and Aging Research,The IMM,The University of Texas Health Science Center at Houston (UT Health), Houston, TX 77030, USA
Center for Regenerative Sports Medicine, Steadman Philippon Research Institute, Vail, CO, USA

Nigel P. Moore
Ubrs GmbH, Postfach, 4058 Basel, Switzerland

Catherine A. Picut
WIL Research, LLC, Hillsborough, NC 27278, USA

Jeffrey H. Charlap
WIL Research, LLC, Ashland, OH 44805, USA

Seong Gu Hwang
Department of Animal Life and Environmental Science, Hankyong National University, Anseong 17579, Republic of Korea

Ho-Sung Lee and Byung-Cheon Lee
Institute for Information Technology Convergence, Division of Electrical Engineering, Korea Advanced Institute of Science and Technology, Daejeon 34138, Republic of Korea

GunWoong Bahng
Department of Mechanical Engineering, The State University of New York Korea, Incheon 21985, Republic of Korea

Vincenzo Giancotti
Department of Life Science, University of Trieste, Trieste, Italy
Trieste Proteine Ricerche, Palmanova, Udine, Italy

Natascha Bergamin, Palmina Cataldi and Claudio Rizzi
Division of Pathology, Azienda Ospedaliero-Universitaria, Udine, Italy

Louise C. Kenny
Department of Obstetrics and Gynaecology, Cork University Maternity Hospital, University College Cork, Cork, Ireland
INFANT Centre, Cork University Maternity Hospital, University College Cork, Cork, Ireland

Katie L. Togher
Department of Obstetrics and Gynaecology, Cork University Maternity Hospital, University College Cork, Cork, Ireland
APC Microbiome Institute, Biosciences Institute, University College Cork, Cork, Ireland
INFANT Centre, Cork University Maternity Hospital, University College Cork, Cork, Ireland

Gerard W. O'Keeffe
APC Microbiome Institute, Biosciences Institute, University College Cork, Cork, Ireland
INFANT Centre, Cork University Maternity Hospital, University College Cork, Cork, Ireland
Department of Anatomy and Neuroscience, University College Cork, Cork, Ireland

Aquiles Sales Craveiro Sarmento, Lázaro Batista de Azevedo Medeiros, Lucymara Fassarella Agnez-Lima and Julliane Tamara Araújo deMelo Campos
Laboratório de Biologia Molecular e Genômica, Departamento de Biologia Celular e Genética, Centro de Bioci^encias, Universidade Federal do Rio Grande do Norte, Natal, RN, Brazil

Josivan Gomes Lima
Departamento de Medicina Clínica, Hospital Universitário Onofre Lopes, Universidade Federal do Rio Grande do Norte, Natal, RN, Brazil

Ranjeet SinghMahla
Department of Biological Sciences, Indian Institute of Science Education and Research (IISER), Bhopal, Madhya Pradesh 462066, India

Julie E.Morgan and Susanna F. Greer
Division of Cellular Biology and Immunology, Department of Biology, Georgia State University, Atlanta, GA 30302, USA

Paula A. Baldión
Grupo de Investigaciones Básicas y Aplicadas en Odontología, Universidad Nacional de Colombia, Bogotá, Colombia

Jaime E. Castellanos
Grupo de Investigaciones Básicas y Aplicadas en Odontología, Universidad Nacional de Colombia, Bogotá, Colombia
Grupo de Virolog´ıa, Universidad El Bosque, Bogotá, Colombia

Myriam L. Velandia-Romero
Grupo de Virolog´ıa, Universidad El Bosque, Bogotá, Colombia

Kazuo Katoh
Laboratory of Human Anatomy and Cell Biology, Faculty of Health Sciences, Tsukuba University of Technology, 4-12-7 Kasuga, Tsukuba, Ibaraki 305-8521, Japan

Masoumeh Rostami and Majid Shahbazi
Medical Cellular and Molecular Research Center, Golestan University of Medical Sciences, Gorgan, Iran

Kamran Haidari
Department of Anatomy, Faculty of Medical Sciences, Golestan University of Medical Sciences, Gorgan, Iran

Norma Estrada
Programa Cátedras CONACyT, Centro de Investigaciones Biológicas del Noroeste, S.C. (CIBNOR), 23090 La Paz, BCS, Mexico

Edwin Velázquez and Felipe Ascencio
Laboratorio de Patogénesis Microbiana, Centro de Investigaciones Biológicas del Noroeste, S.C., 23090 La Paz, BCS, Mexico

Carmen Rodríguez-Jaramillo
Laboratorio de Histología e Histoquímica, Centro de Investigaciones Biológicas del Noroeste, S.C., 23090 La Paz, BCS, Mexico

Saraporn Harikarnpakdee
Cosmeceutical Research, Development and Testing Center, College of Pharmacy, Rangsit University, Pathum Thani 12000, Thailand
Department of Industrial Pharmacy, College of Pharmacy, Rangsit University, Pathum Thani 12000, Thailand

Verisa Chowjarean
Cosmeceutical Research, Development and Testing Center, College of Pharmacy, Rangsit University, PathumThani 12000, Thailand
Department of Pharmaceutical Technology, College of Pharmacy, Rangsit University, PathumThani 12000,Thailand

Peleg Rider and Yaron Carmi
The Department of Pathology, Sackler Faculty of Medicine, Tel-Aviv University, 6997801 Tel-Aviv, Israel

Idan Cohen
Galilee Medical Center, 22100 Nahariya, Israel

Xin Xing, Carmen Mroß, Linlin Hao, Martina Munck, Alexandra Herzog, Clara Mohr, C. P. Unnikannan, Pranav Kelkar, Angelika A. Noegel, Ludwig Eichinger and Sascha Neumann
Institute of Biochemistry I, Medical Faculty, University Hospital Cologne, Joseph-Stelzmann-Str. 52, 50931 Cologne, Germany
Center for Molecular Medicine Cologne (CMMC) and Cologne Cluster on Cellular Stress Responses in Aging-Associated Diseases (CECAD), Medical Faculty, University of Cologne, Cologne, Germany

Index

www.ingramcontent.com/pod-product-compliance
Lightning Source LLC
Chambersburg PA
CBHW080640200326
41458CB00013B/4693